Plant Breeding
in 21st Century

Realted Books

Title	Author
• A Handbook of Laboratory Solutions	*M.H. Gabb*
• Advances in Plant Physiology (Vol. 1-18)	*A. Hemantaranjan*
• An Introduction to Population Genetics Theory	*J.F. Crow*
• Basic Concepts of Plant Biotechnology (With MCQs)	*Vijay Prakash*
• Biofertilizer Technology	*S. Kannaiyan*
• Biology and Biotechnology of Anoxygenic Phototrophic Bacteria	*S. Girisham*
• Biotechnology Emerging Trends	*R.Z. Sayyed*
• Biotechnology of Microbes and Sustainable Utilization	*R.C. Rajak*
• Comprehensive Laboratory Manual of Life Science	*Jyoti Saxena*
• Crop Improvement and Mutation Breeding	*A.K. Sharma*
• CSIR NET: Part A (Hindi)	*Christy Varghese*
• Cytochalasin: Incidence and Biological Activities	*S. Kiran*
• Double Helix: DNA Sanrachna Ki Khoj (Hindi)	*H.C. Dube*
• Fundamentals of Agriculture	*R.L. Arya*
Genetic Improvement of Field Crops	*C.B. Singh*
• Handbook of Mycological Techniques: Identification of Mycotoxigenic Fungi and Mycotoxins	*Rekha Bhadauria*
• Indiras Objective Agricultural Biotechnology *2nd Ed*	*R.L. Arya*
• Indiras Objective Agriculture	*R.L. Arya*
• Laboratory Manual of Microbiology, Biochemistry and Molecular Biology	*Jyoti Saxena*
• MCQs in Plant Breeding Biotechnology and Seed Science	*K. Vanangamudi*
• Methods in Fungal Biology: A Manual of Laboratory Protocols	*A.K. Gautam*
• Microbial Biotechnology for Sustainable Development and Productivity	*R.C. Rajak*
• Mycology and Microbiology: A Textbook for UG and PG Courses	*C. Manoharachary*
• Objective Life Science MCQs for Life Science	*Kailash Choudhary*
• Perspectives in Biotechnology	*S.M. Reddy*
• Plant Biotechnology and Molecular Biology: A Laboratory Manual	*M.S. Punia*
• Plant Molecular Genetics	*S. Chakraborty*
• Plant Reproduction	*T. Pullaiah*
• Plant Tissue Culture: Theory & Practicals *2nd Ed*	*T. Pullaiah*
• Plant Tissue Culture: Theory and Techniques	*Shailesh Kumar*
• Recent Trends in Biotechnology	*M.K. Rai*
• Seed Technology *2nd Ed*	*Dhirendra Khare*

Plant Breeding
in 21st Century

B.D. Singh
N.S. Shekhawat

Published by:

SCIENTIFIC PUBLISHERS (INDIA)
5A, New Pali Road, P.O. Box 91
Jodhpur 342 001, India
E-mail: info@scientificpub.com
Website: www.scientificpubonline.com

Print 2026

Reprinted from the book 'Molecular Plant Breeding' by B.D. Singh & N.S. Shekhawat
(Pages: 544 Page and Year: 2018)

ISBN: 978-93-89061-26-0
eISBN: 978-93-89061-28-4

Printed in India

To

Satish C. Maheshwari

A world-famous scientist
and
an inspiring teacher

Preface

The prehistoric humans domesticated a diverse group of wild plant species, which ultimately gave rise to our present-day crops. Domestication permitted the development of agriculture, which prompted the humans to shift from nomadic to settled life style, and allowed the evolution and refinement of culture and civilization. An inevitable consequence of these developments was the incessant increase in human population, which is over 7.7 billion at present and is expected to cross 9.0 billion by the year 2042. It has long been realized that the crop yields are considerably improved by selecting lines with desirable features for cultivation. In view of this, extensive systematic research activities were devoted to the development of improved crop varieties leading to the formulation of the various schemes and methods of plant breeding. The success of breeding programmes depends mainly on the genetic variation(s) available to the breeder, and the effectiveness of the selection procedures for the characters being improved. The bulk of plant breeding research has been focussed on the development of methods to satisfactorily resolve the above two issues.

Plant breeding has been highly successful in evolving improved crop varieties to provide for the growing needs of the ever expanding human population. But the climatic changes associated with the global warming, and the unfavourable effects of the modern-day agricultural practices on the biotic and abiotic components of the agricultural environment are inevitably adversely affecting crop yields. Therefore, the plant breeders have the dual challenge of developing climate smart crop varieties endowed with higher yielding ability, and resistance to the emerging pathogens and insect pests within progressively shorter time spans than ever before. Thus, plant breeders need to be able to access the desired genes and use them in their breeding programmes, and to devise such breeding strategies that enable them to develop new varieties in as short a period as possible.

Plant breeders are always in search of useful genes that could be deployed in the crop varieties to create in them the desired phenotypes. This search has enabled them to utilize related wild species as donors of valuable traits to many of our crops. But many useful genes remain inaccessible to the breeders due to sexual incompatibilities. The development of recombinant DNA technology provided an opportunity to transfer and express in plants genes from any source and, thereby, create plants with novel traits. The transgenic technology has been considerably refined and used to develop many crop varieties with useful novel traits; a large number of these varieties have been approved for commercialization and are in cultivation since 1996. The transgenic varieties possess such useful traits as disease and insect pest resistance, resistance to drought, improved produce quality, and even higher yields. It is projected that the transgenic technology will enable the development of climate smart crop varieties by bringing together the necessary genes from various sources.

In spite of the reported all-round positive impact, including beneficial effects on the environment, the consumer acceptance of transgenic varieties has been, if anything, lukewarm. The researches directed at the use of DNA for unequivocal identification of human individuals had the pleasant and highly useful side-effect of the development of DNA-based molecular markers. DNA markers have found a variety of applications in the various disciplines of biology, of which perhaps the most notable is that in the field of plant breeding leading to the emergence of marker-assisted plant breeding. Markers have enabled reliable and efficient indirect selection for traits, mapping and selection for quantitative trait loci, and more particularly, selection for various traits at the seedling stage and in off-season crops/greenhouses. Thus, markers allow rapid advance of generations and, thereby, drastically reduce the time required for the development of new crop varieties. Most commercial and many national breeding programmes are using the marker technology to achieve a variety of objectives.

Plant breeding has freely and gainfully utilized the developments in the other disciplines of biology to enrich its arsenal to enhance its capability to development of crop varieties designed to meet the changing requirements of agricultural production systems. The transgenic and marker technologies have emerged as extremely powerful tools for modification of crop genotypes. These technologies are progressively being integrated into breeding programmes. As a result, a modern plant breeder needs to be adequately familiar with the transgenic and marker technologies.

The book *Molecular Plant Breeding in 21st Century* is designed to provide a combined overview of plant breeding tools and techniques along with those of transgenic and marker technologies, each presented in a separate section of the book. This book is intended for graduate students offering an introductory course in plant breeding, and for those pursuing M.Sc. programmes in botany. It would also be useful for the students appearing in various competitive examinations. We have given special attention to keep the text simple and easy-to-understand and enjoyable, but at the same time comprehensive and detailed. Therefore, we made enormous efforts to make the content of this book as pleasant as possible for broad range of readers.

The authors are appreciative of the sincere efforts by Mr. Tanay Sharma and his production team, including the reviewers, editors and the graphics designers, of Scientific Publishers, Jodhpur. The authors like to extend their thanks to all the team for their commendable efforts, as a result of which the book has been brought out speedily in an excellent published form.

With this edition, we aim to serve a wide range of readers' groups with the intention of contributing towards technological advancement in molecular plant breeding. In the end, we would like to express our heartfelt appreciation for our family members, whose affectionate support gave us the courage to undertake this project and the energy to successfully complete the same.

B.D. Singh
N.S. Shekhawat

June 30, 2019
Varanasi, Jodhpur

Contents

Part A
Basics of Plant Breeding

1 Introduction

1.1. INTRODUCTION TO PLANT BREEDING

Plant breeding was described by Nikolai Vavilov, the famous Russian scientist as "evolution directed by man". The green plants are the marvels of nature; use energy from sunlight, carbon dioxide from the atmosphere and water and minerals from the soil to generate the organic materials to support all the living organisms (Biel and Fomina, 2015). Thus the green plants function as self-sustaining and self-propagating machines that convert the abundant, innocuous, and free solar energy into a multiplicity of organic compounds and products that are used by humans, animals, microbes, and parasitic plants. The chloroplast present in the green plant cells is the key to the process of converting the light into chemical energy through the process of photosynthesis. Although, the photosynthetic energy conversion is relatively inefficient (merely ~1.0% of the energy falling on plant is used/converted), still results in massive quantities of global carbon fixation, which is estimated as around 200 billion tonnes per year; this figure that dwarfs the total output from human activities (Galston, 1980). The plant biomass is the source of food, fiber, feed, timber, pharmaceuticals/herbal products and numerous industrial materials. In addition, the green plants provide invaluable ecosystem services including generation of oxygen, and also contribute to the aesthetic values and spiritual satisfaction.

More than 11,000 years ago, Neolithic ancestors of humans made attempts to manage and domesticate plants and animals (and some of the microbes), soils, and to extent even waters (**Neolithic revolution**). This allowed humans to accomplish the transition from hunter-gatherer life-style to the settled (sedentary producer of selected plants and animals) human communities; this transition was the starting point for the development of culture and civilization, including ordered societies and governments. Views on origin of agriculture range from mythological to ecological. With the development of agriculture, large sections of the wild landscapes were converted into crop fields and pastures. Agriculture is one of the major technological innovations of humankind; a form of bio-manufacturing based on plants, and is one of the oldest industries on the planet. Hundreds of wild plants and animals were selected for domestication by the human societies at different locations on the earth and used for bio-manufacturing to serve the local human needs. This can be regarded as the first stage of plant breeding that probably contributed to the characteristic species distribution along the corridors of human migration. Sedentary life style led to reduction in birth spacing, which eventually caused a continued increase in global human population.

1.2. THE GROWING FOOD NEEDS AND DECLINING AGRICULTURAL RESOURCES

The conventional plant breeding techniques applied by humans remained viable for centuries. High yielding/superior varieties developed by humans coupled with the advanced mechanical agricultural processes resulted in production of sufficient food, fiber and other commodities at reasonable costs for supporting the rising population (which is now around 7.3 billion). However, the availability of foods and nutrients is not equal for all the individuals of the world: on one hand we have starving poor humans, while there are overfed rich who waste about 50% of the food. According to the World Bank, approximately 1.2 billion people in absolute poverty live on less than US $ 1 per day, while nearly twice of this number live on less than US $ 2 per day. About 90% of the world's poor (approximately half of the world's population) comprise rural, resource-poor farmers and their families, and land-less poor who depend on agriculture for their livelihood, who live and work in regions that harbor important sources of plant genetic diversity.

Food and Agricultural Organization (FAO) of the United Nations predicts that food production must rise by 70% over the next 40 years to meet the demands of the growing population, which is expected to reach 9 billion by the year 2050 (UNFAO, 2009). The solution to this challenge will be multifaceted and certainly involve changes in what, where and how plant species yielding food and other products are grown and transported, as well as the amount actually consumed relative to what is wasted. Roughly 10% of the known plant species is edible, and of these about 7000 species are cultivated (UNFAO, 2009, 2014). The main factors that contribute to increased agricultural production are the total cultivated area; the inputs like nutrients, irrigation water, plant protection measures, etc. used for crop cultivation; and the genetic potential of the crop varieties that are grown to make the full use of these inputs. The land area cannot be increased; in fact, it is declining due to the diversion of land to fulfill nonagricultural human needs. In addition, the physical and chemical properties of the soils are being adversely affected by the prevalent agricultural practices. Earth's soils have formed by processes that have maintained a persistent and expansive global soil mantle, one that in turn provided the stages for evolution of vast biological diversity. Soils are highly complex ecosystems and are considered as one of the earth's main reservoir of life forms. Human exploitation of the soils systems began thousands of years ago, and made the agriculture a huge success. The vastness of the planet and its soil resources allowed expansion of agriculture and colonization of the favorable regions of planet; the populations moved on to new locations when the soil resources of a given site were depleted. The practices of farming greatly accelerated the rate of soil erosion and depletion of nutrients. The use of chemicals (fertilizers and pesticides) in soils and on plants for intensive agriculture has caused soil-water pollution and affected adversely the soil quality and productivity at large. The chemicals applied in agricultural fields have enormously harmed the pollinator insects.

Civilization's ever-increasing demand for water is a great challenge. Agriculture accounts for 70% of global fresh water use. Therefore, agriculture is a logical target for increased water conservation efforts. The potential of agriculture water use efficiency efforts is large as more than 99% of the irrigation water is lost via transpiration of crops, which is highly correlated with the net incoming radiation (Zamft and Conrado, 2015). Given these challenges/stress in era of global warming, the new breeding strategies have to be adopted for crop water use efficiency and soil conservation. In the past, improvement of crops and agricultural techniques has mainly concentrated on increasing shoot biomass (tuber and root biomass in tuber and root crops) and seed yield, and the relevance of the root (the hidden half) systems for food production has been overlooked. The root systems take care of indispensable plant functions such as uptake of nutrients and water, anchorage in soil/substrate and interaction with symbiotic/ beneficial and antagonistic/pathogenic microbes. It is suggested that the next agricultural revolution will come through root biology. There is now emphasis to breed crops to match their soil-water environment and for improved root phenotypes for better foraging of water and nutrients in stressed soils. Finally, the ability to make the best use of the inputs provided and to tolerate the adverse environmental factors is determined by the genotype of the crop variety being cultivated. Therefore, the development of improved crop varieties with superior performance under the given environment appears to be the major contributor to the enhanced agricultural production in the future.

Since its origin in antiquity, plant breeding has human endeavors/practices of simple discrimination among plants types to select, retain, promote, and cultivate such plants that had the most desirable/advantageous trait combinations (Section 1.3). In the initial stages, plant breeding was essentially an art as its decisions were purely subjective relying on human observation, experience, instinct, and judgment. Although it was a deliberate effort by human to understand nature and manage heredity of plant species to an advantage, most of the changes made in plants were permanent and heritable. Subsequently, the fundamental discoveries in fields of sciences, particularly the life sciences, established the scientific basis for plant breeding. The recent integration of advances in molecular biology, genetics, biotechnology, bioinformatics, and genomics with conventional plant breeding practices have created the foundation for molecular breeding/smart breeding, as an interdisciplinary science that is revolutionizing the 21st Century crop improvement programs. The approaches of ultra-high throughput genome sequencing, metagenomics and modern phenotyping, biosensing and gene/genome imprinting, plant system biology, gene targeted genome editing have opened new avenues for more specific and targeted genetic modifications to supplement plant breeding efforts. The chief objective of plant breeding is to develop improved crop varieties, which may be regarded as products designed to promote human welfare. Therefore, it would be more appropriate to consider plant breeding as a technology that generates useful products in the form of superior plant varieties.

1.3. MAJOR DEVELOPMENTS

Although man does not cause variability (now he/she can do so using various methods) and cannot prevent it, he can select preserve, and accumulate the variations given to him by the hand of nature almost in any way that he chooses, and thus he can certainly produce a great result (Charles Darwin).

Agriculture is the deliberate planting and harvesting of plants (and microbes) and rearing of animals. The human inventions have impacted and continue to impact the society and the environment. Plant breeding has a long history of integrating the latest innovations in biology, genetics and agronomic practices to maximize productivity and enhance improvement. Prehistoric selection for visible phenotype(s) that facilitated harvest and increased productivity led to the domestication of the first crop variety (Harlan, 1992: Moose and Mumm, 2008). Humans had realized long ago that the produce available from the crop field depended both on the plant material grown and the conditions under which it was grown. For this reason, humans selected the best plant types and grew them under the most favorable conditions that could be created for their growth. As a result, efforts were made to modify the genetic makeup of the crop plants to make them more useful to the humans; this activity constitutes **plant breeding**. Plant breeding has progressed through the following four major categories of developments: (i) domestication and introduction (ii) selection, (iii) hybridization, (iv) mutagenesis, polyploidy, and (v) molecular breeding (Fig. 1.1, Table 1.1). These phases not only describe the history of plant breeding, but also the progress of breeding of any individual crop species. **Domestication** (Section 1.4) is the **first** and the most basic phase of plant breeding as it gives rise to the cultivated plant species that could be further modified by breeding activities.

Table 1.1: Major advances in plant breeding

Development phase	Contributed by	Year	Main contribution	Remarks
Phase I: Domestication	Prehistoric humans	Over 11,000 years ago	Bringing selected wild species under cultivation as crop plants	Domestication continues till date and is likely to continue in future
Phase II: Selection		Recorded examples in 19th century	Development of plant varieties with superior performance	Utilized the natural genetic variation already present in the crop species
	Shireff	1873	Only heritable variation responds to selection	Several workers developed by selection improved crop varieties
	Vilmorin	1856	Progeny test	The basis of evaluation of the genetic value of plants
	Nilsson-Ehle and coworkers	1900	Individual plant selection	The present-day, pureline selection
	Johannsen	1903	Pureline theory	The genetic basis of pureline selection
Phase III: Hybridization	Fairchild	1717	Hybrid from carnation x sweet William	Popularly called Fairchilds's mule
	Mendel	1866	The basic laws of inheritance: segregation and independent assortment	Provided scientific basis to plant breeding procedures
	Davenport	1908	Dominance hypothesis of heterosis	Involved in most cases of heterosis
	East	1908	Overdominance hypothesis of heterosis	Existence of heterosis loci demonstrated using molecular markers
	Shull	1909	Use of inbreds for production of hybrid varieties of maize	The basis of development of hybrid varieties
	Karpechenko	1928	Raphano-brassica by crossing radish with cabbage	The first intergeneric hybrid

Development phase	Contributed by	Year	Main contribution	Remarks
	Carlson and coworkers	1972	Somatic hybrid plant from protoplast fusion of *Nicotiana glauca* with *N. langsdorfii*	The first somatic hybrid plant
Phase IV: Mutagenesis and polyploidy			Creation of novel genetic variation	
(i). Mutagenesis	Muller	1927	Mutagenic action of X-rays	X-ray is the most widely used mutagen
	Gustafsson	1929	First mutation breeding programme	First mutant variety released in 1950
	Auerbach and Robson	1946	Mutagenic action of mustard gas	EMS and some other chemical mutagens are widely used
	Larkin and Scowcroft	1981	Somaclonal variation	Some somaclones released as varieties
(ii). Popyploidy	Clause and Goodspeed	1926	Production of the allopolyploid *Nicotiana digluta*	Triticale has developed as a successful crop
	Karpechenko	1928	The intergeneric hybrid *Raphanobrassica*	Being developed as a forage crop
	Blakeslee and Nebel, independently	1937	Chromosome doubling action of colchicine	Widely used for chromosome doubling
Phase V: Molecular breeding			Creation of designer varieties of crops demonstrated	A designer variety has features specified prior to its development
(i). Transgenic technology	Smith and coworkers	1970	Isolation of restriction endonuclease	The basic tool of recombinant DNA technology
	Silber and coworkers	1972	Discovered DNA ligase	The basic tool of recombinant DNA technology
	Jackson and coworkers	1972	Construction of the first recombinant DNA molecule	The basic step of transgenic technology

Development phase	Contributed by	Year	Main contribution	Remarks
	Herrera-Estella and coworkers	1983	The first transgenic plant with kanamycin resistance	The feasibility of creation of 'designer ' crops demonstrated
	Monsanto/ Calgene	1994	Flavr Savr tomato appr-oved for culti-vation in USA	The first designer crop variety
(ii). Marker-assisted plant breeding			Highly effective indirect select-ion for traits	Integrated in many plant breeding programmes
	Botstein and coworkers, and others	1980	Developed the use of RFLPs* for linkage mapping	Facilitated the use of molecular markers in plant breeding
	Concibido and coworkers	1986	Marker-assisted selection (MAS) for resistance to soybean cyst nematode	The first report of MAS
	Monsanto	2006	Maize hybrid variety developed by MAS released for cultivation	Many varieties developed by using molecular markers

*RFLP, restriction fragment polymorphism.

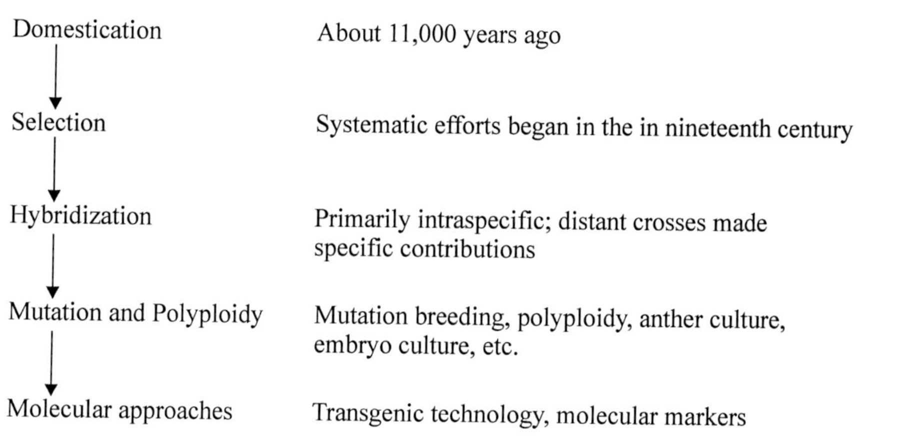

Fig. 1.1. The major developments in plant breeding. The dominant activity of each phase is depicted in the figure, but other activities have continued in every phase of development, following domestication.

The **second phase** of plant breeding comprises planned and conscious **selection** for improved performance. Selection was practiced to develop several useful crop varieties especially in self-pollinated crop species. Notable contributions were made by Knight, Cooper, Le Coutier, Shireff and others. Shireff in 1873 concluded that only heritable variation responded to selection, and that this variation resulted from, in present-day terms, mutation, and segregation and recombination. Vilmorin in 1856 put forth the concept and procedure of progeny test. In 1900, Nilsson-Ehle and coworkers developed the individual plant selection method, i.e., the pureline selection method of today. The concept of purelines was proposed by Johannsen in 1903.

The **third phase** was dominated by hybridization that involves mating of two genetically different plants/lines to bring together in one plant, the *F1* hybrid, the genes present in the two parental plants/lines. Segregation and recombination in *F1* will create genetic variation that is utilized by selection for crop improvement. In 1717, Thomas Fairchild produced the first artificial plant hybrid by crossing carnation with sweet William. This hybrid became popular as the **Fairchild's mule**. Hybridization was used by Knight to develop several varieties of apples, pears, grapes etc. But the rediscovery of the Mendel's work in 1900 provided scientific basis for the breeding methods designed to exploit hybridization for crop improvement. Soon thereafter, detailed observations on inbreeding in maize were independently published by East and Shull in 1908 and 1909, respectively. These studies provided the starter for the development of hybrid varieties. Hybridization continues and is expected to continue to be one of the major approaches for crop improvement; the various aspects of hybridization are discussed in some detail in Section 1.9.

The **fourth phase** in plant breeding comprised the use of various approaches to generate novel genetic variation. The ability of X-rays to induce mutations was discovered by Muller in 1927. Barely two years later, in 1929, Gustafsson and coworkers initiated a breeding program based on induced mutations. A specialized and more recent (beginning during 1980s) approach for mutation breeding is the use of somaclonal variations for crop improvement. **Somaclonal variation** is the genetic variation that is present in cell grown in tissue culture or the plants regenerated from them. Polyploidy has contributed to the evolution of plant species, and some of our important crops are natural allopolyploids, i.e., polyploids produced by chromosome doubling of the *F1* hybrid produced by crossing two species. The first allopolyploid plant species (*Nicotiana digluta*) was synthesized by Clausen and Goodspeed in 1925. The chromosome doubling action of colchicines was independently discovered by Blakeslee and Nebel in 1937. A large number of allopolyploid species have been produced and studied. A new allopolyploid cereal crop triticale has been developed and is being cultivated in some specific areas.

The **fifth phase** is the most recent and the most exciting development; it is generally called **molecular breeding**. The main feature of this phase is the use of molecular biology for achieving plant breeding objectives. There are two main approaches of molecular breeding, viz., genetic engineering or transgenic technology and marker-assisted plant breeding. In the **transgenic technology**, recombinant DNA technology is used to transfer into plants genes from any organism to generate the desired traits. The first genetic engineering experiments were carried out in bacteria in 1973. The technique was soon extended to plants, and the first genetically engineered insect resistant crop varieties were used for commercial cultivated in 1996. The various aspects of transgenic technology are considered in some detail in Chapters 7-10. In case of marker-assisted plant breeding, molecular markers are used for achieving different plant breeding objectives, marker-assisted selection (MAS) being the most widely activity. The first variety developed by MAS was released in 2006 for cultivation in USA. The various aspects of markers and their uses in plant breeding are discussed in Chapters 11, 12 and 13.

The above delineation of the five phases is largely arbitrary. Some of the phases have been named after the predominant activity during the concerned period of time. But some others have been designated on the consideration that the specified technology became available during the specific period and was used in plant breeding although it may not have been the predominant approach. These phases are necessarily overlapping, and at present all the strategies are being used in breeding programmes. It may be pointed out that hybridization continues and is likely to continue to be the predominant breeding approach.

1.4. ACTIVITIES IN PLANT BREEDING

Plant breeding comprises the following five types of activities: (i) creation of genetic variation, (ii) selection of the desirable genotypes, (iii) evaluation of the selected plant/lines to confirm their superior performance and to select one or few best lines for release as new varieties, (iv) seed multiplication of the lines released as varieties, and (v) distribution of their seeds to the farmers for cultivation (Table 1.2). Genetic variation for the traits to be improved must be present in the breeding population. Therefore, creation of genetic variation is the first step in any breeding programme. It can be achieved by a variety of approaches, including somatic hybridization and genetic engineering. In case the desired trait is not available in the germplasm collection of the country, it could be introduced from another country. If the trait is not present in the germplasm of the species, techniques like genetic engineering may be used to generate the same. Sometimes, a new plant species may need to be domesticated to achieve a new objective, e.g., biofuel production.

Table 1.2: Main activities in plant breeding

Activity	Objective	Strategies	Remarks
Creation of genetic variation	Creation of genetic variation in the desired trait	Hybridization, Mutation, Polyploidy, Somatic hybridization, Somaclonal variation, and Genetic engineering	Presence of genetic variation is essential for improvement in the desired trait
Selection	Isolation of a line/ lines with the desired trait combination and superior performance	Phenotypic selection, Marker-assisted selection	Phenotypic selection is effective for qualitative traits, but is much less effective for quantitative traits
Evaluation	Assessment of the superior performance of the selected line(s)	In replicated trials conducted over locations and years	Comprehensive evaluation is necessary for release as a new variety
Multiplication	To produce the required quantities of quality seed	Multiplication in well defined steps to ensure continuous seed supply	The multiplication steps: Nucleus seed Breeder seed Foundation seed Certified seed
Distribution	Supply of certified seed to the farmers	Multiple agencies are involved in seed sale and supply	The role of private sector is increasing with time

Selection aims to identify and isolate the plant(s) having the desired trait combination from a genetically variable population. This is the most critical step, and its effectiveness determines the success of the breeding programme. Selection is based on trait phenotype, but molecular markers (*See* Chapter 11) can be used for marker-assisted selection (MAS; Chapter 13). MAS is particularly useful when the phenotypic evaluation of the trait presents problems or yields less than 100% reliable results. The objective of **evaluation** is to confirm the superiority of the selected lines, and to identify the best performing line. Initially, the lines and two or more of the popular varieties (used as checks) are evaluated at a single location. The lines performing better than checks are included in multi-location trials for three or more years. A selected line performing better than the checks may be released as a new variety for commercial cultivation. The seed on the new variety is multiplied through well defined steps. The new variety is popularized among the farmers and its seed is sold to the cultivators through various channels.

1.5. DOMESTICATION

All crops have been derived from wild plant species that were selected by the pre-historic humans for growing them under human supervision. The process of bringing wild plant species under human management is called **domestication**. Domestication began over 12,000-10,000 years ago at Fertile Crescent and elsewhere. Cereals and legumes were among the first to be domesticated. During domestication, the plant populations would have been subjected to both artificial (human-mediated) and natural selections. **Selection** refers to differential reproduction rates of individuals with different genotypes. As a result, the domesticated populations diverged from their wild relatives to give rise to our present-day crop species. Domestication continues as an important activity in plant breeding, e.g., the efforts for domestication of plant species suitable for biofuel production.

It is believed that the domestication of the crops took place in the following six regions of the world: (i) Mesoamerica, (ii) Southern Andes, (iii) Near East, (iv) Africa, (v) South East Asia, and (vi) China. It may be pointed out that these regions are geographically diverse, and that the crop domestication in these regions occurred independently of each other. Yet the set of traits selected for during domestication in many different crops are remarkably similar; these traits are often called **domestication syndrome traits**. For example, increased seedling vigor, increased self-pollination and loss of seed dormancy has occurred in many diverse crops like sunflower, tomato, mungbean, *Brassica juncea* etc. Similarly, vegetative reproduction has been adopted in crops like sugarcane, potato, sweet potato, cassava etc. The following traits have been selected against during domestication: shattering of pods, spikes etc., dormancy, toxins, tall plants (in many crops) etc. On the other hand, the following traits seem to have been favoured during domestication: shorter life cycle, taller plants (in some crops), larger grains/fruits, higher economic yields, asexual reproduction, polyploidy etc.

There are degrees of domestication of species. Species that become completely domesticated are unable to survive when re-introduced into the wild. This is so because of the selection process that drives domestication removes natural adaptive features and mechanisms of the plants that are essential for survival in the wild but not required if cultivated/managed as crop. Maize as cultivated crop plant depends on human aid for survival as there is no intrinsic mechanism for natural dispersion. The suite of phenotypic changes that transform wild plants into food crops are known as **domestication syndrome**. Rapid progress in archaeobotany and ethanobotany has contributed to the understanding of crop domestication. The new discoveries in genetics, molecular biology, genomics and bioinformatics are providing new potent tools for crop improvement applications. These tools allow plant breeders to engineer crops to make them more suitable for human needs, and capable of giving higher yields with the minimal environmental impact. This is considered to constitute **super-**

domestication, e.g., in tomato, banana and rice. For example, the modern tomato fruit is 100-times larger than that of its ancestors due to domestication and improvement focused on two independent sets of quantitative trait loci (QTL).

The genetic variation that was utilized by selection during domestication had originated by mutation, segregation and recombination following natural hybridization, interspecific hybridization. Most of the useful mutations were *micromutations*, each of which produces small effect on the phenotype. Accumulation of micromutations has been the main cause of modifications during the domestication of many crops like barley, rice, peas, *Phaseolus* spp. etc. But some mutations produce large individual phenotypic effects; they are termed as **macromutation**. Macromutations have been important for the domestication of some crops like maize, cabbage, cauliflower etc. The maize differentiated from the grassy pod corn due to a single macromutation. Similarly, cabbage, cauliflower, broccoli and Brussel's sprouts differ from each other for merely a few major genes. Interspecific hybridization may have contributed to the development of few crops like maize. In contrast, many of our crops, e.g., banana, wheat, groundnut, cotton, tobacco are polyploids.

1.6. GENETIC RESOURCES

Genetic resources or **germplasm** of a crop species may be describes as the sum total of the genetic information present in the species and its wild relatives. The genetic information comprises all the alleles of all the genes present in the given crop species and its related species. The germplasm, in practical terms, consists of the following materials: (i) varieties in cultivation, (ii) obsolete or old varieties, (iii) breeding lines, (iv) special genetic stocks, (v) land races or local varieties, (vi) wild forms and (vii) wild relatives. A **wild form** is the wild species from which the given crop has been directly derived by domestication. A **wild relative** is one of the other species that are related to the crop species by descent during the evolution. The first four categories of the materials broadly constitute the **adapted** or **elite germplasm**, while the remaining materials comprise the **unadapted germplasm**. The germplasm originating in the country under consideration are called **indigenous germplasm**, while those from another country are referred to as **exotic germplasm**.

The germplasm is thus the chief source of desirable alleles and genes that are used to construct the improved varieties through hybridization. The germplasm accessions that easily hybridize with the crop species and yield fertile hybrids are grouped into the **primary gene pool** (GP1). Generally, GP1 germplasm comprises the strains of the concerned crop species and is the most extensively used. The **secondary gene pool** (GP2) accessions present difficulties in hybridization with the crop species, and the hybrids are partially fertile. GP2 serves as a valuable source of useful genes and alleles. The hybridization between members of the **tertiary gene pool** (GP3) with gene transfers into the crop species generally requires special techniques and great amount of work.

The genetic variability present in the crops and their relatives is consistently declining due to various human activities; this is termed as **genetic erosion**. The variability in the cultivated germplasm is declining due to widespread adoption of the improved varieties. It is ironical that the improved product derived from the genetic variability represented in the germplasm is adversely affecting the genetic variability itself. The wild forms and the wild relatives of the crop species are threatened by extension of cultivation and grazing into the forests, developmental activities like railways, roads, industrial areas, hydroelectric projects etc., and even introduction of some species that invade the natural habitats. In view of the above, it is necessary to conserve the germplasm of crop species to ensure their availability in the future.

The wild forms and wild relatives can be conserved in the areas where they occur by protecting them from human interference; this is called **in situ germplasm conservation**. The protected areas are known as **gene sanctuaries** or **biosphere reserves**. They are best established in the centres of origin of crop species. In India, 14 biosphere reserves have been established. The germplasms present in gene sanctuaries continue to evolve in response to the natural forces. Still it is generally very difficult to protect these areas from human interference. At the same time, the cultivated germplasm cannot be conserved in gene sanctuaries.

The cultivated germplasm has to be collected and conserved as germplasm collections, often called **gene banks**. This is known as **ex-situ germplasm conservation**. Even the wild forms and wild relatives of crops collected and conserved in gene banks. The germplasm can be conserved in one of the following five ways: (i) stored as seeds, (ii) grown in the fields as plants, (iii) maintained as slow growing shoot cultures, (iv) kept frozen in liquid nitrogen at -196°C as cell and organ cultures or even embryos, and (v) stored as DNA fragments as cosmid clones or phage lysates. Storage as seed is used wherever it is feasible. This is because seed storage is easy, safe, and is amenable to rapid use after storage. Most crop species produce **orthodox seed**, which can be dried to 5% or lower moisture content. These seeds (with 5% moisture) are stored for long periods (20-100 years depending on species) at -20°C; this constitutes **base collection**. The seeds in an **active collection** are stored at a temperature close to 0°C for 10-15 years. The **working collection** is maintained at ~15°C for up to 5 years. Government of Norway has established a **Global Gene Vault** at Svalbard, near North Pole in 2008. In India, Defense Research and Development Organization (DRDO) has set up similar **Gene Vault** at Chang La in Ladakh in 2009. The genetic resources of the asexually propagated species are maintained plants in the field, as slow-growth shoot cultures or as cryopreserved cells, tissues and organs including embryos. The genetic material from the extinct or nearly extinct species can be stored as cosmid clones or phage lysates (Table 1.3). The large gene banks having germplasm from all over the world are recognized as world collections (Table 1.4). In India, the germplasm are maintained as base collections at the National Bureau of Plant Genetic Resources (NBPGR), New Delhi. In 2006, the NBPGR had 339,194 accessions of germplasm of different crop species.

Table 1.3: A summary of the various approaches for germplasm conservation

Approach	Accessions maintained as	Application	Remarks
In situ conservation	Growing plants in reserves protected from human interference	Germplasm of wild forms and wild relatives	Established in centres of origin/diversity; difficult to maintain
Ex situ conservation	Plants, seeds, shoot cultures etc. in gene banks	Cultivated germplasm, wild forms, and wild relatives	Established in places with appropriate facilities
(i) Seed banks	Seeds	All species producing orthodox* seeds	The most preferred approach
a. Base collection	Seeds with 5% moisture at -20°C	The ultimate source of active collections	Disturbed only for regeneration** and germination tests
b. Active collection	Seeds with 5% moisture at close to 0°C	Used for evaluation, multiplication and distribution	Periodically replaced by accessions from the base collection
c. Working collection	Seeds with 10% moisture at ~ 15°C	Used in breeding programmes	Maintained by the breeders
(ii) Field banks	Plant maintained in the field	Asexually propagated species	Requires large areas and considerable resources
(iii) Slow-growth cultures	Slow growing shoot cultures	Asexually propagated species	A substitute for field banks; requires tissue culture facility; in use for some species
(iv) Cryopreservation	Cells, tissues, organs, e.g., embryos, kept at -196°C in liquid nitrogen	Asexually propagated species	Requires considerable expertise and sophistication; in use for some species
(v) DNA banks	DNA fragments as cosmid clones or phage lysates	Extinct or nearly extinct species; a source of desired DNA sequences	Highly sophisticated; needs considerable resources

*Seeds that can be dried to 5% or lower moisture level without any adverse effect on their viability.

**Growing the stored seeds in the field, and using the freshly harvested seeds for further storage.

Table 1.4: World collections of germplasm. The various international crop research institutes, in addition, maintain large collections of germplasm of the concerned crops

Crop	World collection located at
Coffee	Ethiopia, Africa
Groundnut	Bambey, Senegal
Potato	Cambridge, UK
Sugarcane	Canal Point, Florida, USA
Sweet potato	New Zealand
Small grain crops	Beltsville, USA
All crops	Institute of Plant Industry, Russia
All crops	Royal Botanic Gardens, Kew, England
All crops	Svalbard Global Seed Vault*, near North pole

* Popularly known as doomsday vault. It has more than 230,000 accessions stored in the permafrost zone near North Pole to safeguard against natural and human-mediated disasters.

The germplasm stored in gene banks has to be collected by exploration or introduction (**collection of germplasm**). All germplasm is obtained by **exploration**, i.e., trips for collection of germplasm. Explorations generally cover the areas of centres of origin/diversity, and in areas where the diverse genotypes of the crops are expected to be found. Centres of origin are the areas where the crop species are believed to have originated. These areas show great plant diversity, and often termed as **centres of diversity**. Initially, only eight centres of origin were proposed, but the total number is now 11 (China, Hindustan, Central Asia, Asia Minor, Mediterranean, Abyssinia, Central America, Peru, Brazil-Paraguay, Chile and USA). Similarly, 12 megacentres of plant diversity have been recognized; these generally overlap the centres of origin. Within each centre, there are relatively much smaller areas that show very high variability; these areas are called **microcentres**. Microcentres are the first to be explored for germplasm collection.

The germplasm accessions have to be evaluated for the various traits of interest to enable their judicious utilization (**evaluation**). This multidisciplinary activity is the most critical and expensive step in germplasm conservation. Each germplasm accession is given a unique identification number, and all the data available for it are stored in a suitable format that facilitates its retrieval when required. This information is published by the gene banks as **catalogues** (**cataloguing**). The germplasm accessions are stored, their germination is periodically assessed and the accessions are regenerated when germination reaches a critical level. **Regeneration** consists of growing the accessions in the field and using their fresh seeds for further storage. The germplasm accessions are multiplied and distributed to the research workers as per their research needs. The germplasm accessions can be used for selection of suitable lines for direct release as new varieties, as parents in hybridization or in other numerous types of investigation (**utilization**). In India, large number (nearly 200) varieties have been developed through selection in germplasm accessions; over 50 varieties have been developed in rice alone.

1.7. MODES OF REPRODUCTION

The mode of reproduction is the chief factor affecting the choice of breeding methods and the end products of breeding of crop species. **Mode of reproduction** may be defined as the method of production of new individuals from pre-existing or parental individuals. The crop species reproduce by asexual or sexual methods. The **asexual reproduction** is based on mitosis and does not involve meiosis and gametic fusion. The progeny plants may develop from vegetative parts of the parental plants (**vegetative reproduction**) or they may arise from seeds in which embryos develop without fertilization (apomixis). Many of the crop species are vegetatively propagated, and several of them also show sexual reproduction. Apomixis is found in many cultivated species, and most of them reproduce sexually as well. The asexually reproducing species are highly heterozygous and they show severe inbreeding depression. Any plant with desirable features can be asexually multiplied to give rise to a new clone that may be used as a variety. A **clone** comprises the asexual progeny of a single plant. As a result, all the individuals of clone have identical genotype, and the variation in their phenotypes is purely due to the environmental factors. But the traits present in two different clones can be combined into a single clone only when the concerned crop also reproduces sexually. The breeding of apomictic crop species requires especially designed breeding schemes.

In case of **sexual reproduction**, haploid male and female gametes fuse to produce diploid zygotes that develop into embryos and, ultimately, plants. The male and female gametes are produced in the anthers (a component of stamens) and ovules (present inside the ovaries), respectively. Several specialized cells, called **microspore mother cells**, in each anther undergo meiosis to produce haploid microspores or pollen grains. Two rounds of mitosis in the pollen grains produce to sperms, the male gametes. But a single specialized cell, called **megaspore mother cell**, in each ovule divides meiotically to yield four haploid cells one of which functions as the megaspore. Three rounds of mitotic division in this cell produce eight nuclei, one of which functions as egg cell (forms the zygote) while two nuclei fuse to generate the secondary nucleus (produces endosperm). The mature anthers dehisce to liberate mature pollen grains. Many of the pollen grains land on the stigma of the same or some other flowers; this is called **pollination**. The pollen grains germinate, and the pollen tubes grow through the style, enter the ovules. In the end, they enter the embryo sacs, and liberate two sperms in each embryo sac. One sperm fuses with the egg cell to produce diploid zygote that gives rise to embryo and, in the end, progeny plant. The other sperm fuses with the secondary nucleus to produce the triploid endosperm that provides nutrition to the developing embryo. The ovule develops as seed, which is a hardy organ and used for propagation of crops.

When pollen produced by a flower pollinates the same flower, it is called **self-pollination**. Self-pollination is promoted by non-opening of the flowers, opening of the flowers only after pollination has occurred; due to the anthers being located on top/around the stigma or the anthers and stigma being enclosed in a flower part like petals. The self-pollinated crops (Table 1.5) show very high amount of self-pollination, and less than 5% cross-pollination; some genotypes of some crop species may show very little cross-pollination. Self-pollination leads to a rapid increase in homozygosity, and the level of heterozygosity decreases to 50% of that in the previous generation. Therefore, natural populations of such species are composed of homozygous genotypes. These species show little or no inbreeding depression but may show considerable extent of heterosis (Section 1.10). Therefore, the cultivated varieties in these crops are generally homozygous lines, called **purelines** (Section 2), but hybrid varieties are also used.

Table 1.5: Some examples of crops belonging to the self-pollinated, cross-pollinated and often cross-pollinated categories

Mode of pollination		
Self-pollinated crops	**Often cross-pollinated crops**	**Cross-pollinated crops**
Wheat	Maize	Sorghum (jowar)
Rice	Pearl millet	Cotton
Barley	Alfalfa (*Medicago sativa*)	Pigeon pea (arhar)
Pea	Cabbage	Safflower
Groundnut	Carrot	*Brassica juncea*
Chickpea	Onion	Triticale
Soybean	Sunflower	Tobacco
Tomato	Castor	Jute (*Corchorus olitorius*)
Brinjal	Apple	Broad bean (*Vicia faba*)
Linseed	Mango	

In case of cross-pollination, pollen produced by a flower pollinates other flowers. In nature, the pollen is transferred from one flower to the other by wind, insects or water (in case of water plants). A variety of mechanisms promote cross-pollination, e.g., unisexual flowers, stamens maturing before (**protandry**) or after (**protogyny**) the pistils mature self-incompatibility etc. The cross-pollinated crops (Table 1.5) show very high frequency of cross-pollination, but they may exhibit up to 10% self-pollination. Cross-pollination promotes heterozygosity so that cross-pollinated populations are highly heterozygous. These crops show some to considerable amount of inbreeding depression, and substantial amount of heterosis. Therefore, the varieties used in these crops are highly heterozygous (open-pollinated, synthetic and composite varieties; Section 3). The hybrid varieties (Section 2) are the F_1 generations of crosses and show the maximum possible level of heterozygosity.

Many sexually reproducing crop species are called **often cross-pollinated species**; these species may show more than 5% and up to 30% cross-pollination. The genetic make-up of these crop species is intermediate between that of self- and cross-pollinated crops. Therefore, the breeding methods designed for either self-pollinated or cross-pollinated species can be applied in these crops. Further, both pureline and hybrid varieties are used in often cross-pollinated species.

In conclusion, mode of reproduction determines the genetic constitutions of the populations, the prevalent nature of gene action, the ease in controlled pollination, stability of the varieties after release, the method of germplasm conservation, the breeding schemes, and the types of varieties used for cultivation (Table 1.6).

Table 1.6: Relevance of the mode of reproduction to the breeding approaches and other features of the crop species

Feature	Asexual reproduction	Cross-pollination	Self-pollination
Genetic constitution	Highly heterozygous	Highly heterozygous	Homozygous
Inbreeding depression	Severe	Low to severe	Little to none
Heterosis	Considerable	Considerable	Considerable
Nature of gene action	Mainly nonadditive	Mainly nonadditive	Mainly additive
Ease in controlled pollination	Sexual reproduction suppressed	Easy out-crossing; selfing requires effort	Selfing effortless; cross-pollination requires considerable effort
Stability of varieties after release	Stable (as clones)	Hybrid variety can be grown for one generation; other varieties remain genetically stable*	
Germplasm conservation	In field as plants, slow-growth shoot cultures, cells/tissues/organs frozen in liquid nitrogen	In seed banks	In seed banks

* The synthetic, composite and open-pollinated varieties can be grown for many generations provided contamination from foreign pollen is prevented.

1.8. QUALITATIVE AND QUANTITATIVE TRAITS

A character or trait is any morphological, anatomical or biochemical feature of an organism. A given trait of an organism would be either a qualitative or quantitative trait, depending on its mode of inheritance and the effects of the environment on its expression. A **qualitative trait** is governed by one or few genes producing large effects, is little affected by the environmental variations, and shows discrete variation (distinct groups). The genes governing such traits are called **major genes** or **oligogenes**; therefore, these traits are also called **oligogenic traits**. The **quantitative traits**, on the other hand, are governed by several genes with small effects, are markedly affected by the environment, and show continuous variation. The genes governing such traits are called polygenes so that these traits are also known as **polygenic traits**. The current distinction between qualitative and quantitative traits is based on the effect of the environment on the trait phenotype (Table 1.7). Therefore, a trait that is markedly affected by the environment is considered as a qualitative trait even when it is governed by a single gene with major effect.

Table 1.7: The main characteristics of qualitative and quantitative traits

Type of trait	Distribution in F_2	Percent phenotypic effect due to genotype	Gene governing the trait
Qualitative trait	Discrete	100	Oligogene
Quantitative trait	Continuous	10 and above	Major QTL*
Quantitative trait	Continuous	<10	Minor QTL

* A QTL is a genomic region that has genes involved in the control of a quantitative trait.

1.8.1. Inheritance of Qualitative Traits

The two fundamental laws of inheritance were first described by Mendel.in 1865 (published in 1866) based on his studies on qualitative traits. The discussion of these laws would be facilitated by using some specific terms as defined here. A character is presumed to be governed by a gene located at a specific site in a chromosome. A gene has two (or more) alternative forms called **alleles**; each allele governs one form of the trait in question. Generally, one of the alleles is dominant and the other one is recessive. The dominant allele is able to express itself in the heterozygote, while the recessive allele is unable to do so; this situation is described as **complete dominance**. A heterozygote has two alleles of the concerned gene, but a homozygote has a single allele in two copies. The alleles of a gene may show **incomplete dominance** (the F_1 is intermediate between the two homozygotes), **co-dominance** (both the alleles are expressed) or **overdominance** (F_1 is superior to both the homozygotes). The first law, the law of segregation, of Mendel specifies that the two alleles of a gene present in the F_1 hybrid separate from each other and pass into different gametes. As a result, each gamete has only a single of a gene, and 3:1 ratio is obtained in the F_2 generation when the alleles of the gene show complete dominance. According to the second law, the **law of independent assortment**, when two genes governing two different traits segregate together, their segregation is independent of each other. As a result, four different gametes containing the four possible combinations of the alleles of the two genes are produced, and in the F_2 generation a 9:3:3:1 ratio is obtained. It is now known that a single character may be governed by two or more genes. Further, these genes may affect the expression of each other in a variety of ways; this is called **gene interaction** or **epistasis**. The different types of gene interactions give rise to the various modifications of the typical F_2 ratios. In addition, in many cases, a single gene affects more than one trait; this is known as **pleiotropy**. The pleiotropic effects of a gene can be enhanced or reduced by the genetic background in which it is placed. **Genetic background** comprises all the genes present in the nucleus of an individual, except the gene in question.

It is now well known the genes located close to each other in the same chromosome tend to be inherited together; this is known as **linkage**. But new combinations of linked genes, i.e., recombination, are produced by crossing over, which exchange the homologous segments of chromatin between homologous chromosomes. The frequency of recombination between linked genes is generally proportional to the distance between them in the chromosome. This property has been used to construct linkage maps of genes. A linkage map is a linear representation of the genes present in a chromosome in the order and at distances derived from the frequencies of recombination between them. The genes may be linked in coupling phase or in repulsion phase. In **coupling phase**, the dominant alleles of two genes are linked together. In case the dominant alleles produce favourable effects, coupling phase linkage is advantageous since selection for the favourable allele of one gene will also select for that of the other gene. But in case of **repulsion phase linkage**, the dominant allele of one gene is linked with the recessive allele of the other gene. The repulsion phase linkage will be undesirable in all such cases where the recessive allele of a gene produces detrimental effects.

The expression of qualitative traits is little affected by the environment. But the expression of the mutant alleles of many oligogenes is markedly affected by the environment; this sometimes produces variable phenotypes of the otherwise oligogenic trait. In addition, some phenotypic expression of some traits requires some specific environment; such traits are known as **threshold characters**. For example, the expression of a disease resistance gene can be assayed only when the lines are exposed to the concerned pathogen under conditions suitable for the development of the concerned disease. Similarly, drought resistance can be selected only under the conditions of drought. Therefore, often the selection for threshold traits presents problems since creation of the appropriate conditions may not be easy.

1.8.2. Inheritance of Quantitative Traits

In 1908, Nilsson-Ehle proposed the **multiple factor hypotheses** to explain the inheritance of quantitative traits. According to this hypothesis, quantitative traits are governed by several genes that have small individual effects and act in an additive manner. It was also assumed that all the genes, called **polygenes,** affecting a trait have equal effect on the trait. But it is now recognized that polygenes show dominance as well as epistatic gene action in addition to the additive effects. It was not possible to follow the inheritance of quantitative traits using the techniques developed for the analysis of the qualitative traits. Therefore, methods based on statistical procedures were developed for their analyses. Further, the development of molecular markers (Chapters 13, 14 and 15) has allowed the identification of genomic regions involved in the control of various quantitative traits; such genomic regions are known as **quantitative trait loci (QTLs).** A single QTL may contain one or, more often, more genes. Many QTLs have been cloned and it has been found that the genes present in them perform a variety of functions, and many of them encode transcription factors. QTL analysis has revealed that in most of the cases, a quantitative trait is governed by one or few genes with large effects and a large number of QTLs each producing a small effect on the trait. It may be pointed out that the continuous variation of quantitative traits is produced due to the following two factors: (i) involvement of several polygenes (or QTLs as they are identified today) in the control of each trait, and (ii) the effect of the environment on the phenotypic expression of the polygenes/QTLs. The expression of QTLs is also affected by the genetic background, which also contributes to the continuous distribution of quantitative traits.

The phenotypic expression (P) of a quantitative trait in an individual is affected by the genotype (G) the individual for the trait, the prevailing environment (E) as well as the interaction between G and E (G x E). The presence of G x E interaction signifies that the relative performance of different genotypes changes in different environments. Thus the phenotypic expression of a trait in a genotype can be depicted as follows.

$$P = \mu + G + E + G \times E$$

where, μ is the population mean. One of the crucial questions, therefore, relates to the relative contribution of the G and E (including G x E) to the P. One approach uses the phenotypic variance estimates for partitioning into components due to G, and E effects. However, the estimation of variance due to G x E effects requires specific experimental designs and considerable effort. The relative importance of the variance components due to G and E is easily revealed by the estimate of heritability estimate for the trait, which is estimated as follows.

$$\text{Heritability (\%)} = [Vg/(Vg + Ve)] \times 100$$
$$= (Vg/Vp) \times 100$$

where, *Vg*, *Ve* and *Vp* are the estimates of genotypic, environmental and phenotypic variances, respectively. This estimate of heritability is known as **broad sense heritability**. It indicates the proportion of phenotypic variance for the concerned trait produced by genotype. The genotypic variance can be divided into additive, dominance and epistatic components. Of these components, only the additive variance and additive x additive epistatic components are expressed in the homozygous genotypes. Therefore, these components are known as **fixable gene effects**. In contrast, the dominance and the other components of the epistatic variance are expressed only in heterozygous genotypes. Therefore, these components are called **nonfixable gene effects**. Thus, in case of segregating generations the broad sense heritability is appropriate. In these generations, the appropriate estimate of heritability is the ratio of additive genetic variance (*Va*) to the phenotypic variance.

$$\text{Heritability (narrow sense) (\%)} = (Va/Vp) \times 100$$

It may be pointed out that selection for a trait is based on its phenotype. The estimate of heritability indicates the effectiveness of the phenotype based selection. A high heritability estimate, e.g., 60% or more, indicates that selection for the trait should be quite effective. But a low estimate of heritability, say, <40%, suggests that selection for the trait may have poor effectiveness.

1.9. ACHIEVEMENTS

Plant breeding has been highly successful in developing high yielding crop varieties adapted to the various agroclimatic conditions. These varieties are developed by using a variety of tools and breeding methods, which are discussed in adequate detail in the Chapters 2-6. The improved varieties, in addition to being high yielding, are resistant to many diseases and insect pests, and often possess desirable consumer qualities. For example, the semidwarf varieties of cereals like wheat and rice initiated the **green revolution** that made India self-sufficient in food grains. These varieties are resistant to lodging, responsive to fertilizer application and high yielding. The semidwarf wheat varieties are generally resistant to the three rusts and other major diseases of wheat. Similarly, the rice varieties are resistant to many of the prevalent disease. These wheat and rice varieties are also photoperiod insensitive and, as a result, suitable for late planting/planting in nontraditional areas. Another excellent example is the noblization of sugarcane. The Indian canes (*Saccharum barberi*) are hardy but give poor cane and sugar yields. The noble canes (*Saccharum officinarum*), on the other hand, are high yielding and have high sugar content, but they not adapted to the climatic conditions of North India and are susceptible to diseases and insect pest. Therefore, the noble canes were crossed with *S. barberi* and other *Saccharum* species to develop the present-day sugarcane varieties, which are known as **noblized canes**. The noblized sugarcane varieties show increased resistance to diseases and insect pests, greater adaptation to the environmental conditions and, above all, higher yield that the noble canes themselves.

The development of hybrid varieties is an important contribution of plant breeding. Heterosis or hybrid vigor is expressed in almost all the crops, including the self-pollinated ones, and it is best exploited by hybrid varieties. The hybrid varieties were first developed in maize and used in USA beginning in the year 1922. In India, work on hybrid variety development in maize began in 1957, and the first hybrid varieties were released in 1961. Soon thereafter, in 1964, hybrid varieties of sorghum and pearl millet were released for commercial cultivation. One of the notable developments in hybrid variety development in India is the use of hybrid varieties in cotton. The first hybrid cotton variety, H4, was developed in 1970; it was a hybrid between two strains of *Gossypium hirsutum*. Subsequently, hybrid varieties were developed by crossing strains of *G. hirsutum* with those of *G. barbadense*, and within the *desi* cottons. Initially, the hybrid seed was produced by hand emasculation followed by hand pollination, but at present cytoplasmic-genetic male sterility is used for this purpose. Hybrid varieties have also been developed in several other crops like pigeon pea, rice, sunflower, etc. and in several vegetable crops.

1.10. UNDESIRABLE CONSEQUENCES OF PLANT BREEDING

Human altered the ecosystems they encountered as they began inhabiting the planet. Inventions and adaptations of agricultural practices involved removal of natural flora and wide-spread habitat destruction. The simplification of biodiversity favored monoculture in many crops. Tomato and its processed products are one of the most widely consumed fruits. Its domestication for example has resulted in the loss of some 90% of the genetic and chemical diversity of wild relatives. Only less than 5% of the genetic variation of wild relatives of crops is present in domesticated/super-domesticated species due to selection of preferred genotypes in the existing germplasm (a process known as **domestication syndrome**).

Plant breeding does not always lead to only desirable modifications in the crop plants and their populations. It does produce some undesirable effects, e.g., genetic erosion, genetic vulnerability, susceptibility to otherwise minor diseases and yield plateau. The great success of the improved crop varieties is one of the main reasons for genetic erosion particularly in the cultivated plant species. The improved varieties are highly homogeneous and, in the case of self-pollinated crops, highly homozygous. They have replaced the genetically variable land races (old local varieties) of crops leading to loss of genetic variability present in those varieties. In addition, a single improved variety is usually grown on large areas. Therefore, when this variety becomes susceptible to a disease, the whole area occupied by the variety is affected leading to considerable economic losses. In addition, the genetic base of the crops tends become increasingly with the level of crop improvement since the cultivated varieties of a crop have one or more parents in their ancestry. As a result, they are more similar to each other than they would be if they did not share ancestor. This genetic similarity leads to genetic vulnerability, i.e., susceptibility of most of the cultivated varieties of a crop to a disease. For example, the early pearl millet hybrids in India had either Tif 23A or Tif 23D2A as their female parent. These male sterile lines were highly susceptible to downy mildew and ergot. As a result, all the pearl millet hybrid varieties were susceptible to these diseases and had to be withdrawn from cultivation. Subsequently, hybrids resistant to these diseases were developed.

Sometimes, high yielding varieties show unusual susceptibility to an otherwise minor disease of the concerned crop. This is because the plant breeding efforts are focused on the development of superior varieties with resistance for important diseases for the crop. For example, the wheat variety WL711 was an excellent wheat variety, which showed severe infection by Karnal bunt caused by *Tilletia indica*. This wheat variety had to withdrawn from cultivation within a very short period after its release. In addition, the breeding efforts in all crops sooner or later reach a stage when yield increases are not easily achieved, and a plateau is reached. This situation has been experienced in crops like rice and wheat and novel approaches are devised to break this plateau. For example, yield genes may be introgressed from related wild species to increase the opportunities for yield improvement of the concerned crops.

1.11. LIMITATIONS OF PLANT BREEDING METHODS

There are two critical requirements for improvements in the various characteristics of the crops. The first requirement is the presence of the desired genes/alleles in the breeding populations or the germplasm collection. It may be emphasized that in every crop species, the desired genes/alleles for many important traits are not available in the lines/accessions of the crops. Breeders have attempted to introduce the genes from related wild species and even other crop species (Sections 2.6, 2.7). Since this requires sexual compatibility between the concerned species, some to considerable difficulties are encountered in these efforts and sometimes somatic hybridization may have to be used (Sections 2.7). But often the desired gene may be present in species that cannot be crossed with the concerned crop species. In such cases, recombinant DNA technology or transgenic technology could be used to isolate the desired genes and to transfer them into the concerned crop species. The various aspects of transgenic technology are discussed in some detail in the Chapters 7 to 10. The transgenic approach has been quite successful in the development of crop varieties with the desired features, and in 2012 these varieties were grown in more than two dozen countries in over 170 million hectares.

The second requirement is a dependable technique for a reliable identification of the desired genotype from among the large number of genotypes present in the segregating generations of the appropriate crosses. This technique, in addition to being reliable, has to be easy, fast and inexpensive, and allow the evaluation of the target traits and allow the prediction of the performance of the progeny from the selected plants. In addition, accessing the desirable genes present in related species and eliminating the risk of linkage drag often encountered in introgression of desirable traits from unadapted germplasm are the other important challenges. Finally, the breeding process has to be accelerated to enable the development of new varieties to meet the challenges posed by the growing human population and the abiotic and biotic stresses associated with the climate change. The molecular marker technology facilitates almost all of the above objectives. This technology is being widely used by the private sector seed companies and is being uses by several national programmes as well. In India, some varieties developed by using molecular markers have been released for commercial cultivation. The various aspects of molecular markers and their use to achieve plant breeding objectives are discussed in the Chapters 11-13. The last chapter of the book deals with variety release, seed multiplication and intellectual property rights issues relevant for plant breeding and transgenic technology.

1.12. CONCLUSIONS

Our crop plants were derived from wild weedy species by the process of domestication that began over 10,000 years ago and was rather quite successful. Domestication new plant species continues even today to develop crops for some specific purposes. After domestication, crop plants were subjected to selection to utilize the genetic variation present in their populations. A variety of approaches, including interspecific hybridization, somatic hybridization, mutagenesis, polyploidy were devised to create new genetic variation for crop improvement. The new genetic variation was used by devising clever breeding schemes designed to make the best use of the genetic variability. The plant breeding efforts have been very successful in developing a large number of improved varieties in every crop species. However, plant breeding has generated some undesirable consequences as well, e.g., genetic erosion, genetic vulnerability, susceptibility to otherwise minor diseases and yield plateau. Plant breeders are always developing strategies to minimize the consequences of these undesirable effects, e.g., collection and conservation of plant genetic resources to minimize the adverse effects of the genetic erosion. The two major limitations of plant breeding methods relate to the availability of the desired genes in the breeding population, and the selection on the basis of plant phenotype. The transgenic technology was developed to make it possible to use any gene from any organism to be used for crop improvement. Similarly, molecular marker technology has been refined to enable efficient indirect selection for genomic regions involved in the control of the desired crop traits. Crop varieties developed by these two modern technologies are already being cultivated. It is expected that these and other novel technologies being developed for application in plant breeding will make crop improvement methods more effective and precise. These and future developments in the tools and techniques of plant breeding may be expected to enable humans to face the dual challenges of increasing human population and the adverse impacts of climate change.

PRACTICE QUESTIONS

1. Define plant breeding. Discuss its relevance to food security and climate change.
2. Plant breeding is a process that proceeds in distinct steps. Briefly describe the different steps in plant breeding programmes, and discuss their relevance.
3. Briefly describe the major developments in plant breeding since the domestication of certain wild species as crops.
4. What do you understand from genetic resources? Briefly describe their collection and conservation, and discuss their relevance to crop improvement.
5. Briefly describe the different modes of reproduction of crop plants, and discuss their significance in plant breeding.
6. Briefly describe the inheritance of qualitative and quantitative traits. Highlight the chief differences in the inheritance of these two types of traits and their implications in plant breeding.
7. Highlight the significant achievements of plant breeding. Discuss the limitation of conventional plant breeding methods and examine the undesirable effects of plant breeding activities.
8. Plant breeding has brought about remarkable improvements in the features of the crop plants, but it has also produced some undesirable side effects. Discuss this statement in the light of the available information.
9. Write short notes on the following: (i) plant breeding, (ii) domestication, (iii) germplasm or genetic resources, (iv) germplasm collection, (v) germplasm conservation, (vi) asexual reproduction, (vii) sexual reproduction

RELEVANT LITERATURE

1. Mehta TR 1986. Early history of plant breeding in India. Indian J. Genet. 46: 1-11.
2. Simmonds NW 1973. Plant breeding. Phil. Trans. Royal Soc. London B 267: 145-156.

3. Simmonds NW 1979. Principles of Crop Improvement. Longman, London and New York.
4. Simmonds NW 1991. Bandwagons I have known. TAA Newsletter Dec. 1991: 7-10.
5. Singh BD 2015. Plant Breeding: Principles and Methods, 10th edn. Kalyani Publishers, Ludhiana, New Delhi.
6. Zamft BM. and Robert J. Conrado 2015. Engineering plants to reflect light: strategies for engineering water-efficient plants to adapt to a changing climate. Plant Biotechnology J.13: 867–874.
7. Biel K and I. Fomina 2015. Benson-Bassham-Calvin cycle contribution to the organic life on our planet. Photosynthetica, 53 (2): 161-167, 201.
8. UNFAO, 2009. The State of Food and Agriculture, 2009. Food and Agriculture Organization of the United Nations Rome, 2009.
9. UNFAO, 2014. The State of Food and Agriculture, 2014. Food and Agriculture Organization of the United Nations Rome, 2014.
10. Galston, AW 1980. GREEN WISDOM. The Inside story of Plant Life. Basic Books, Inc., Publishers, New York. p. 8.
11. Harlan, JR 1992. Crops and Man, 2nd Ed., American Society of Agronomy, Madison, Wisconsin, USA.
12. Moose, SP and Rita H Mumm 2008. Molecular Plant Breeding as the Foundation of 21st Century Improvement Plant Physiology. 147: 969-977.

■■■■

2 The Relevance of Genetics and Genomics in Plant Breeding

2.1. INTRODUCTION

Plant breeding aims to improve the various characteristics of crop plants so as to make them more useful to humans. It is universally accepted that the expression of a trait (phenotype) of an organism is governed by one or more genes (genotype), and that gene expression is influenced by the prevailing environment. Thus, phenotype (P) of a trait is produced by the actions of the relevant genotype (G) and the prevailing environment (E); this can be expressed as follows: $P = G + E$. In addition, there is an interaction (G x E interaction) component of the phenotype. Thus, $P = G + E + (G \text{ x } E)$,

In general, the contribution of the environment is more important in the case of quantitative traits than that for qualitative traits (Section 1.8, Table 1.7). The genes themselves are made up of deoxyribonucleic acid (DNA), which is present mainly in the nuclei and, to a small extent, in the chloroplasts and mitochondria. The inheritance of nuclear genes, or simply genes, was first elucidated by Gregor J. Mendel in 1866. The genes present in the mitochondria and chloroplasts generally show uniparental transmission. This inheritance pattern is usually called **cytoplasmic inheritance**, and was first reported in 1908 independently by Correns and Bauer. The genes are organized in chromosomes that make up the genomes of the organisms. The phenotypic expression of genes is mediated by ribose nucleic acid (RNA) and protein syntheses, and a variety of biochemical reactions catalyzed by the proteins. In this chapter, we shall consider the salient features of qualitative and quantitative inheritance, genome organization, gene expression and their relevance to plant breeding

2.2. GENES AND CHROMOSOMES

In *classical terms*, **gene** is a functional unit that governs the development of a character of an organism. But it is well known that a trait is not produced by a single gene alone, and that each gene codes for one RNA molecule, which generally specifies a protein or more specifically a polypeptide. Thus, *in the current perspective*, a **gene** is a segment of DNA (including the regulatory sequences at its both the ends) that encodes a polypeptide or an RNA molecule (product). Some genes, however, are known to encode two or more polypeptides that are produced either by alternative splicing of their RNA products or by processing of the polypeptides encoded by them. A polypeptide generates a functional protein molecule either alone or in association with some other polypeptide(s). A given protein may serve an enzymatic, structural, regulatory or hormonal function in the cells and tissues. Thus the various proteins encoded by the different genes of an organism perform the vast array of cellular functions, which ultimately give rise to the variety of traits of the organisms. Each gene has two or more alternative forms called **alleles**, which generally give rise to the different phenotypes of the concerned traits.

It is universally accepted that genes are located in chromosomes. Further, in plants, the genes are present in nuclei, mitochondria and chloroplasts. The genes present in nuclei are called **nuclear genes** or, simply, **genes**, while those located in mitochondria and chloroplasts are known as **plasmagenes**. The **chromosomes** present in nuclei are linear bodies composed of DNA and proteins (mainly histones) and they stain deeply with basic dyes. A chromosome has one centromere, which divides the chromosome into two arms; each arm has one telomere and contains several specialized sequences that function as origins replication, i.e., they function in the initiation of DNA replication. Each chromosome, generally, has two chromatids, called **sister chromatids**, which are produced by replication of a single parental chromatid. The chromosomes become much shorter and thicker during some stages of cell division and can be easily seen (after proper staining) with the help of light microscope. In a diploid plant species, each somatic cell has two copies of each chromosome. Therefore, the somatic chromosome complement of a species is represented by $2n$. In contrast, the gametes of a species have only one copy of each chromosome, and the gametic chromosome complement is denoted by n. In a diploid species, the n chromosomes constitute the genome of the species. Each member of a genome is distinct from the others in terms of gene content and, in many cases, morphological features. Since the genes are located in chromosomes, each somatic cell has two copies of each gene, while the gametes have only single copies of the genes. Thus, the behavior of genes during inheritance is the same as that of the chromosomes. Each gene occupies a fixed position, called **locus**, in a specific chromosome. These genes show typical Mendelian inheritance, and they are the chief determinants of the plant growth, development and performance.

The chromosomes present in the mitochondria and chloroplasts, in contrast, are circular in shape and are composed primarily of DNA. In addition, the chromosome is present in multiple copies in each mitochondrion and chloroplast. The genes present in these chromosomes show cytoplasmic inheritance (Section 2.7), and they contribute to the performance of plants in many ways.

2.3. CELL DIVISION

The **cell** is the unit of structure and function of all organisms. Further, all cells originate from pre-existing cells by a process called cell division. In **cell division**, the cytoplasm and the chromosomes of an existing cell, called **parent cell**, divide into two new cells, termed as **daughter cells**, which are comparable with each other. There are two types of cell division, viz., mitosis and meiosis. During **mitosis**, the two chromatids of every chromosome of a cell separate and move to the opposite poles. As a result, the two poles receive the same number and type of chromatids as those present in the parent cell. Thus the two daughter cells produced by mitosis in a parental cell are identical with each other and with the parent cell in terms of their chromosome complement. Each of these chromatids later undergoes replication to generate chromosomes composed of two chromatids each. As a rule, each mitotic division is preceded by replication of the chromosomes (each chromosome represented by a single chromatid). As a consequence, the chromosome of the somatic cells is maintained as $2n$. In case of plants, mitosis occurs in root and shoot tips, young leaves, flower buds and cambium. It also occurs in the microspores and megaspores during gameto-genesis. Mitotic cell division occurs in all somatic tissues, and is the basis of vegetative propagation as well as asexual reproduction, including some forms of apomixes. In view of the above, the progeny produced by asexual or vegetative reproduction of a single parent plant have the same genotype as the parent plant, and they are identical with each other. Such progeny together are called **clone**. Therefore, the phenotypic variation present within a clone is generated by the environment, does not have any genetic basis, and is non-heritable.

In case of **meiosis**, a single cell divides two times one after the other, and there is no DNA or chromosome replication between these two divisions. As a result, each parent cell produces four daughter cells, each of which has only one-half (that is, n) the number of chromosomes present in the parent cell, i.e., $2n$. In the prophase (zygotene) of the first division of meiosis, the two homologues of each chromosome pair and form n number of bivalents. At anaphase, the two homologues of each chromosome separate and move to the opposite poles so that each pole receives only one member of each homologous pair. The second meiotic division follows soon, during which one chromatid each from all the chromosomes moves to the opposite poles. As a result, four haploid nuclei are produced at the end of the second meiotic division, and each of these nuclei has only n number of chromosomes. In case of plants, meiosis occurs in anther and ovule cells, which are involved in spore production. The spores ultimately give rise to gametes that are the basis of sexual reproduction. Meiosis leads to the phenomena of segregation and independent assortment that generate genetic variation in the sexual progeny. During pachytene stage of the prophase of first meiotic division, crossing over takes place. In **crossing over**, there is an exchange of strictly homologous parts between the two members of the homologous chromosome pair forming a bivalent. This leads to recombination between linked genes.

2.4. INHERITANCE OF QUALITATIVE TRAITS

In 1866, Mendel proposed the two fundamental laws of inheritance on the basis of his work with peas. These laws were rediscovered in 1900, which sparked research activities in this new area, which was later (in 1905) termed as **genetics** by Bateson. A discussion of the two fundamental laws of inheritance, applicable to both qualitative as well as quantitative traits, would be greatly facilitated by using some specific terms as defined here. A character is presumed to be governed by a gene that is located at a specific site, called **locus**, in a chromosome. A gene has two (or more) alternative forms called **alleles**. Each allele of a gene governs one form of the character or trait specified by the gene in question. Generally, one of the alleles is dominant and the other/others is/are recessive. A **dominant** allele is able to express itself in the heterozygote, while a recessive allele is unable to do so; this situation is described as **complete dominance**. A **heterozygote** has one copy each of two different alleles of the concerned gene. In contrast, a **homozygote** has two copies of a single allele the gene in question. The alleles of a gene may show **incomplete dominance** (the F_1 is intermediate between the two homozygotes), **co-dominance** (both the alleles are expressed in the heterozygotes) or **overdominance** (the F_1 is superior to both the homozygotes).

2.4.1. Segregation and Independent Assortment

The first law proposed by Mendel is the **law of segregation**. It specifies that the two alleles of a gene present in the F_1 hybrid separate from each other and pass into different gametes. As a result, each gamete has only a single allele of the gene, and the F_1 produces two types of gametes in equal frequency. A random union among the male and female gametes of the F_1 produces three genotypes in 1:2:1 ratio, and a 3:1 phenotypic ratio is obtained in the F_2 generation when the alleles of the gene show complete dominance (Fig. 2.1, Table 2.1). But in cases of incomplete dominance and over-dominance, phenotypic 1:2:1 ratio would be obtained in the F_2 generation. Segregation occurs due to meiosis. It may be pointed out that the two alleles of a gene are located at the same locus in the homologous chromosomes. The homologous chromosomes pair during prophase and move to the opposite poles during the anaphase of first division of meiosis. This leads to the separation of the two alleles of a gene, and their inclusion in different gametes. Segregation leads to production of genetic variation in the sexual progeny. All the genes that are present in heterozygous condition in a plant segregate and produce the homozygous genotypes of the concerned genes, which generate genetic variation in the sexual progeny.

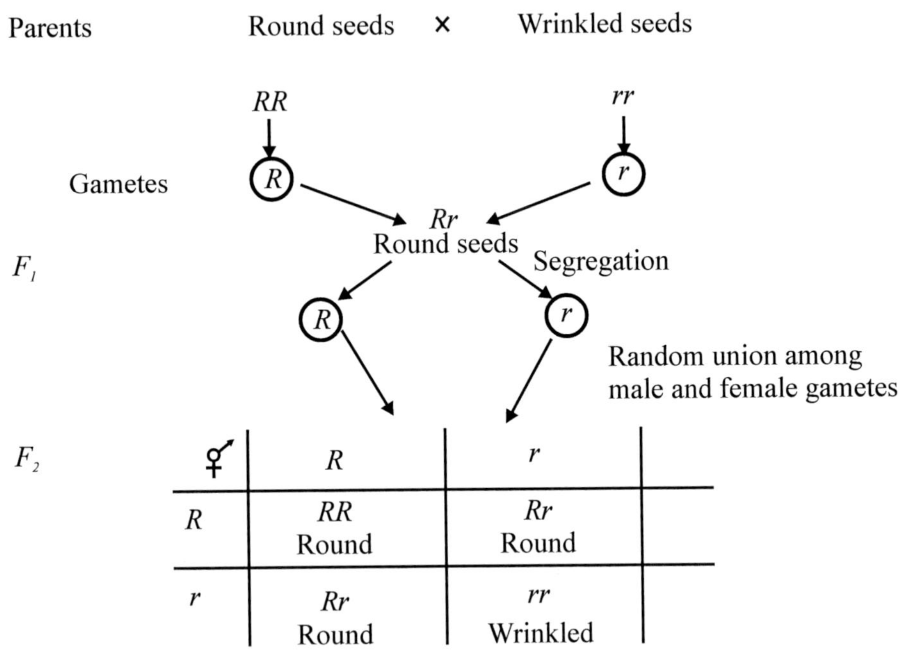

Fig. 2.1. Inheritance of seed shape in pea. Round and wrinkled seeds are determined by the dominant allele *R* and the recessive allele *r*, respectively. The parents are homozygous *RR* and *rr*, and their F_1 is *Rr* (heterozygous). The alleles *R* and *r* pass into different gametes produced by the F_1. A random union among the male and female gametes produces 1 *RR,* 2 *Rr* and 1 *rr* progeny in F_2, which gives 3 : 1 phenotypie ratio (Table 2.1)

Table 2.1: The genotypes and phenotypes of the progeny produced due to segregation of the two alleles, *R* and *r*, of the gene governing seed shape in pea (see, Fig. 2.1)

Genotype	Frequency	Phenotype	Remarks
RR	1	Round	Genotypic ratio: 1:2:1
Rr	2	Round	Phenotypic ratio: 3:1
rr	1	Wrinkled	

According to the second law, the **law of independent assortment**, when two genes, each governing a different trait, segregate together, their segregation is independent of each other. As a result, the two alleles of one gene combine freely with the two alleles of the other gene. This would generate equal frequencies of four different combinations of the alleles of the two genes. Thus independent segregation of two alleles produces four types of gametes in equal frequencies in the F_1 generation. A random union among these male and female gametes would produce nine different genotypes and four different phenotypes in 9:3:3:1 ratio in the F_2 generation, when there is complete dominance at the two loci (Fig. 2.2, Table 2.2). Independent assortment always occurs when the two genes are located in different chromosomes. This is because the two members of one chromosome pair combine randomly with the two members of the other chromosome pair during the first anaphase of meiosis. Therefore, two alleles of the genes present in these chromosomes also combine randomly leading to their independent assortment. However, genes located in the same chromosome tend to be inherited together but this tendency declines as the distance between the

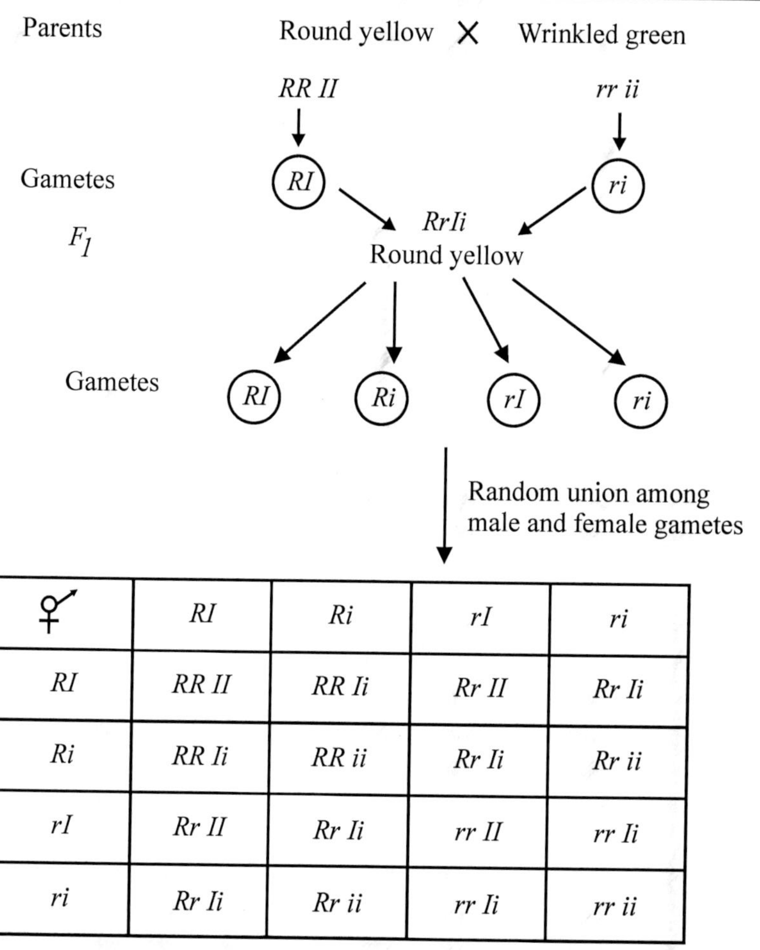

♀ ♂	*RI*	*Ri*	*rI*	*ri*
RI	*RR II*	*RR Ii*	*Rr II*	*Rr Ii*
Ri	*RR Ii*	*RR ii*	*Rr Ii*	*Rr ii*
rI	*Rr II*	*Rr Ii*	*rr II*	*rr Ii*
ri	*Rr Ii*	*Rr ii*	*rr Ii*	*rr ii*

Fig. 2.2. Independent assortments for genes governing seed shape and cotyledon colour in pea. Alleles *R* and *r* govern seed shape, while *I* and *i* specify cotyledon colour. The alleles *R/r* and *I/i* combine independently (= randomly) to produce four different types of gametes in F_1. These gametes unite randomly to produce 9:3:3:1 phenotypic ratio in F_2 (Table 2.2)

two genes increases. Therefore, when two genes are located in a chromosome at a relatively longer distance from each other, they tend to assort independently. Independent segregation produces new combinations of genes and, as a result, character phenotypes as compared to those present in the parents that were crossed to produce the F_1.

Table 2.2: The genotypes and phenotypes produced by independent segregation for seed shape and cotyledon colour in pea (Fig. 2.2). Seed shape (round and wrinkled) is governed by the alleles *R* and *r*, while cotyledon color is determined by the alleles *I* and *i*

Genotype	Frequency	Phenotype	Remarks
RR II	1	Round yellow	Genotypic ratio: 1:2:2:4:1:2:1:2:1
RR Ii	2	Round yellow	Phenotypic ratio: 9:3:3:1
Rr II	2	Round yellow	
Rr Ii	4	Round yellow	
RR ii	1	Round green	
Rr ii	2	Round green	
rr II	1	Wrinkled yellow	
rr Ii	2	Wrinkled yellow	
rr ii	1	Wrinkled green	

2.4.2. Gene Interaction and Pleiotropy

It is now well recognized that a single gene alone is not likely to specify the development of a character. In fact, many different genes participate in the development of every character. But often a single gene may influence the phenotype of a given character to such a large extent that it appears to determine the given trait. The phenotypes of many characters are known to be governed by two or more genes. Further, these genes generally affect the expression of each other in a variety of different ways; this is called **gene interaction** or **epistasis**. The different types of gene interactions give rise to the various modifications of the typical F_2 ratios. Some of the common interactions between two genes are summarized in Table 2.3. An example of gene interaction is provided by the production of hydrocyanic (HCN) acid in *Trifolium* spp. The HCN production is known to be governed by two genes, *L* and *H*. The genes *L* and *H* show complementary gene interaction. Gene *L* encodes an enzyme that produces a cyanogenic glucoside from a substrate. Similarly, the enzyme encoded by gene *H* converts this cyanogenic glucoside into HCN. As a result, the dominant alleles of both the genes *L* and *H* are required for the production of HCN (Fig. 2.3).

The occurrence of gene interactions has the following consequence relevant to plant breeding: (i) production of di- or tri-hybrid ratios in F_2 for a single trait in the place of the simple monohybrid ratio of 3:1 or 1:2:1, which may be sometimes be confusing; (ii) the need for transfer of two or more gene for improvement in a single trait; and (iii) a delay in the isolation of homozygous lines due to the involvement of more than one gene.

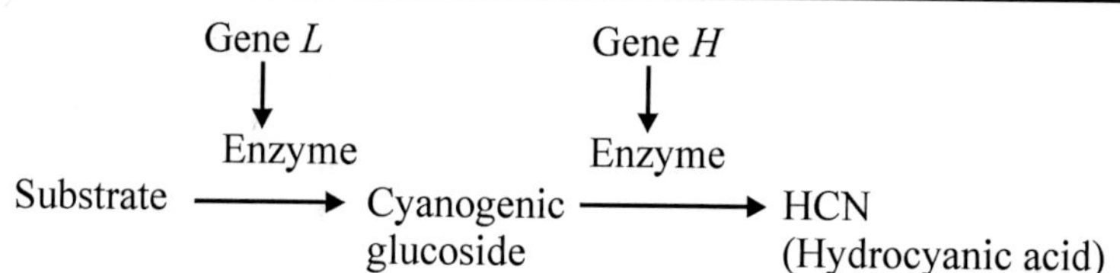

Fig. 2.3. Production of HCN (hydrocyanic acid) in *Trifolium* sp. is governed by genes *L* and *H*. These genes show complementary interaction since dominant alleles at both the loci are essential for HCN production.

Table 2.3: Some of the common types of interactions between two genes governing a single trait

Gene interaction	**F_2 ratio**	**Remarks**
Duplicate	15:1	Identical phenotypes produced by dominant alleles at one or both the loci
Complementary	9:7	At least one dominant allele at both the loci needed for production of the phenotype
Supplementary	9:3:4	The phenotype produced by the dominant allele of one gene is modified by the dominant allele of the other gene
Inhibitory	13:3	Expression of the dominant allele of one gene is inhibited by the that of the other gene
Masking	12:3:1	The phenotype produced by the dominant allele of one gene cannot be observed due to the masking effect of the phenotype produced by the dominant allele of the other gene
Polymeric	9:6:1	The dominant alleles of the two genes produce identical phenotype when they are present with the recessive allele of the other gene. In contrast, the phenotype due to the dominant alleles at both the loci is much more intense
Additive	1:4:6:4:1	The alleles at the two loci do not show dominance. Each gene produces small phenotypic effect, and the effects of the two genes are cumulative. Basis of multiple factor hypothesis (Section 2.4)

In general, one gene governs a single trait. But in many cases, a single gene affects more than one trait; this is known as **pleiotropy**. For example, the wheat gene *Lr34* affects yellow rust resistance, stem rust resistance, powdery mildew resistance, in addition to determining leaf rust resistance and leaf-tip necrosis. The pleiotropic effects of a gene are determined by comparing the phenotypes of near-isogenic lines (NILs). NILs are a pair of lines that have the same genetic constitution, except for the gene under consideration. NILs are generally produced by using backcross programme to transfer the dominant (or recessive) allele of the concerned gene in the genetic background of a line having the recessive (or dominant) allele of the gene. Pleiotropic condition can be produced by the following two situations: (i) pleiotropic action of a single gene, and (ii) effects of other genes tightly linked to the gene in question. It is generally not easy to distinguish between the above two causes of pleiotropic condition.

It may be pointed out that the undesirable effects associated with the pleiotropic condition can be removed by breeding when the pleiotropic condition is created by linked genes. But the undesirable effects generated by true pleiotropic gene action cannot be easily removed. It is noteworthy that the true pleiotropic effects of a gene can be enhanced or reduced by the genetic background in which it is placed. Therefore, the pleiotropic gene can be transferred into a genetic background in which its undesirable effects are mitigated. When the pleiotropic effects of a gene are desirable, it is a sort of bonus since more than one trait can be improved by transferring a single gene.

2.4.3. Genetic Background

Genetic background comprises all the genes present in the nucleus of an individual, except the gene in question. Genes present in the genetic background influence the level of expression of the other oligogenes and quantitative trail loci (QTLs) present in the nucleus. As a result, the intensity of the phenotype produced by a given oligogene/QTL in modified by these genes. For this reason, these genes (present in the genetic background) are often termed as **modifying genes** or, simply, **modifiers**. For example, the *opaque 2* (*o2*) mutant allele of maize has the valuable feature that it increases the lysine and tryptophan contents of the storage proteins of maize kernels. But it has the undesirable pleiotropic effects of producing unattractive opaque and soft kernels. It has been found that modifying genes present in the genetic background reduce the opaqueness as well as softness of the kernels. In fact, a suitable combination of modifiers leads to the production of normal looking hard kernels in *o2o2* homozygous plants. Therefore, *adequate care should be taken to select for the appropriate genetic background when genes are being transferred especially from the unadapted germplasm*.

2.4.4. Linkage

It is now well known the genes located close to each other in the same chromosome tend to be inherited together; this is known as **linkage** (Fig. 2.4). But new combinations of linked genes, i.e., **recombinants**, are produced by crossing over. **Crossing over** is an exchange of homologous segments of chromatids between the two members of a homologous chromosome pair (Fig. 2.5). Crossing over occurs during the pachytene stage of prophase of the first meiotic division. The frequency of recombination between linked genes is generally proportional to the physical distance between them in the chromosome. This property has been used to construct linkage maps of genes. A **linkage map** is a linear representation of the genes present in a chromosome, and the order of the genes in the map is presumably the same as that in the chromosome. The order of genes in the map and the distances between the neighboring gene pairs is derived from the frequencies of recombination between these genes.

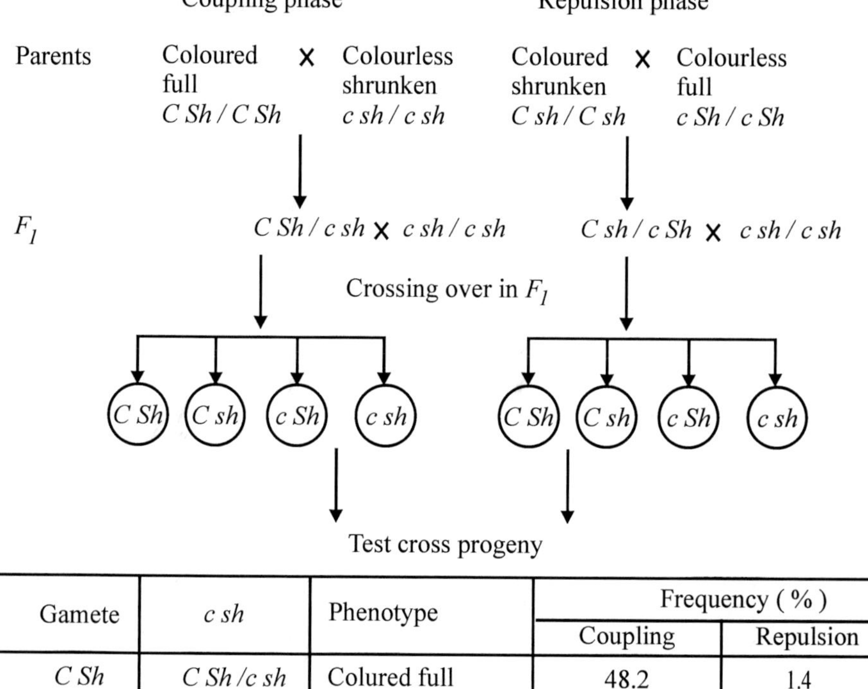

Gamete	*c sh*	Phenotype	Frequency (%)	
			Coupling	Repulsion
C Sh	*C Sh /c sh*	Colured full	48.2 (Parental)	1.4 (Recombinant)
C sh	*C sh /c sh*	Colured shrunken	1.7 (Recombinant)	47.9 (Parental)
c Sh	*c Sh /c sh*	Colourless full	1.8 (Recombinant)	49.1 (Parental)
c sh	*c sh /c sh*	Colourless shrunken	48.3 (Parental)	1.5 (Recombinant)

Fig. 2.4. Linkage between genes *C* and *Sh* of maize. Linked genes are depicted as *C Sh/ C Sh* in place of *CC ShSh*. In coupling phase, dominant alleles of the two genes are linked together. But in repulsion phase, dominant allele of one gene is linked with the recessive allele of the other gene. The net effect of linkage is as follows: the parental allelic combinations are far more frequent in the gametes than their new combinations (recombinants). Therefore, parental trait combinations are much more frequent than the two new trait combinations in the test cross progeny.

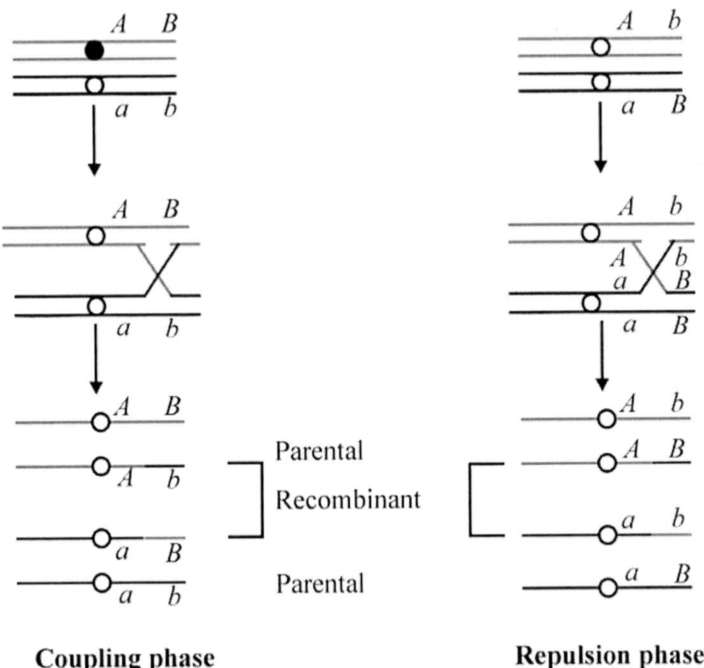

Fig. 2.5. Crossing over produces recombinant gametes. The recombinant gametes have new combinations of the alleles of the two linked genes. In case of coupling phase, *Ab* and *aB* are recombinant. But in the case of repulsion phase, *AB* and *ab* are recombinant.

The genes may be linked either in coupling phase or in repulsion phase (Fig. 2.4). In **coupling phase**, the dominant alleles of two genes are linked together. As a result, the frequency of individuals with the dominant alleles of both the genes is much more than expected in the testcross and other progeny. In case the dominant alleles produce favorable effects, coupling phase linkage is advantageous since selection for the favorable allele of one gene will also select for that of the other gene. But in case of **repulsion phase linkage**, the dominant allele of one gene is linked with the recessive allele of the other gene. Therefore, the frequency of individuals with the dominant alleles of both the genes is much lower than expected in the testcross and other progeny. The repulsion phase linkage will be undesirable in all such cases where the recessive allele of the linked gene produces detrimental effect. It may be pointed out that the phase of linkage does not affect the frequency of recombination between the linked genes.

The main effect of linkage is that it reduces the frequencies of new allelic combinations of the linked genes in comparison to those produced by independent assortment. Therefore, much larger segregating generations are needed for the recovery of recombinant types in comparison those that would be required if the concerned genes were segregating independently. In view of the above, linkage between desirable alleles of two genes greatly facilitates selection for them in the segregating generations. In such cases, one needs to select for only one of the traits, and the other trait would be automatically selected for. This would be particularly true for the traits governed by genes that are tightly linked with <1% recombination between them. The marker-assisted selection is, in fact, based on this expectation. However, tight linkage between the desirable allele of one gene with the undesirable allele of another gene is a nuisance since it would be very difficult to recover a recombinant having the desirable alleles of both the genes. Thus, linkage between the desirable allele of one gene and the undesirable allele of another gene slows down the progress of crop improvement programmes. Therefore, this phenomenon is called **linkage drag**. Linkage between the genes governing two different quantitative traits is one of the chief reasons for genetic correlation between these traits. In addition, repulsion phase linkage tends to increase the estimate of dominance variance and to reduce that of additive genetic variance. Repulsion phase linkage between genes affecting the same quantitative trait also mimics overdominance.

2.4.5. Threshold Characters

The expression of qualitative traits is generally little affected by the environment. But the expression of mutant alleles of many oligogenes is markedly affected by the environment; this sometimes produces variable phenotypes of the otherwise oligogenic traits. In addition, the phenotypic expression of some traits requires some specific environment; such traits are known as **threshold characters**. For example, the expression of a disease resistance gene can be seen only when the plants/lines are exposed to the concerned pathogen under conditions suitable for the development of the concerned disease. Similarly, drought resistance can be evaluated and selected for only when the plant population is subjected to appropriate drought conditions. Therefore, often the selection for threshold traits presents problems since creation of the appropriate conditions for their reliable expression may not be easy.

2.4.6. Maternal Effects

The development of many traits in higher plants is affected by the genotype of the female parents of the individual plants. This phenomenon is known as **maternal effect**. Some of the examples of the traits showing maternal effect are seed size, seed protein content, etc. As a result of the maternal effect, the F_1 progeny from reciprocal crosses generally show marked differences for such traits. These differences, however, disappear in the F_2 generation. But the magnitude of phenotypic variation for such traits present in F_2 is much smaller than that observed in the F_3 generation of the same cross. Therefore, maternal effect would tend to reduce the effectiveness of selection for the concerned traits in the segregating generations of crosses.

2.5. INHERITANCE OF QUANTITATIVE TRAITS

In 1908, Nilsson-Ehle proposed the **multiple factor hypotheses** to explain the inheritance of quantitative traits. According to this hypothesis, quantitative traits are governed by several genes that have small individual effects, and these effects are additive in nature (Fig. 2.6, Table 2.4). The genes having this type of action are called **polygenes**. One allele of a polygene produces effect on the trait (termed as **positive allele**), while the other allele does not produce any effect (called **negative allele**). The alleles of polygenes are presumed to show additive gene action and a lack of dominance. It was also assumed that all the polygenes affecting a trait have equal effect on the phenotype of that trait. In the case of polygenic inheritance, such individuals may appear in the F_2 generation that surpass the parental values for the trait in question. This phenomenon is termed as **transgressive segregation**. This type of segregation would be observed whenever the two parents involved in the cross have positive alleles of different genes for the trait in question. For example, a cross between two strains of wheat with the genotypes $R_1R_1\ r_2r_2$ and $r_1r_1\ R_2R_2$ (both the parents have medium red seeds) would produce $R_1R_1\ R_2R_2$ and $r_1r_1\ r_2r_2$ plants in the F_2 generation. These plants will produce dark red seeds and white seeds, respectively, which represent transgressive segregation. It may be pointed out that the seed color in wheat is not a true quantitative trait since it shows discontinuous variation in contrast to the continuous variation exhibited by the true quantitative traits. Subsequent studies by a number of workers have clearly shown that true quantitative traits are also governed by polygenes, and that the multiple factor hypothesis is able to explain their inheritance.

Table 2.4: A summary of the F_2 progeny produced from the cross between dark red x white seeded varieties of wheat

Genotype	Frequency	Number of R_1 and R_2 alleles	Units of phenotypic effect	Phenotype (= seed color)	Frequency
$R_1R_1\ R_2R_2$	1	4	4	Dark red	1
$R_1r_1\ R_2R_2$	2	3	3	Medium-dark red	4
$R_1R_1\ R_2r_2$	2	3	3	Medium-dark red	
$R_1r_1\ R_2r_2$	4	2	2	Medium red	6
$R_1R_1\ r_2r_2$	1	2	2	Medium red	
$r_1r_1\ R_2R_2$	1	2	2	Medium red	
$R_1r_1\ r_2r_2$	2	1	1	Light red	4
$r_1r_1\ R_2r_2$	2	1	1	Light red	
$r_1r_1\ r_2r_2$	1	0	0	White	1

Phenotypic ratio: 1:4:6:4:1

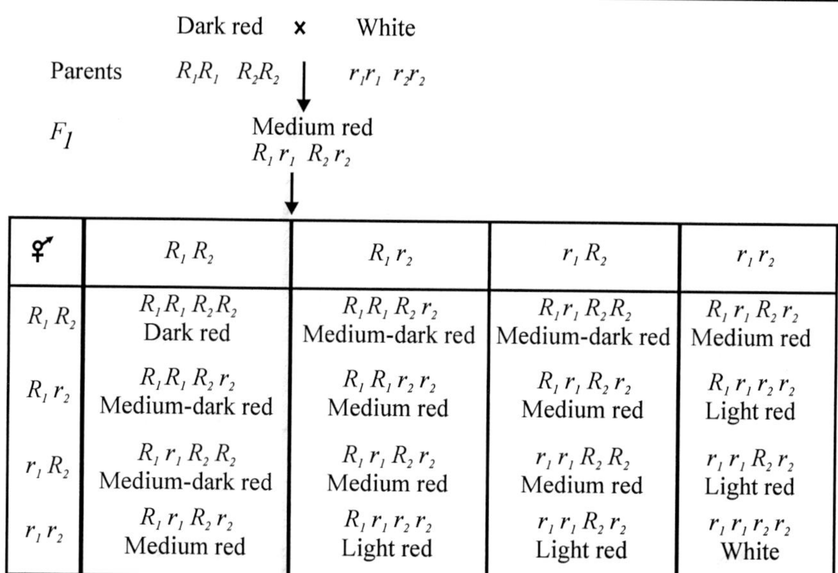

Dark red × White

Parents R_1R_1 R_2R_2 r_1r_1 r_2r_2

F_1 Medium red R_1r_1 R_2r_2

⚥	R_1R_2	R_1r_2	r_1R_2	r_1r_2
R_1R_2	$R_1R_1R_2R_2$ Dark red	$R_1R_1R_2r_2$ Medium-dark red	$R_1r_1R_2R_2$ Medium-dark red	$R_1r_1R_2r_2$ Medium red
R_1r_2	$R_1R_1R_2r_2$ Medium-dark red	$R_1R_1r_2r_2$ Medium red	$R_1r_1R_2r_2$ Medium red	$R_1r_1r_2r_2$ Light red
r_1R_2	$R_1r_1R_2R_2$ Medium-dark red	$R_1r_1R_2r_2$ Medium red	$r_1r_1R_2R_2$ Medium red	$r_1r_1R_2r_2$ Light red
r_1r_2	$R_1r_1R_2r_2$ Medium red	$R_1r_1r_2r_2$ Light red	$r_1r_1R_2r_2$ Light red	$r_1r_1r_2r_2$ White

Fig. 2.6. Inheritance of seed colour in wheat explained by the multiple factor hypothesis. Seed colour is assumed to be governed by two genes R_1 and R_2. The alleles R_1 and R_2 produce colour, while the alleles r_1 and r_2 do not. These two loci do not exhibit dominance. The effects of the R_1 and R_2 alleles are cumulative. As a result, the intensity of seed colour depends on the total number of R_1 and R_2 alleles present in a given seed (Table 2.4)

2.5.1. Quantitative Trait Loci

It was not possible to follow the inheritance of quantitative traits using the techniques developed for the analysis of the qualitative traits. Therefore, methods based on statistical procedures were developed for their analyses. Further, the development of molecular markers (*See* Chapters 13-15) has allowed the identification of genomic regions involved in the control of various quantitative traits; such genomic regions are known as **quantitative trait loci (QTLs)**. A single QTL may contain one or, more often, more genes. Many QTLs have been cloned and it has been found that the genes present in them perform a variety of functions, and many of them encode transcription factors. QTL analysis has revealed that in most of the cases, a quantitative trait is governed by one or few genes with large effects and a large number of QTLs each producing a small effect on the trait.

2.5.2. Continuous Variation in Quantitative Traits

It may be pointed out that the continuous variation is the main feature of quantitative traits. Further, the frequency distribution for any quantitative trait in a population, e.g., F_2, generally tends to follow the bell-shaped normal distribution curve. The continuous variation in quantitative traits is produced due to the following three factors: (i) involvement of several polygenes or QTLs in the control of each trait, (ii) the effect of the environment on the phenotype produced by these polygenes/QTLs, and (iii) the modifying effects of the genetic background. The number of phenotypes in F_2 would be twice the number of genes affecting the quantitative trait and segregating in F_1 if there were complete dominance at all the loci. Thus, the number of phenotypes produce by segregation of n genes will be $2n$. In case there was no dominance at these loci, the number of phenotypes would be $2n+1$. As the number of different phenotypes for a trait increases, the distribution for the trait tends to become continuous. Thus, with 12 polygenes controlling a trait, the number of different phenotypes produced will be 25, and the distribution would be nearly continuous even with 100% correspondence between trait the phenotype and the genotype. It is now well known that the expression of polygenes is markedly affected by the environmental factors. The environmental factors may contribute, in case of some traits, up to 50% of the trait phenotype. The net effect of the environmental influence is to increase the range of phenotypes produced by a single genotype. As a result, the phenotypes produced by different genotypes tend overlap each other, which makes the trait distribution continuous. It has been estimated that a 50% contribution of the environment to the phenotype would make the distribution of a trait continuous even if it were governed by a single gene. In addition, the expression of QTLs is affected by their genetic background, which also contributes to the continuous distribution of quantitative traits.

2.5.3. Types of Gene Effects

It was originally proposed that polygenes produce small individual effects, which are additive in nature. However, it is now well recognized that polygenes show dominance (Fig. 2.7) as well as epistatic gene actions in addition to the additive effects (Table 2.5). In case of epistatic gene action, the phenotypic expression of one gene is affected by the genotype of the second locus as well. The epistatic gene action has been divided into the following three types: additive x additive, additive x dominance, and dominance x dominance. The **additive x additive interaction** denotes the effects of homozygous genotypes at one locus on the expression of homozygous genotypes at the other locus. Similarly, the **additive x dominance epistasis** refers to the effects of homozygous genotypes at one locus on the expression of heterozygous genotype at the other locus. Finally, the **dominance x dominance gene interaction** signifies the influence of heterozygous genotype at one locus on the expression of heterozygous genotype at the other locus (Table 2.5).

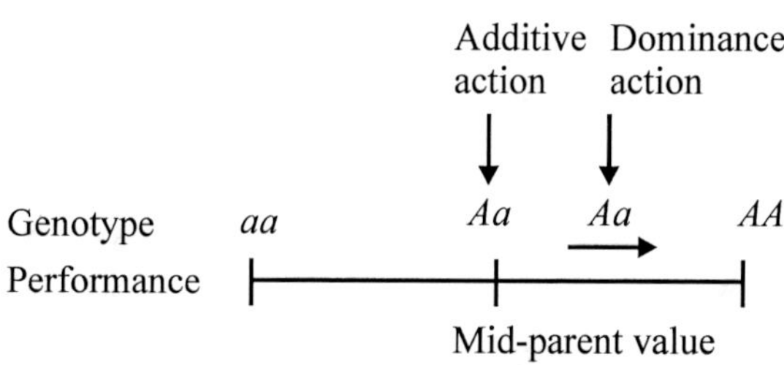

Fig. 2.7. The effect of the type of gene action on performance of the heterozygote for a gene in comparison to those of the two homozygotes. In case of strictly additive gene action, the performance of heterozygote *Aa* equals the average of performances of the homozygotes *AA* and *aa* (= the mid-parent value). But when dominance gene action is present, the performance of *Aa* deviates from the mid-parent value toward that of, say, *AA*

Table 2.5: A simplified representation of the different types of gene effects in the control of quantitative traits

Gene effect	Expressed in genotypes	Explanation	Fixable/ nonfixable	Remarks
Additive	*AABB*, *AA bb*, *aa BB*	The phenotypic effects of *A* and *B* alleles are cumulative	Fixable	Expressed in purelines; effectiveness of selection depends on this gene effect
Dominance	*Aa Bb*, *Aa BB*, *AA Bb*	The heterozygote phenotypes deviate from the mid-parent values	Non-fixable	Expressed in heterozygous genotypes; contributes to heterosis
Epistasis		Phenotypic expression at one locus affected by the genotype at the other locus		
Additive x additive	*AA BB*, *AA bb*, *aa BB*	Interaction between the homozygous genotypes at the two loci	Fixable	Expressed in purelines; effectiveness of selection depends on this gene effect
Additive x dominance	*AA Bb*, *Aa BB*	Interaction between the homozygous genotypes at one locus and the heterozygous genotype at the other locus	Non-fixable	Expressed in heterozygous genotypes; contributes to heterosis
Dominance x dominance	*Aa Bb*	Interaction between the heterozygous genotypes at the two loci	Non-fixable	Expressed in heterozygous genotypes; contributes to heterosis

It would be seen that the additive and additive x additive gene effects relate to the homozygous genotypes at the concerned loci. Therefore, these gene effects would be expressed in purelines that are homozygous. The homozygous state for a gene can be considered as a **fixed state** since there can be no segregation for this gene. For this reason, the additive and additive x additive gene effects are often referred to as **fixable gene effects**. These gene effects are the only ones that are relevant to the performance of pureline varieties, and would be utilized by selection during the development of purelines. On the other hand, the dominance, additive x dominance, and dominance x dominance gene effects are termed as **nonfixable gene effects** because their expression requires at least one of the concerned loci to be in the heterozygous state, i.e., non-fixed state. Therefore, these gene effects are not relevant for the performance of pureline varieties. However, they are involved in the expression of heterosis, and are important for the performance of the hybrid varieties.

2.5.4. Variance Components

The phenotypic expression (P) of a quantitative trait in an individual is affected by the genotype (G) of the individual for the trait under consideration, the prevailing environment (E) as well as the interaction between G and E ($G \times E$). The presence of $G \times E$ interaction signifies that the relative performance of different genotypes changes in different environments (Fig. 2.8). Thus, the phenotypic expression of a trait in a genotype can be depicted as follows.

$$P = \mu + G + E + G \times E$$

where, μ is the population mean. One of the crucial questions, therefore, relates to the relative contributions of the G and E (including $G \times E$) components to the P. A number of experimental approaches have been developed for the estimation of the G, E and $G \times E$ components to the phenotypic variance. The estimation of

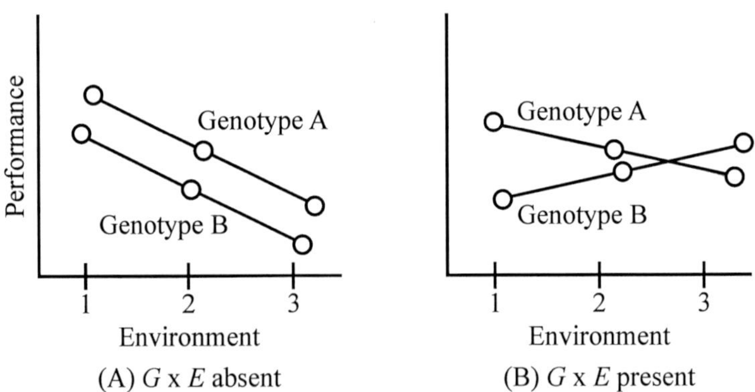

Fig. 2.8. A simple representation of the effect of $G \times E$ interaction on performance of two different genotypes, A and B. When $G \times E$ interaction is absent, the environmental effects on their performances are similar. But when $G \times E$ interaction is present, the genotypes respond differently to the three environments.

phenotypic variance and partitioning of this estimate into components due to *G*, and *E* is relatively simple. However, the estimation of variance component due to *G* x *E* requires specific experimental designs and considerable effort. Further, a variety of mating designs have been proposed to partition the *G* component into additive, dominance and epistatic components. These types of studies have shown that in general the additive component is much more important in the control of most quantitative traits than are dominance and epistatic components.

2.5.5. Heritability and Genetic Advance

The variance components due to *G* and *E* can be easily estimated by evaluating a large number of homozygous genotypes in a replicated trial. The data on the concerned quantitative trait collected from this trial are analyzed according to the experimental field design used for the trial (Table 2.6). The analysis of variance table provides the estimates of the mean sum of squares (MS) due to genotypes and due to error. The error MS is taken as the estimate of variance due to the environment ($\sigma^2 e$ or Ve). The estimate of the genotypic variance (Vg) is obtained from the estimate of the MS due to genotypes as given below.

$$\text{Expectation of MS due to genotypes} = \sigma^2 e + R\sigma^2 g$$

where,

σ^2 is the variance due to error (=Ve),

R is the number of replications in the trial, and

$\sigma^2 g$ is the variance due to genotype (Vg).

Therefore, Vg = (MS due to genotypes – MS due to error) / R

The relative importance of the variance components due to *G* and *E* is easily represented as the estimate of heritability estimate for the trait, which is estimated as follows.

$$\text{Heritability (\%)} = [Vg/(Vg + Ve)] \times 100$$
$$= (Vg/Vp) \times 100$$

where, Vg, Ve and Vp are the estimates of genotypic, environmental and phenotypic variances, respectively. This estimate of heritability is known as **broad sense heritability**. It indicates the proportion of phenotypic variance for the concerned trait produced by genotype. The genotypic variance can be divided into additive, dominance and epistatic components. Of these components, only the additive and additive x additive epistatic variance components are expressed in the homozygous genotypes. Therefore, these components are known as **fixable gene effects**. In contrast, the dominance and the other components of the epistatic variance are expressed only in heterozygous genotypes. Therefore, these components are called **nonfixable gene effects**. Thus, the broad sense heritability estimates are not appropriate in the cases of segregating generations. In such cases, the appropriate estimate of heritability is the ratio of additive genetic variance (*Va*) to the phenotypic variance.

$$\text{Heritability (narrow sense) (\%)} = (Va/Vp) \times 100$$

It may be pointed out that selection for a trait is based on its phenotype. The estimate of heritability indicates the effectiveness of phenotype-based selection. A high heritability estimate, e.g., 60% or more, indicates that selection for the trait should be quite effective. But a low estimate of heritability, say, <40%, suggests that selection for the trait may have poor effectiveness.

The population obtained after selection would show a shift in the mean performance of the character being selected for; this change would obviously be in the direction of selection. The change in performance of the selected progeny in relation to the original population is described as **genetic advance under selection** The extent of this change will depend mainly on the following factors: (i) the intensity of selection, (ii) the amount of phenotypic variation for the trait present in the population subjected to selection, and (iii) the heritability of the trait. The genetic advance under selection (*Gs*) is considered in a greater detail in Section 3.9.1.

Table 2.6: Analysis of variance for a quantitative trait measured on *G* number of genotypes grown in a trial with *R* number of replications according to the randomized block design

Source of variation	D.F. (degrees of freedom)	Mean sum of squares (MS)	Expectation of MS
Genotype	*G*-1	*MSg*	$\sigma^2 e + R\sigma^2 g$
Replication	*R*-1	*MSr*	$\sigma^2 e + G\sigma^2 r$
Error	(*G*-1)(*R*-1)	*MSe*	$\sigma^2 e$
Total	(*GR*)-1		

Where, *G* is the number of genotypes evaluated in the trial, *R* is the number of replications, $\sigma^2 e$ is the variance due to error, $\sigma^2 g$ is the variance due to genotype, and $\sigma^2 r$ is the variance due to replications.

2.6. POPULATION GENETICS

The methods of classical analyses and the inferences derived from studies based on them are applicable to self-pollinated populations. But many species are cross-pollinated and their populations are considered as random mating populations. In a **random mating population**, a male gamete has equal opportunity for uniting with any female gamete produced by the population. In other words, a male individual has equal opportunity of mating with any female individual of the population. Therefore, any generation of such a population is produced by random union among a random sample of male and female gametes produced by the previous generation of the population. Therefore, the genotypes of particular individuals comprising the population are of little significance in determining the genetic makeup of the population over generations. In fact, the genetic composition of these populations can be best described as well as studied in term of the gene and genotype frequencies. A specialized branch of genetics, called **population genetics**, has been developed for the study of such populations. The salient features of the genetic behavior of random mating populations are described in some detail in Section 3.9.2.

2.7. CYTOPLASMIC INHERITANCE

As stated in Section 2.2, most of the genes are located in chromosomes present in the nuclei of plant cells. The principles and the processes of inheritance discussed in the foregoing sections relate to the genes present in nuclei. But many genes, called **plasmagenes**, are present in the chromosomes of mitochondria and chloroplasts. The plasmagenes control some important plant characteristics like male sterility (specified by mitochondrial plasmagenes), leaf variegation (due to plastid plasmagenes), etc. The plasmagenes do not follow the inheritance patterns exhibited by the nuclear genes. In contrast, they show an inheritance pattern (Fig. 2.9) that is entirely distinct from that of the nuclear genes; this inheritance is called **cytoplasmic inheritance**. The salient features of cytoplasmic inheritance are as listed below.

1. The F_1 generally has the phenotype of the female parent (Fig. 2.9).

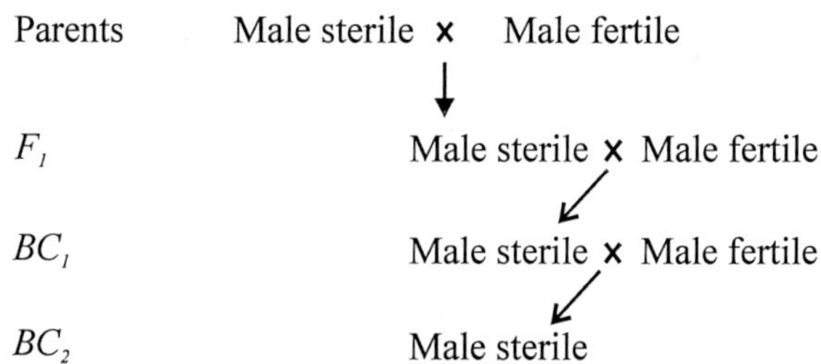

Fig. 2.9. Inheritance of cytoplasmic male sterility. The F_1 is male sterile. The backcross (*BC1*, and *BC2*) generations are also male sterile. Further, the backcross progeny do not show segregation for male sterility. Since the F_1 and later generation progeny are all male sterile, they have to be backcrossed to the male fertile parent (or any other male sterile line) to produce progeny.

2. As a result, traits governed by plasmagenes show reciprocal differences, i.e., the F_1s from reciprocal crosses show different phenotypes for the trait.
3. There is no segregation in the F_2 and the later generations of the cross.

4. Rarely, segregation may occur for plasmagenes. Whenever this happens, the segregation ratios are irregular and markedly different from those for the nuclear genes.
5. Plasmagenes show segregation during mitotic cell division, which is in contrast to the behavior of nuclear genes.
6. In a vast majority of crop plants, plasmagenes from the female parent are transmitted to the progeny, i.e., they show maternal transmission (Fig. 2.9, 2.10). But in conifers, plastids from the male parent are transmitted, i.e., they show paternal transmission.
7. In some crops, plasmagenes from both the parents are transmitted to the progeny. For example, in the case of *Medicago* spp., the plasmagenes present in the plastids are transmitted from both the parents. Even in other crop species like tobacco, a very low frequency (<1%) of paternal transmission of the plasmagenes does take place.

The maternal transmission of the plasmagenes occurs because of the following mechanisms as discussed below.

1. The egg cells produced by plants have large amounts of cytoplasm and many plastids and mitochondria. In contrast, the generative cells produced during the first mitotic division of the microspores receive several mitochondria, but almost no plastids.
2. Further, the mitochondria are lost from the generative cells as they mature into sperms.
3. Even when generative cells receive both plastids and mitochondria, they degenerate subsequently during their maturation into sperms.
4. In any case, only the nucleus of the sperm is able to gain entry into the egg cell during fertilization, and almost the whole of the cytoplasm, and the plastids and mitochondria present in it are left out (Fig. 2.10). It may be pointed out that none of these mechanisms is 100% effective. As a result, there is a very low frequency of paternal transmission of plasmagenes. In many cases, the expression of plasmagenes is affected by nuclear genes; this leads to nuclear cytoplasmic-interaction. For example, the expression of cytoplasmic male sterility is affected by restorer (*R*) genes present in the nucleus. Generally, the *R* allele of this gene suppresses the effects of male sterile cytoplasm, and leads to the male fertile phenotype. This feature of cytoplasmic male sterility is of great relevance in the use of male sterility for hybrid seed production (Section 3.3.2).

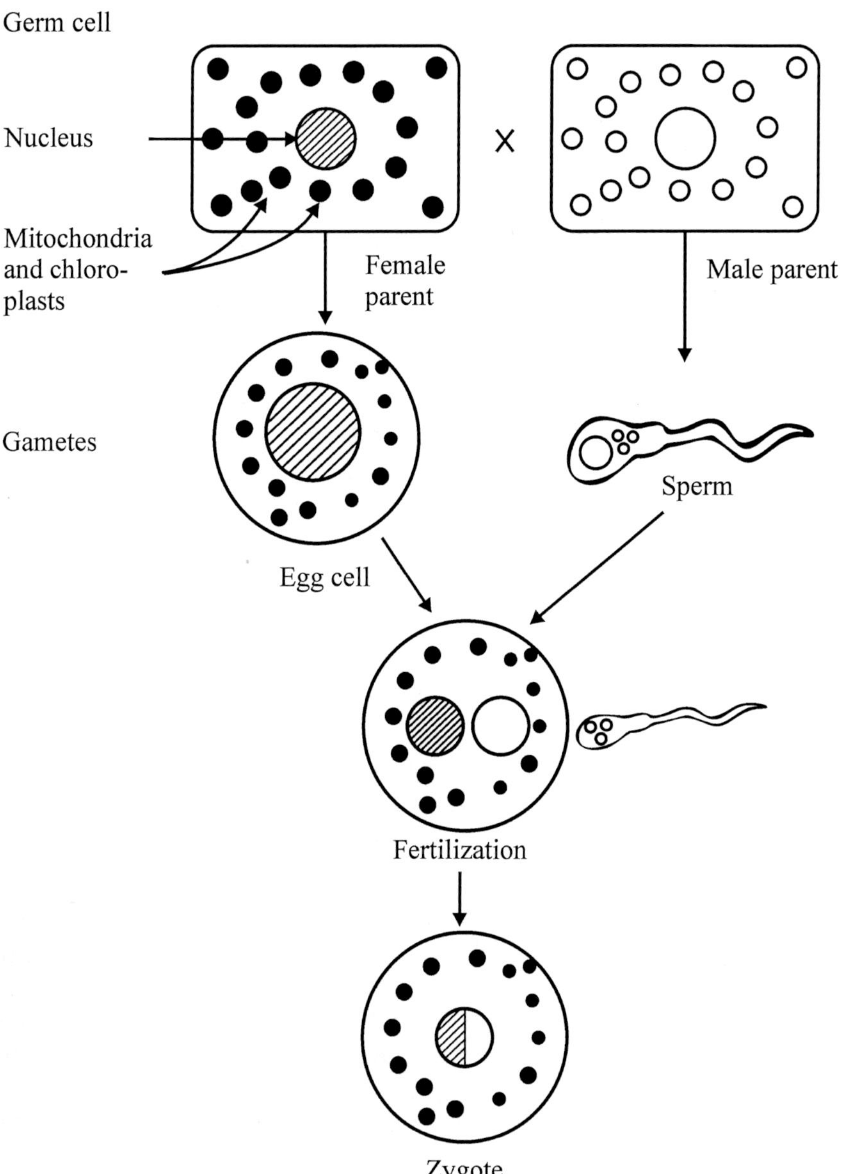

Fig. 2.10. Explanation for cytoplasmic inheritance. The egg cell receives a large amount of cytoplasm and many plastids and mitochondria. In contrast, sperms have little cytoplasm and few plastids and mitochondria. At the time of fertilization most of the sperm cytoplasm, and the mitochondria and plastids present in it are left out of the egg cell. As a result, the zygote contains almost exclusively the maternal plastids and mitochondria.

2.8. MOLECULAR GENETICS

In the foregoing sections, it has been presumed that genes affect the development of trait phenotypes. It is now universally accepted that genes are made up of DNA, and that they are located in nuclei and organelles of cells. It is well recognized that the various cellular processes are catalyzed by proteins that function as enzymes. The DNA must somehow specify these cellular processes to be able to regulate the trait phenotypes. In this context, the DNA must perform the following two functions: (i) a high fidelity self-propagation so that the genetic information can be passed on to the succeeding generations without alteration, and (ii) regulation of/participation in cellular processes to be able to govern the development of trait phenotypes. In this section, an attempt is made to give a general outline of the molecular events that allow the genes to propagate themselves as well as to direct and regulate the cellular processes so that they are able to produce the phenotypes specified them.

2.8.1. DNA Replication

The DNA molecule is a double helix. Each DNA strand is composed of four different nucleotides that are generally represented simply as A, T, C and G. These nucleotides are linked with each other by sugar-phosphate-sugar linkages. The DNA molecules propagate themselves by semiconservative replication, in which each of the two old strands serves as template for the synthesis of a new strand. As a result, the two progeny DNA molecules produced by this type of replication comprise one old and one newly synthesized strand each (Fig. 2.11).

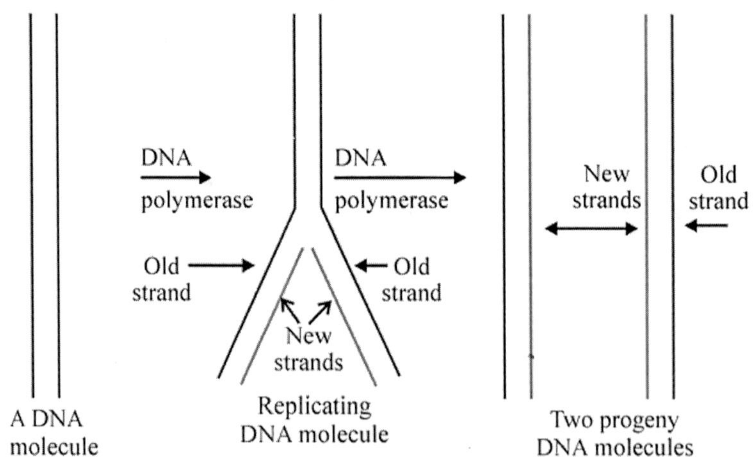

Fig. 2.11. A highly simplified schematic representation of DNA replication. The DNA strands are shown as straight lines. In fact, the two strands of a DNA molecule are twisted into a right-handed double helix. The replication reaction is catalyzed by the enzyme DNA polymerase. Thus, each progeny DNA molecule has one old and one newly synthesized strand.

During synthesis of the new strand, the nucleotides of the old strand serve as the template, and the nucleotide added to the new strand must be complementary to that present in the old strand. For example, if A is present in the old strand, the nucleotide added to the new strand opposite to this nucleotide has to be T. Similarly, a C will be added in the new strand opposite to a G in the old strand. This process is called **DNA replication**, and is catalyzed by the enzyme DNA polymerase. This enzyme commits a low level of error, but it has proofreading capacity so that it examines each newly added nucleotide, and replaces the incorrect nucleotide with the correct one.

2.8.2. Transcription

The gene expression involved in the development of trait phenotypes comprises two distinct processes, viz., transcription and translation. **Transcription** consists of the use of DNA sequence of a gene as template to produce RNA copies of the concerned genes (Fig. 2.12). Transcription begins in the promoter region of the gene, and is catalyzed by the enzyme RNA polymerase that is assisted by several proteins called **transcription factors**. The base sequence of the DNA serves as

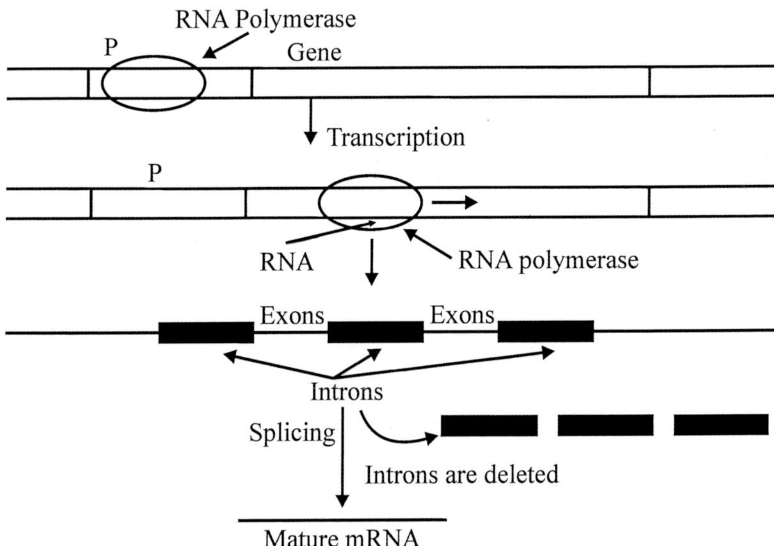

Fig. 2.12. A highly simplified schematic representation of production of messenger RNA (mRNA) from protein encoding genes. The base sequence of the gene serves as template for RNA synthesis; this constitutes transcription. Transcription is catalyzed by RNA polymerase, which is assisted by several proteins called transcription factors. Transcription begins from the promoter region of the gene. The RNA transcripts of plant genes generally have noncoading regions called introns. The introns are removed by a process called splicing to yield mature messenger RNA (mRNA). P, promoter.

template for synthesis of the RNA transcript. The plant genes generally comprise exons (expressed sequences) and introns (interrupting sequences), while the mRNA (messenger RNA) from these genes consists of only exons. Therefore, the primary transcripts from the protein encoding genes of plants are processed to precisely remove all the introns and to join all the exons in the proper sequence to produce the mature mRNA. This type of RNA transcript processing is called **splicing**. Generally, the RNA transcripts from one gene are processed to yield a single type of mRNA. But in case of some genes the RNA transcripts from a single gene are processed in two or more different ways to generate more than one different type of mRNA molecules (**alternative splicing**). Plant cells produce a variety of other RNA molecules, including ribosomal RNA (rRNA), transfer RNA (tRNA), microRNA, etc. by transcription of the concerned genes.

2.8.3. Translation

The mRNA is used as template for the production of polypeptides, this process is called **translation**. In translation, the base sequence of an mRNA molecule is used as template for conversion of this information into the amino acid sequence of the polypeptide encoded by the concerned gene. Translation involves ribosomes that associate with the mRNA and make it available in the condition suitable for translation. The tRNA molecules bring one amino acid at a time to the ribosome where this amino acid is added to the growing polypeptide chain. The amino acid sequences of polypeptides are specified by the base sequences of the concerned mRNAs. One amino acid is specified by a set of three mRNA bases called **codon**. Each amino acid is encoded by at least one distinct codon, but some amino acids are specified by more than one codon. Each tRNA molecule has three bases that comprise its **anticodon**, which is complementary to a specific codon of the mRNA. A given tRNA always attaches to that particular amino acid for which it carries the anticodon. Thus base-pairing between the codon of mRNA and the anticodon of tRNA determines the amino acid that will be added in response to a specific codon of the mRNA. At any given time, two tRNA molecules can pair with the mRNA region associated with the ribosome. When the first two tRNAs pair with the mRNA, the amino acids attached with them become linked together by a peptide bond (Fig. 2.13). The mRNA now moves one codon, the first tRNA is ejected from the ribosome and a new tRNA come to the ribosome. In this way, the entire mRNA sequence is copied into a polypeptide. The polypeptides ultimately become organized into functional molecules called proteins. A protein molecule may be composed of a single polypeptide (*monomeric protein*) or more than one polypeptide (*multimeric protein*). Further, a multimeric protein molecule may contain two or more copies of a single polypeptide (*homomultimeric protein*) or of two or more different polypeptides (*heteomultimeric protein*), each of which isencoded by a different gene.

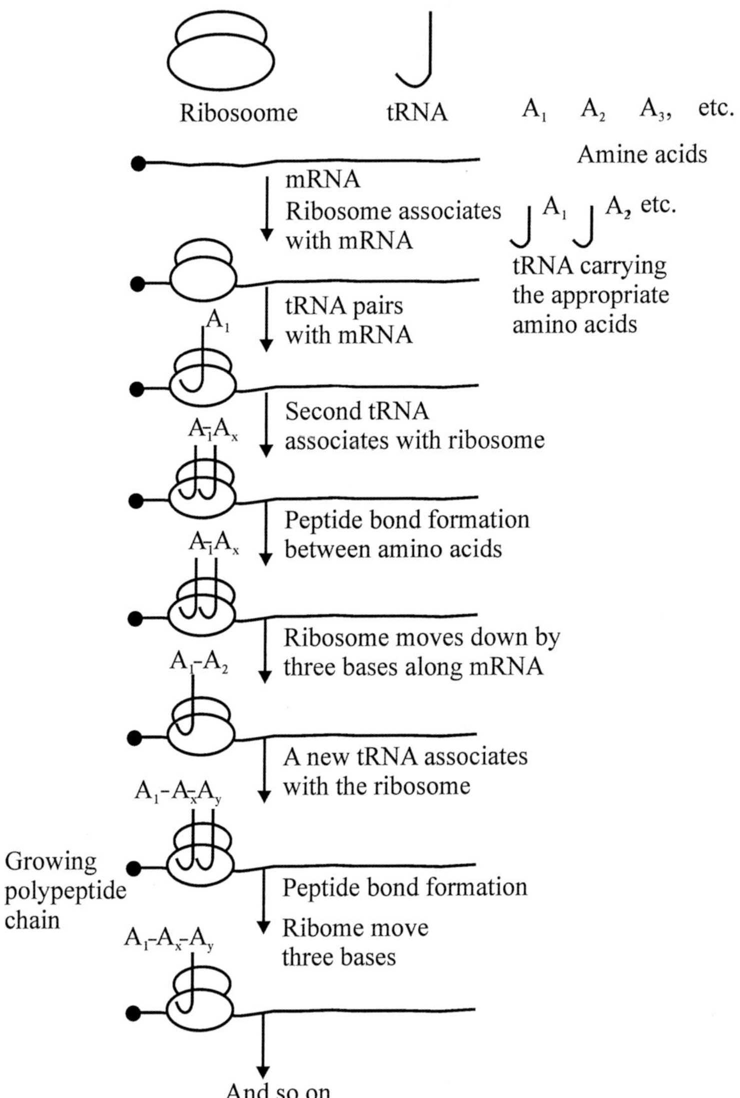

Fig. 2.13. A simplified schematic representation of translation. Each tRNA carries the appropriate amino acid. Base pairing between mRNA codons and tRNA anticodons ensures that the amino acid sequence of the polypeptide is the same as that specified by the base sequence of mRNA. After peptide bond formation, the ribosome moves 3 bases down along the mRNA. This ejects the freed tRNA and makes the next colon of mRNA available for translation. Several different proteins participate in and facilitate the translation process. The required energy is provided by GTP (guanosine triphosphete). tRNA, transfer RNA, mRNA, messenger RNA.

2.8.4. Regulation of Gene Action

Gene expression is mainly regulated at the level of transcription, but regulations during translation and after translation are also important. Transcription regulation is mediated by a class of proteins called **regulator proteins**, which bind to specific DNA sequences that are often located within or around the promoter regions of the concerned genes. Binding of some of the regulator proteins to DNA represses the transcription of the neighbouring gene; these proteins are called repressors and the phenomenon is known as **repression of gene expression**. On the other hand, the binding of another group of proteins to DNA activates gene expression; these proteins are called **activators**, and the phenomenon is known as **activation of gene expression**. In case of prokaryotes, gene regulation involves both repression and activation of transcription, but repression seems to be predominant. On the other hand, *transcription regulation in plants is primarily based on transcription activation.*

2.8.5. Biosynthetic Pathways

The proteins serve a variety of functions in the cells and tissues. Most of the proteins are enzymes and catalyze biochemical functions. Proteins also participate in the organization of cellular structures like membranes, ribosomes, etc., and function in the regulation of gene expression. In addition, proteins perform many other functions, e.g., they function as hormones, chaperones to facilitate proper protein folding, carriers of molecules and ions, food reserves, etc. The enzymes catalyze biochemical reactions and are the basis of all metabolic processes of cells. All the enzymes involved in a metabolic process, e.g., biosynthesis of tryptophan (Fig. 2.14), function in a specific sequence and give rise to a metabolic pathway. In such a pathway, the reactions occur in a chain so that the product of the previous reaction serves as the substrate for the next reaction. Each reaction is catalyzed by a single enzyme, and the different enzymes are encoded by specific genes. At the end of the pathways, specific metabolites are produced. These metabolites in conjunction with the various proteins ultimately determine the various cellular features that give rise to the various phenotypes of the organisms.

2.9. EPIGENETICS

The classical view considers heritable part of the phenotypic variation to be due to genetic differences among the individuals/lines under consideration. Genetic differences arise from alterations in gene functions as a result of changes in gene structure; these changes are reflected in the DNA sequences of the concerned genes. Thus **genetics**, in simple terms, is the study of heritable changes in gene function arising due to alterations in the DNA sequences of the concerned genes. However, there is considerable evidence that many changes in gene function do not involve alterations in DNA sequences of the concerned genes. These changes in gene function are transmitted through mitosis and meiosis, and some of them are stable over several generations. Such heritable changes in gene function are referred to as **epigenetic changes**. The study of heritable changes in gene functions without any change in the DNA sequences of the genes involved is called **epigenetics**.

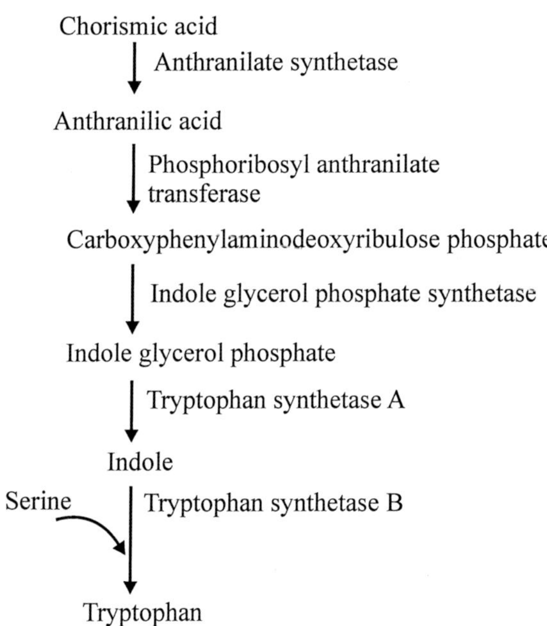

Fig. 2.14. The pathway for tryptophan biosynthesis in *E. coli*. The reactions occur in a chain so that the product of the previous reaction is substrate for the next reaction. Each reaction is catalyzed by an enzyme, which is encoded by a single gene. For example, the enzyme anthranilate synthetase is encoded by the gene *trpE*, while tryptophan synthetase A and tryptophan synthetase B are encoded by genes *trpA* and *trpB*, respectively.

Epigenetic changes in gene function are produced by one or more of the following modifications: (i) DNA methylation/demethylation, (ii) histone modification coupled with chromatin condensation/relaxation, and (iii) RNA-induced interference (RNAi). DNA methylation was the first discovered heritable epigenetic change. Mostly the C in the CpG and CpNpG (N is any of the four nucleotides) sequences is methylated; this represses transcription. DNA methylation promotes histone methylation and deacetylation, which lead to more compact chromatin formation. DNA methylation plays an important role in several biological processes, including differentiation and development. DNA methylation is catalyzed by a family of conserved DNA methyltransferase or

DNA methylase enzymes. The DNA demethylation takes place by removal of the methylated nucleotide and is catalyzed by the enzyme DNA glycosylase.

Histone modification was the second epigenetic change to be discovered. It involved methylation, acetylation, phosphorylation, ribosylation, and uniquitination of the histones, which together constitute the histone code of the chromatin. The targets of modification are Lys (lysine), Arg (Arginine) and Ser (Serine) amino acid residues. Histone modifications are involved in the conversion of euchromatin into heterochromatin leading to gene silencing and that of heterochromatin into euchromatin resulting in the activation of genes silenced by heterochromatin formation (Table 2.7). For example, methylation of Lys at position 9 of the H3 histone blocks transcription initiation of the concerned gene. In addition, histone modifications can attract DNA methylases to initiate methylation of C residues of DNA.

Table 2.7: The salient features of the epigenetic modifications

Epigenetic change	Molecular changes	Effects on genome organization and gene function
DNA methylation	Methylation of cytosine	Repression of transcription, and promotion of more compact chromatin formation
	Demethylation of cytosine	Promotes gene expression and less compact chromatin formation
Histone modification	Methylation of H3 and H4 histones at specific Lys residues	Heterochromatin formation and gene silencing
	Deacetylation	
	Methylation of histone H3 at Lys at position 4	Euchromatin formation and gene expression
	Acetylation	
RNA interference (RNA_i)	Small RNAs mediate silencing of genes that encode mRNAs; these RNAs are complementary to a part of the concerned the genes	The genes complementary to the interfering RNA are silenced; many types of endogenous interfering RNAs; they are also produced by transgenes

RNAi was discovered from analysis of gene silencing, particularly cosuppression, in transgenic plants. When an additional copy of chalcone synthase gene (involved in flower color production) was expressed in *Petunia* to enhance flower color, surprisingly the expression of this gene was completely suppressed leading to white flowers. A detailed analysis of this phenomenon revealed RNAi, which has now been shown to be an important mechanism of gene regulation in eukaryotes. The RNAi is involved in gene regulation and is based on endogenous interfering RNAs, which are of several types, including microRNA. The interfering RNA is complementary to the mRNA encoded by the gene regulated by it. This RNA pairs with the concerned mRNA to form double-stranded RNA, which is cleaved, and the short double-stranded fragments so produced ultimately lead to gene silencing (Fig. 2.15). RNAi is central to the epigenetic modifications as it regulates both DNA methylation and histone modification. RNA interference is widespread in eukaryotes, and is believed to have evolved as a defense mechanism against viruses, transposable elements and other repeat elements; it was only later recruited to regulate endogenous genes, particularly those concerned with plant development.

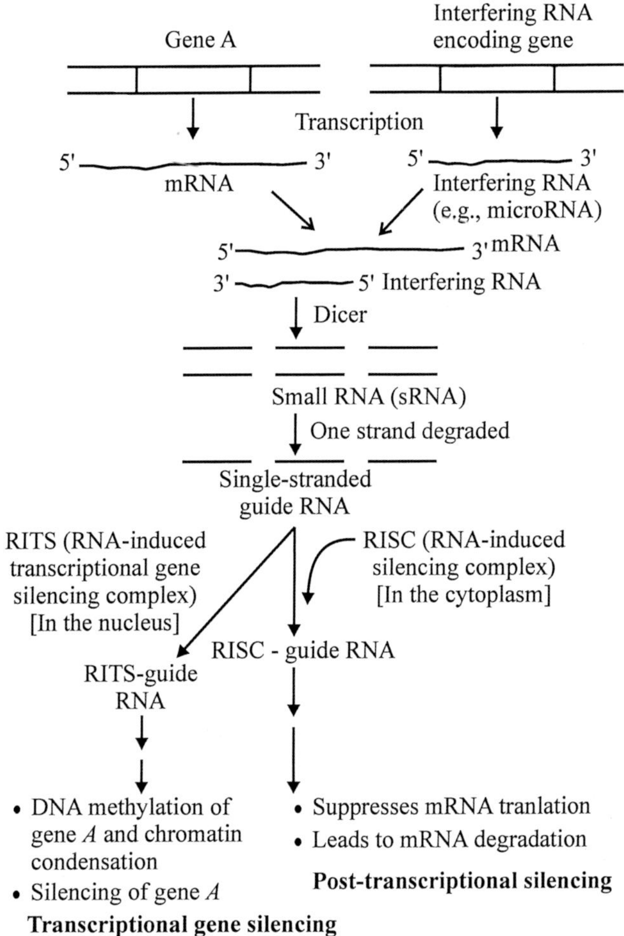

Fig. 2.15. A simplified schematic representation of RNA interference (RNAi). The interfering RNA is complementary to the mRNA encoded by the concerned gene, i.e., the gene that is silenced. The interfering RNA pairs with the mRNA to form double-stranded RNA (dsRNA). The dsRNA is cut into pieces of 21-23 nucleotides, which ultimately become single-stranded. These singles-stranded guide RNAs associate with RISC (in the cytoplasm), or RITS (in the nucleus) and lead to gene silencing.

Various epigenetic modifications of gene function vary in stability; some changes have been stable for very long periods of time. For example, the *Peloric* phenotype of flower in *Linaria vulgaris,* generated by methylation of the *LYC* locus, has been stable for over 250 years. Some well known phenomena involving epigenetic modifications are paramutation, parental imprinting, vernalization, inhibitory gene interaction, etc. In addition, several phenomena of considerable importance in crop improvement, such as polyploidy (both allo- and auto-polyploidy), interspecific hybridization, heterosis, etc., involve epigenetic modifications. The epigenetic information systems are modulated by environmental factors, and they provide a channel by which the environment exerts such influences that often lead to heritable changes in genome organization and gene function. This role of the environment runs contrary to the classical concept wherein environmental factors are considered to only provide a passive milieu for proper gene function, and to have no influence either on genome organization or heritable changes in gene function. DNA methylation and histone modifications interact with and reinforce each other. The epigenetic information systems can be exploited to generate basic information in molecular biology and to contribute to crop improvement endeavors; the various possibilities of this aspect are only beginning to emerge.

2.10. GENOMIC IMPRINTING

In simple terms, the **genomic imprinting** or simply **imprinting** is the mechanism by which the expression of a set of genes contributed by the male gametes is suppressed, i.e., the genes are silenced, while the expression of a different set of genes is suppressed in the case of female gametes (Fig. 2.16). Thus, the expression of the genes subjected to genomic imprinting depends on the sex of the parent that contributed the concerned genes. The phenomenon of genomic imprinting was first described for the gene encoding IGF-II (insulin-like growth factor II) in case of mice. The suppression of gene expression or gene silencing involves DNA methylation of the concerned genes as well as histone modification. The genes subjected to genomic imprinting, in general, are those that are essential for early embryo development. As a result, the zygote contains at least one non-silenced functional copy of all the genes required for embryo development. The set of genes silenced in the genome contributed by the sperms is functional in the genome provided by the egg cell, while those suppressed in the egg cell genome are fully functional in the genome of the sperms. Therefore, the zygote is able to embark on embryo development.

The methylation pattern is erased during the embryo development, and is reset later in the cells destined to produce male and female gametes. The methylation pattern in the male gametes is the same that is specific for the sperms, while that in the egg cells is specific for the egg cells. In plants, the genomic imprinting is quite flexible in the case of embryo development so that apomictic embryos develop in many species. In addition, certain treatments like low or high temperature exposure are able to induce embryo development from pollen grains of cultured anthers and egg cells of unfertilized ovaries. However, genomic imprinting is quite rigid in case of endosperm development. As a result, fertilization is essential for the development of the endosperm. Genomic imprinting is believed to be the reason for the abortion of endosperms in case of certain interspecific crosses.

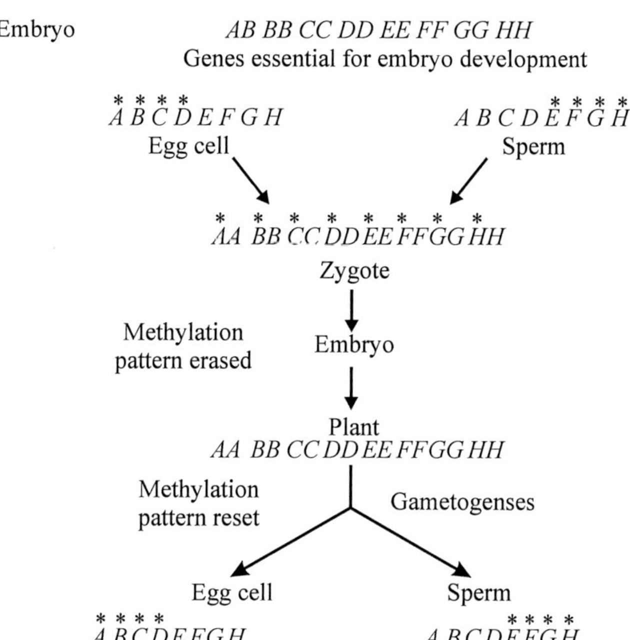

Fig. 2.16. A highly simplified schematic representation of genomic imprinting in plants. Let us suppose that genes *A, B, C, D, E, F, G* and *H* are essential for embryo development. In the egg cells, genes *A, B, C* and *D* are silenced by methylation (denoted by *A**, *B**, *C**, and *D**), while in the sperms genes *E* , *F, G,* and *H* are silenced (depicted as *E**, *F**, *G**, and *H**). These genes become demythlated during the embryo development. But during gamete development, the methylation pattern is reset so that the genes *A, B, C* and *D* become methylated in the egg cells, and the genes *E, F, G,* and *H* become methylated in the sperms.

2.11. GENOME EDITING

Genome editing, gene editing, and gene-targeting, etc. comprise the induction of random and/or pre-determined genetic changes in the genes/genome of an organism. Genome editing uses genetically engineered nucleases, called **site-directed nucleases (*SDN*)**, for induction of the genetic changes at pre-decided locations in the genomes/genes of the selected organism. The SDNs can also be used to integrate transgenes at the pre-determined locations in the genome. The SDNs are designed to bind sequences of the genomic DNA close to which the genetic changes are to be induced; this is the function of the DNA binding domains of these nucleases. Therefore, the DNA binding domain of every SDN has to be specifically designed keeping in view the DNA sequence close to the site of the targeted genetic change. Once a SDN binds to DNA, its nuclease domain induces a double-strand break close to the binding site. The repair of these double strand breaks generates both random mutations as well as pre-specified mutations in the target genes. These breaks are also used for integration of transgenes at the sites of the breaks in the genome.

There are primarily the following four types of SDNs: zinc-finger nucleases (ZFNs), meganucleases (homing endonucleases), transcription-activator like effector nucleases (TALENs), and clustered regularly interspaced short palindromic repeat (CRISPR) system. ZFNs were the first SDNs to be designed and used for genetic modification. They have a DNA binding domain comprising 3-4 zinc fingers, and the nuclease domain of the FokI restriction enzyme. The nuclease action of FokI cleaves the DNA 9 or 13 nucleotides downstream of the DNA binding site, and the sequence at the cleavage site is irrelevant (Fig. 2.17).

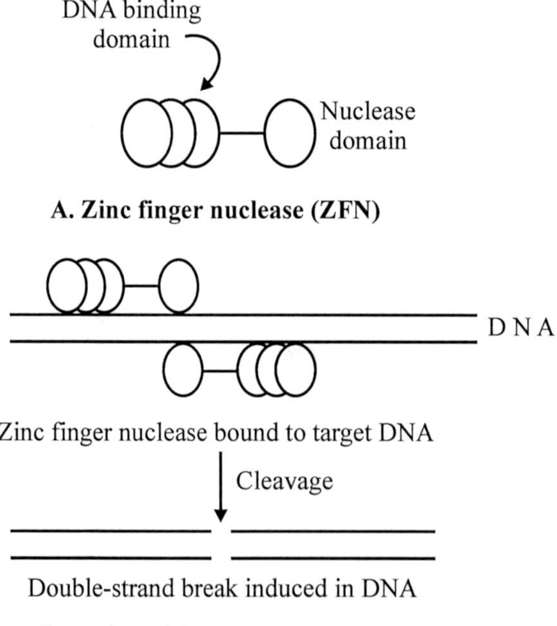

Fig. 2.17. A-ZFN molecule has a DNA binding domain and a nuclease domain. The DNA binding domain is designed to bind specific sequences of DNA. The ZFN binds to target DNA as a dimer and induces a double-strand break between the binding sites. ZFN, zinc finger nuclease.

But the ZFNs are difficult to design, and only a limited flexibility is available in terms at the sequences of the binding sites. The CRISPR system is, on the other hand, is a simple and highly flexible SDN system. As a result, this system has emerged as the most widely used genome editing approach within a short span of about three years. A CRISPR SDN molecule has a guide RNA (gRNA) molecule and the Cas9 (CRISPR-associated nuclease 9) endonuclease. The gRNA has two functional regions, one of which pairs with the target DNA, while the other region associates with Cas9. Thus, guide RNA pairs with the desired target DNA and places the Cas9 nuclease in the vicinity to induce a double strand break.

The SDNs can be delivered either as proteins or as DNA directly into plant protoplasts. Alternatively, their protein sequences can be converted into DNA sequences, and suitable gene constructs can be created to encode the SDNs. These gene constructs are integrated into appropriate expression vectors, which can then be introduced into plant cells. In the end, plantlets are regenerated from the cells/tissues treated/transformed with the SDN constructs. The regenerated plantlets are then screened to isolate plants having the desired mutant phenotype. The mutant plants can also be selected using a suitable molecular technique. The mutational changes in the DNA are random (DNA repair by non-homologous end joining) (Fig. 2.18) in case a donor DNA is not applied with the SDN. Alternatively, a donor DNA can be provided with the SDN; in this case, the mutational change would be specified by the sequence of the donor DNA, which is used for repair by homologous recombination (Fig. 2.19). In case a transgene were provided with the SDN, repair by homologous recombination would lead to integration of the transgene at the site of double-strand break. The sequences flanking this transgene must be homologous to the sequences surrounding the double-strand break to allow homologous recombination. The mutation frequency induced by the SDN technology may be up to 10% or higher in case of plants, and up to 4% of these may be due to homologous recombination. Thus, SDN technology consists of the following four steps:

1. Designing of a SDN that binds to the desired site in the target gene, and producing the gene construct encoding this SDN.
2. Introduction of this gene construct into plant cells; if desired, a donor DNA may also be delivered.
3. Induction of the SDN in the transformed cells, followed by production of double strand breaks, DNA repair and the induction of mutations or transgene integration.
4. Regeneration of plants, and their screening leading to the isolation plants with the desired mutant phenotype.

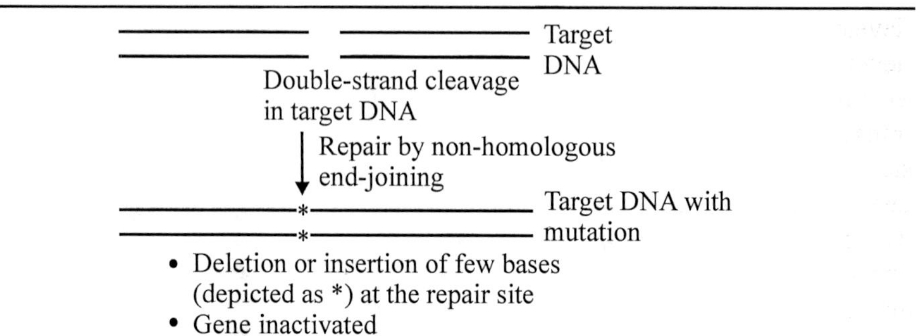

Fig. 2.18. Induction of random mutations in target genes by site-directed nucleases.

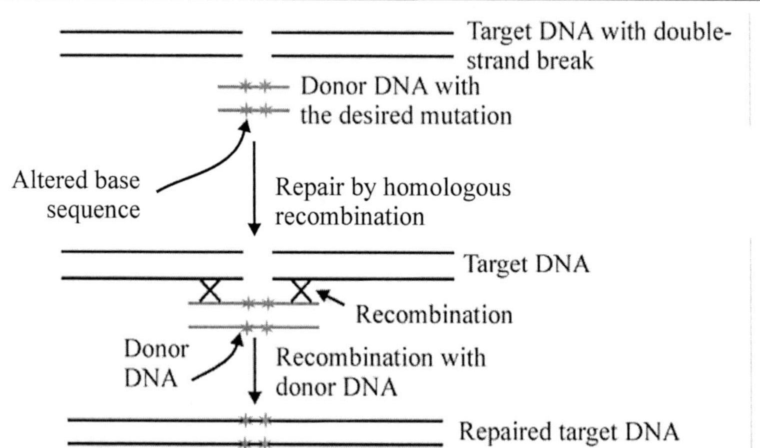

Fig. 2.19. Induction of specified mutations at the target site. The SDN is used along with a donor DNA. The donor DNA is homologous to the region in which mutation is to be induced, and contains the desired mutated sequence. The double-strand break is repaired by homologous recombination. As a result, the mutant sequence contained in the donor DNA becomes integrated into the target genomic DNA. SDN, site-directed nuclease.

2.12. GENOMIC RESOURCES

In simple terms, **genome** represents the total genetic information contained in the nuclear and organellar chromosomes (chromosomes present in plastids and mitochondria) of an organism. **Genomics**, therefore, comprises the analysis of the organization and the function of complete genomes of the organisms. This analysis includes sequencing, genome annotation, analysis of proteomes, transcriptomes as well as various metabolic pathways and the interactions among these pathways. **Transcriptome** is the sum total of different types of RNA produced in various tissues, and the quantities of each of the RNA species. Therefore, the transcriptome would vary with the tissue type, the stage of development and the prevailing environment. Therefore, **transcriptomics** is devoted to the study of transcriptome, and includes the determination of expression levels of different RNA species, splicing patterns of primary RNA transcripts, etc. in various cell/tissue types, during different developmental stages, and under different environmental conditions. The term **proteome** denotes to the full set of proteins present in a cell type; the proteome composition will depend on the developmental stage and the environmental conditions to which the cells are exposed. **Proteomics**, therefore, is the study of proteome, and includes the study of 3-D structures of proteins and protein complexes present in the cells, the expression patterns of the proteins, and the protein networks operating in living cells. Genomics may be divided into two subgroups, viz., structural and functional genomics. The determination of complete genome sequence of an organism, and the complete set of different proteins encoded by its genome constitutes **structural genomics**. However, **functional genomics** analyzes the patterns of gene expression in an organism, and the different types of metabolic pathways present in the organism and the interactions among these pathways. **Metabolome** consists of all the metabolites produced in a cell.A **metabolite** represents the end product of a metabolic pathway. The composition of metabolome would also differ depending on the cell/tissue type, the stage of development and the prevalent environment. Therefore, **metabolomics** is a systematic analysis of the profiles of various metabolites in various cells/tissues of an organism at different stages of development and under different environmental conditions.

The **genomic resources** of a species consist of all the available information about the structure and function of the genome of the concerned species. Therefore, the genomic resources include genetic maps, physical maps, genome sequences, annotations of the genome sequences, expressed sequence tags, transcriptome analysis, proteomics data, and even the available information on the metabolomics of the species. An extremely large amount of data is being generated in the different areas of genomics. The pace of data generation has been tremendously accelerated by the genome sequencing projects, and the associated developments in the technologies and bioinformatics tools required for the various genome wide studies. The genomics data are now available for many different plant species, but those for *A. thaliana* and rice are far more complete than those for others. The various genomic features of *A. thaliana* and rice have established them as the model dicot and monocot plant species, respectively. The sequencing of the genomes of *A. thaliana* and rice has been completed, and their genomes have been almost completely annotated. In addition, the functions of most of their genes have been determined by appropriate experimental procedures. In view of the above, the genomes of *A. thaliana* and rice, and any other plant species for which extensive genomic data have been generated, are used as reference in various studies, e.g., gene/gene family identification,

identification of orthologous sets of genes that are conserved across various species, etc.

A comparison among the molecular marker linkage maps revealed that the gene order is more or less similar in the related species, and even in species belonging to thc same family; this is termed as **synteny** or **colinearity**. But this similarity is much less at the genome sequence level due to small chromosomal aberrations, including insertions and deletions. The synteny relationships can be used to devlop molecular markers in a species with limited genomic resouces on the basis of the genomic information of a related species. The analysis of genome sequence of a species with suitable bioinformatics tools would enable the discovery and identification of molecular markers like simple sequence repeat (SSR), conserved orthologous set of genes (COS) and single nucleotide polymorphism (SNP) markers. The transcriptome data can be analyzed to develop a library of ESTs (expressed sequence tags), which represent a part of the transcribed region of the concerned genes. The EST databases can be analyzed to identify expression QTLs (eQTLs). The eQTLs are those genomic regions that regulate the levels of expression of some other genes present in the genome. The study of eQTLs is expected to help identify such genes and *cis*-acting DNA sequences that participate in the regulation of expression of various genes. Many workers are investigating the feasibility of using the levels of different metabolites as markers for predicting the performance of different genotypes/lines. Efforts are also being made to evaluate the usefulness of the metabolite levels as selection criteria for the isolation of superior genotypes/lines.

PRACTICE QUESTIONS

1. Discuss the relevance of genetics and genomics in plant breeding.
2. Explain the inheritance of quantitative traits and production of continuous variation in the phenotypic expression of these traits.
3. Explain the meaning of independent assortment. How does linkage differ from independent assortment and what is the significance of linkage in plant breeding.
4. Explain the meaning of polygenes. Describe the fixable and nonfixable components of genetic variance, and discuss their relevance in plant breeding.
5. Give a brief account of the processes of propagation of genes, gene expression and its regulation, and the control of biochemical pathways.
6. Discuss the significance of genomic imprinting and epigenetics in plant breeding.

7. White short notes on the following: (i) heritability, (ii) genome editing, (iii) genomic imprinting, (iv) cytoplasmic inheritance, (v) genomic resources, (vi) epigenetics, (vii) threshold characters, (viii) multiple factor hypothesis, (ix) segregation, (x) linkage, (xi) genetic background.

RELEVANT LITERATURE

1. Tabata S 2002. Impact of genomics approaches on plant genetics and physiology. J. Plant Res. 115: 271-275.
2. Morgante M and S Salamini 2003. From genomics to breeding practice. Curr. Opin. Biotechnol. 14: 214-219.
3. Singh BD 2009. Genetics, 4th edn. Kalyani Publishers, Ludhiana, New Delhi

■■■■■

3 Tools of Plant Breeding

3.1. INTRODUCTION

Plant breeding began with the pre-historic humans as a domestic activity, and has progressed to the stage of assimilating the sophisticated molecular marker and the transgenic technologies, and gene/genome editing. But the objective of the plant breeding activities throughout has been the genetic modification of crop plants so as to make them more useful to the humans. As noted in the first chapter, there are two critical requirements for crop genotype modification, viz., the desired gene/allele should be available and usable in breeding programmes, and a suitable technique/method should be used for identification of the desired genotypes from the breeding populations. The plant breeders, therefore, use a variety of tools and breeding strategies to accomplish their objectives. In this chapter, the various tools used by the breeders are considered in some detail, while the various breeding schemes are described in the following four chapters. The tools for plant breeding are considered to include the sources of various genes and the genetic mechanisms that facilitate breeding activities and help achieve the various objectives.

3.2. INTRODUCTION TO GENETIC RESOURCES

> *"The greatest service which can be rendered to any country is to add a useful plant to its culture".* – **Thomas Jefferson,** the third president of USA

The process of bringing and growing a new plant genotype or species into an area where it was not grown before constitutes **plant introduction**. This plant material may come from a different region of the same country or from another country. In the early days, plants were introduced by invaders, settlers, traders, travelers, etc. Many of our important crops like maize, pigeon pea (arhar), sesame, sorghum (jowar), mustard, tomato, potato, groundnut (peanut), etc. are due to past introductions; soybean and sugar beet are examples of introductions during the 20th century. At present, all plant introductions must be routed through NBPGR, New Delhi. The procedure to be followed for plant introduction is outlined in next page.

1. The concerned individual/institution may request the provider of the plant material to send the same to NBPGR. Alternatively, the request may be submitted to the NBPGR for procuring the desired germplasm. The germplasm are obtained as gift, in exchange for germplasm, are purchased or are collected by explorations. During 1976 to 2005, NBPGR introduced 20,22,022 germplasm entries. It may be noted that introduction of certain plant materials in India is prohibited, e.g., *Theobromo* spp. from Africa, Ceylon and West Indies; sugarcane from Australia, Fiji, Philippines, etc.
2. The introduced material is subjected to quarantine. **Quarantine** is a set of procedures that is followed to prevent the entry of insect pests, pathogens, and weeds along with the introduced plant materials. The introduced germplasm is kept in isolation during the quarantine. Quarantine for propagules of all agricultural and horticultural species is carried out by NBPGR, New Delhi and its regional station, Hyderabad. The introduced germplasm is examined closely using suitable techniques, including X-ray screening. The bureau makes every effort to free the germplasm from contaminations. In case the efforts fail, the introduced material is destroyed. The process of quarantine takes a minimum of three weeks. During 1956-2005, quarantine prevented the entry of 12 new pathogens, and many new pathotypes of several pathogens in the country.
3. The germplasm that is cleared by quarantine is sent to the concerned person/institution. It is important that the process of introduction is initiated well in advance so that the material reaches at least two months before it is needed by the individual/institution making the request.

The introduced germplasm may be an entirely new crop species or a new varieties of an extinct crop species. It may be released for cultivation as a new variety (**primary introduction**) or it may be subjected to selection or used in hybridization for the development of new varieties (**secondary introduction**). Many introductions have been directly used as new varieties, e.g., Sonora 64, Lerma Rojo wheat; TN1, IR8, IR28, etc. rice and some varieties of many other crops. Similarly, many introductions have been used for creation of new varieties through selection (e.g., Kalyan Sona, Sonalika wheat; Jamnagar Giant, Improved Ghana pearl millet; Pusa Lal, Pusa Sunehari sweet potato, etc.) or hybridization (e.g., almost all the present-day semi-dwarf wheat and rice varieties; all the sugarcane varieties; Pusa Ruby, Pusa Early Dwarf tomato, etc.). Thus, introduction provides entirely new crop species, and superior new varieties with minimum demand on the various resources. In any case, it makes available new genetic variability that can be used in breeding programmes. A crop species may be taken to a new area for cultivation because this area is free from a threatening disease/insect pest. But new pathogens, insect pests and weeds, and new pathotypes/biotypes of existing pathogens/insect pests may be brought into the country if adequate care is not taken during the quarantine. Further, some introduced plant species may become problematic weeds, e.g., *Lantana* sp., and some others may threaten the ecological balance, e.g., *Eucalyptus* spp.

3.3. POLLINATION CONTROL

Controlled pollination is critical to almost all plant breeding schemes. **Controlled pollination** involves pollination of a given line (used as the female parent) only with the pollen from a specified line (used as the male parent). The mechanisms that allow controlled pollination or pollination control are particularly required for hybrid seed production because self-pollination can be achieved with relatively much less effort. There are two chief mechanisms that ensure controlled cross-pollination, viz., self-incompatibility and male sterility.

3.3.1. Self-Incompatibility

Self-incompatibility (**SI**) is the inability of functional pollen grains to effect fertilization on self-pollination. It was first described by Koelreuter in the middle of 18th century. In some cases, fertilization may take place, but the embryos abort at an early developmental stage. Self-incompatibility can be divided into the following types after Lewis (1954): (1) heteromorphic, (2) homomorphic, (2a) sporophytic homomorphic, and (2b) gametophytic homomorphic systems.

The heteromorphic system of SI

In the **heteromorphic SI system**, the flowers of different SI groups differ in the lengths of their styles and stamens. For example in *Primula*, the flowers are either of pin (long style, short stamen) or thrum (short style, long stamens) type (Table 3.1). The only compatible mating is between pin and thrum flower types. The SI reaction is governed by the locus *S*, which has five tightly linked genes (CCM^T, GLO^T, CYP^T, PUM^T and KFB^T). The recessive allele *s* is, in fact, a mutation that involves the deletion of all the five genes present in the *S* locus. The CYP^T gene encodes an enzyme that degrades brassinosteroids. This gene is expressed only in the styles so that *Ss* plants have short styles, while the *ss* plants have long styles. It is suggested that one of the genes controls the SI reaction of style, while another gene controls the SI reaction of pollen grains. In some plant species, the system is relatively more complex, e.g., styles of three different lengths are present. The SI reaction shows sporophytic control. This system is found only in few crop species like sweet potato and buckwheat.

Table 3.1: Salient features of the heteromorphic self-incompatibility system of *Primula.*

Flower type	Style	Stamen	Genotype	Pollen	SI reaction	
					Pollen	Style
Pin	Long	Short	ss	s	s	s
Thrum	Short	Long	*Ss*	*S* and *s*	S	S

The homomorphic-gametophytic system of SI

In the **homomorphic SI system**, there is no association between flower morphology and SI reaction. The SI reaction may be controlled by the genotypes of pollen grains themselves (**gametophytic system**) or by the genotypes of the sporophytes (the plant) that produced them (**sporophytic system**). The homomorphic-gametophytic system system was first described by East and Magelsdorf in 1925. The SI reaction is generally controlled by a single locus, e.g., in *Nicotiana, Solanum, Trifolium*, etc. This is the most widespread and the ancestral system. But in some species two, three or even more loci are involved. The SI reaction effectively prevents fertilization on self-pollination, but sib pollination may lead to fertilization. Further, all the plants are heterozygous at the *S* locus. Sometimes, polyploidy may lead to loss of incompatibility. In general, pollen germinates, but the growth of pollen tube is inhibited in the style (Table 3.2). In species where more than one gene is involved, the SI reaction shows complex control, and polyploidy has no effect on incompatibility. In the style, both the alleles of the *S* locus express themselves, i.e., the SI reaction generally shows codominant control.

The homomorphic-sporophytic system of SI

In the **homomorphic-sporophytic system** of self-incompatibility, the SI reaction of pollen grains is determined by the genotypes of the plants that produces them. In addition, flower morphology is not associated with the SI reactions of pollen and styles. In this system, the SI reaction is governed by a single locus (the *S* locus) with multiple alleles (Table 3.2). This system of SI was first described by Hughes and Babcock in 1950. The sporophytic system is found in radish, diploid *Brassica* crops and in *Sinapis*. The *S* locus alleles generally show complex dominance relationships. In addition, the *S* alleles show codominance and competition. Generally, pollen grains fail to germinate on incompatible stigma, and often this failure is due to lack of pollen hydration after pollination. Even when the pollen grains germinate, they fail to enter stigma and there is callose deposition in the stigma as well as the pollen tubes.

The molecular mechanism of SI

In all the SI systems, each *S* locus has at least two genes. One of these genes encodes a male-specific determinant protein, which is produced either in the anther tapetum (in case of the sporophytic system) cells or in the pollen grains/tubes (in case of the gametophytic system). The other gene encodes a female-specific determinant protein that is expressed either in the stigma (in the sporophytic system) or the style (in the gametophytic system) cells. The male and female determinant proteins interact in a highly specific manner. This interaction initiates a phosphorylation cascade, RNA degradation or Ca^{2+}-mediated signaling cascade that generates the SI response (Table 3.2).

Table 3.2: The salient features of the gametophytic and sporophytic systems of SI

Feature	Gametophytic system		Sporophytic system
	Monofactorial	**Bifactorial**	
Number of loci	One	Two	One
Number of alleles	Multiple (often >50)	Multiple (often >20)	Multiple (often >50)
Control of pollen SI reaction	Pollen genotype	Pollen genotype	Sporophyte genotype
Pollen coating	Simple	Complex	Complex
Pollen germination	Slow	Fast	Fast
Stigma type	Wet	Dry	Dry
Site of inhibition	Style	Stigma	Stigma
Pollen when liberated from anthers	Binucleate	Trinucleate	Trinucleate
Male determinant*	SLF; predominantly in pollen grains	Unknown**	SP11; predominantly expressed in stigma papilla cells
Female determinant*	S-RNase; mostly located in the upper region of style	S-protein; mostly in the style**	SRK; mainly expressed in anther tapetum cells
Mechanism of SI reaction	RNA degradation in incompatible pollen	Ca^{2+}-mediated signaling cascade	Phosphorylation cascade

* Male and female determinants are proteins specific to the pollen and stigma, respectively. These proteins interact in a highly specific manner to initiate the SI reaction.

** The mechanism observed in *Papaver*, in which three loci control the SI reaction.

Elimination and suppression of SI

Self-incompatibility can be eliminated to produce self-fertile forms by one of the following approaches briefly described below.

1. Chromosome doubling generally eliminates single gene determined gametophytic SI.
2. Mutations in the *S* locus produce S_f alleles that specify self-fertility. Selection for S_f alleles is relatively easier in the gametophytic than in the sporophytic SI system.
3. Finally, alleles for self-compatibility may be transferred from a related species. Elimination of SI will be useful when self-fertile lines of a self-incompatible crop species are to be developed.

But for several purposes, e.g., production of self-incompatible inbred lines, it would be desirable to temporarily suppress the SI reaction. A temporary suppression of SI reaction is often referred to as **pseudofertility**. The most successful techniques for temporary suppression of SI reaction are as follows.

1. Immature buds (2-4 days before anthesis or the first openonig of the flowers) may be pollinated with mature pollen (**bud pollination**). This is the most practicable and successful method in both gametophytic and sporophytic systems.
2. A part or the whole of the stigma and even a part of the style may be removed, and pollen dusted onto the cut end. This method is quite successful in the sporophytic system.
3. In some species, a brief exposure to a high temperature induces pseudofertility, but its effectiveness is species- and genotype-dependent.
4. The sporophytic SI is suppressed by increased CO_2 concentration as well as high humidity. High humidity can be produced simply by enclosing the flowers or even the inflorescence in polythene or cellophane bags.
5. In case of *Brassica* spp., spraying the flowers with 5-10% salt solution for 3-5 days is reported to suppress SI.

Relevance of SI in plant breeding

Since SI effectively prevents selfing, it can be used for hybrid seed production in one of the following ways.

1. Two self-incompatible, but cross-compatible, lines may be inter-planted. The seeds produced by both the lines will be hybrid seed.
2. A self-incompatible line may be inter-planted with a self-compatible line. In this case, the seed produced by the self-incompatible line would be hybrid seed. But the seed produced on the self-compatible line will be a mixture of selfed and hybrid seed.

3. In addition, feasible schemes for the production of double cross and triple cross hybrids have been devised (Banga and Banga 1998). The gametophytic system has been used for hybrid seed production in clover (*Trifolium* spp.), while the sporophytic system has been employed in brassicas. The use of SI for hybrid seed production presents several problems, e.g., hand pollination for production and maintenance of inbreds, depression in SI reaction and selection for self-fertility by the continued selfing of parental lines, generation of new SI reactions by inbreeding in the gametophytic systems, adverse effects of environmental factors on the SI reaction, etc.

3.3.2. Male Sterility

In **male sterility**, the pollen grains are nonfunctional, poorly developed or not produced at all. In some cases, the androecium itself is either malformed or itlack normal anthers. The male sterility may be governed by a nuclear gene (genetic male sterility, GMS), the cytoplasm (cytoplasmic male sterility, CMS), or an interaction between the nuclear and cytoplasmic genes (cytoplasmic-genetic male sterility, CGMS). In addition, male sterility may be induced by a chemical (chemically-induced male sterility) (Table 3.3).

Table 3.3: Salient features of different male sterility systems in plants

Male sterility system	Genetic basis	Inheritance pattern	Used in hybrid seed production	Remarks
Genetic	Recessive alleles of single nuclear genes	Typical Mendelian	In a limited number of crops; two-line system: male sterile and pollinator lines	Elimination of the male fertile plants from the MS line is the chief problem
(i). Temperature-sensitive	The male sterility allele expresses only when the temperature is above a critical point	Typical Mendelian	Very convenient to use in crops where available; two-line system: male sterile and pollinator lines	Regions with the critical temperature regimes should be available
(ii). Photoperiod-sensitive	The male sterility allele expresses only when the photoperiod is above a critical point	Typical Mendelian	Very convenient to use in crops where available; two-line system: male sterile and pollinator lines	Regions with the critical photoperiod regimes should be available
(iii). Transgenic	Transgene introduced by genetic engineering	Typical Mendelian	Usable in hybrid seed production	Systems developed for easy use in hybrid seed production

Male sterility system	Genetic basis	Inheritance pattern	Used in hybrid seed production	Remarks
Cytoplasmic	Mitochondrial genome	Cytoplasmic or maternal	Limited	The hybrids are male sterile
Cytoplasmic-genetic	Mitochondrial genome; male sterility restored by a nuclear gene	Cytoplasmic and nuclear gene interaction	Widely used in many crops; three-line system: male sterile, maintainer and restorer lines	Many of the CMS systems are not usable due to some undesirable effects on hybrid performance
Chemically induced	Treatment with a chemical that produces male sterility	Nonheritable	Used in some crops; two line system: male sterile and pollinator lines	Not widely used

Genetic male sterility

Genetic male sterility is governed by nuclear genes. In general, a single recessive gene, *ms*, governs GMS. But in most crop species a very large number of different genes (over 70 genes in maize, over 60 genes in tomato, etc.) produce GMS. The *ms* alleles are produced by spontaneous or induced mutations. Generally, GMS is not sensitive to environmental factors. But some forms of GMS are sensitive to temperature (*temperature-sensitive GMS, TGMS*), while some others are sensitive to photoperiod (*photoperiod-sensitive GMS, PGMS*) as well. In case of *TGMS*, the *ms* allele expresses and causes male sterility when the temperature is above a threshold point; below this point, the *ms ms* plants are fully fertile. Similarly, the *ms* allele in the case of *PGMS* expresses itself under long-day photoperiods, while the *ms ms* plants are fully fertile under short-day conditions (Fig. 3.1). In case of rice, the effect of photoperiod is produced only when the prevailing temperature is within a critical range. The temperature and photoperiod effects are produced during specific periods of floral development. In case of rice TGMS, the critical developmental stage is between formation of PMCs and meiosis, while that for PGMS lies between the secondary rachis branching and PMC formation. TGMS and PGMS are known in several crop species, e.g., rice, tomato, wheat, etc.

The use of GMS in hybrid seed production is problematic since the maintenance of male sterile line involves successful roguing of male fertile plants. However, GMS has been used for hybrid seed production of castor in USA and of pigeon pea in India. But the use of TGMS and PGMS lines eliminates this bottleneck. The TGMS/PGMS lines are multiplied in an area where the *ms* allele in unable to express itself. As a result, the TGMS/PGMS lines are fully male fertile and produce selfed seed. The hybrid seed production, however, is done in an area where the *ms* allele expresses itself so that the TGMS/PGMS line is male sterile. Therefore, the seed produced on the TGMS/PGMS line would be hybrid seed produced by pollination by the male parent. Thus, the TGMS and PGMS systems have given rise to two-line hybrid seed production systems (Fig. 3.1). The TGMS/PGMS line is multiplied in a location where the *ms* gene is not expressed due to the temperature/photoperiod effect. The hybrid seed production, however, is done in such areas where the *ms* gene is expressed so that the TGMS/PGMS line does not produce functional pollen. As a result, the seed produced by this line is hybrid seed.

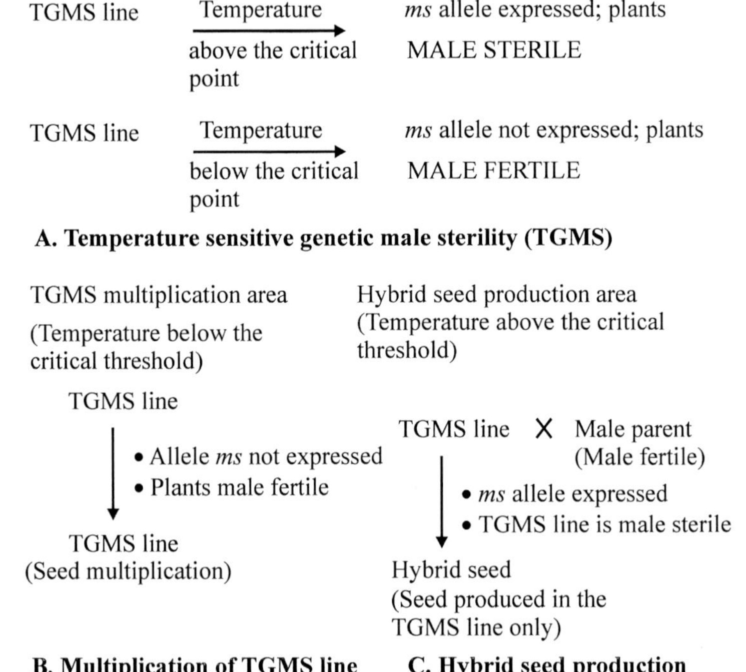

Fig. 3.1. The temperature-sensitive male sterility (TGMS) system and its use in hybrid seed production. The critical temperature must occur during the period from PMC (pollen mother cell) formation to meiosis in order to be effective in producing male sterility. A. Temperature sensitive genetic male sterility. B. Multiplication of TGMS line. C. Hybrid seed production. TGMS, temperature-sensitive genetic male sterility.

Genetic male sterility can be produced by introducing and expressing transgenes that interfere with pollen development and maturation; this type of male sterility is referred to as **transgenic male sterility**. The *Barnase/Barstar* system was the first successful example of this type of male sterility. The bacterial gene *Barnase* encodes an RNase. This gene was transferred into tobacco and expressed in the tapetum cells leading to their degeneration and, as a result, male sterility. It may be pointed out that tapetum cells provide the nutrition necessary for pollen development and maturation. The gene *Barstar* encodes a specific inhibitor of the Barnase RNase. Therefore, when a line with *Barstar* is crossed with a *Barnase* male sterile line, the hybrid tapetum cells express both Barnase and Barstar proteins. As a result, the Barnase RNase becomes inactive and the tapetum cells continue to function normally. Therefore, the hybrid plants are male fertile. The chief limitation of this system is that the male parent of the hybrid has to have the *Barstar* gene so that the F_1 hybrid is male fertile. In India, the *Barnase/Barstar* system has been used to develop male sterile lines and male fertility restorer lines (*Brassica juncea*).

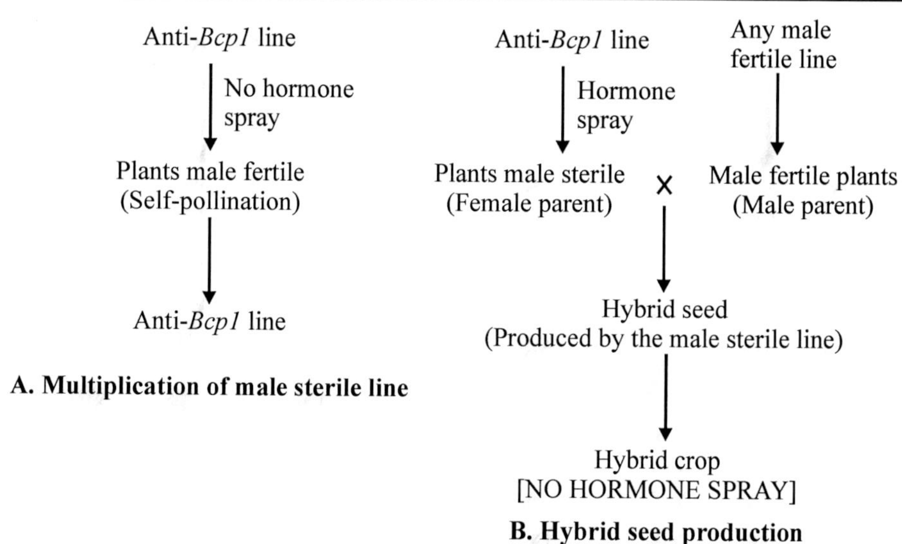

Fig. 3.2. The male sterility system based on the Anti-*Bcp1* (antisense construct of the *Bcp1*) gene. The promoter of this gene is hormone-responsive. A. The Anti-*Bcp1* line is multiplied without hormone spray. The Anti-*Bcp1* is not expressed, and the line is male fertile. B. When a line having Anti-*Bcp1* is sprayed with the hormone, Anti-*Bcp1* is expressed. This blocks the expression of normal *Bcp1* gene, and the plants become male sterile.

Attempts to develop more convenient to use transgenic male sterility have been successful. One such system uses antisense gene construct of the *Bcp1* gene (anti-*Bcp1*). The *Bcp1* gene is expressed in tapetum and pollen grains and is essential for normal pollen grain development. This antisense construct is driven by its own promoter that has been modified to function as a hormone-inducible promoter. When the anti-*Bcp1* gene is expressed in a plant it prevents the expression of the normal *Bcp1* gene in this plant. Therefore, this plant produces nonfunctional pollen grains. Plants having two copies of this anti-*Bcp1* gene are used as the male sterile line. The plants of this line will be male fertile so long as they are not strayed with the concerned hormone. The male sterile lines are multiplied by growing them without the hormone spray (Fig. 3.2). But these plants will be male sterile if they were sprayed with the concerned hormone since they will be unable to produce the Bcp1 protein. This male sterile line can be pollinated by any male fertile line to produce hybrids. The hybrid plants will be male fertile as long as the hybrid plants are not sprayed with the hormone.

Cytoplasmic-genetic male sterility

Some cases of male sterility are caused by changes in the mitochondrial genome. In such cases, male sterility shows cytoplasmic inheritance. Therefore, this type of male sterility is called **cytoplasmic male sterility (CMS)**. The CMS lines are maintained by crossing them with respective male fertile lines, called **maintainer lines** or **B lines** (Fig. 3.3A). Each CMS line has a specific maintainer line that has the same nuclear genotype as the CMS line. It should be noted that when a given

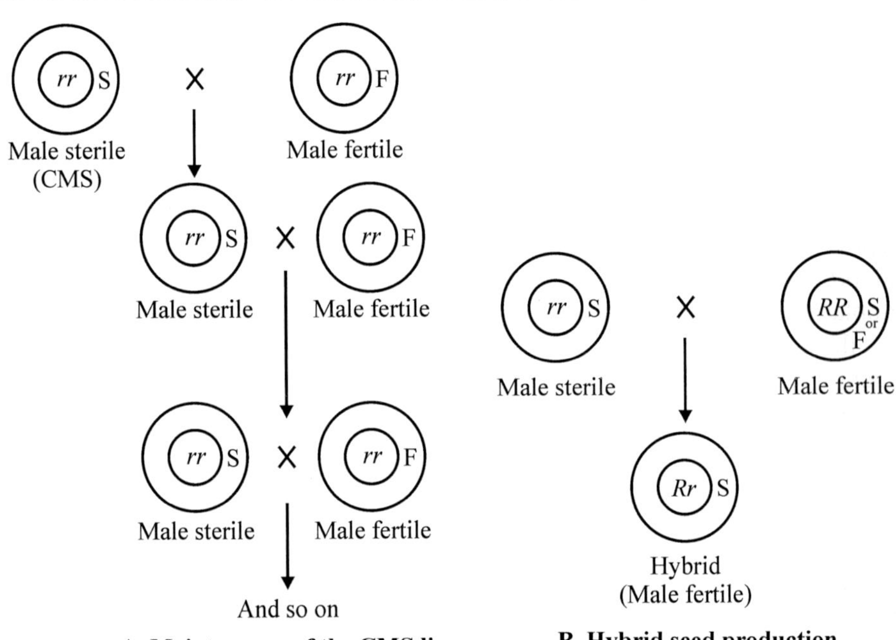

Fig. 3.3. Cytoplasmic male sterility (CMS). The male sterility is produced by mutant mitochondria located in the cytoplasm (denoted by S). The CMS line (called *A line*) is maintained by crossing it with a male fertile line called *B line* or *maintainer line*. The A and B lines have the rr nuclear genotype. Some lines have the dominant gene *R* that restores male fertility in the CMS lines. When such a restore line (*R line*) is crossed with a CMS line, the hybrid seed is obtained and the hybrid plants are male fertile.

CMS line or A line is maintained by crossing it with a single male fertile line, the nuclear genotype of the A line will become almost identical with that of the B line after 6-8 generations. It may be pointed out that both the A and B lines are homozygous for the nuclear gene *r* (restorer gene). The **restorer gene** negates the effects of the sterility producing cytoplasm and, thereby, restores fertility in the CMS lines. Some lines on the other hand, have the dominant allele (*R*) of this gene; these lines are called **restorer lines** or **R lines**. When a restorer line is crossed with a CMS line, the seed produced on the CMS line will be hybrid seed. The hybrid seed will be heterozygous for the *r* gene (*Rr*) and, as a result, will give rise to male fertile plants (Fig. 3.3B). When restorer gene/genes is/are available for a CMS system, it is generally known as **cytoplasmic-genetic male sterility (CGMS)** system. The CGMS system is available in many crops (Table 3.4), and is the most widely used pollination control strategy for production of hybrid seed. In addition, a photoperiod sensitive CGMS has been described in wheat. In this wheat CGMS system, the male sterility is produced in longer photoperiods, while shorter photoperiods completely suppresses the expression of male sterility.

The CMS may involve changes in the genome organization or gene expression of mitochondria. In most of the CMS systems, there is production of new, often abnormal, RNA transcripts that are not produced by the male fertile plants with the normal cytoplasm. The restorer genes of several systems encode proteins of the Pentatrico Peptide Repeats (PPR) family. The restorer genes suppress the production of abnormal transcripts associated with the male sterility and promote the production of normal transcripts. Many of the male sterile cytoplasms were discovered as mutant cytoplasms in the concerned crop species, e.g., the CMS-S, CMS-C and CMS-T in maize. In several other crop species, mutant cytoplasms are not available. The CMS systems in such crops have been developed by transferring the cytoplasm from a related species. For example, the cytoplasmic male sterile lines of pigeon pea have been developed by transferring the cytoplasm from the related species *Cajanus cajanifolium* (Table 3.4). In addition, ethidium bromide has been used to induce mutations in the mitochondrial genomes to produce CMS lines in several plant species like petunias, pearl millet, barley, etc. Once a CMS line is obtained in a crop species, the male sterility can be transferred into different lines of the crop. This is done by crossing the CMS line (A line) with the selected male fertile line (line C) into which the male sterility is to be transferred. The F_1 from this will be male sterile. The F_1 and the subsequent generations are backcrossed with the male fertile line C. After 6-8 backcrosses, the male sterile progeny will be the male sterile version of line C; the line C will be used as the maintainer line for propagation of the new male sterile line.

Table 3.4: Selected examples of cytoplasmic-genetic male sterility systems in crops

Crop	Male sterility cytoplasm	Restorer gene	Remarks
Maize	CMS-C*	*Rf*$_4$	Now used commercially
	CMS-S*	*Rf*$_3$	Used commercially
	CMS-T*	*Rf*$_1$ and *Rf*$_2$ (complementary action)	Most commonly used; reduces yield by 2-4%, susceptibility to Southern leaf blight
Wheat	*Triticum timopheevii*	*Rf*$_1$ and *Rf*$_2$ from *T. timopheevii*	Used commercially
	Aegilops crassa	*Rfd1* or multiple *Rf* genes	Photoperiod sensitive; longer day lengths produce male sterility
Cotton (*Gossypium hirsutum*)	*G. harknesii*	Gene *Msc* from *G. harknesii*	Used commercially; may lead to up to 50% lower yields
Chilies	Chilies*	Available	Used commercially
Pearl millet	Tifton (A_1)*	Available	Used commercially
Brassica napus	CMS-Ogu**	From wild radish	Used commercially
Brassica juncea	CMS-Ogu**	Restorer genes available	Used commercially
Pigeon pea	A_4 (from *Cajanus cajanifolium*)	Available	Commercially used
Sorghum	Milo or A_1*	*Msc*	Used commercially
Rice	CMS-WA (wild abortive)**	Four genes; *Rf-WA*$_1$ is the strongest	Used commercially
Onion	CMS-S (alloplasmic)**	available	Used commercially
Sunflower	PET-1 (from *Helianthus petalaris*)	*Rf*$_1$ and *Rf*$_2$ (polymeric gene action)	Used commercially

* Mutant cytoplasm from the crop species concerned.
** Cytoplasm from a related species.

The critical step in the commercial use of a CMS system is the discovery of suitable restorer lines. The CMS line is crossed with a large number of male fertile lines, and male fertility of the F_1s is scored. The lines producing fertile F_1s will have the restorer gene and can be use as restorer lines. The restorer gene, once discovered, can be transferred into any given line as follows. The restorer line is crossed with the concerned male sterile line. The F_1 will be fertile due to the restorer action of the restorer gene. It may be pointed out that in this *R* gene transfer scheme, the cytoplasm from the CMS line serves as the selection environment for the restorer gene during the backcross programme. The F_1 is used as the female parent in a cross with the line (say, line D; used as the male parent) into which the restorer gene is to be transferred. The male fertile plants in the progeny are selected and crossed with the line D (used as the male parent); this step is repeated 6-8 times. In the end, the male fertile plants in the progeny are selected, they are self-pollinated, and their seeds are harvested separately and grown in individual plant progeny rows. The progeny row having all male fertile plants and nearly identical to the line D in appearance is selected for multiplication. This line is the restorer version of line D and it can be used as the male parent in a hybrid seed production programme.

For hybrid seed production, the CMS line is used as the female parent and a restorer line is used as the male parent. The F_1 hybrid produced in this way will be heterozygous for the *R* gene, and will be fully male fertile (Fig. 3.3B). The CGMS system has been extensively used in hybrid seed production. The chief limitation of this system is that the male parent of the hybrid must be a restorer line. Therefore, the male parents of promising hybrids have to be first converted into restorer lines by backcross programmes. In addition, the male sterile line has to be maintained by crossing it with the respective maintainer line, which is identical with the male sterile line in appearance. The maintainer seed often contaminates the male sterile line seed, which reduces the proportion of hybrid seed in the seed lot. All the male sterile cytoplasms reduce hybrid performance and many of them generate other detrimental effects as well. For example, the most successful male sterility cytoplasm, the CMS-T in maize, reduces yields by up to 2-4% and leads to susceptibility to the Southern leaf blight. In many CMS systems, fertility restoration is not satisfactory, many others show unstable expression, and some show spontaneous reversion to male fertility. These problems have prevented the use of many CMS systems in hybrid seed production programmes. The problems associated with the CGMS systems might be minimized by one or more of several approaches, including search for new CMS sources, attempts at defect correction, use of alternative methods of producing male sterility and/or pollination control, etc.

Chemically induced male sterility

An alternative approach to produce male sterility is the use of certain chemicals that prevent the production of normal pollen grains; these chemicals are called **pollen suppressants** or **chemical hybridizing agents (CHAs)**. The CHAs effectively induce male sterility so that the treated plants can be used as the female parent in a hybrid seed production programme. An ideal CHA should have certain important properties, including no adverse effect on female fertility, economical and easy to use, and absence of the CHA from the hybrid seed. Some examples of CHAs are zinc methyl arsenate (MG1), sodium methyl arsenate (MG2), ethephon or ethrel, GA_3, RH0007 (Hybrex), etc. MG2 application as foliar spray (usually, at 150 mg L^{-1}) 5 days before heading produces 99% male sterility in rice. It has been extensively used for rice hybrid seed production in China. It suppresses respiration in anthers leading ultimately to male sterility. Similarly, Hybrex causes 95-100% male sterility in a wide spectrum of wheat varieties; it has been used for production of hybrid wheat in USA since 1982. Hybrid seed production based on CHAs is more complex and demanding than that based on GMS/CGMS. The use of CHAs allows any line to be used as the female or the male parent of the hybrid. Further, the hybrid seed production is based on only two lines, and these lines are readily maintained as any other line of the concerned crop. Successful use of CHAs requires a set of environmental conditions, e.g., temperature between 26 and 28°C in the case of rice, lack of rain and adequate sunshine, etc. In addition, the CHAs have to be applied at precise developmental stages, and complete male sterility may not be produced leading to the production of some mount of selfed seed. Many of the CHAs are toxic to plants and animals and some of them may be carried over in the hybrid seed.

Other mechanisms of pollination control

Some of our crops are monoecious, i.e., both male and female flowers are produced on the same plant. Some mutant lines in such crops may produce only female flowers, such lines are called **pistillate lines,** and the condition is referred to as **pistillate condition**. The pistillate lines can be used as female parents in hybrid seed production programmes. In case of castor pistillate condition is produced by a single recessive gene *n*; the *nn* pistillate lines are known as **N type pistillate lines**. This pistillate system is similar to genetic male sterility and offers the same advantages andlimitations as the latter. Efforts to develop temperature-sensitive N type pistillate lines of castor have been successful; these lines are called **NES pistillate lines**. These lines produce both male and female flowers when the temperature during flowering is below 35°C, but they produce only female followers at temperatures higher than 35°C. In addition, the pistillate condition in some castor lines is governed by polygenes. These lines, called **S type pistillate lines**, tend to produce male flowers as well in the later inflorescences provided the temperature during flowering is above 40°C. In India, both S and NES type pistillate lines have been used for hybrid seed production in castor.

Apomixis is a form of asexual reproduction, in which embryo development in seeds takes place without fertilization. Apomixis occurs in many crop species in low frequencies, and these species generally reproduce sexually; this is known as **facultative apomixis**. When haploid cells of a normal embryo sac develop as embryos, haploid plants are produced. But when the embryos develop from diploid cells, diploid plants identical in genotype to the parent plant are produced. Therefore, a suitable apomixis system might allow fixation of heterosis by production of seed in the F_1 and the subsequent generations via the apomictic route. In most known cases, one or two recessive or dominant genes are involved in the control of apomixis. But it appears that these genes are most likely complex loci and each of them consists of several tightly linked genes. New apomictic lines can be developed by gene transfer from wild species, induction of mutations, and selection for apomictic genotypes in the segregating generations of interspecific crosses. Attempts to develop obligate apomictic types have not been successful in any of the grain crops. In contrast, efforts to develop sexually reproducing genotypes in obligate apomictic forage grasses have been much more successful. These sexual forms have permitted the development of effective breeding schemes suitable for their improvement. For example, when apomixis is dominant over sexual reproduction, a sexually reproducing line may be used as a female parent in a cross with an apomictic line. The F_1 hybrid will be apomictic and all the progeny from the F_1 and the subsequent generations will be identical to the F_1 hybrid (Fig. 3.4). This would lead to permanent fixation of heterosis. It has been achieved in case of some forage grasses. Even when apomixis is facultative, two facultative apomictic lines may be crossed to produce a facultative apomictic F_1 hybrid, called **vybrid**. A vybrid can be propagated by harvesting the seeds from F_1 like plants; this would lead to partial fixation of heterosis.

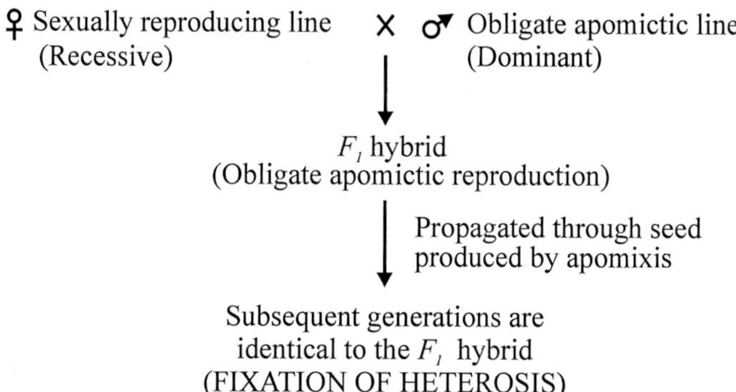

Fig. 3.4. Obligate apomixis leads to fixation of heterosis. The apomictic reproduction is dominant over sexual reproduction so that the F_1 and the subsequent generations reproduce apomictically.

3.4. HYBRIDIZATION: INTRASPECIFIC AND INTERSPECIFIC

Hybridization is the mating of two different plants or lines with different genotypes. This involves placing of the mature pollen grains from one of the plants/lines (called the **male parent**) onto the receptive stigmas of the other plant/line (known as the **female parent**). The first artificial plant hybrid was produced by Fairchild in 1717, after which it was used by several researchers to develop varieties of several fruit trees and for scientific studies. *The chief objective of hybridization is to create genetic variation by generating new* *combinations of the alleles of different genes present in the parents involved in hybridization.* Hybridization has been and remains the chief method of creation of genetic variation, which is achieved as follows. When two homozygous parents with different genotypes are hybridized or crossed, their alleles are brought together in the progeny produced by the cross, i.e., the F_1 progeny. The F_1 plants, therefore, are heterozygous for all the genes for which the two parents differ. The F_1 plants are allowed to self-pollinate or they are intermated to produce the F_2 generation. Segregation of and recombination among the genes heterozygous in the F_1 generation will create new allelic combinations and genetic variation in the F_2 generation. Selfing in the F_2 generation yields F_3 generation, that in F_3 generation gives the F_4 generation, and so on.

The F_2 and the later generations are called **segregating generations** since they exhibit genetic variation as a result of segregation. Selection of plants with the desired combination of traits may be done during the segregating generations to develop new varieties superior to the two parents involved in the cross for one or more traits, especially yield. This method is called **transgressive breeding** and is achieved by pedigree method or one of its modifications. The F_1 plants and the subsequent progeny may be repeatedly crossed with one of the parents; this is called **backcrossing**. Backcrossing allows the transfer of one or few characters from the other parent into the parent to which the F_1 and the later generations are backcrossed. This approach is called **combination breeding** and is represented by the backcross method of breeding. The F_1 generation itself can be used a new variety, called **hybrid variety**; this is the basis of **heterosis breeding**.

3.4.1. Intervarietal Hybridization

The hybridization is called intervarietal or interspecific depending on the taxonomic relationship between the parents involved in the given cross. In **intervarietal hybridization**, also called **intraspecific hybridization**, the parents of the cross belong to a single species. But when the genotypes from two different species are crossed, it is known as **interspecific, wide** or **distant hybridization**. The intervarietal crosses are easy to make and are the most commonly used in crop improvement programs. But in few cross combinations, problems like low seed set, hybrid necrosis and even hybrid mortality may be encountered. The intervarietal crosses generally involve two parents (**simple cross**), but crosses with three or more parents (**complex cross**; Fig. 3.5) are becoming increasingly common particularly in highly improved crops like wheat.

3.4.2. Procedure for Hybridization

The first and the most critical step in hybridization programme is the selection of parents to be crossed. One of the parents of each cross is used female, while the other is used as male parent. It is essential that self-pollination is prevented in the female parent. This is achieved by either removing the anthers or inactivating the

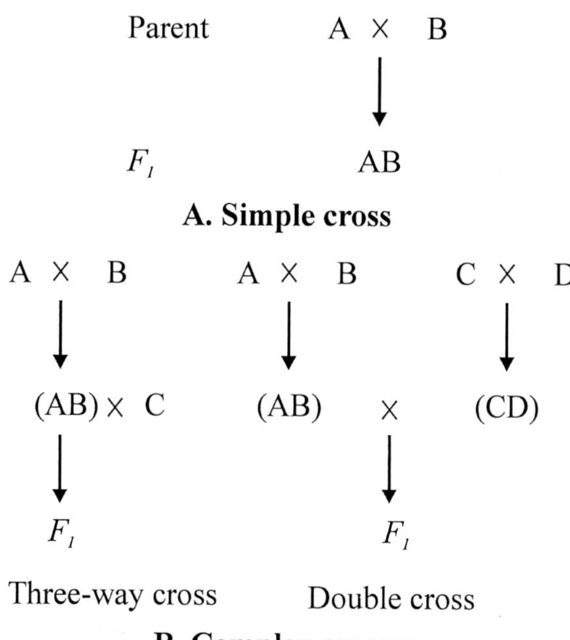

Fig. 3.5. Examples of (A) simple and (B) complex crosses. A three-way cross involves three parents, while four parents are used to generate a double cross. The generation obtained at the end of making a complex crosses is called F_1 generation. But the F_1 generation from a complex cross will be genetically variable due to segregation and recombination in the F_1s (e.g., AB and CD) crossed to generate the complex cross.

pollen grains of the flowers of the female parent before anthesis or the first opening of the flowers; this is called **emasculation**. One common method of emasculation (Table 3.5) is hand emasculation that is suitable for crops with relatively larger flowers. The flowers expected to open the next morning are selected and emasculated, generally, between 4 and 6 PM. The flowers are carefully opened and all the anthers are removed. Care is taken to prevent dehiscence of the anthers and damage to the pistil. The emasculated flowers are enclosed in suitable bags to prevent pollination by unintended pollen. These flowers are pollinated by pollen grains collected from the flowers of the male parent the morning after emasculation, and enclosed with the same bags. The seeds are harvested, stored and grown in the next season to raise the F_1 progeny. The F_1 generation will be heterozygous for all the genes that differ in its two parents.

Table 3.5: A summary of the different methods of emasculation

Emasculation method	Procedure in brief	Remarks
Hand emasculation	Anthers removed with the help of forceps	Between 4-6 PM in the flowers expected to open the next morning; all the anthers must be removed and the ovary must not be damaged*
Suction method	Suction is used to suck the anthers and the pollen grains	Used in the morning around the time the flowers open; up to 15% self-pollination*
Hot water method	Flowers treated with hot water, e.g., 10 min at 40-44°C for rice	Treatment prior to anthesis; water temperature and duration of treatment critical*
Alcohol treatment	Flowers/inflorescence immersed in alcohol, e.g., in sweet clover, 57% alcohol for 10 seconds, then rinsed with water	Treatment time critical as even few seconds longer treatment will greatly reduce seed set; selfing less than 1%*
Cold treatment	Exposure of flowers to low temperature, e.g., 0-2°C for 15-24 hr in wheat	Less effective than hot water*
Chemical emasculation	Application of chemical hybridizing agents	See, Section 3.3.2**
Genetic emasculation	Selfing prevented by genes governing male sterility or self-incompatibility	See, Section 3.3.1, 3.3.2**

* Not suitable for hybrid seed production. Hand emasculation has been used for hybrid seed production in some crops like cotton.
** Used for hybrid seed production.

3.4.3. Consequences of Hybridization

There will be segregation and recombination in the F_1 generation to produce many new gene combinations in the F_2 generation. If the F_1 plants were heterozygous for n genes, there will be 2^n different types of gametes in the F_1 generation, 3^n different genotypes, and 2^n and 3^n different phenotypes (assuming complete and partial dominance, respectively) in the F_2 generation. The number 2^n also represents the total number of different homozygous genotypes obtainable from the concerned F_1. When the F_1 and the subsequent generations are subjected to selfing, the level of heterozygosity in a generation declines to half of that in the previous generation, and the amount of homozygosity increases correspondingly. The level of heterozygosity declines at the same rate when the F_1 and the subsequent generations are repeatedly backcrossed to one of the parents (Table 3.6). It may be pointed out that heterozygotes often tend to be favored in nature. In such cases, the rate of increase in homozygosity may be lower than expected. In any case, the frequency of plants completely homozygous for all the genes increases much more slowly than the level of homozygosity, and it can be estimated by the following formula.

$$\text{Frequency of completely homozygous plants} = [(2^m\text{-}1)/2^m]^n$$

where, m is the number of generations of self-pollination, and n is the number of segregating genes. In this case, linkage among the segregating genes will reduce the number of genes segregating independently. This will increase the frequency of completely homozygous plants in comparison to the expected frequency.

Table 3.6: Effect of continued selfing (or repeated backcrossing to one of the pare-nts) of the F_1 and the subsequent generations on the levels of heterozygosity and homozygosity

Generation	Generations of self-pollination	Level (%) of heterozygosity	Level (%) of homozygosity
F_1*	0	100	0
F_2 (BC_1)**	1	50	50
F_3 (BC_2)	2	25	75
F_4 (BC_3)	3	12.5	87.5
F_5 (BC_4)	4	6.3	93.7
F_6 (BC_5)	5	3.2	96.8
F_7 (BC_6)	6	1.6	98.4
F_8 (BC_7)	7	0.8	99.2

* The heterozygosity and homozygosity are expressed in terms of only those genes that differ between the two parents of the F_1 hybrid. Since the F_1 generation is heterozygous for all such genes, it is said to be 100% heterozygous.

** The F_1 and the subsequent generations are backcrossed to one of the parents of the hybrid.

3.4.4. Distant Hybridization

In **distant hybridization**, parents from different species are crossed. Generally, one of the parents in such crosses is a crop species, and the objective of such crosses is to transfer some desired genes or the cytoplasm to the crop species. In general, interspecific hybridization is more problematic than intervarietal hybridization, and the difficulty increases with the taxonomic distance between the parents being crossed. Hybridization between species differing in ploidy level is generally more problematic than that between species of the same ploidy level. The use of growth regulators have allowed many interspecific crosses to be made, and even immunosuppressive compounds have been used for this purpose. In many interspecific crosses, e.g., in potato, unreduced gametes may be produced by one or both the parents. This greatly facilitates interspecific hybridization, and the recovery of interspecific hybrids. In some cases, two species (say, species A and B) may not cross with each other, but they both may hybridize with another species C. In such cases, species C can be crossed with species A, and the F_1 hybrid is then hybridized with species B. The species C in this case is called **bridge species** since it serves as a bridge and allows gene transfers from species A into species B (or vice-versa).

The difficulties in distant hybridization may arise due to failure of fertilization, or the failure of the hybrid zygote/embryo to develop. The latter problem may arise from lethal genes, elimination of chromosomes from one of the parents, disharmony between the genomes of the two parents or, quite often, endosperm abortion. In case, there is no fertilization, one of several treatments may be tried or somatic hybridization (Section 3.5) may be attempted. The problem of endosperm abortion is routinely overcome by excising the young embryo and culturing it on a suitable culture medium (**embryo rescue**). The endosperm balance number (EBN) hypothesis has been proposed to explain endosperm abortion. *EBN* is the relative strength of a genome to support endosperm development in interspecific crosses. When the EBNs of the parental genomes of a cross are different, the endosperm tends to be aborted during development. The EBN seems to be related to **genomic imprinting**, which is the silencing of different sets of genes in the male and female gametes. Genomic imprinting ensures that only the zygote, and not the gametes, develops into an embryo. In some interspecific crosses, the hybrid seedlings/plants may be weak, abnormal or sterile, or they may even die. These phenomena are generally called hybrid incompatibility. The genes involved in hybrid incompatibility have diverse functions, and in some cases they can be components of the resistance response to diseases.

The interspecific hybrids generally show partial to complete sterility. In most cases, the sterility is due to reduced chromosome pairing between the chromosomes of the two paremtal species. The chromosome numbers of such F_1 hybrids can be doubled by colchicine treatment; the resulting plants would be allopolyploid, they would constitute a new species, and would generally be at least partially fertile. But in some cases, the hybrid sterility may be due to specific genes or produced by an incompatible cytoplasm. Segregation in interspecific hybrids produces considerable genetic variation, but most of the segregants may be either weak or undesirable form a plant breeding point of view. Therefore, most interspecific crosses are used for transfer of either some desired genes or the cytoplasm from a related species into a crop species. But there are some instances of the development of new varieties from interspecific crosses, e.g., from *Oryza sativa* x *O. perennis*, or of using an interspecific hybrid itself as a new variety, e.g., hybrids from *Gossypium hirsutum* x *G. barbadense*. Further, distant crosses can be used to create a new species, e.g., synthesis of *Triticale* (Table 3.7).

Table 3.7: Applications of distant hybridization in crop improvement

Application	Approach	Remarks
Alien addition/substitution lines	Repeated backcrossing to the crop species and selection of the plants with the desired chromosome constitution	Generally used for basic studies and gene/chromosome segment transfers
Gene transfer	Repeated backcrossing to the crop species and selection of the desired trait	Generally for disease resistance, but genes for other traits, including yield and quality have been transferred
Cytoplasm	Repeated backcrossing to the crop species, which is used as the male parent	To create cytoplasmic male sterility in the crop species
Development of new varieties	Pedigree method or a variation thereof; selection for desirable features in the segregating generations	Some examples, e.g., in rice, cotton, etc.
Hybrid varieties	The F_1 generation is used as hybrid variety	In cotton (*G. hirsutum* x *G. barbadense*); sugarcane varieties are complex interspecific hybrids
New crop species	Chromosome doubling of the F_1, followed by hybridization and selection	Triticale has been developed as a new cereal crop
Widening of the genetic base	New versions of our allopolyploid crops are created from their parental species	Creates new genetic variation in the allopolyploid crop species

3.4.5. Somatic Hybridization

Production of hybrid plants by fusion of protoplasts derived from somatic cells is called **somatic hybridization**. A **plant protoplast** is a cell without its cell wall and is surrounded by plasma lemma. The protoplasts can be isolated from nearly all types of cells. But generally leaf mesophyll cells are used for protoplast isolation from dicot species, while embryogenic cell cultures are employed in the case of monocot plants. The mesophyll cells are treated with a mixture of the enzymes pectinase (0.1 to 1%) and cellulase (1-2%) for several hours (Fig. 3.6A). The protoplasts have to be protected from bursting by keeping them in a suitable osmoticum (~500-800 mmol L^{-1} sorbitol or mannitol plus 50-100 mmol L^{-1} $CaCl_2$). The net charge of the plasma lemma is negative; therefore, plant protoplasts do not fuse by themselves, and their fusion has to be induced by a suitable treatment. The protoplasts of the two fusion parents, i.e., the species involved in somatic hybridization, are mixed in equal proportion. This protoplast mixture is treated with 28-50% PEG (polyethylene glycol) for 15-30 min followed by gradual washing to remove the PEG. Protoplast fusion occurs during washing. Alternatively, the protoplast mixture is exposed to a high voltage (800-1000 V/cm) DC current for a brief period of time (~ 60 μs; Fig. 3.6B).

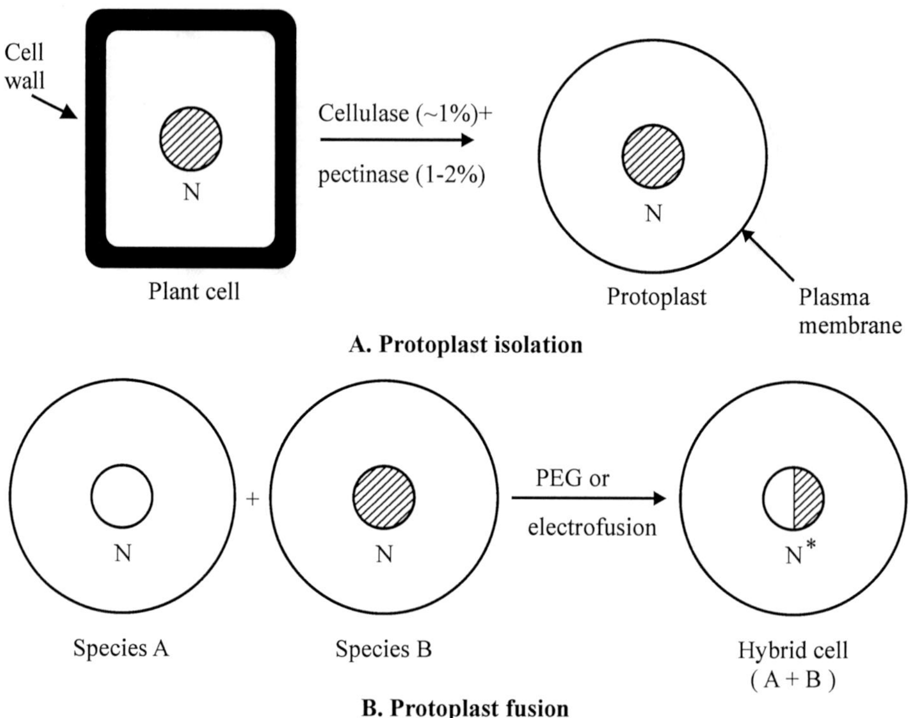

Fig. 3.6. A schematic representation of (A) protoplast isolation, and (B) protoplast fusion. N, nucleus; PEG, polyethyene glycol; N*, hybrid nucleus.

The fusion products are cultured, and the hybrid cells are selected using a suitable strategy. One of the simplest and generally applicable approaches consists of the culture of all the protoplasts after the fusion treatment, and regeneration of plantlets from them. The regenerated plants are then analyzed for morphological features, chromosome constitution, and presence of specific molecular markers for the identification of hybrid plantlets from those of the two parental species. Protoplast fusion can be used to produce **symmetric hybrids** that have the full chromosome complements of both the parental species. These hybrids can be used to widen the genetic base of existing allopolyploid species by their synthesis from the concerned parental species. The somatic hybrid can be backcrossed to the parental crop species to transfer desired genes or the cytoplasm form the other fusion parent into the crop species. The symmetric hybrids may themselves have superior features and be used in crop improvement programmes. In many species combinations, most of the chromosomes of one of the fusion patents are eliminated so that the hybrid has only one to few chromosomes from this species; these hybrids are called **asymmetric hybrids**. These hybrids are useful for gene transfers from the species whose chromosomes are eliminated into the other species. Finally, the protoplasts of one species may be fused with cytoplasts of the other species to produce cybrids. Cytoplasts are protoplasts without nuclei so that **cybrid** cells have nuclei of only one parent and the cytoplasm from both the parents. In fact, all kinds of somatic hybrid cells have cytoplasms form both the parents. However, all kinds of somatic hybrid plants ultimately have chloroplasts from only one of the parents; this is true for the mitochondria as well. But a single somatic hybrid plant may have chloroplasts and mitochondria from either the same parent or the two organelles may be from different parents.

The somatic hybridization approach has been the most successful in the members of the families Cruciferae and Solanaceae. Somatic hybridization has been used to develop at least some varieties (e.g., DelGold and Dr. Chang varieties of tobacco in Canada) and useful breeding materials (e.g., in *Brassica* spp. and potato).

3.5. HETEROSIS AND INBREEDING DEPRESSION

In simple terms, **heterosis** is the superiority of the F_1 generation from a cross to both the parents of the cross in terms of yield (almost always in the plant breeding context) and/or some other trait (Fig. 3.7). In general, heterotic hybrids show faster growth, flower earlier, and show higher biomass than their parents, but their harvest indices are generally comparable. Heterosis is almost universal, is commercially used worldwide, and has been one of the most widely studied phenomena. But the genetic and molecular bases of heterosis have not been fully resolved. Several different terms have been used in literature to describe heterosis, but some of them refer to different methods of estimating heterosis. Some of these terms have been listed in Table 3.8, and their meanings and uses are also indicated. Generally, heterosis is manifested as an increase in vigor, growth rate, yield, etc. It may be pointed out that in terms of practical usefulness, the only relevant measure of heterosis is **economic**, **standard** or **useful heterosis**; this denotes the superiority of the F_1 hybrid over the best variety, including hybrid varieties, of the concerned crop.

Some important features of heterosis are as follows: (i) Heterosis is usually seen in the F_1 generation from crosses between genetically distinct lines. (ii) Heterosis may be observed for some traits of a given F_1 hybrid, but not for others. (iii) A number of different loci seem to be involved in the expression of heterosis. (iv) A variety of different mechanisms have been proposed to explain heterosis. (v) There is evidence that different genes involved in heterosis may affect different characters in the heterotic hybrids. (vi) The genetic basis of heterosis may differ from one crop to the other. Finally, (vii) the extent of heterosis varies with the crop species; for example, it is much higher in maize than in other crops.

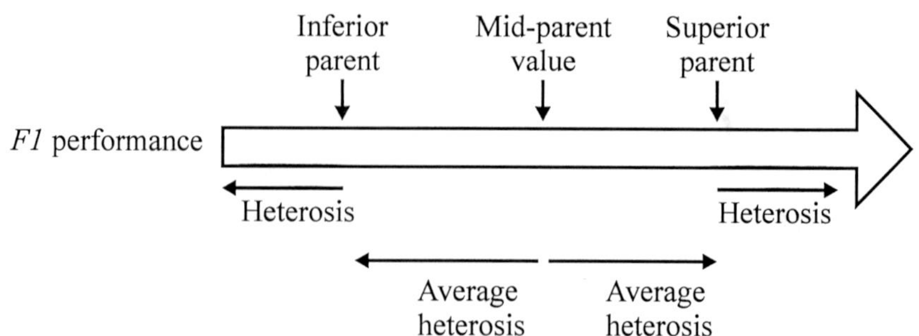

Fig. 3.7. A representation of heterosis and average heterosis. Mid-parent value is the average of mean performance of the two parents.

Table 3.8: Some of the terms in used to describe heterosis

Term	Meaning	Remarks
Heterosis (Syn., true heterosis, euheterosis, heterobeltiosis, hybrid vigor)	F_1 mean either higher than the superior parent or lower than the inferior parent	Heterosis is the most widely used term (the synonyms are not in common use)
Average heterosis (Syn., mid-parent heterosis, relative heterosis)	F_1 mean is either superior or inferior to the mid-parent value for the trait*	Used in quantitative genetic studies
Economic heterosis (Syn., standard heterosis, useful heterosis)	F_1 mean is higher than that of the currently best variety, including hybrid variety, of the crop	This estimate of heterosis is of commercial application

* Mid-parent value is the average of the mean values of the two parents of the concerned F_1.

A phenomenon closely related to heterosis is inbreeding depression. **Inbreeding depression** is the reduction in vigor and reproductive capacity as a result of inbreeding. **Inbreeding** is the mating between individuals related by descent. One of the chief consequences of inbreeding is that there is increase in homozygosity. Generally, inbreeding depression is observed for traits related to fitness. The degree of inbreeding depression observed in different plant species varies from very low (in self-pollinated species) to very high (in asexually propagated species). Most likely the degree of inbreeding depression depends on the level of inbreeding to which the populations of the concerned species have been subjected in the past. For example, populations of carrot and alfalfa are seldom exposed to inbreeding, and they show severe inbreeding depression. In contrast, cucurbit populations are generally subjected to inbreeding, and they show little inbreeding depression. Thus, it seems that inbreeding depression is associated with homozygosity, while heterosis is linked to heterozygosity. As a result, heterosis and inbreeding depression are generally regarded as the opposite sides of the same coin.

3.5.1 Genetic Basis of Heterosis

The different hypotheses proposed to explain heterosis and inbreeding depression are based on either (i) specific genes or (ii) genome-wide mechanisms (Kaeppler 2012; Table 3.9). There are three hypotheses based on specific genes: (i) dominance, (ii) overdominance, and (iii) epistasis hypotheses. In 1908, Davenport proposed **dominance hypothesis**, which states that the dominant alleles of genes mask the harmful effects of their recessive alleles, which accumulate in random mating populations (Section 3.10.2). Inbreeding will lead to homozygosity for the recessive alleles; as a result, they will be expressed and cause inbreeding depression. In the same year, East and Shull proposed that heterosis is the result of superiority of the heterozygotes to the homozygotes at certain loci; this is called **overdominance hypothesis**. The above two hypotheses usually lead to similar expectations, but there are some notable differences. For example, as per dominance hypothesis, an inbred homozygous for dominant alleles of all the concerned genes should perform as well as the heterozygote for these genes. However, according to the overdominance hypothesis, the above heterozygote should be superior to the homozygotes (Fig. 3.8). Therefore, the isolation of homozygous lines as good in performance as the F_1 hybrids should be possible if heterosis were due to dominance gene action, but it will not be possible in case overdominance gene action were involved. The epistasis hypothesis of heterosis was proposed by Gowen (1952). According to this hypothesis, heterosis is the result of interations between the genes affecting the concerned traits.

Cross			Heterosis	
Parent 1	Parent 2	F_1	Dominance	Over-dominance
AA BB cc dd ×	*aa bb CC DD*	*Aa Bb Cc Dd*	Yes	Yes
		Cross I		
AA BB cc dd ×	*aa bb cc dd*	*Aa Bb cc dd*	**No**	**Yes**
		Cross II		
AA BB cc dd ×	*AA BB cc dd*	*AA BB cc dd*	No	No
		Cross III		

Fig. 3.8. Dominance and over-dominance hypotheses lead to different expectations in some of the crosses. For example, F_1 from cross II is heterozygous for the genes *A* and *B*. Since the dominant alleles *A* and *B* are present in parent I, the F_1 will not be heterotic according to the dominance hypothesis. But it will show heterosis according to the over-dominance hypothesis due to heterozygosity for the genes *A* and *B*.

Table 3.9: A summary of the different genetic mechanisms proposed for the explanation of heterosis and inbreeding depression (Based on Kaeppler 2012 and other sources)

Genetic mechanism	Explanation	Remarks
Specific gene mechanisms:	Specific genes involved in heterosis	The first genetic mechanisms to be proposed
(i) Dominance	Inbreeding depression due to homozygous recessive alleles; masking of deleterious effects of recessive allelea by their dominant alleles leads to heterosis	Seems to be the major contributor to heterosis in most of the cases
(ii) Over-dominance	Heterozygotes for some of the genes are superior to their two homozygotes	Repulsion phase linkage between two genes with dominant action would mimic overdominance
(iii) Epistasis	One gene governing a trait affects the action of another gene specifying the same trait	Epistasis would also mimic overdominance
Genome-wide mechanisms:	Mechanisms based on the entire genome and not on specific genes	Can explain heterosis where the specific gene mechanisms prove inadequate
(i) Gene dosage balance*	Inbred genomes show gene presence/absence and copy number variations; hybrids would show better dosage balance for such genes	Inbreds would show extreme dosages for the genes affected by presence/absence and/or copy number variations
(ii) Epigenetic mechanisms	Stable DNA methylation and histone modification	In maize, rice, etc. DNAs of inbreds are more methylated than those of hybrids
(iii) Increased energy efficiency hypotheses	Selective protein synthesis and metabolism in the hybrids	Faster cell division and growth and higher yields of hybrids

* The dosage effects produced over the whole genome accumulate to produce heterosis.

The evidence from a wide range of experiments reveals the following:

1. The heterozygotes for at least a small number of genes are superior to the respective homozygotes.
2. In many cases, the heterozygote superiority may be due to repulsion phase linkage (it gives rise to *pseudo*-overdominance; Fig. 3.9) or epistatic gene action or both (*see* Singh and Singh for more detailed discussion).
3. The dominance gene action seems to be the chief contributor to heterosis.

   (But overdominance and epistatic gene actions are also involved in the expression of heterosis).
4. The relative contributions of the above three types of gene actions differ from one crop species to the other, for different crosses of the same crop, and even for different traits in the same cross.

5. In some cases, dominance, over-dominance and epistasis hypotheses fail to explain heterosis.

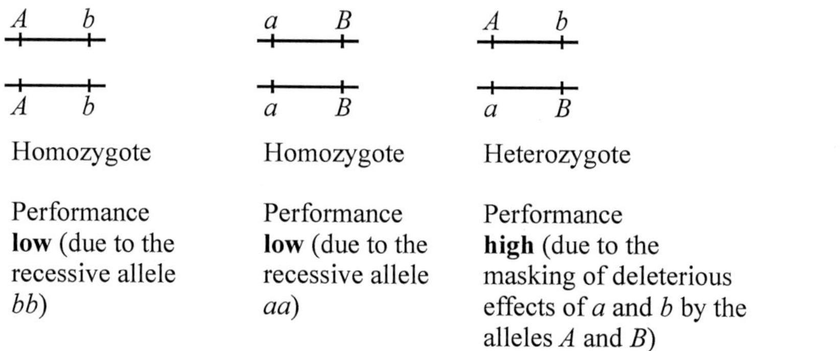

Fig. 3.9. Repulsion phase linkage (*a* linked with *B*, and *b* linked with *A*) mimics over-dominance; this phenomenon is known as pseudo-over-dominance.

The following genome-wide mechanisms have been put forward to explain heterosis in such cases: (i) gene dosage balance, (ii) epigenetic mechanisms, including DNA methylation, and (iii) increased energy efficiency hypotheses (Kaeppler 2012, Schnable and Springer 2013). These mechanisms would produce heterosis for all the traits of a given cross and to a similar degree.

3.5.2 Molecular Basis of Heterosis

At least the following nine types of mechanisms have been proposed to account for heterosis as discussed below:

1. The F_1 produces intermediate amount of a biochemical, which may be the optimum level for the species.
2. The protein products from the different alleles of the concerned genes have different properties/functions so that the F_1 can perform the functions of both the alleles of each gene.
3. The F_1 may produce 'hybrid' products like 'hybrid' transcription factors.
4. The two alleles of the concerned gene may produce effects in two different tissues; in such cases, the effects will be produced in both the tissues of the F_1 generation.
5. The F_1 may show enhanced levels of expression of some the genes.
6. Expression of a larger number of genes may be obtained in the F_1 than in the parents.
7. The F_1 may show a superior gene dosage balance.
8. F_1 may exhibit an increased energy efficiency, or
9. The F_1 hybrid may possess and increased connectivity in the metabolome networks.

It may be pointed out that the above mechanisms involve the coding regions of genes. But some mechanisms, listed here, are based on changes in the noncoding regions of genes.

For example, mutations in the promoter region of a gene responding to a transcription factor, and in the coding region of the gene encoding this transcription factor would produce heterozygote superiority.

3.5.3. Commercial Exploitation of Heterosis

Heterosis is a universal phenomenon that has been observed in every crop species studied so far. Heterosis is best exploited by cultivation of the F_1 hybrid itself as a variety. Such varieties are called **hybrid varieties**; these varieties are discussed in some detail in Chapter 5. Heterosis can also be utilized by cultivation of synthetic and composite varieties, but these varieties exploit only a part of the heterosis. The use of hybrid varieties makes it necessary to produce the F_1 hybrid seed in large quantities every year since the farmers will have to plant new hybrid seed in every crop season. Therefore, hybrid seed production has to be undertaken on a large scale, and a suitable method of ensuring controlled pollination has to be available. In addition, the parents of the hybrid varieties have to be inbred/homozygous lines so that they can be maintained over the years and the same hybrid variety is produced form them every year. These activities require considerable breeding effort and hybrid seed production itself is a demanding task. Therefore, the cost of hybrid seed is generally much higher than that of other variety seed. As a result, the level of heterosis expressed in the particular hybrid should be reasonably high to enable the farmers make substantial profit after paying for the hybrid seed and any other additional inputs that may be required for raising the hybrid crop.

The hybrid varieties have been quite successful in both cross- and self-pollinated crop species. The hybrid technology was developed in USA for application to maize. The maize varieties in common use up to 1930 were open-pollinated varieties; they yielded from ~12 to ~20 quintals per hectare (q/ha). During 1930s, double cross hybrid varieties of maize were introduced. The yields of these varieties increased steadily from ~15 q/ha in 1935 to ~34 q/ha during 1960s. The single cross hybrids of maize were introduced around this time, and maize yields continued to increase from ~39 q/ha in 1960 to ~75 q/ha in 1990. Clearly, the introduction of hybrid varieties in maize enabled the phenomenal increase in the yielding ability of the crop. The data on the performance of hybrid varieties of other crops in the different countries of the world confirm the unquestionable superiority of hybrid varieties, although the increases in yields are not as spectacular as those in maize.

3.6. POLYPLOIDY

The somatic cells of a species generally contain two copies of each chromosome, and this chromosome complement is designated as $2n$. A species that has two copies of a single genome is called **diploid**, and its somatic chromosome number is denoted by $2x$. A **genome** comprises a set of chromosomes that differ from each other in gene content and, often, morphology. A change in the chromosome number from either $2n$ or $2x$ is known as **heteroploidy**. A large number of different terms are used to describe the various types of heteroploidy; the terms in common use are briefly described in Table 3.10 and some of them are discussed below.

Table 3.10: A brief description of the terms used to describe changes in chromosome numbers of various species

Term	Type of change	Symbol	Remarks
Heteroploid	Change from $2x$*	-	-
(i) *Aneuploid*	Few chromosomes more or less than $2n$*	$2n \pm$ few	Generally weak, used for genetic and cytogenetic studies and for gene transfers
Nullisomic	One pair of chromosomes missing	$2n - 2$	Survive only in highly polyploid species
Monosomic	One chromosome missing	$2n - 1$	Survive in polyploid species and few diploid species
Trisomic	One chromosome extra	$2n + 1$	Survive in diploid species
Tetrasomic	One pair of chromosomes extra	$2n + 2$	Survive in diploid species
(ii) *Euploid*	Genome constitution other than two copies of a single genome	-	Useful in basic studies and in plant breeding
Monoploid	One copy of a single genome	x	Sterile, chromosome doubling produces homozygous plants
Haploid	Gametic chromosome complement of the species	n	Sterile, chromosome doubling produces homozygous plants
Polyploid	More than two copies of one genome or two copies each of more than one genome	-	Two types: autopolyploid and allopolyploid
(a) *Autopolyploid*	More than two copies of a single genome	-	Less successful than allopolyploids
Autotriploid	Three copies of a single genome	$3x$	Highly sterile
Autotetraploid	Four copies of a single genome	$4x$	Partial sterility, larger size
Autohexaploid	Six copies of a single genome	$6x$	Partial sterility, larger size
(b) *Allopolyploid*	Two copies each of more than one genome**	-	Many of our crops are allopolyploids
Allotetraploid	Two copies each of two genomes	$2x_1 + 2x_2$*	Tobacco, *Brassica juncea, B. napus, etc.*
Allohexaploid	Two copies each of three genomes	$2x_1 + 2x_2 + 2x_3$	Bread wheat, oats

* $2n$, the somatic chromosome complement of a species whether diploid or polyploid
n, the gametic chromosome complement of a species whether diploid or polyploid
x, the basic or genomic chromosome number of a species
x_1, x_2, etc., distinct genomes from different species
** $2x_1$, $2x_2$, etc., in general, this condition is found, but other situations may also occur

3.6.1. Aneuploidy

A change of one or few chromosomes with reference to the $2n$ chromosome complement of the concerned species is called **aneuploidy**. A **nullisomic** individual ($2n$ - 2) has one pair of chromosomes missing from the $2n$ complement (Fig. 3.10), while a **monosomic** ($2n$ - 1) has a single chromosome less than the somatic number. A **trisomic** ($2n$ + 1) has one chromosome in addition to the somatic complement, while a tetrasomic ($2n$ + 2) has one chromosome pair in addition to the somatic complement. Aneuploids are produced spontaneously in low frequency from normal disomic ($2n$) plants. A relatively higher frequency of aneuploids is produced in the progeny of translocation heterozygotes, asynaptic and desynaptic mutants, and tetrasomic plants. But the best source of aneuploidy plants are autotriploid plants (Section 3.6.4). The behavior of chromosomes during mitosis and meiosis is summarized in Fig. 3.10. The chromosome pairing and distribution in aneuploids, except in nullisomics, is relatively irregular so that they produce considerable frequency of aneuploid (mainly, n + 1, n - 1) gametes.

Aneuploids are generally weaker than normal disomic ($2n$) plants. Nullisomics survive only in highly polyploid species like bread wheat and oats, while monosomics do not survive in diploid species. A complete series of nullisomics (nullisomics for all the n chromosomes) is available in bread wheat and oats, while full series of monosomics in available in several species, including wheat, tobacco, and oats. Monosomics of wheat are often comparable to normal disomic plants in morphology. In diploid species like barley, maize, pea, tomato, etc. only trisomics are known to survive, and a full series of trisomics are available in several species. Trisomics vary considerably in vigor, but some trisomics, e.g., trisome 5 (trisomic for chromosome 5) of maize, are comparable to the normal diploid plants.

In view of the above, aneuploids cannot be directly used for the development of new varieties. But they are quite useful for many types of basic studies that considerably facilitate breeding programs. Aneuploids enable the determination of effects of loss or gain of specific chromosomes on plant morphology and performance. They permit the localization of different linkage groups to specific chromosomes. The homeologous relationships among the chromosomes of allopolyploid species are determined by studies with aneuploids. **Homologous chromosomes** are identical with each other in their gene contents and, in general, morphology. In contrast, **homeologous chromosomes** are similar in their gene contents and, in general, morphology, but they come from different parental species of the concerned allopolyploid species. Aneuploids are also useful in the identification of chromosomes involved in translocations, and they allow the production of substitution lines. A **substitutions line** has one chromosome pair from a different variety or species; in the latter case, it is called **alien substitutions line**. Finally, investigations with aneuploids have allowed the discovery of certain novel genetic systems. For example, the *ph* gene responsible for suppression of homeologous chromosome pairing in wheat was discovered from analysis of the nulli-5B (nullisomic for chromosome 5 of the B genome of bread wheat) line.

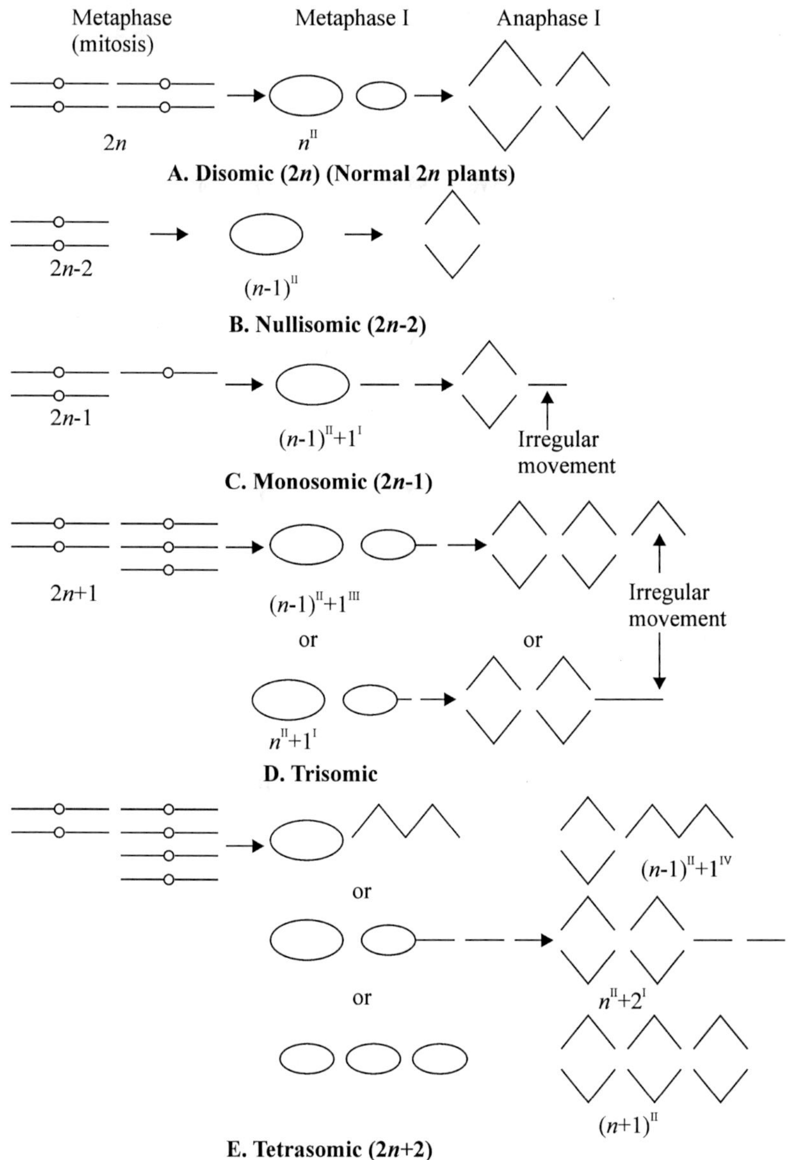

Fig. 3.10. A schematic representation of the chromosome constitution of certain aneuploids, and the behavior of chromosomes during meiosis. For simplicity, only two chromosome pairs of the somatic complement are shown here. Nullisomics survive only in highly polyploid species. n^{II}, *n* bivalens; 1^{I}, one univalent; I^{III}, one trivalent; 1^{IV}, one quadrivalent.

3.6.2. Euploid Variation

In case the somatic chromosome complement of a plant/line/species differs from the $2x$ state by whole genome(s), it is called **euploid variation**. The **$2x$ state** is the presence of two copies of a single genome. The euploid state is represented by haploidy and polyploidy; the latter can be either autopolyploidy or allopolyploidy. In all these cases, the change in chromosome numbers is related to the $2x$ rather than the $2n$ state, and the change involves the complete genome. Further, the genomes involved may be either identical, i.e., from the same species, or different from each other, i.e., belong to different species. A brief description of the different common euploid states is given in Sections 3.7.3 - 3.7.5.

3.6.3. Haploids

An individual with the gametic chromosome complement (n) of the given species is called **haploid**. Haploids of a diploid species are called **monoploids**. Haploids are produced spontaneously in low frequencies. In some species like maize some inducer strains induce haploid production when they are used as males in crosses; these haploids are generally derived from the female gametes. In several species like potato, tobacco, wheat, barley, rapeseed, etc. haploids are produced in high frequencies when their anthers are cultured on suitable tissue culture media; these haploids develop from pollen grains. Production of haploid plantlets from cultured anthers is based on the pioneering work of Guha and Maheshwari (1964) on *Datura* anther cultures. Haploids can also be produced from cultured unfertilized ovaries of several crops, including sugar beet, barley, rice, etc. Some interspecific crosses also yield haploid progeny due to chromosome elimination of one of the parental species. For example, barley and wheat haploids are recovered from their crosses with *Hordeum bulbosum*, and of wheat can be obtained from wheat x maize crosses.

Haploids are much weaker than normal disomic ($2n$) plants. Further, they are highly sterile because each chromosome has only a single copy, which form univalent during meiosis. Therefore, they can be maintained only through asexual propagation or else they will have to be produced afresh at every occasion. Haploids can be subjected to colchicines treatment for doubling their chromosome number; this produces **doubled haploids** (DH). The DH plants are homozygous at all loci and are fully fertile. The doubled haploids can be used in breeding programs to develop new varieties. In fact, many varieties of rice, barley, tobacco, etc. have been produced by isolating doubled haploids from the F_1 hybrids obtained from suitable crosses. The doubled haploids can also be used for hybrid variety development. In crops like potato, which behave like autotetraploids, haploid plants are fully fertile. Therefore, breeding and selection are carried out at the haploid stage, and DHs are produced from the selected plants. These DH lines are developed as new varieties. This breeding approach is called **analytical breeding**. In Asparagus, DH male plants are YY; when they are crossed with XX females, they produce all male progeny. The male asparagus plants are more desirable than female plants since their spears are more delicious.

3.6.4. Autopolyploidy

When the somatic cells of a plant contain more than two copies of a single genome, it is known as **autopolyploid**. The most common types of autopolyploids are triploids (3 copies of the genome), autotetraploids (4 copies of the genome) and, in some cases, autohexaploids (6 copies of the genome). The autopolyploids are produced as a result of chromosome doubling, which occurs spontaneously in low frequencies. In some plant species, e.g., in species of Solanaceae, decapitation leads to regeneration of adventitious shoot buds, some of which may have doubled chromosome number. Some of the plants regenerated from cell/tissue cultures may be polyploids. In addition, several physical and chemical agents induce chromosome doubling. Of these agents colchicine treatment is the most effective in chromosome doubling. Colchicine alkaloid is obtained from autumn crocus (*Colchicum autumnale*); it is readily soluble in cold water and alcohol. Generally, colchicine is applied to growing shoot tips as 0.1 to 1.0 percent aqueous solution two times a day for a few days. Alternatively, it can be applied at 0.5% to 1.0% in lanoline paste two or three times in a week. Colchicine blocks spindle formation during mitosis, which leads to chromosome doubling in the affected cells.

Autopolyploids generally have larger cell size, including stomatal guard cells and pollen grains. They usually have larger leaves, flowers, fruits and seeds, which are, in general, less in number than those of diploids. Autopolyploids are generally slower in growth and later in flowering. They show reduced fertility, which can be improved considerably by selection for increased fertility. Autopolyploids generally show increased vigor and vegetative growth, but their water content is higher than that of diploids. **Autotriploids** show irregularities during meiosis, and produce various types of aneuploid progeny. Generally, they are highly sterile, but autotriploids of some species like spinach are highly fertile. Autotriploids are readily produced by crossing a diploid strain with an autotetraploid strain. **Autotetraploids** show quadrivalent formation during meiosis, and much lower fertility than diploids. Selection for increased fertility in autotetraploids is successful and has been improved in autotetraploids of several crop species. Most of the progeny of autotetraploids are tetraploid, but some progeny have other chromosome numbers also.

Autopolyploidy has contributed to evolution of plant species, but its direct contribution to the evolution of new plant species appears to be much less than that of allopolyploidy. Some of our crop species like alfalfa, banana, coffee, groundnut, etc. are believed to be autopolyploids. But recent studies using genomic in situ hybridization (GISH) have shown coffee and groundnut to be most likely allotetraploids (Table 3.11). These findings raise the possibility that the other species believed to be autopolyploids might turn out to be allopolyploids when they are intensively investigated. Perhaps the indirect contribution of autopolyploidy to the evolution of many of the present-day diploid species is more important. Many diploid plant species are believed to have undergone one or more rounds of whole genome duplication (WGD), i.e., chromosome doubling. This WGD was followed by genome reduction by loss of genetic material that was associated with extensive genomic rearrangements, i.e., translocations, inversions, deletions, etc. As a result, the diploid states of these species were ultimately restored (Fig. 3.11), but their genomes were reorganized and modified. *Rice in believed to have undergone two or three rounds of WGD followed by genome reduction.*

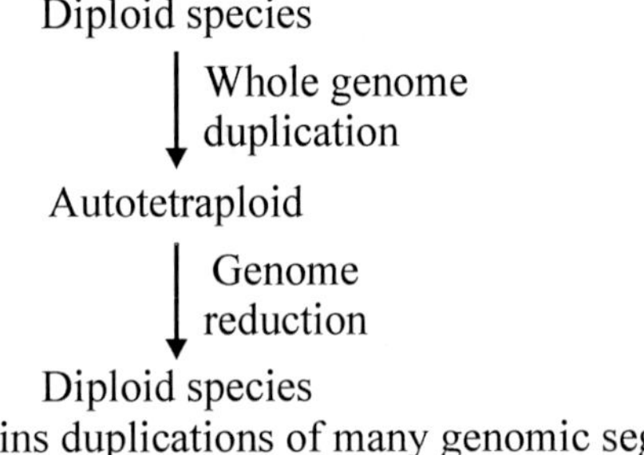

(The genome now contains duplications of many genomic segments)

Fig. 3.11. Role of autopolyploidy in the evolution of many present-day diploid species. Rice is proposed to have undergone two or three rounds of whole genome duplication, followed by genome reduction. In genome reduction, there is progressive loss of the genetic material till the species becomes essentially diploid.

Table 3.11: A list of the crop species that are considered to be autopolyploid

Crop	Somatic chromosome complement (2*n*)	Remarks
Alfalfa (*Medicago sativa*)	32 (4*x*)	-
Banana (*Musa sapientum*)	33 (3*x*)	-
Coffee (*Coffea arabica*)*	44 (4*x*)	Parental species: *C. congensis* and *C. eugenioides*
Peanut (*Arachis hypogaea*)*	40 (4*x*)	Parental species: *A. villosa* and *A. ipaensis*
Peppermint (*Mentha piperita*)	- (3*x*)	-
Potato (*Solanum tuberosum*)	48 (4*x*)	-
Sweet potato (*Ipomoea batatas*)	90 (6*x*)	-

* Recent evidence from GISH (genomic in situ hybridization) indicates these species to be allopolyploids rather than autopolyploids.

Autopolyploidy has found some useful applications in crop improvement. Autotriploid watermelons are grown commercially in Japan and USA. The triploid watermelons are produced by crossing selected diploid lines with tetraploid lines. Their fruits re seedless, but some normal sized seeds may be produced, and some fruits may be hollow. These problems are minimized by planting improved diploid pollinator strains in the fields of the triploid plants. Similarly, triploid sugar beets are being cultivated in Europe and Japan since 1970s. Triploid sugar beets produce larger roots and higher yields per unit area than diploids. The triploid varieties of sugar beet are generally mixtures of triploid (most of the plants), diploid (small proportion of plants) and other ploidy level plants. In India, a triploid clone of tea has been released for commercial cultivation. Autotetraploids have been produced and studied in a large number of crop species. Autotetraploids can be useful in crop improvement both directly (as new varieties) and indirectly (useful in breeding activities). Autotetraploidy has been used to overcome self-incompatibility in some species, including *Brassica* spp., *Petunia*, white clover, etc. Some autotetraploids may have superior quality characteristics, e.g., tetraploid maize has 43% higher carotenoid content. Autotetraploids have been used as varieties in some forage crops, including red clover, berseem and ryegrass. Many varieties of ornamental plants (e.g., the geranium variety Velvet Rose) and medicinal plants (e.g., variety HMT-1 of *Hyoscyamus niger*, Sugandha of *Vetiveria zizonoides*) are autotetraploids. Autotetraploid varieties have been used in rye (*Secale cereale*), e.g., Double Steel, Tetra Petkus, etc., but they are no longer in cultivation. Efforts are being made to develop disease resistant tetraploid varieties of banana.

The experience with autopolyploids allows the following conclusions: it is more likely to be successful in species (i) having lower chromosome numbers than in those with higher numbers, (ii) that are cross-pollinated, and (iii) that are cultivated for their vegetative parts, e.g., forage crops, ornamentals, etc.

3.6.5. Allopolyploidy

An **allopolyploid** generally has two copies of two or more distinct genomes. The production of an allopolyploid involves the following two steps: (i) hybridization between two distinct species, (ii) chromosome doubling of the F_1 hybrid, and (iii) diploidization of the alloplyploid so produced (Fig. 3.12). Since the genomes of the two parental species are likely to be distinct, there will be little or no pairing between their chromosomes, and the F_1 hybrid will be generally sterile. In nature, doubling of chromosome numbers in the hybrids occurs spontaneously either in somatic cells due to irregular mitosis, or due to the production of unreduced gametes. The plants produced after chromosome doubling will have two copies of each of the chromosomes of the two genomes; these plants are, therefore, called **amphidiploids** and constitute a new plant species. The newly produced amphidiploids show irregularities during meiosis and are partially sterile. Natural selection would favor increased regularity in meiosis and improved fertility.

Therefore, genes and gene combinations that promote diploid like regular meiosis and complete fertility are selected for. This process is termed as **diploidization**. Diploidization is promoted by selection following hybridization among the different plants of the new allopolyploid. In case of bread wheat (6*x*), a single gene *ph* is responsible for the diploid like behavior of wheat. This gene suppresses pairing between homeologous chromosomes so that only homologous chromosomes pair and the hexaploid wheat plants behave as diploids. Similar genetic systems are found in other the natural allopolyploid species like oats, tobacco, *Festuca*, etc.

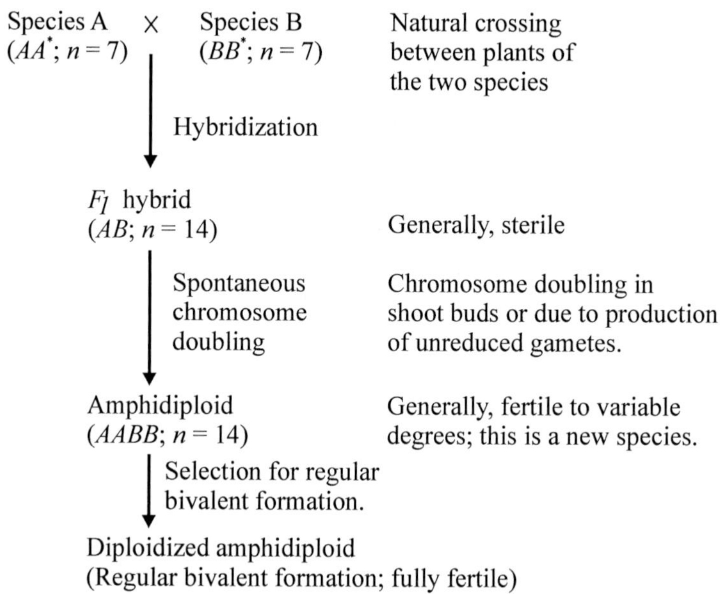

Fig. 3.12. A generalized scheme for the production of allopolyploid species. **A* and *B* represent the two different genomes present in species A and B, respectively.

A new allopolyploid has a combination of the morphological and physiological features of the parental species. But it is difficult to predict which parental features will be combined in the allopolyploid. Genome-wide gene expression analyses reveal that different diploid species of the same genus show different levels of expression of the same genes; these genes are said to be **differentially expressed**. In some cases, up to 50% of the genes may show differential expression. The allopolyploid between two diploid species would show average expression for some of the differentially expressed genes. But the expression of other differentially expressed genes in the hybrid may either be reduced or even suppressed or they may be expressed at higher levels than the average value for the two parents. Which genes will show average, reduced or increased expression cannot be predicted; this may be the reason for the unpredictable features of the allopolyploids. Generally, allopolyploids are hardier than their parental species. The distribution of natural allopolyploids is generally different from those of its parental species. Allopolyploids may produce some new metabolites, exhibit increased resistance to biotic and/or abiotic stresses, and may even possess some novel morphological features. Many natural allopolyploids, e.g., many grass species, are apomictic and do not reproduce sexually; most likely this is due to the near complete sterility of the interspecific hybrids.

Table 3.12: A list of some allopolyploid crop species and their parental diploid species

Crop	Somatic chromosome complement (2*n*)	Diploid parental species
Cotton (*Gossypium barbadesnse*)	52 (4*x*)	*G. herbaceum* ssp. *afircanum* and *G. raimondii**
Cotton (*Gossypium hirsutum*)	52 (4*x*)	*G. herbaceum* ssp. *afircanum* and *G. raimondii**
Mustard (*Brassica juncea*)	36 (4*x*)	*B. nigra* and *B. campestris*
Oats (*Avena sativa*)		
Rapeseed (*Brassica napus*)	38 (4*x*)	*B. oleracea* and *B. campestris*
Tobacco (*Nicotiana tabacum*)	48 (4*x*)	*N. sylvestris* and *N. tomentosa*
Tetraploid wheat (*Triticum turgidum*)	28 (4*x*)	*T. monococcum* ssp. *urartu* and an unknown *Triticum* species
Bread wheat (*Triticum aestivum*)	42 (6*x*)	*T. monococcum* ssp. *urartu*, an unknown *Triticum* species, and *T. tauschii*

* It is proposed that there was a single event of polyploidization, and the different polyploid species have originated due to divergence at the polyploid level.

The importance of allopolyploidy in evolution was first proposed by Winge in 1917. According to an estimate, about one-third of the Angiosperm species are polyploids, and the vast majority of them are allopolyploids. Thus, allopolyploidy has been much more important in the evolution of new plant species, and a larger number of our crops are allopolyploids (Table 3.12) than autopolyploids. The evolutionary history of many important allopolyploid species has been determined by using a variety of techniques, including hybridization with the suspected parental species and the allopolyploid species, use of DNA markers and GISH. The allohexaploid bread wheat is proposed to have evolved as follows. First, *Triticum monococcum* var. *urartu* (the contributor of *A* genome of wheat) hybridized with an unknown diploid species (the contributor of *B* genome of wheat) to produce a hybrid that was presumably sterile. Chromosome doubling of this hybrid gave rise to *Triticum turgidum*, i.e., the tetraploid wheats.

Subsequently, *T. turgidum* is believed to have crossed with *Triticum tauschii* (the contributor of *D* genome of wheat). Chromosome doubling of the hybrid from this cross gave us the present-day hexaploid wheats (*Tritium aestivum*; Fig. 3.13). Some of our important allopolyploid crops evolved in a similar manner and their most likely diploid parental species are listed in Table 3.12.

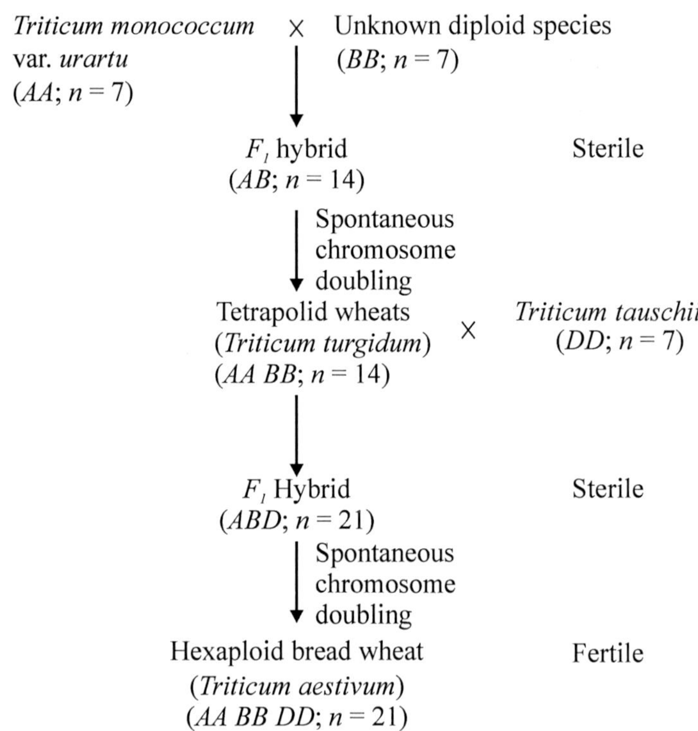

Fig. 3.13. A schematic representation of the evolution of bread wheat. The donor of B genome is an unknown diploid species.

Allopolyploidy has been used for crop improvement in the following three important ways: (i) as bridging species, (ii) as new crop species, and (iii) for creation of genetic variation in the existing allopolyploid crop species. Gene transfers into crop species from related species is often hampered by the sterility of their F_1 hybrids. In such cases, the chromosomes of the F_1 hybrids are doubled to produce amphidiploids. Each of these amphidiploids is a new plant species, and they show enough fertility to allow backcrosses with the concerned crop species for the desired gene transfers. Thus, the amphidiploids serve as **bridging species** in these cases. There are many examples of such gene transfers from related species into crop species. Some amphidiploids may have enough desirable features to be used as a new crop species. But newly synthesized amphidiploids have several undesirable features, including irregular chromosome behavior and variable sterility. Hybridization and selection at the amphidiploid level would greatly improve their performance and some of them may succeed as new crop species. Triticale is the most successful synthetic allopolyploid so far. It is generally produced by producing amphidiploids from crosses between tetraploid wheat (*Triticum turgidum*) and diploid rye (*Secale cereale*; Fig. 3.14). Triticales possess the yielding ability and grain quality of wheat and tolerance to adverse environment of rye. The newly synthesized triticales were low yielding, showed high sterility, poor seed set, etc. But the triticales developed after 50 years of intensive breeding are high yielding so that triticale varieties are grown in certain areas in several countries, including Germany, India and Poland. Some other successful synthetic allopolyploids are raparadish and *Festulolium*. *Raparadish* is the amphidiploid from the cross between radish and *Brassica rapa* (Syn., *B. campestris*), and is grown commercially as a forage crop in Europe. Similarly, *Festulolium* is the amphidiploid from the cross between *Lolium perenne* or *L. multiflorum* with *Festuca pratensis*. These amphidiploids are used as forage crops in Europe and USA. Finally, the variation present in some of our allopolyploid crops like *Brassica napus* is limited. New genetic variation in such crops can be generated by synthesizing the concerned amphidiploid species from their respective parental species. This approach is being followed to widen the genetic base of *B. napus*.

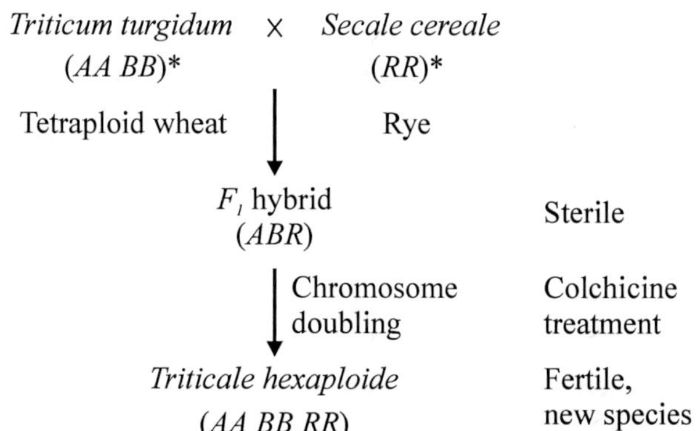

Fig. 3.14. The scheme for synthesis or creation of *Triticale hexaploide*. *The letters within parenthesis depict the genomic constitution of the species.

3.7. MUTATION BREEDING

A sudden heritable change in a characteristic of an organism is called **mutation**. Mutations can be classified in a variety of ways. A common way based on the type of change involved in mutations is as follows. A **gene** or **point mutation** is produced by a change in the base sequence of a gene, including its regulatory sequences. A **plasmagene mutation**, on the other hand, involves change in mitochondrial or chloroplast genome, and shows cytoplasmic inheritance pattern.

Many mutations in eukaryotes are produced due to changes in chromosome structure and even in number; these mutations are called **chromosomal mutations**. It is generally not easy to distinguish between gene and chromosomal mutations especially in case of eukaryotes.

3.7.1. Molecular basis of Gene Mutations

The gene mutations can be produced by base substitution, base deletion, base addition, transposition, and trinucleotide repeat expansion; in addition, epigenetic changes like DNA methylation, and RNA interference (RNAi) can produce heritable changes in gene function. Base substitutions involve replacement of a base by another base, e.g., A in place of G, in a gene (Fig. 3.15A,B). Most of the base substitutions are produced due to errors committed by DNA polymerase during DNA replication, but some other mechanisms are also involved. Base addition and base deletion lead to changes in all the subsequent codons if the number of bases added or deleted were not a multiple of three (Fig. 3.15C).

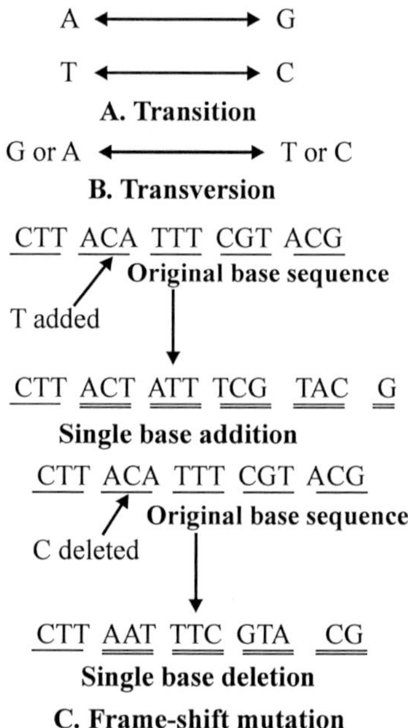

Fig. 3.15. Molecular bases of mutation. (A) Transition involves replacement of one purine by the other or of one pyrimidine by the other. (B) Transversion is substitution of a purine by a pyrimidine and vice-versa. C. The frame shift mutations are produced by base addition (single base addition shown here) or base deletion (single base deletion shown here) when the number of bases added or deleted is not a multiple of three. The codons located downstream of the site of addition/deletion are changed as a result. The changed codons are underlined, by two lines.

Transposition is the movement of transposable elements from one location in the genome to another genomic site; this would generate disruption of the genes within which the transposable elements happen to integrate. Some heritable changes are produced by relatively stable methylation of the promoter regions of genes. The inhibitory genes provide an example of RNAi; these genes are generated by the presence of inverted copies of the genes whose actions they inhibit (through RNAi). Both these changes are called **epigenetic changes** as they produce heritable changes in gene function and phenotype, but the base sequences of the concerned genes remain unchanged.

3.7.2. Characteristics of Mutations

Mutations are random, recurrent, generally recessive, and mostly harmful to the organisms. They can originate without any specific treatment (**spontaneous mutations**) or can be induced by treatment with some physical or chemical agents called **mutagens** (**induced mutations**). Most mutations reduce survival (**lethal**, **sublethal**, **subvital mutations**), but some mutations do not affect survival or may even increase it (**vital mutations**). Mutation breeding generally makes use of those mutations that do not reduce survival and have some beneficial effect on the concerned plants; the frequency of such mutations is around one in 800 mutations, i.e., about 0.1% of the total mutations. Many spontaneous mutations have been used to develop new varieties. But since the frequency of induced mutations is very low (10^{-5} or less), mutation breeding relies on mutations induced by chemical or physical mutagens. In general, the features of induced and spontaneous mutations are comparable, except for the very high frequency of induced mutations.

3.7.3. Mutagens

Induction of mutations a by specific treatment is called **mutagenesis**, while the use mutations for crop improvement is known as **mutation breeding**. Several chemical and physical agents generate mutagenic effects, and some of them have been widely used for mutagenesis (Table 3.13). The physical mutagens are different kinds of radiations, including ultraviolet radiation. Fast neutrons, X-rays and gamma-rays are ionizing radiations, and highly penetrating in biological materials. The primary effect of these radiations is **ionization**, which is the removal of an electron from an atom, which makes the nucleus positively charged. The ions produced by radiations in the biological materials, including DNA, generate the direct effects of radiations. In addition, ions lead to the formation of free radicals, which are responsible for the indirect effects of radiations. **Free radicals** are electrically neutral molecules each of which has an unaired electron in its outer orbit. Free radicals are extremely short-lived and highly reactive. They are responsible for a fair proportion of the genetic effects produced by radiations. The genetic effects of radiations are produced by their effects on DNA. These effects include modification, e.g., deamination, of bases, loss of a base, single- and double-strand breaks and cross-linking of DNA strands. Mutations are produced by deamination of bases, and dimerization of bases present in the same strand. The UV radiation is nonionizing, and causes excitation of atoms. In case of **excitation**, an electron is pushed to an outer orbit representing a higher level of energy. Excitation increases the reactivity of the affected atoms and molecules. UV radiations produce their genetic effects mainly through the production of thymine dimers, and deamination of DNA bases. But UV rays can penetrate only few cell layers. Therefore, they can be used to irradiate only pollen grains, which can be used for pollination. But this approach presents practical problems, and UV rays are seldom used for mutation induction in plants.

Table 3.13: The mutagens widely used for mutation breeding

Mutagen	Features	Remarks
1. Physical mutagens		Various kinds of radiation
(a). Fast neutrons	Densely ionizing, particulate, uncharged, highly penetrating in biological tissues	Produced in thermal reactors
(b). X-rays	Sparsely ionizing, nonparticulate, 0.001 – 10 Â wavelength, high energy radiation produced by X-ray tubes	Mutagenic effects produced mainly by ionization
(c). γ-rays	Sparsely ionizing, nonparticulate, 0.001 – 10 Å wavelength, high energy radiation, generally produces by radioactive decay of ^{60}Co	Mutagenic effects produced mainly by ionization
(d). Ultraviolet rays	Nonionizing, nonparticulate, wavelength of 100 – 3,900 Å, most effective wavelength 2,540 Å, limited penetration in biological tissues	Mutagenic effects mainly due to thymine dimer formation
2. Chemical mutagens		
(a). Ethylene imine (EI)	Alkylating agent, causes extensive cross-linking of DNA, breaks in DNA backbone, alkylation of N at position 7 in guanine	Mainly G:C to A:T and A:T to G:C transitions
(b). Ethyl methane sulfonate	Alkylating agent, causes extensive cross-linking of DNA, breaks in DNA backbone, alkylation of N at position 7 in guanine	Perhaps the most commonly used chemical mutagen in plants
(c). Sodium or potassium azide	Used as NaN_3 or KN_3, mutagenic effect produced by azidoalanine that is produced in the treated cells, causes base substitutions	Perhaps the least dangerous and the most efficient mutagen

Among the various chemical mutagens, ethylene imine (EI), ethyl methane sulfonate (EMS), and sodium or potassium azide are the most commonly used. EI and EMS are called **alkylating agents** since they transfer alkyl groups (methyl, ethyl, etc. groups) to DNA bases; this is known as **alkylation**. It is extremely important that the solutions of alkylating agents are freshly prepared, since storage leads to the production toxic but non-mutagenic hydrolysis products. Alkylating agents cause cross-linking of DNA, chromosome breaks as well as gene mutations. Generally, they alkylate the nitrogen at position 7 of guanine, to a less extent the nitrogen at position 3 of adenine, and rarely the N of cytosine. Alkylation of guanine causes it to mispair with thymine to generate G:C to A:T transition. Similarly, alkylation of thymine leads to A:T to G:C transition. Inside the cells, sodium/potassium azide is converted to azidoalanine, which is the real mutagenic metabolite. Azide has been shown to be mutagenic in several crop species, and it is perhaps the least dangerous and the most efficient mutagen. However, azide is a known respiratory poison, and it can cause some other problems. The azide solution (10^{-4} – 10^{-3} M sodium azide) is prepared in phosphate buffer. The seeds should be soaked in water for 4 – 12 hr before they are treated with the azide at 20 – 25°C.

3.7.4. Mutation breeding

A mutation breeding program should have clearly defined objectives, and should be well planned. Generally, the variety selected for mutagen treatment should be the best variety of the concerned crop unless the objectives of the program necessitate the use of some other variety. In sexually propagated species, ordinarily seeds are used for mutagen treatment, while in asexually propagated crops buds or cuttings are used. In case the whole plants are to be irradiated, it should be done when the plants are in flowering stage; however, this would require a gamma-garden facility. Chemical mutagens are best applied to seeds, but they have been used with cuttings as well. The mutagen dose used for treatment should be such that it induces the maximum frequency of mutations and causes the minimum killing in the M_1 generation, i.e., the generation obtained from seeds/cuttings subjected to mutagen treatment. This dose will depend on the mutagen and the treatment conditions, but many feel it will be close to LD_{50} (lethal dose 50). The dose of mutagen that would kill 50% of the treated individuals is called $\mathbf{LD_{50}}$. The LD_{50} of the same mutagen varies with the plant species. Therefore, for each new species the LD_{50} is experimentally determined. The dose of mutagen can be changed by altering the intensity (physical mutagens) or the concentration (chemical mutagens) of the mutagen or the duration of mutagen treatment. In case of chemical mutagens, the seeds are generally presoaked for few hours to initiate metabolic activities. After the mutagen treatment, the seeds are washed in running tap water for about 4 hr and then planted in the field. But, generally, dry seeds are irradiated and then planted in the field. It has been estimated that ~6,000 seeds should be treated to obtain about 10 desired mutants. Therefore, as many seeds should be treated as is permitted by the facilities available to the researcher. In any case, it is desirable that about 500 fertile plants should be available in the M_1 generation for each mutagen dose.

The selfed or clonal progeny of the M_1 plants are known as M_2, those of M_2 are called M_3 and so on. About 20 seeds for each M_1 plant should be harvested separately and 2-4 thousand individual plant progeny rows are grown in M_2. The plants are regularly observed, and clearly identifiable and desirable mutant M_2 plants may be selected and multiplied for evaluation. Further, all other plants that are suspected to be mutant for a desirable trait are harvested separately. Individual plant progenies are grown from the selected plants, and the plants are subjected to careful scrutiny. Mutant progenies are selected, harvested in bulk since most of them would be homogeneous, and used for yield testing. The parent variety and other popular varieties are used as checks in the trials. This approach is suitable for mutation breeding designed to improve a qualitative trait. This approach is modified as follows for the improvement in quantitative traits. The M_1 and M_2 generations are raised as above. In the M_2 generation, vigorous, fertile and normal looking plants are selected, and their seeds are harvested separately. In M_3, ~1000 or so individual plant progenies from the selected M_2 plants are grown along with the parent variety (used for comparison). In M_3, the individual plant progenies are carefully observed, and plants or progenies superior to the parent variety for the desired trait are selected. In M_4, individual plant progenies from the selected M_3 plants are grown, superior progenies are identified, harvested in bulk, and used for yield testing in M_5. In case, some M_3 progenies were homogeneous for a mutant phenotype, they could be used in yield testing in M_4 generation itself.

It may be pointed out that care should be taken to prevent out-crossing in the M_1 generation even in the case of self-pollinated crops. Further, mechanical mixtures should be avoided throughout the experiment to eliminate the risk of contamination from other genotypes of the crop. The mutants may be either released directly as a new variety or they may be used as parents in hybridization programmes to develop improved varieties. Mutation breeding is ideally suited to improve a specific weakness of an otherwise excellent variety. To achieve this objective, efforts were made to achieve directed mutagenesis. Ideally, **directed mutagenesis** should produce mutations in sufficiently high frequency only I n the specified genes/traits. Directed mutagenesis could not be achieved with the conventional mutagens. But the development of site-directed nucleases (SDNs) has made it feasible to induce mutations in the selected genes in relatively high frequencies (Section 2.11). These approaches are being rapidly refined, but they already appear to be potent technologies for site-directed mutagenesis.

3.7.5. Contributions of mutation breeding

Mutation breeding has been extensively used for the development of new improved varieties. The first mutant variety was the cotton variety MA-9 developed in 1948 in India; it was developed by X-ray irradiation and had increased tolerance to drought. Since then a large number of mutant varieties have been developed. In the year 2012, the total number of mutant varieties released for commercial cultivation stood at 3,200 in over 200 different crop species (*http://mvgs.iaea.org*). The bulk of these varieties have been developed in Asia, followed by Europe and North America; Latin America, Africa, etc. account for merely <5% of the varieties. Most of the mutant varieties have been developed in food crops, followed by flowers and ornamentals, while only few varieties have been developed in forage, fruit and industrial crops. Several mutant varieties have made noteworthy contributions either directly by being cultivated in large areas or indirectly by serving as sources of desirable traits for which they have been widely used as parents in hybridization programmes (Table 3.14). The mutant varieties developed in India in several food crops are listed in Table 3.15.

Table 3.14: Some examples of widely used varieties developed by mutation breeding

Crop	Mutant variety	Improved trait	Remarks
Barley	Golden promise[1]	Straw strength, yield, malting quality	Extensively used in hybridization programmes throughout Europe to develop many varieties
	Diamant[1]	Grain-to-straw, yield, lodging resistance	
Rajma (*P. vulgaris*)	Michelite[1]	Bush type	Extensively used in hybridization programmes in USA
	NEP-2[1]	White seed color	
Sunflower	Pervenets[2]	High oleic acid and low linolenic acid contents	Extensively used for the development of high oleic acid hybrids
Peppermint	Todd's Mitcham[1]	Tolerant to *Verticillium* wilt	Bulk of global mint oil production
Grapefruit	Star Ruby[1]	Improved red color of fruit flesh	Occupy 75% of grapefruit area in Texas, USA
	Rio Red[1]		
Pear	Gold Nijiseeiki[3]	Improved resistance to black spot disease	Additional income of ~50 million US dollars in Japan

[1]Radiation induced; [2]Chemical mutagen-induced; [3]Chronic irradiation in a gamma-field

Table 3.15: A selected list of mutant varieties developed in India

Crop	Mutant varieties
Brassica juncea	Kranti, RLM198, RLM514, RLM619
Chickpea (gram)	BGM408, BGM413, BGM417, RSG2, RS11
Cotton (*G. hirsutum*)	MCU7, MCU10
Cotton (*G. herbaceum*)	Indore-2, DB-3-12
Groundnut	TG17, TG19A*, TG1 (Vikram), TG13A*, TG14*, TG15*, TG16*, TG18*, TG19*, BP1, BP2, CO2
Jute	JRO3690, TKJ40 (Mahadev), JRO412, JRO514
Mungbean	Pant Mug-2, Co-4, TAP-7
Pea	Hans
Rice	Jagannath, Prbhavati, HUR-36, Gautam, Shyama, GEB24, Biraj, CNM6
Sugarcane	Co8152, Co8153
Tobacco	Jayasri, Bhavya
Urdbean	Co4, TAU-1
Wheat	NP836, Pusa Lerma, Sharbati Sonora

* Derived from a cross in which one or both the parents were mutants

3.8. SOMACLONAL VARIATION

The genetic variation present in plant tissue/cell cultures or the plants regenerated from them is called **somaclonal variation**. Plant tissue and cell cultures are generally derived from somatic cells, and these cells divide mitotically during the culture period. Therefore, all the cells present in given culture are expected to have the same chromosome constitution as well as genotype. However, chromosomal changes and genetic variation do arise in the tissue/cell cultures. These genetic changes may arise due to one or more of the following: (i) gene mutations due to the errors committed by DNA polymerase during DNA replication, (ii) gene amplification (increase in the number of copies of a gene in a single genome), (iii) gene deamplification (reduction in the number of copies of a gene in a single genome), (iv) increased transposition of transposable elements during the tissue culture, (v) mutations in the genomes of chloroplasts and/or genomes, and (vi) changes in chromosome number or structure.

Somaclonal variants for agronomic traits, including yield are generally isolated by the screening procedure. In this approach, a large number of plantlets are regenerated from cell cultures, and they are transferred into the field. The plants are selfed and the progeny are grown in, preferably, individual plant progeny rows. These progeny plants are evaluated for the various agronomic traits and compared with those of the parent variety; any plant showing a desirable change is identified and selected. The selfed progeny from the selected plants are grown in individual plant progeny rows and desirable progenies/plants are selected. The selected lines are evaluated in replicated yield trials and superior variants may be released as new varieties. This approach has been used to develop some useful varieties in several crop species (Table 3.16). Alternatively, selection may be applied to cell cultures to select for variant cells, e.g., cells resistant to a toxin produced by a pathogen causing a disease. In case the toxin is involved in disease development, the toxin resistant cells would give rise to plants that would be resistant to the concerned pathogen. This approach can be used for selection for resistance to abiotic stresses like salt stress, cold stress etc. as well. The technique shows some to considerable success in the selection of resistant cells, and the cell selection efforts have resulted in the development of some useful somaclonal variants, e.g., in case of *Mentha* sp.

Table 3.16: A list of some somaclonal variants released as new varieties

New variety (somaclone)	Crop	Remarks
Sigma	Alfalfa	Improved yield and disease resistance
Pusa Jai Kisan	*Brassica juncea*	Increased (17.4%) yield
Vegisnax	Carrot and celery	Crunchier and more palatable than the parental varieties
Bio-13	*Citronella java*	More oil and citronellal yield
Velvet rose	Geranium	Sturdiness, vigor, attractiveness
Andro	Linseed	Salt tolerant, immune to rust
Ono	Sugarcane	Resistant to Fiji disease
Scarlet	Sweet potato	Darker and more stable skin color

3.9. SELECTION

In population genetics, **selection** is defined as differential rates of reproduction for different genotypes of the population. But in plant breeding context, **selection** may be defined as identification and isolation of individual plants or groups of plants with the desired phenotype and raising the next generation from the seeds produced by the selected plants. *Selection, as a rule, is based on phenotype.* Therefore, the effectiveness of selection depends on the heritability of the trait under selection. In addition, the consequences of selection will depend on whether the selected plants are homozygous or heterozygous and the mode of reproduction of the concerned crop species. An idea of the homozygous/heterozygous state of the selected plant and the relative contributions of the genotype and the environment to the trait phenotype can be obtained by progeny test. The **progeny test** is comprises the estimation of worth of a plant by the performance of its progeny. This test was developed by Vilmorin and is an integral part of the most breeding methods. If a plant were heterozygous for the genes governing a trait, its progeny will show variation for the trait. The converse will be the case if it were homozygous. Similarly, the progeny will be close the parent in terms of the trait phenotype if there were little influence of the environment. The difference between the parent plants and their progenies will increase with an increase in the contribution of the environment to the trait phenotype. The consequences of selection are further discussed in the following sections with reference to the mode of reproduction of the concerned crop species.

3.9.1. Selection in Self-Pollinated Crops

The populations of self-pollinated crops are highly homozygous. Heterozygous plants are produced only by chance out-crossing or planned hybridization carried out by the breeders. Even these heterozygous plants will ultimately give rise to homozygous genotypes as a result of continued self-pollination (Table 3.6). Therefore, the varieties used for commercial cultivation are generally homozygous genotypes. Selection in a homozygous population for a trait that is genetically variable will be effective, i.e., it will lead to a change in the concerned character in the direction of selection. Generally, individual plants are selected; their seeds are harvested separately and grown in separate rows called **individual plant progeny rows**. Thus, each individual plant progeny will be progeny of a single homozygous self-pollinated plant; such progenies are called **purelines.** The concept of purelines, generally called **pureline theory**, was developed by Johannsen in 1903 based on his selection for seed size in French bean (*Phaseolus vulgaris*). The plants within a pureline will have the same genotype as that of the parent plant. As a result, there will be no genetic variation within a pureline. Further, the phenotypic variation present in a pureline will be due to the environment, and selection within a pureline will be ineffective. However, purelines do become genetically variable with time due to one or more of the following: mechanical mixture, natural hybridization, spontaneous mutations and chromosomal aberrations. Therefore, selection in old purelines is often effective and several improved varieties have been developed by such selection.

The population obtained after selection is expected to show a change in the character being selected for. This change will be in the direction of selection. The extent of this change will depend on the intensity of selection, the amount of phenotypic variation for the trait present in the population subjected to selection, and the heritability of the trait. The magnitude of change expected from selection can be estimated as **genetic advance under selection (Gs)** by the following formula.

$$Gs = k.\sigma p.H$$

where, k is selection differential, and σp and H are the phenotypic standard deviation and heritability, respectively, for the trait under selection. **Hertitablity** is the ratio of genetic variance to the phenotypic variance for the concerned trait in the given population. This can be expressed as follows.

$$H = \text{Genetic variance } (\sigma^2_g) \text{ / Phenotypic variance } (\sigma^2_p)$$

The estimates of σ^2_g and σ^2_p can be obtained from data from replicated trials using appropriate populations. The above estimate of H is used for the estimation of Gs. But H can also be expressed in percentage.

$$H \text{ (in \%)} = (\sigma^2_g / \sigma^2_p) \times 100$$

The value of k is expressed in terms of the standard deviation units; as a result, this value depends only on the **intensity of selection** (Table 3.17). The intensity of selection is the ratio (in percent) of the number of plants/lines selected for raising the next generation to the total number of plants/lines present in the population subjected to selection. The Gs estimate has the same unit as the mean for the concerned character. It may be pointed out that when the Gs is estimated for a segregating population, the narrow sense estimate of H should be used in the above equation. Thus, Gs is the expected advance in the mean genotypic value of the selected population as compared to that of the original population that was subjected to selection.

Table 3.17: The value of k for some selected intensities of selection.

Selection intensity (%)	k
1	2.64
2	2.42
5	2.06
10	1.76
20	1.40

3.9.2. Selection in Cross-Pollinated Crops

The populations of cross-pollinated crops are considered as random mating populations. In a **random mating population**, a male gamete has equal opportunity for uniting with any female gamete produced by the population. In other words, a male individual has equal opportunity of mating with any female individual of the population. Therefore, any generation of such a population is produced by random union among random samples of male and female gametes produced by the previous generation of the population. If a gene has two alleles, say, A and a, with the frequencies p and q, respectively, a random union among the male and female gametes will produce progeny with the genotypes AA, Aa and aa in the frequencies p^2, $2pq$ and q^2, respectively (Fig. 3.16). It can be shown that this generation of the population will also produce the gametes A and a in the frequencies p and q, respectively. Further, random union among these gametes will yield the same three genotypes with the same frequencies as listed above (Fig. 3.16). Thus, the gene and genotype frequencies in a random mating population will remain constant generation after generation if there were no selection, mutation, migration or random drift. Therefore, the genotype frequencies p^2, $2pq$ and q^2 for a gene with two alleles represent equilibrium frequencies of the three genotypes. This is generally called **Hardy-Weinberg equilibrium** since this conclusion was reached independently by Hardy (in 1908) and Weinberg (in 1909).

Allele	A	a
Frequency	p	q
Condition	$p + q = 1.0$	

Random union among gametes :

♀ ♂	A (p)	a (q)
A (p)	AA p^2	Aa pq
a (q)	Aa pq	aa q^2

Genotype frequency : p^2 AA, $2pq$ Aa, q^2aa

Gametes produced by these genotypes :

$$A = p^2 + pq$$
$$= p\,(p{+}q) \qquad (p \text{ taken as common})$$
$$= p \qquad (p + q = 1)$$

Similarly, $a = q$

Fig. 3.16. Gene and genotype frequencies in a random mating population for a gene with two alleles, say, A and a. The frequency of union between gametes will be the product of frequencies of these gametes. All the gametes produced by p^2 individuals with AA genotype will be A. But only half of the gametes produced by $2pq$ Aa individuals will be A.

The Hardy-Weinberg equilibrium is disturbed by selection, mutation, migration or random drift. These factors are called evolutionary factors since the gene and genotype frequencies must change from the equilibrium state for evolution to take place. **Migration** is the movement of some individuals from one population into another and participation in the reproduction of the latter. The extent of change in gene frequency due to migration will depend on the magnitude of the difference between the gene frequencies of the two populations, and the proportion of the migrant individuals in the new population. In plant breeding situations, migration is represented by varietal crosses. The **random drift** or **genetic drift** is the random change in gene frequency of a population due to sampling error. The ultimate effect of random drift is that the frequency of one allele becomes zero and that of the other allele becomes one, i.e., the other allele becomes fixed. The extent of random drift increases as the population size decreases, and in small populations all the genes may, ultimately, become fixed. Since breeding populations are generally small, some amount of random drift is bound to occur in them. Mating between individuals related by descent is called **inbreeding.** Inbreeding reduces the proportion of heterozygotes and increases that of the homozygotes in the population. The extent of inbreeding occurring in a population depends primarily on the population size, and the rate of decrease in heterozygosity is $1/2N$, where N is the number of plants in the population. But in plant species in which self-pollination cannot take place, this rate is slightly lower [$= 1/2N + 1)$].

Selection is defined as differential reproduction rates of different genotypes. In plant breeding situations, only the selected genotype is allowed to reproduce, and the other genotypes are not allowed to contribute to the next generation. Let us suppose we wish to select plants with the genotype *AA*. In case there was no dominance, the heritability was 100%, and there was no cross-pollination, all the selected plants will be *AA* and their progeny will also be *AA*. But in practice, dominance is likely to be present, heritability for quantitative traits will be less than 100%, and cross-pollination is predominant. Therefore, selection cannot be expected to eliminate an allele from a population; it can be expected to increase the frequency of the selected allele, but it would not lead to fixation of the allele. In general, the rate of change in the gene frequency will depend on the intensity of selection, the degree of dominance and the frequency of the selected allele in the population. The rate of change in gene frequency increases with gene frequency till 0.3, 0.5 and 0.7 for dominant, additive and recessive alleles, respectively, beyond which it declines. In case selection favors the heterozygotes, both the alleles will be maintained in the population.

Since selection is always based on the phenotype, selection of the extreme phenotype will increase the frequency of the desirable alleles in the population. The rate of change in the frequency of the desirable allele of each gene will become smaller as the number of genes governing the trait becomes larger. The progress under selection will be hampered by non-additive gene action, low heritability and the limits to permissible selection. Permissible selection intensity is the minimum proportion of population that must be selected to maintain the population size. Thus, permissible selection intensity will depend on the rate of reproduction of the crop species. For example, if each plant produced 20 seeds, a minimum of 5% of the population (= 100/20) has to be selected to maintain the population size. In case, each plant produced 100 seeds, only 1% of the plants could be selected without reducing the population size.

The type of response to selection depends on the genetic control of the concerned trait. When the character is governed by few genes with large effects and several genes with small effects, there will be rapid gain for some generations, followed by slow gain for several generations. Studies with molecular markers reveal that most quantitative traits show this type of genetic control. The rapid gain period is due to larger increase in the gene frequencies of the genes with large effects. The slow gain, on the other hand, will be due to the progressively smaller change in the frequencies of the major effect genes and the increase in the frequencies of the genes with small effects. In case of some traits, there is slow gain for a long period of time. For example, selection for high protein and high and low oil contents in the Burr's White open-pollinated variety of maize has been effective for over 80 generations. Such traits are presumed to be governed by a large number of additive action genes, each of which has small effect size. For example, it was estimated that a minimum of 69 genes are involved in the low oil line, while 173 genes were estimate for the high protein line (Dudley and Lambert 1992). In case of some traits like grain yield, there may be little or no gain in response to selection particularly when the selection scheme is not carefully designed and the experiments are not properly executed. The lack of response to selection in such cases may be due to low heritability of the concerned traits.

Many quantitative traits show the following response to selection: rapid gain, followed by a plateau, again followed by rapid gain. This type of response was first reported by Mather and Harrison for the trait abdominal bristle number in *Drosophila melanogaster*. The genetic explanation for this type of response is as follows. In natural populations, the mean value for a quantitative trait is the optimum value for the trait. Therefore, selection will favor the individuals having the mean value for the trait. However, the population has to maintain genetic variability for the trait in order to meet the long-term evolutionary requirements. These two opposing demands can be met by a genetic organization that attains the optimum phenotype of the trait, but retains considerable amount of variability. One highly effective scheme is based on repulsion phase linkage between the genes governing the trait (Fig. 3.17). The initial rapid gain phase represents the selection for linkage blocks that would obviously be small in number. The plateau will be reached when nearly all the linkage blocks would have become heterozygous, and after this there will be no further gain due to selection. In Figure 3.17, the four genes linked in repulsion phase are assumed to show additive gene action. Therefore, the highest value for the trait will be obtained when each of such linkage blocks would have become heterozygous. Crossing over between the linked genes would generate linkage blocks with three and four positive alleles linked together. Selection will now act on these new linkage blocks (generated by recombination) and will lead to a change in the trait phenotype in the direction of selection. This gain would continue for several generations since crossing over between the linked genes would occur at relatively low frequencies over generations. In addition, random mating would continue to generate new combinations of linkage blocks that would respond to selection as well produce new gene combinations due to crossing over.

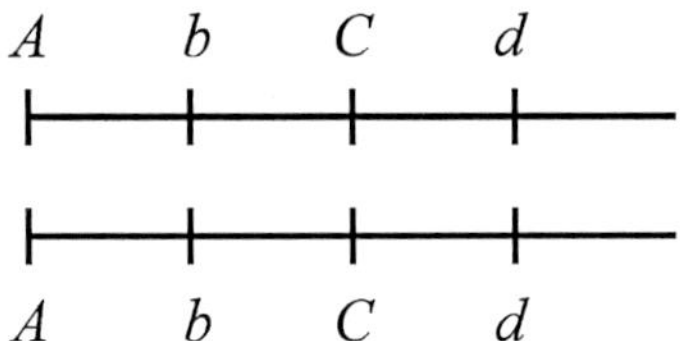

A. Repulsion phase linkage (gene arrangement)

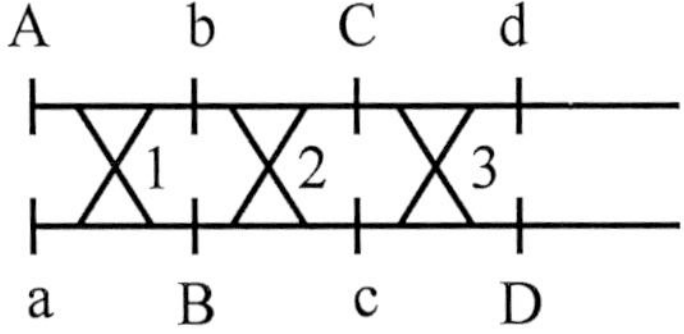

B. Heterozygote

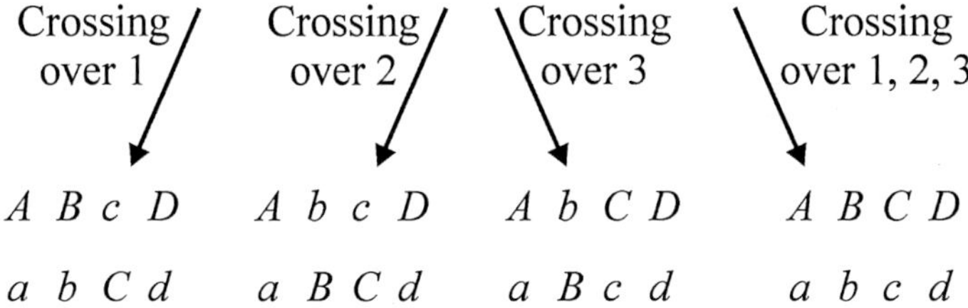

C. Gene arrangement after crossing over

Fig. 3.17. Gene organization in natural populations. Genes *a, b, c* and *d* govern the same trait, and show additive gene action. The alleles *a, b, c* and *d* are negative alleles, while alleles *A, B, C* and *D* are positive alleles. The arrangement *A-b-C-d* produces a linkage block with two positive alleles. Linkage blocks with 3 and 4 positive alleles can be produced only after 1 and 3 crossing overs, respectively.

PRACTICE QUESTIONS

1. Explain the meaning of genetic resources. Briefly describe the procedure of plant introduction and discuss its relevance in plant breeding.
2. Briefly describe the various types of self-incompatibility, and discuss its relevance in plant breeding.

3. Define male sterility and differentiate it from self-incompatibility. Briefly describe the various types of male sterility and discuss its usefulness in plant breeding.
4. Define hybridization and briefly describe its various types. Explain the consequences of hybridization and discus its significance to crop improvement.
5. Define heterosis and inbreeding depression. Explain the genetic and molecular bases of heterosis, and discuss its exploitation in crop improvement.
6. Briefly describe the various types of polyploidy and discuss their role in crop improvement.
7. Define autopolyploidy and allopolyploidy, and discuss their roles crop improvement and evolution.
8. Define mutation; explain its molecular basis, and the role in crop improvement.
9. Define selection; discuss the effects of selection in self- and cross-pollinated crops.
10. Write short notes on: (i) genetic resources, (ii) plant introduction, (iii) quarantine, (iv) self-incompatibility, (iv) molecular basis of self-incompatibility, (v) male sterility, (vi) cytoplasmic-genetic male sterility, (v) chemical hybridizing agents, (vi) pistillate condition, (vii) apomixis, (viii) hybridization, (ix) heterosis, (x) inbreeding depression, (xii) somatic hybridization, (xiii) polyploidy, (xiv) aneuploidy, (xv) doubled haploids, (xvi) mutagens, (xvii) mutation breeding, (xviii) somaclonal variation, (xix) selection.

RELEVANT LITERATURE

1. Dudley JW and Lambert RJ 1992. Ninety generations of selection for oil and protein in maize. *Maydica* 37: 81-87.
2. Hardy GH 1908. Mendelian proportions in a Mixed Population. *Science* **28**: 49–50. doi:10.1126/science.28.706.49.
3. Kaeppler S 2012 Heterosis: many genes, many mechanisms – end of the search for an undiscovered unifying theory. *ISRN Botany*, doi:10.5402/2012/682824
4. Lewis D. 1954. Comparative incompatibility in angiosperms and fungi. *Adv. Genet.* **6**: 235-285.
5. Mather K and Harrison BI 1949. The manifold effects of selection. *Heredity* 3: 1-52, 131-162.

6. McClure B 2016. The genetic basis of heterostyly. Nature Plants. doi: 10.1038/NPLANTS.2016.184
7. Singh BD and Singh AK 2015. Marker-Assisted Plant Breeding: Principles and Practices. Springer, New Delhi
8. Weinberg W 1908. Über den Nachweis der Vererbung beim Menschen. Jahreshefte des Vereins für vaterländische Naturkunde in Württemberg **64**: 368–382.
9. Winge Ø 1917 The chromosomes: their numbers and general importance. Compt. Rend. Trav. Lab. Carlsberg, 13: 131-275.
10. Gowen JW ed 1952. Heterosis. Iowa State College, Ames, Iowa, USA.
11. Schnable PS and Springer NM 2013. Progress toward understanding heterosis in crop plants. Annu Rev Plant Biol 64:71–88.
12. Guha, S and Maheshwari, SC (1964) *In vitro* production of embryos from anthers of *Datura. Nature.* **204**: 497.

■■■■

4 Breeding of Self-Pollinated Crops

4.1. INTRODUCTION

The populations of self-pollinated crops are naturally homozygous since self-pollination causes rapid reduction in heterozygosity and corresponding increase in homozygosity. Therefore, populations of self-pollinated crops generally comprise homozygous genotypes. Similarly, their improved varieties are, usually, homozygous genotypes or purelines; most varieties consist of single purelines, while some comprise a mixture of purelines. Self-pollinated crops also show considerable amount of heterosis. Therefore, in several such crops hybrid varieties are also used for commercial cultivation. In this chapter, we will consider in some detail the methods of developing both pureline and hybrid varieties in self-pollinated crops.

4.2. METHODS OF SELECTION

A population of self-pollinated crop may contain genetic variation. Such populations will be local or *desi* varieties of crops and old pureline varieties that have become genetically variable due to mutation, natural outcrossing and/or mechanical mixture. Such genetically variable populations can be used for either mass selection or pureline selection to develop new varieties with improved performance.

4.2.1. Mass Selection

In **mass selection** a large number of plants with improved characteristics but similar in phenotype are selected from a genetically variable population. The seeds from all the selected plants are composited, i.e., mixed together, to constitute the new variety. In general, there is no progeny test of the selected plants. Allard (1960) proposed that progeny test should be done as it might improve the performance of the new variety. The selection would be based on easily observable characters, e.g., plant height, grain characteristics, disease resistance, shattering resistance etc. Care is taken to select plants that are similar in appearance so that the new variety developed by selection appears homogeneous, i.e., the plants appear similar to each other. In case, progeny test is to be done, the seeds from about 200 to 2000 or 3000 selected plants are harvested separately, and individual plant progenies are grown in the second year. The progenies are generally evaluated visually, inferior and variable progenies are rejected. Seeds from the remaining superior progenies that are similar in appearance are composited to produce the improved version of the population (Fig. 4.1).

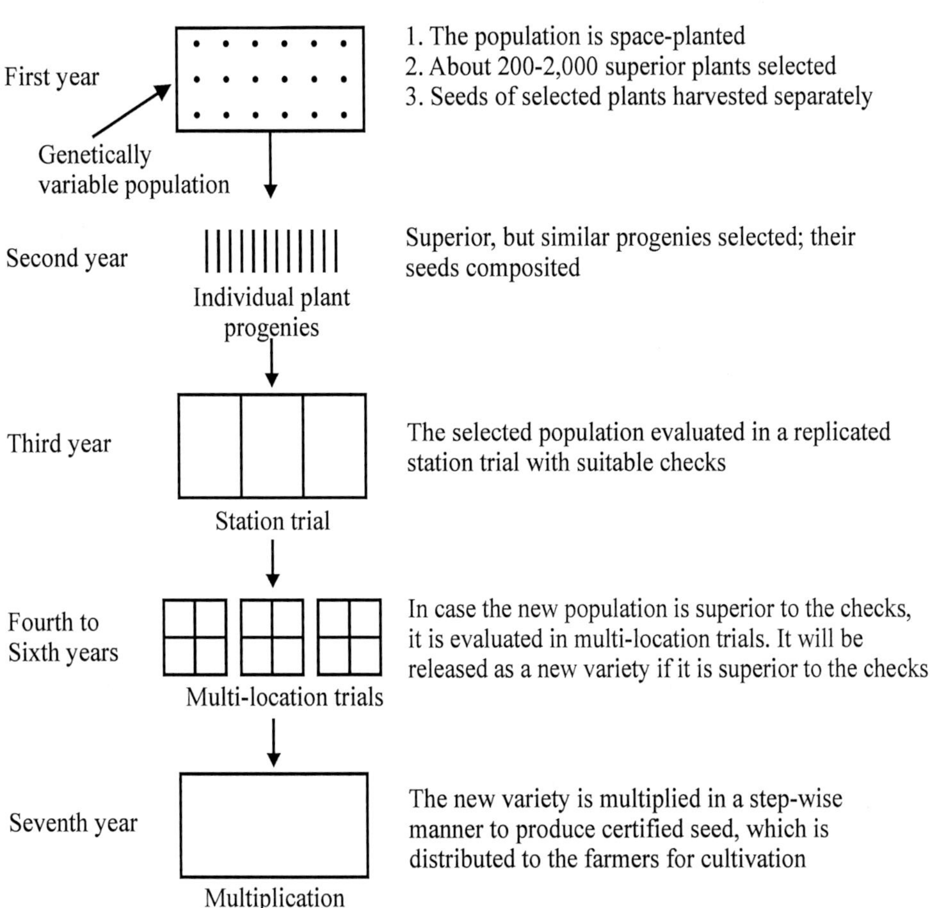

Fig. 4.1. A scheme for mass selection based on progeny test. Progeny test is not necessary, but would be helpful in the identification of superior progenies. The activities of the first and the second years are used for purification of pureline varieties. The station trial, multi-location trials and seed multiplication steps are common to all breeding methods, but their details vary with crops (for more details, see, Chapter 15). This scheme becomes pureline selection if superior progenies are selected in the second year and their seeds are harvested separately. In the third and later years, superior lines are used for yield trials.

The selected population is evaluated in a replicated trial along with suitable checks; this trial is called **station trial**. The best variety or the best two or three varieties of the concerned crop are used for comparison with the selected population or line; these varieties are called **checks**. In case, the selected population is superior to the checks in yield and other characteristics, it is advanced to multi-location trials. The **multi-location** trials are conducted under the All India Coordinated Project for the concerned crop. These projects are run mostly at agriculture universities and Indian Council for Agricultural Research (ICAR) institutes, and are funded and coordinated by ICAR, New Delhi. The trials are usually conducted for three years at several locations and include appropriate checks. If the selected population (or line) is superior to the checks, it will be released as a new variety (Fig. 4.1), and this release will be notified. The seed of new variety will be multiplied in a stepwise manner to make the certified seed of the variety available to the farmers for cultivation (for some more details, see Chapter 15). The steps of station trial and multi-location trials are common to all the breeding methods, and for all the crops. However, their details vary from one crop to the other.

Mass selection would be useful in improving the local/*desi* varieties or land races of crops since they are genetically variable. But pureline selection is better suited for this purpose and is almost exclusively used for variety development. At present, mass selection is used for maintaining the genetic purity of the pureline varieties since purelines tend to become variable with time (Section 3.10.1). For this purpose, the steps of the first two years given in Fig. 4.1 are implemented. In the first year, plants similar to the given pureline variety are selected and their seeds are harvested separately. During the next year, individual plant progenies are grown and all the progenies identical to the given pureline variety are selected and their seeds are composited to give the purified version of the pureline variety. Mass selection is simple, easy to practice, and produces a variety that is expected to retain the adaptability of the original variety. Nevertheless, the variety developed by mass selection is not completely uniform and its performance would be lower than that of the best pureline variety that can be developed from the same population.

4.2.2. Pureline Selection

The **pureline selection** method had been designed to select the best pureline from a genetically variable population and to use it as a pureline variety. A **pureline variety** is obtained from a single homozygous plant of a self-pollinated species, and is maintained by self-pollination. Therefore, all the plants of such a variety have the same genotype, and any phenotypic variation present among them is only due to environmental factors and/or mutations. Individual plant selection had long been used by farmers. But the present form of pureline selection utilizes the concepts of progeny test and purelines developed by Vilmorin (during 1850s) and Johannsen (in 1903) respectively. This method was refined to its present form by the workers at the Svalof Experiment Station of the Swedish Seed Association. In the pureline selection method, a genetically variable population is space-planted, and a large number (200 – 2000 or 3000) of plants are selected on the basis of their superior phenotype for easily observable traits like plant height, disease resistance, flowering time, etc. The seeds from the selected plants are harvested separately, and individual plant progenies are grown in the next crop season. The progenies are evaluated visually, and inferior progenies are rejected. Segregating progenies will be generally rejected unless they have some highly desirable features. In the latter case, superior plants may be selected from such progenies and their progenies evaluated in the next year. Several superior progenies are selected and their seeds are harvested separately. The selected progenies are evaluated in a replicated trial with suitable checks, and one or few superior progenies are identified. These progenies are then evaluated in multi-location yield trials, and a pureline superior to the checks would be released as a new variety (Fig. 4.1).

Table 4.1: An illustrative list of some varieties developed in India by pureline selection, pedigree selection, and backcross methods of breeding in selected self-pollinated crops

Crop	Variety developed		
	Pureline selection	**Pedigree method**	**Back cross method**
Barley	C251, 50; K12		
Brassica juncea	Varuna, Krishna, Shekhar, Seeta, RH30, Durgamani	Prakash, Vaibhav, Bardan	
Cotton (*G. hirsutum*)	Gadag 1, CC2, Buri 107	Laxmi, LK861, LAM389	170-Co2, 134-Co2M
Linseed	Chambal		
Mungbean	T1, B1		
Pea	T163, BR2, No.43, Harbhajan, RP3	Rachna, Pant Pea 5, HUP5, DMR11, VL1	
Pigeon pea	Malviya Vikalp	Prabhat, Mukta, Manak, Pusa 84, Sharda, T21	
Rice	Mtu1, 2, 7; Ar1, Br2, 3; Patni 6, T1, 3, 22, 29, etc.	Jaya, Ratna, Swarna, Samba Mahsuri, PB-1, etc.	Peta, Sabarmati, Jamuna
Tobacco (*N. tabacum*)	NP28, 63, 70	Godavari, Bhagya-lakshmi, Bhavya	
Urdbean	T9, 27, 122; Naveen, Kulu 4		
Wheat	NP4, 6, 12; Pb 8, 8A, 11; C13 (later K13), K46, 53, 54	K65, 68; NP50, 120, 125 and the NP series varieties; Malviya 12, WL711, KRL1-4	Kharchia 65, NP852, HUW234, K8027, HW2004, KSML3*, MLKS11*, KML7406*, Sonalika Multiline-1*

* Multiline varieties

Pureline selection is used to improve local/*desi* varieties, introduced varieties, or old pureline varieties that have become genetically variable. This method can achieve the maximum possible improvement in a genetically variable population. The pureline varieties are uniform and are preferred by consumers and farmers. They are also easily identifiable for seed certification. But the breeder has to devote more time for pureline selection, and the pureline variety may not have the wide adaptability of the original variety/population. A large number of pureline varieties have been developed in most of our crops; some examples are listed in Table 4.1. A relatively recent example of a variety developed by pureline selection is the pigeon pea pureline variety Malviya Vikalpa. This variety was selected from material collected from the Mirzapur district of Uttar Pradesh, and was released for commercial cultivation in 1999.

4.3. HYBRIDIZATION

The genetic variation present in the populations of a self-pollinated crop is soon exhausted by pureline selection. Therefore, it becomes imperative to generate new genetic variation to enable further improvement. Hybridization (Section 3.4) is the most commonly used method for generating genetic variation in breeding populations. A **breeding population** is created by a breeder for the purpose of using a breeding method for developing a new improved variety. Hybridization brings the allelic combinations present in the two parents together in the F_1 hybrid. As a result, the F_1 hybrid is heterozygous for all the genes for which its parents have different alleles. The F_1 hybrid and its subsequent generations may be maintained by continued selfing to generate F_2, F_3, F_4, etc. generations. These generations are called **segregating generations**, and would show genetic variation produced by segregation and recombination of the genes present in heterozygous state in the F_1 generation. The segregating generations are generally handled according to the pedigree scheme but often they are carried forward by the bulk or the single seed descent (SSD) scheme. The F_1 and the subsequent generations can be repeatedly backcrossed to one of the parents of the F_1; this scheme is the basis of the backcross method of breeding. Sometimes the plants in the F_2 and a few of the subsequent generation are subjected to intermating before the population is used for, usually, pedigree selection. Finally, the F_1 generation may itself be used for commercial cultivation as a hybrid variety to exploit heterosis.

4.4. PEDIGREE METHOD

The **pedigree method** is so named because maintenance of a pedigree record is an essential component of this breeding scheme. A **pedigree record** is a document containing the relationships of each plant selected in F_2 and the individual plant progenies grown in the subsequent generations. A simple representation of a F_5 progeny in the pedigree record may be as follows: 1527-

36-12-4. These numbers are interpreted as follows: F_5 progeny obtained from the plant number 4 selected from the F_3 progeny number 12 derived from the plant number 36 selected in the F_2 generation of the cross number 27 made in 2015. The pedigree record contains brief notes on the distinguishing features of each progeny in every generation. Pedigree record is often useful in the elimination of redundant progenies: in later generations (F_5, F_6, etc.) if two or more lines are similar in appearance and they were selected from the same progeny of the previous generation, only one of them may be retained. These records may also provide a general idea about the inheritance of characters of interest.

4.4.1. The procedure of Pedigree Method

In the pedigree method, individual plants are selected and individual plant progenies are grown in each generation from F_2 to F_4 or F_5 till the progenies become almost homogeneous. In the F_1 generation, about 20-30 plants are space-planted and their seeds are harvested in bulk to raise a space-planted F_2 population of 2,000-10,000 plants. In F_2, 100 – 500 plants with superior phenotype are selected, and their seeds are harvested separately (Fig. 4.2).

In F_3, individual plant progenies are raised from the seeds of selected F_2 plants, and superior plants are selected particularly from superior progenies. Individual F_4 progenies are raised from the seeds of selected plants, and the emphasis is on selection of desirable plants from superior progenies. The individual plant progenies are grown in F_5, and the progenies are evaluated visually for their superior features and homogeneity. Each superior homogeneous F_5 progeny is harvested as bulk, and in F_6 multiple rows are grown from seeds of the selected progenies and the plots are evaluated visually. A limited number of superior homogeneous progenies are selected and their seeds are harvested as separate bulks. In F_7, a replicated yield trial is planted along with suitable checks. The progenies outperforming the checks are advanced to multi-location trials. The progeny that performs better than the checks in these trials will be released as a new variety. Its seed will be multiplied in a step-wise manner so that certified seed is made available to the farmers for commercial cultivation.

4.4.2. Selection of Parents

Selection of the parents to be crossed is the most critical step of a breeding program. All the characters that the new variety is proposed to have must be present in one of the parents. Selection of parents for qualitative trait improvement is rather simple, but that for improving qualitative traits is problematic because of the complex inheritance pattern and less than 100% heritability. In general, the selection is based on the performance of the parents themselves, but the performance of their crosses (based on F_1, F_2 or F_3 performance data, or data from advanced generation lines or doubled haploids) are also used. Several approaches for using the data on parental performance for

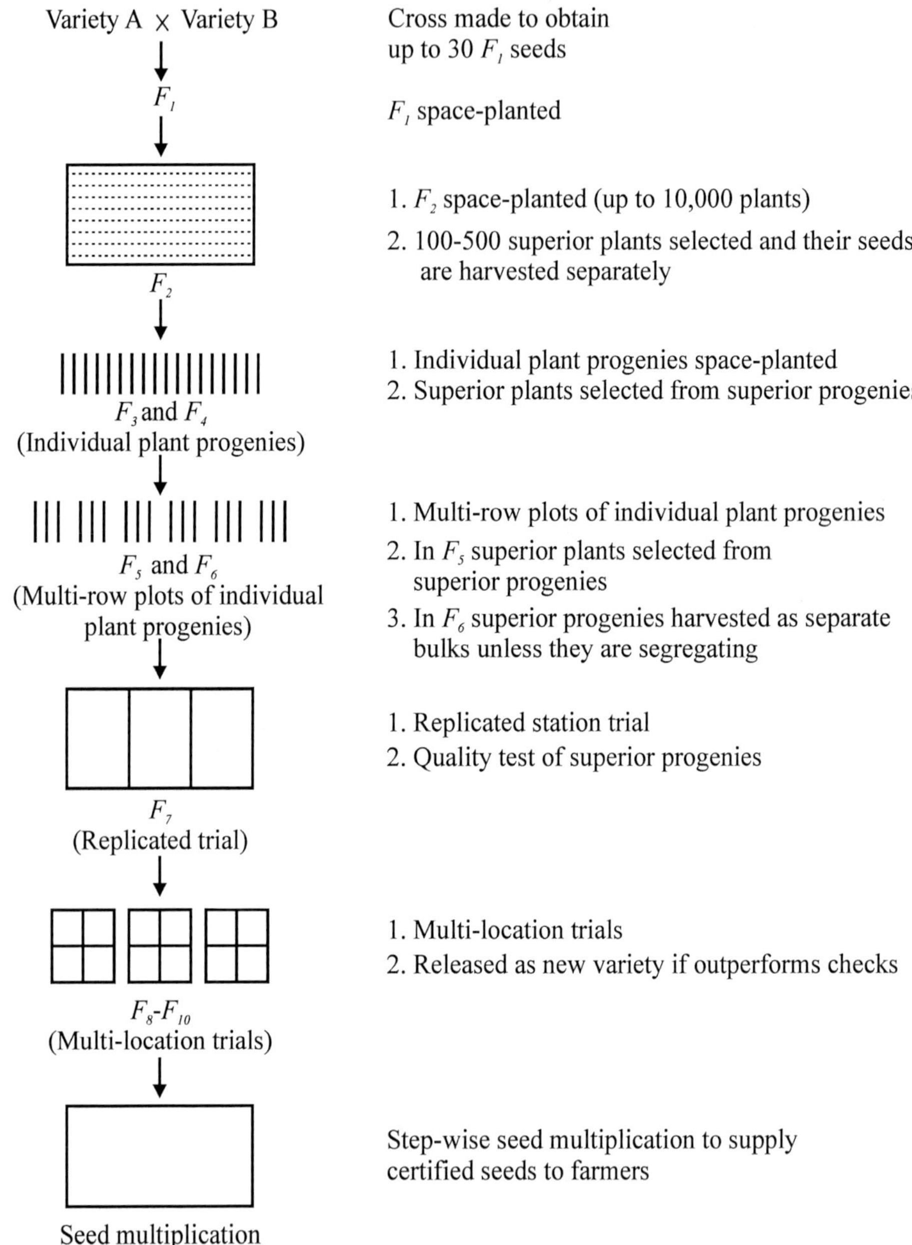

Fig. 4.2. A simplified schematic representation of the pedigree method. Many breeders harvest superior progenies as separate bulks in F_4 and conduct preliminary yield trials in F_5. If required, the progenies may be further purified before multi-location trials.

the selection of suitable parents have been proposed. Each of these approaches offers some advantages, but suffers from some weaknesses. For example, the midparent values for yield and other quantitative traits are good indicators of parental usefulness in some cases, but not in others. It has been suggested that the parents to be used in hybridization programmes should be first selected on the basis of their performance for the traits of interest. Then multivariate analysis should be used to group them in clusters on the basis of genetic distance, and parents should be selected from different clusters. Finally, the selected parents should have small coefficient of parentage, i.e., should not be closely related. This approach might be better than the one based on the parental performance and the midparent value. The future gains justify the additional work involved in the parent selection process. The availability of molecular markers linked to important agronomic traits, including disease resistance can be gainfully used to identify parents with the desired genes and/or QTLs (quantitative trait loci).

4.4.3. Selection in the Segregating Generations

In case of pedigree method, the selection of individual plants is based on visual evaluation of the phenotypes of plants. Visual selection would be quite effective for qualitative traits and highly heritable quantitative traits, but it would not be effective for traits like yield that have complex inheritance and low heritability. In addition, a breeder generally makes several different crosses, which would differ in their potential to yield superior segregant progenies. It would be highly desirable to identify the more promising crosses so that the breeder could devote more time to them. Generally, this decision is also made on the basis of visual evaluation of F_2 or a later segregating generation. Efforts have been made to design suitable schemes for the selection of promising crosses, and of superior F_2 and later generation plants.

There are two approaches to identify superior crosses: (i) evaluation of F_2, F_3 or F_4 bulks, and (ii) evaluation of F_3 or F_4 individual plant progenies in replicated trials. Evaluation of bulks is useful in the identification of inferior crosses since crosses with poor performance are less likely to yield superior performing progenies. The evaluation of F_3 or F_4 progenies is based on around 100 random individual plant progenies from each cross. A desirable cross should have high estimates for mean performance of the progenies, genetic variance among them, and expected genetic gain under selection. Generally, this approach generates more reliable information about the potential of crosses than the evaluation of cross bulks. In either case, the information is more reliable if the trials are conducted at more than one location. These approaches require additional cost and effort, but help eliminate poor crosses that would save resources as well as the breeder's time. At CIMMYT, Mexico, wheat crosses are regularly evaluated in F_2 and inferior crosses are rejected.

The selection of individual plants during segregating generations is generally based on visual observation. But visual selection for traits like yield is almost ineffective. One of the approaches designed for more effective selection in the segregating generations is based on the evaluation of a larger number of F_3 or F_4 individual plant progenies isolated from the cross of interest. The progenies are evaluated in an unreplicated trial, preferably at more than one location, for yield and other traits of interest. The selection of superior progenies is on the basis of their comparison with contiguous control or moving average. In the case of **contiguous control**, a check variety is planted after every 4, 6 or 8 progeny rows. The performance of a progeny is compared with that of the nearest check. The progeny will be selected if it is superior to the check. But in the **moving average** method, the performance of a progeny is expressed as percent of the average performance of 7 progenies, viz., three progenies on either side of the progeny in question and the progeny itself. The performance of the first three and the last three progenies in the trial plot will be compared with the average of the progeny itself and the six progenies nearest to it. The two approaches appear to be comparable with each other in effectiveness. But it is doubtful if the additional effort and the expenditure involved is justified by the limited improvement in the effectiveness in comparison to that of the visual selection.

Some other possible alternatives to visual selection are the grid method of selection (Selection 5.2.1), the honeycomb design, and indirect selection for yield and other low heritability traits. In case of **indirect selection**, an easily scorable trait with high heritability and exhibiting high correlation with the trait, for which selection is to be done, is used as the basis of selection. For example in case of maize, selection for yield on the basis of number of ears per plant and the ear height has been found to be more effective than direct selection for yield itself. The chief limitation of this approach is that there are few traits that can be used for reliable indirect selection for the desired traits. The development of molecular markers has provided abundant marker polymorphism that can be used for reliable indirect selection for yield and many other traits that have either low heritability or a direct selection for which is not feasible for one reason or the other (Chapter 14).

4.4.4. Merits and Limitations

In this method, the plant breeder has the maximum opportunity for using his skill and judgment for the selection of superior progenies and plants. It is well suited for the improvement of easily observable and highly heritable characters. At the same time, transgressive segregants for yield and other quantitative traits can also be recovered. A **transgressive segregant** is either superior to the superior parent or inferior to the inferior parent for a quantitative trait. Therefore, this method can be used to correct some specific defects of a superior variety (**combination breeding**), as well as to improve the performance of this variety by recovering some superior transgressive segregants (**transgressive breeding**). The maintenance of accurate pedigree records takes time. In addition, the selection of large numbers of individual plants in each generation takes considerable time of the breeder. Finally, the effectiveness of visual selection in the segregating generations is largely dependent on the skill and experience of the breeder.

4.4.5. Achievements of Pedigree Method

Pedigree method has been the most extensively used breeding scheme for handling of the segregating generations from crosses in self-pollinated crops. A large number of pureline varieties have been developed in many crops, including rice, wheat, barley, pea, chickpea, groundnut, sorghum, etc. (Table 4.1). These varieties represent improvements in yield, quality, disease resistance and many other agronomic characteristics. For example, the basmati rice variety Pusa Basmati 1 (PB-1) was developed from the cross Pusa 150 x Karnal Local, and released for cultivation in 1989. PB-1 is the first semidwarf variety of Basmati rice, and it is photoinsensitive and high yielding (6-7 tons/ha). PB-1 grains are extra long slender and aromatic, and they have improved cooking quality as well. For a long time, this was the most popular variety of Basmati rice: during the period 1995 – 2007, it accounted for 60% of the total Basmati rice export, and its contribution was valued at rupees 24,031 crores. Similarly, the wheat variety KRL1-4 was developed from a cross between Kharchia 65 (salt tolerant) and WL711 (high yielding, excellent agronomic traits). KRL1-4 is salt tolerant, has amber colored, hard, medium-bold grains, and can yield up to 20 – 30 q/ha in saline/alkaline soils.

4.5. BULK METHOD

In the **bulk method** of breeding, seeds from all the plants in F_2 and later generations are harvested and their seeds are mixed or bulked to raise the next generation (Fig. 4.3). At the end of the bulking period, a number of superior plants are selected, and their seeds are harvested separately. Individual plant progenies are grown in the next generation, progenies with desirable features are selected, and the seeds from each progeny are harvested in bulk since the progenies would have become reasonably homogeneous. The progenies are evaluated in a replicated station trial and then in multi-location trials, and the progeny performing better than the checks is released as a new variety. The seed of the new variety is then multiplied for distribution to the farmers. In the F_2 and the later generations, a fairly large number of plants (up to 30-50 thousand) can be grown since they are grown in bulk. The period of bulking would vary depending on the breeding objective. Further, mass selection may be practiced for such traits as seed size, plant type, plant height, flowering time, etc. during the bulking period.

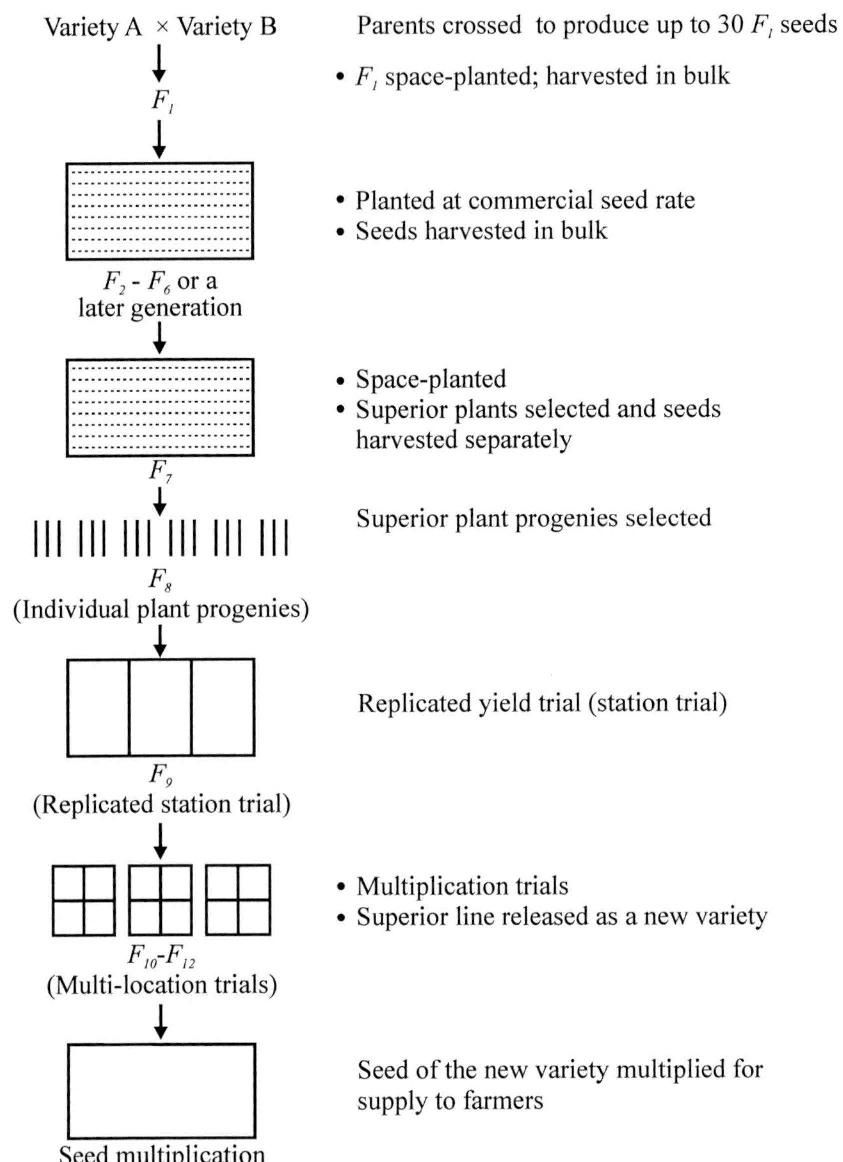

Fig. 4.3. A simplified schematic representation of the bulk method of breeding

The bulk method can be used for one of the following purposes: (i) isolation of homozygous lines, (ii) waiting for the opportunity for selection, and (iii) to allow natural selection to improve the genetic composition of the bulk population. The scheme presented in Figure 4.3 is for the isolation of homozygous lines. In such cases, the bulking period is short of 6-7 generations, after which purelines are isolated from the population, which are evaluated for release as new variety. In case selection is to be made for a trait that requires a specific environment, e.g., lodging resistance, cold tolerance, etc., the population could be maintained as bulk till the occurrence of the concerned environmental condition. After this, the population may be handled as per the pedigree method of breeding. But the bulks are maintained for long periods in time, viz., 20-30 generations or even more, when the objective is to allow natural selection to act on the segregating populations. Natural selection is likely to exert only a small effect up to F_6 or so since the population will have considerable heterozygosity; it would be more effective during the later generations. In any case, mass selection during the bulking period for traits like larger seed size, dwarf plant type, etc. would be desirable since these traits are likely to be selected against by natural selection. Further, disease epidemics may be created to eliminate susceptible genotypes and favor the disease resistant segregants.

In general, the survival of a genotype would depend on the number of seeds/plant it produces, and the proportion of plants obtained from these seeds that reproduce: the higher these values, the greater will be its survival in mixtures. There is considerable evidence that natural selection does change the genetic composition of mixtures of genotypes. The evidence for this has been obtained from the following types of studies as discussed below.

1. When mixtures of several varieties are grown at several locations, one or few varieties become predominant with time and the inferior varieties are eliminated. Different varieties tend to predominate at different locations. Often the ability to survive in mixtures may not be related to agronomic performance.

2. In populations derived from hybrids, such as composite crosses, agronomically inferior alleles of genes are quickly eliminated from the population. When the populations are maintained at two or more locations, the survival of alleles may be markedly affected by the locations.

3. The performance of hybrid populations tends to improve with the generation of bulking. For example, the yield of barley Composite Cross V increased by 21% over 24 generations of bulking. Thus, the frequency of higher yielding genotypes tends increase in the hybrid populations in response to natural selection acting on the bulk populations.

4. The performance of purelines isolated from the bulk populations tends to increase with the duration of bulking.

Thus, it may be concluded that long-term bulking could help create populations that are more likely to yield purelines with superior performance than the early segregating generations or the short-term bulks. A modification of the bulk procedure, called **selected-bulk method**, is used for wheat breeding at CIMMYT, Mexico. In this method, F_2 is space-planted, 10 best plants are selected, and their seeds are bulked. The F_3 to F_5 generations are grown at commercial seed rate, 30-60 best spikes are selected, and their seeds are composited. In F_6, up to 100 best spikes are selected, the seeds are harvested separately, and individual plant progenies are grown in F_7. The superior progenies are selected and evaluated in replicated trials.

The bulk method has not been widely used for crop improvement, but CIMMYT, Mexico regularly uses modifications of the bulk procedure. In India, *B. juncea* variety Narendra Rai was developed by the bulk method. The bulk method of breeding is inexpensive, simple and convenient since it requires little attention from the breeder. Artificial selection may be used to increase the frequency of desirable alleles, e.g., for seed size, dwarf plant type, etc. In long-term bulks, natural selection may increase the likelihood of isolation of superior purelines from the bulked population. But this method takes much longer than pedigree method to develop a new variety particularly when long-term bulks are maintained. The breeder has little opportunity to use his skill for the selection of superior plants, and off-season crops and greenhouses cannot be used to advance the segregating generations.

4.6. SINGLE SEED DESCENT METHOD

The **single seed descent (SSD) method** is a simple modification of the bulk method of breeding. In this method, a single seed is harvested from each plant in the F_2 and the subsequent generations, the seeds are mixed together, and the next generation is grown as bulk. In F_5 or F_6 generation, a large number (up to 500) of individual plants is selected, and individual plant progenies are grown in the next generation. The progenies are evaluated and superior progenies are selected and evaluated in replicated trials (Fig. 4.4). The harvesting of one seed from each plant minimizes the effects of natural selection and there is artificial selection only at the end of the SSD procedure. The SSD procedure is used to rapidly advance the generation of hybrid populations. By using off-season nurseries and greenhouse facilities one could grow up to three generations in a single year. Thus, under ideal conditions, one could raise the F_6 generation in the third crop season and use it for individual plant selection. The SSD procedure is used for the isolation of homozygous lines to be used as new improved varieties. This simple procedure is becoming increasingly popular with the breeders and several varieties have been developed in crops like soybean, yarldlong beans, etc. There is some evidence that the effectiveness of the SSD procedure for the identification of superior purelines may be comparable to that of the pedigree method. The SSD procedure is also used for the construction of a mapping population called recombinant inbred lines (RILs; Chapter 13).

The SSD procedure advances the generation with the maximum possible speed, requires little, space and resources, and makes the maximum use of off-season and greenhouse facilities. It ensures that the population retains a random sample of the homozygous genotypes derived from the F_2 population. But it does not allow any selection, except at the end, and there is plant loss in each generation. In crops like pulses, plant loss may become a serious limitation of the SSD procedure. In such cases, a single pod may be selected from each plant, and the bulked seed may be used to raise the next generation.

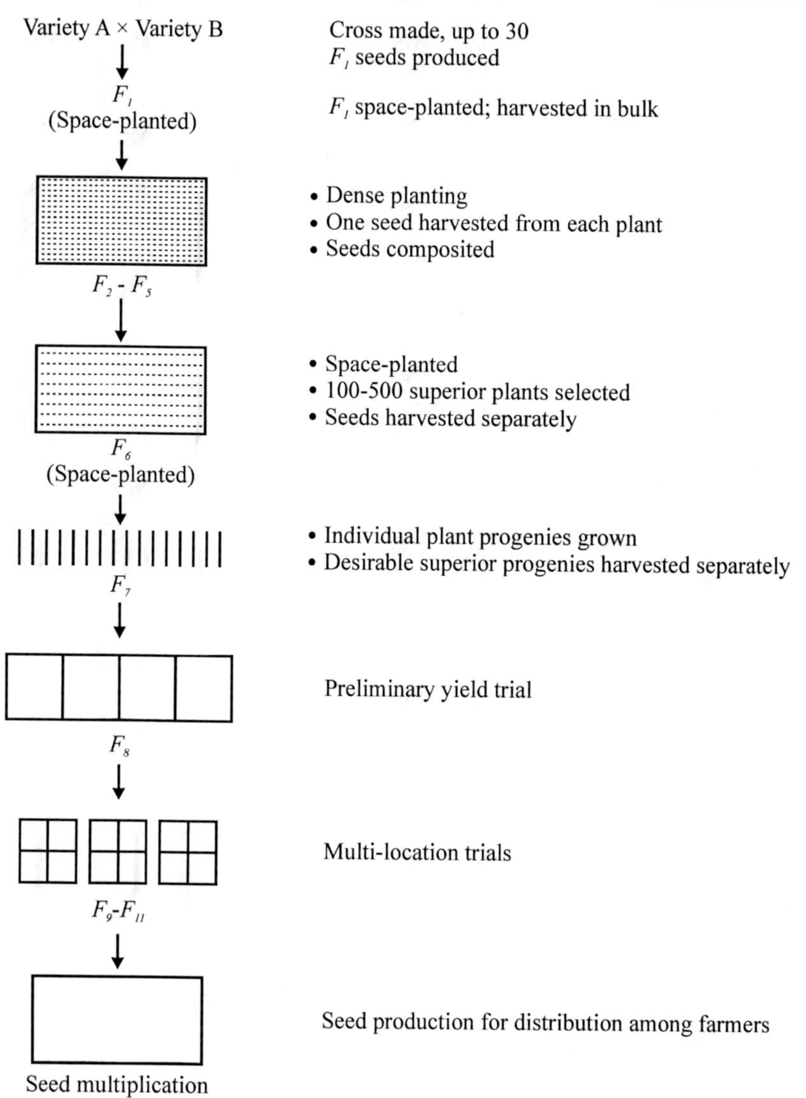

Fig. 4.4. A simplified schematic representation of the single seed-descent (SSD) procedure used for variety development. In case a population of recombinant inbred lines (RILs; see Chapter 13) is to be developed, seeds from all the plants in F_5 or F_6 (or a later generation) are separately harvested and individual plant progenies are grown in F_6 or F_7. Each progeny is a recombinant inbred line, and all the RILs together constitute the RIL population.

4.7. USE OF HAPLOIDS FOR RAPID ISOLATION OF HOMOZYGOUS LINES

A **haploid plant** has the gametic chromosome complement (*n*) of the species. As a result, haploids have only a single copy of each chromosome of the genome. Further, they have only one copy of each gene, i.e., they have only a single allele of each gene. The haploid plants themselves are weak and sterile. But their chromosome numbers can be doubled by a suitable treatment like treatment with an appropriate concentration of colchicine. The plants generated by chromosome doubling of a haploid plant will have two copies of each chromosome, and they are called **doubled haploid (DH) plants**. As a result, the doubled haploid plants will be homozygous for all the genes. In case of many crop species like, wheat, rice, barley, tobacco, *Brassica* spp., haploid plants can be obtained in large numbers by the procedure of anther culture. Anther culture involves the culture of anthers having pollen grains around the binucleate stage on suitable culture medium. The haploid nuclei of pollen grains divide and give rise to haploid plants. In some other crops, haploid plants can also be produced from interspecific crosses. For example, haploids of wheat and barley are produced by crossing them to *Hordeum bulbosum.*

In a breeding programme, the desired parents are crossed to produce the F_1 generation. Haploid plants are produced from the F_1 generation either by anther culture or by interspecific hybridization. These haploid plants will differ in their genotype as a result of the segregation and recombination in the F_1 plants. The chromosomes of the haploid plants are now doubled and the DH plants so obtained are grown to maturity. The seeds from all the DH plants are harvested separately and grown in the next generation in individual plant progeny rows. Each progeny row will be a pureline. The progeny rows are subjected to careful selection for the desirable features and the best DH lines will be used for evaluation in station and multi-location trials. Several varieties of rice, wheat, barley, triticale, tobacco, etc. have been developed following this approach.

4.8. BACKCROSS METHOD

Backcross is the cross between F_1 hybrid and one of its parents. The **backcross method** involves hybridization of the F_1 hybrid and the subsequent backcross generations to one of the parents of the F_1 hybrid; this parent is called the **recurrent parent (RP)** of the **recipient parent**. The RP is an excellent variety of the concerned crop that is popular among the farmers, but is deficient in, usually, a specific trait. The other parent is selected for the high intensity of the trait that is deficient in the RP, and its yield and other traits are not of great importance. This parent is generally called the **donor parent (DP)** or **nonrecurrent parent**. The scheme of backcross breeding for the transfer of a dominant trait, e.g., disease resistance, from the DP into the RP is as follows. The RP is crossed with

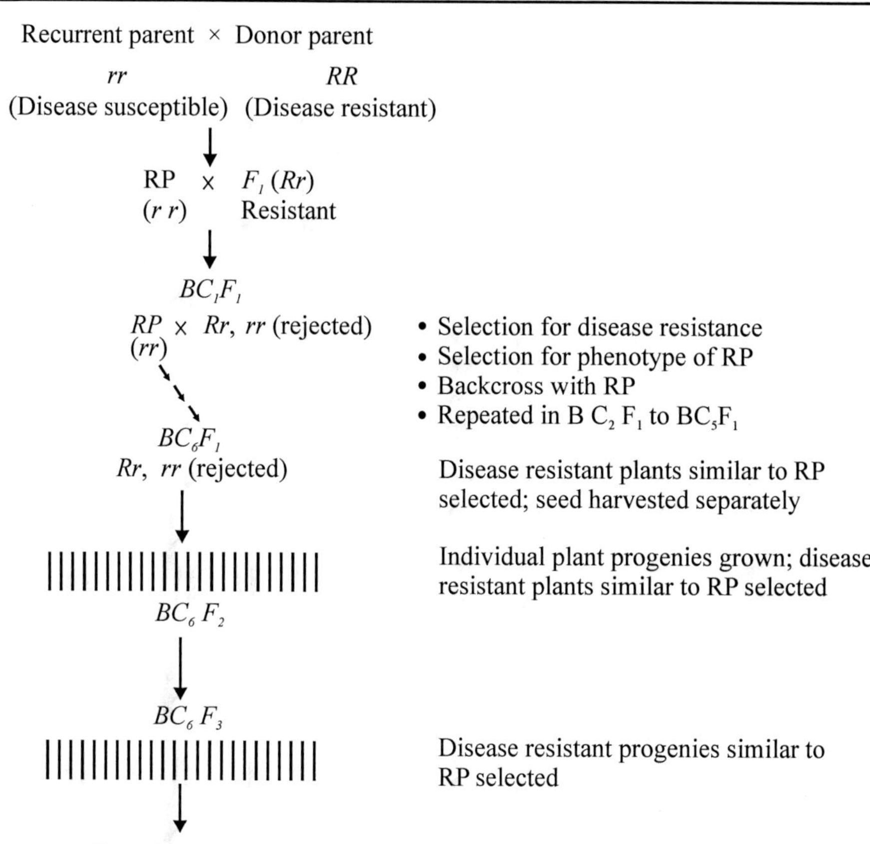

Fig. 4.5. A simplified representation of the backcross scheme for transfer of disease resistance specified by a dominant gene. *RP*, recurrent parent; DP, donor parent; BC_1F_1, backcross 1 F_1; BC_6F_2, backcross 6 F_2, etc.

the DP (used as the male parent), and the F_1 plants are backcrossed to the RP. This backcross generation is denoted as BC_1F_1; it would show segregation for the disease resistance and for the other traits that differ between the DP and the RP. Artificial disease epidemic is created and plants resistant to the disease and similar in phenotype to the RP are selected, and backcrossed to the RP to obtain the BC_2F_1 generation. The steps of BC_1F_1 are repeated in this and the later backcross generations. In BC_5F_1 or BC_6F_1 generation, the disease resistant plants almost identical to the RP in phenotype are selected, their seeds are harvested separately, and individual plant progenies are grown in the next (BC_5F_2 or BC_6F_2) generation. Individual plants resistant to the disease and most similar to the RP are selected and individual plant progenies are grown in the next generation. Plant progenies that are homogeneous, disease resistant, and almost identical to the RP are selected and their seeds are composited to generate the improved version of RP. The 'Improved RP' is evaluated in replicated yield trials and released for commercial cultivation if it performs at least as well as the RP and is resistant to the concerned disease (Fig. 4.5).

The scheme for the transfer of a recessive trait, e.g., disease resistance, is slightly different from the above scheme (Fig. 4.6). Since resistance to the disease is recessive, all the plants in the backcross (BCF_1) generations will be susceptible to the disease. Therefore, the BC_1F_1 and the BC_3F_1 generations are selfed to produce the BC_1F_2 and the BC_3F_2 generations, respectively. In these generations, disease tests are done to select resistant plants, which are backcrossed to the RP. In the other backcross generations, the progeny are not selected for disease resistance (Fig. 4.6). However, in case molecular markers linked to the disease resistance gene are used for selection, the heterozygous (*Rr*) disease susceptible plants can be selected in every backcross generation and used for backcrossing. In this case, there will be no need to raise the BCF_2 generations. In all the backcross generations, plants similar to RP are selected. The BC_5F_1 generation is selfed, and disease test is done in the BC_5F_2 generation. The seeds from disease resistant plants are harvested separately, individual plant progenies are grown in BC_5F_3, all the progenies resistant to the disease and similar to the RP are harvested and bulked to create the improved version of RP. The new version of RP is evaluated in replicated trials, and is released as the improved RP provided it performs at least as well as the RP in other agronomic traits and is superior to RP for resistance to the concerned disease. This scheme can be used for the transfer of any trait governed by the recessive allele of a gene.

Often, two or more traits/genes need to be transferred into a single RP. In case, these traits/genes are present in a single DP, the backcross scheme is similar to that followed for the transfer of a single trait, except that in each BCF_1 or BCF_2 generation plants with all the genes being transferred are selected and used for backcrossing. But when each gene is to be transferred from a different DP, the RP may be first improved by transferring one gene. Then the improved version of the RP is used for the transfer of the second gene, and so on (stepwise transfer). This approach would take a very long time to transfer multiple genes. Alternatively, the RP may be crossed with one DP, the F_1 so obtained is crossed with the second DP, the F_1 hybrid from this crossed with the third DP, and so on till all the DPs are involved in the cross. The complex hybrid is then repeatedly backcrossed to RP, and selection is practiced for all the genes (simultaneous transfer). In this approach much larger backcross population would be needed. In the simultaneous but separate transfer approach, the same RP is used in a separate backcross programme with each of the DPs to transfer a single gene from each of the DPs. In the end, as many improved versions of the RP as the DPs used in the programme are obtained. These versions are hybridized together to produce a complex cross, the complex hybrid and its progeny are subjected to selfing, and homozygous lines carrying all the genes being transferred are isolated. *This seems to be the most useful and practicable scheme of multiple gene transfers from two or more DPs.*

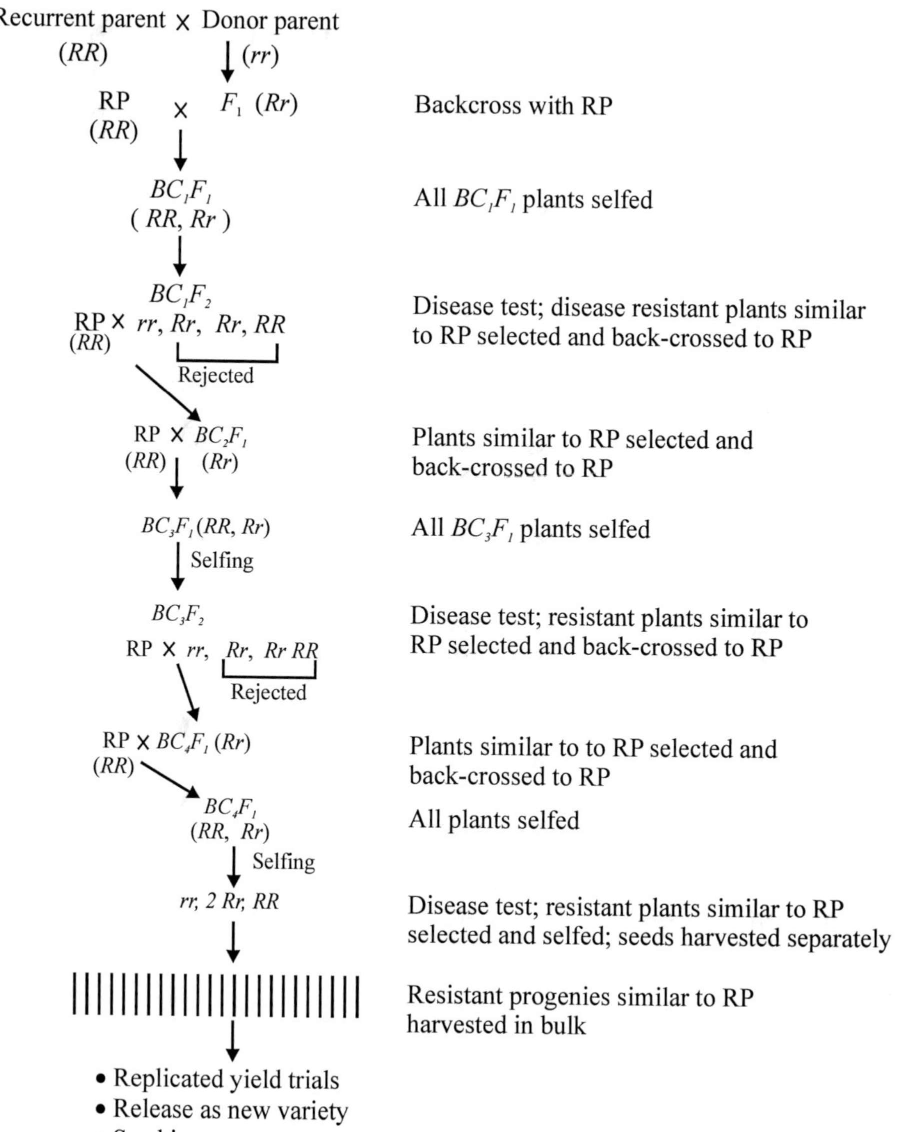

Fig. 4.6. A simplified scheme for transfer of a recessive trait, *e.g.*, disease resistance, by back-cross breeding. RP, recurrent parent; DP, donor parent; BC_1F_1, BC_1F_2, etc., backcross 1 F_1, back-cross 1 F_2, etc.

Repeated backcrossing of the F_1 and the backcross progenies with the RP leads to a rapid decline in the degree of heterozygosity at the same rate as does continued selfing (Table 3.6). At the same time, the genes from the DP are replaced by those from the RP at the same rate as the decline in heterozygosity. Therefore, strict and effective selection for the gene being transferred from the DP has to be practiced during the backcrossing to retain it in the backcross population. If this selection were not done, this gene will be replaced by the allele of this gene from the RP. Further, the genotype of the backcross progeny rapidly converges onto that of RP, except for the gene being transferred. As a result, the backcross progeny become increasingly similar to that of RP. This similarity can be increased by a deliberate selection for the RP phenotype. Molecular markers distributed throughout the genome can be used for a much more effective selection for the RP genetic background (Chapter 14).

The backcross method can be used for the transfer of simply inherited traits from one variety of the crop into another. In fact, this is the most suitable method for gene transfers from unadapted germplasm like land races or local varieties and from related species into the crop varieties. In such gene transfers, the performance of the purelines developed at the end of the backcross programme may be reduced due to tight linkage of some undesirable genes with the transferred gene; this phenomenon is called **linkage drag**. The backcross method can also be used to transfer quantitative traits provided the trait has high heritability. Backcross method is the only breeding method that allows the transfer of cytoplasm from one variety of crop into another variety or from a related species into a variety of the crop species. For this purpose, the variety/species from which the cytoplasm is to be transferred is used as the female parent, and the F_1 and the subsequent generations used for backcrossing are also used as the female parent. This method is also used for **germplasm conversion**. In this process, photoperiod sensitive germplasm lines that cannot be used in crossing programmes because they do not flower under the prevailing day-length conditions are made photoinsensitive by transferring the trait from other germplasm lines.

The backcross method can be modified to combine bulk of the genes from the RP with a reasonable proportion of genes from the DP. This is done by making only one or two backcrosses to the RP and then handling the backcross progeny as per the pedigree scheme. This modification can be used only when the DP not only has the desired trait in high intensity, but it also has some other desirable traits that can improve the performance of the RP. The new variety developed by this modified method is likely to be substantially different from the RP so that it would to designated as a new variety.

The backcross method generates pairs of near isogenic lines (NILs) that are used for mapping the genes that were transferred by the backcross programme. A pair of NILs comprises RP as one line and the line developed by the backcross programme (denoted by RP') as the second member of the pair. The two members of a NIL pair will differ for the gene transferred from the DP as well as some genes that are linked to this gene and some other unlinked genes that are retained in the RP genome by chance. Thus, the product of a backcross programme is almost always the RP improved by the transfer of the concerned gene. As a result, the outcome of a backcross programme of breeding is reliably predictable, and always the same.

The backcross method has been widely used for the transfer of chromosomes, genes, and cytoplasms from one variety into another and from a related species into a crop species. It is extensively used for the development of new cytoplasmic male sterile (CMS) and restorer lines. It has also been used for the improvement of popular varieties of various crops by transferring desirable genes like those for disease resistance. Some examples of the varieties developed by this method are as follows: Vijapla, Vijai, Digvijaya and Kalyan varieties of *Gossypium arboreum* and Kharchia 65, NP852, HUW234, HW2004, etc. varieties of wheat and so on. This method allows the improvement of otherwise excellent varieties of crops for specific traits, which is not possible with any other breeding method. It enables gene and cytoplasm transfers from related species much more efficiently than by the other methods. However, the performance of the new variety developed by this method is the same as that of RP, except for the trait improved by the gene transfer. In almost each generation, crosses with the RP have to be made produce the backcross progeny. Finally, by the time the RP is improved by the gene transfer, it may have been displaced from cultivation by a superior new variety of the crop.

4.9. MULTILINE VARIETIES

A multiline variety is a mixture of several purelines that are similar in phenotype, but differ for the gene specifying vertical resistance to a given disease. The concept of multiline varieties was put forward by Norman E Borlaug in 1959. Such a multiline variety can be developed by using an excellent variety of the concerned crop as RP in separate backcross programmes with different DPs, each of which has a different resistance gene for the concerned disease. Each backcross will yield a modified version of the RP having one gene for resistance to the disease. The seeds of these purelines can be mixed in equal proportions to produce a multiline variety. The number of component lines should be reasonably large, e.g., 15-20 according to Borlaug. A multiline variety allows the vertical resistance genes to mimic the desirable features like durability of resistance of horizontal resistance (Chapter 6). Even when a component line of a multiline variety becomes susceptible to the disease, it will constitute only a small proportion of the plants in the field. As a result, the loss caused by the disease will be relatively small. Once a susceptible line is discovered, it is replaced by a new component line that has a different gene for resistance to the disease, and shows resistant response to the prevailing races/ pathotypes of the pathogen. However, production and maintenance of the component lines of multiline varieties is a time-taking job, and their seed certification presents problems. A limited number of multiline varieties have been produces and used in cereals like oats and wheat, but these varieties have rarely been very popular. For example, three multiline varieties of wheat, viz., KSML3, MLKS (both based on the popular variety Kalyan Sona), and Sonalika Multiline 1 (based on the popular variety Sonalika) were produced in India, but they did not become popular.

4.10. POPULATION BREEDING APPROACH

In this approach, a number of outstanding plants are selected from the F_2 generation of a cross and the selected plants are mated in pairs or according to some other mating scheme (Fig. 4.7). The progenies from these matings are grown, outstanding plants are selected and they are intermated as in the F_2 generation. This step may be repeated as many times as is considered desirable by the breeder. Thus, in the population breeding approach, in F_2 and the subsequent generation, a phenotypic selection followed by intermating among the selected plants scheme is implemented. This phenotypic selection-intermating scheme is similar to the phenotypic recurrent selection used in the cross-pollinated crops (Section 5). Therefore, one round of selection followed by intermating among the selected plants is generally referred to as one selection cycle. At the end of the selection-intermating phase, the progenies from the crosses made in the previous generation are handled according to the pedigree method to isolate superior purelines from the population. The population breeding approach allows for the accumulation of superior combinations of those alleles that are favored by the phenotypic selection. It also allows creation of new allelic combinations by intermating among the selected plants, which brings together the alleles present in two different segregants into a single progeny. Intermating also restores heterozygosity at the concerned loci so that recombination among linked genes can continue for several generations. This is expected to break down the undesirable linkages and generate new combinations of favorable alleles.

Experiments with soybean, wheat, oats, tobacco, etc. show that this breeding scheme is effective in improving yield and yield traits in self-pollinated crops, and the quantum of improvement increases with the number of selection-intermating cycles. More importantly, the amount of genetic variability did not decline with the number of selection cycles. One of the limitations of this approach is that several crosses have to be made in each selection cycle. In addition, the level of improvement generated by this scheme depends primarily on the effectiveness of phenotypic selection for the traits in question. This problem can be overcome by using molecular markers linked to the concerned genes and/or QTLs for selection of the desirable plants. This approach is being extensively used by multinational seed companies. In India, this approach has been used in cotton (*G. hirsutum* and *G. barbadense*), and the varieties JK97, JK119, Sharda and Abadhita-1 selected from these materials have been released for commercial cultivation.

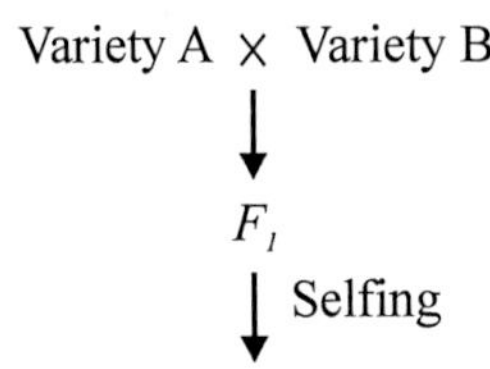

- F_2 is space-planted
- Outstanding F_2 plants are selected and mated in pairs or according to some other mating scheme
- Progenies from matings are space-planted
- Outstanding plants selected and inter-mated

↓ May be repeated one or more times

↓

- Progenies from matings space-planted
- Population handled as per pedigree scheme
- Purelines isolated and evaluated for release as new varieties

Fig. 4.7. A simplified schematic representation of population breeding approach for improvement of self-pollinated crops. This scheme is essentially phenotypic recurrent selection and is being used for improvement of crops like soybean, wheat, etc.

4.11. RAPID ISOLATION OF HOMOZYGOUS LINES (HYBRID SORTING)

In self-pollinated crops, homozygous purelines are used as varieties. The production of homozygous lines from the F_1 hybrids requires selfing for 6-7 generations (= years). This period can be reduced to merely one year by producing haploid plants from the F_1 hybrids, and doubling the chromosome numbers of these plants to obtain completely homozygous **doubled haploid (DH) plants**. The DH plants are selfed and their seeds are harvested separately. In the next crop season, individual plant progenies are grown from the seeds of the DH plants. The progenies are evaluated and superior progenies are selected; the seeds from each of the selected progenies are harvested separately. The selected progenies are evaluated in replicated trials, and the progeny that outperforms the checks is released as a new variety (Fig. 4.8). The haploid plants can be produced by anther culture (wheat, barley, tobacco, etc.), culture of unfertilized ovaries (being used in *Allium* sp. and *Beta* sp.), or hybridization with a suitable species (with *Hordeum bulbosum* for haploid production in barley and wheat) or pollination with a special haploid-inducing pollinator strain (e.g., in case of maize, inbreds like 'RSW'). The chromosome numbers of the haploid plants are generally doubled by colchicine treatment. According to an estimate, about 300 varieties belonging to 12 different crops have been developed using the DH technology. Out of these varieties, more than 100 varieties have been developed in barley alone. The DH approach saves 3-4 years in the development of homozygous lines. But it requires considerable technical expertise, and reliable facilities for plant tissue culture and dependable greenhouse facility for raising of the haploid plants and the production of DH plants. In addition, production of haploid plants in sufficiently large numbers is not feasible in many crops.

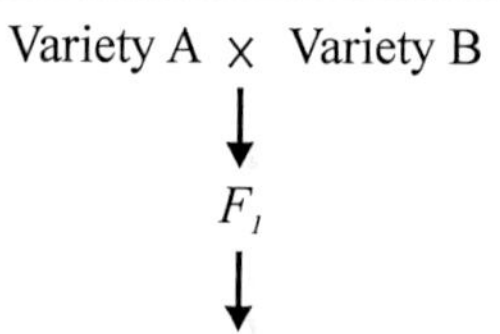

- Haploid plants produced by anther/ovary culture or hybridization with a suitable species/strain
- Chromosome number of haploids doubled by colchicine treatment to produce doubled haploid (DH) plants.
- DH plants self-pollinated; seeds harvested separately

↓

- Progenies from all DH plants grown
- Superior progenies selected and harvested separately

↓

- Selected progenies evaluated in replicated trials
- Outstanding progeny released as new variety
- Seed of new variety multiplied

Fig. 4.8. A schematic representation of production of doubled haploid (DH) plants and their use in breeding of self-pollinated crops.

4.12. HYBRID VARIETIES

The self-pollinated crops show considerable heterosis that is sufficient to warrant the use of hybrid varieties for commercial cultivation. A **hybrid variety** is the F_1 generation from a cross between two or, sometimes more, parents that is used for commercial cultivation. The only issue with the hybrid varieties the production of large quantities of hybrid seed with the minimum cost. The hybrid seed production is greatly facilitated by a suitable pollination control measure like cytoplasmic male sterility. The use of hybrid varieties is discussed with the help of the basmati rice hybrid Pusa RH-10, which matures in 110-115 days and gives, yields of 6.5 tons/ha. The first step in hybrid variety development is to identify a cross combination that gives outstanding performance mainly in terms of yield. One of the parents of this cross is converted into male sterile line by transferring a suitable male sterility cytoplasm. In this case, the CMS line Pusa6A was produced by transferring the wild abortive (WA) male sterility cytoplasm. The original male fertile version of this CMS line, viz., Pusa6B, is used as maintainer of the CMS line. The other parent of the cross is converted into a restorer line by transferring the appropriate restorer gene into it. The line PRR76 carries the restorer gene *Rf1* that restores fertility in the presence of WA cytoplasm.

The male sterile line Pusa6A is propagated by planting it in isolation with the maintainer line Pusa6B, and harvesting the seeds from the plants of line Pusa6A. The seeds produced by the Pusa6B plants will be the maintainer line seeds. For hybrid seed production, the CMS line Pusa6A is planted with the restorer line PRR76 in isolation (Fig. 4.9). The seeds produced by the Pusa6A plants will be hybrid seed. In case of some crops like tomato, tobacco, etc. it may be economical to produce hybrid seeds by hand pollination since each fruit in these crops produces a large number of seeds. It may be pointed out that hand pollination was used to produce hybrid cotton seed till a useful CMS system became available. Hybrid varieties are being used in a large number of self-pollinated crops, e.g., tomato, rice, brinjal, etc.

Pusa 6A × Pusa 6B
CMS line | Maintainer line
(WA cytoplasm) ↓ (Male fertile)
Pusa 6A
CMS line

A. Multiplication of the female parent of Pusa RH10

Pusa 6A × PRR 76
CMS line | Restorer line
(Female parent) ↓ (Male parent)
Pusa RH10
(Male fertile)

B. Production of hybrid (Pusa RH10) seed

Fig. 4.9. A schematic representation of hybrid seed production in rice. The restorer PRR 76 has the restorer gave *Rf1*. The CMS line Pusa 6B has the WA (wild abortive) male sterility cytoplasm. Pusa RH10 matures in 110-115 days and yield 6.5 tons/ ha. CMS, cytoplasmic male sterile.

4.13. PARTICIPATORY CROP IMPROVEMENT

In case of **participatory crop improvement**, the farmers participate in local crop improvement programmes, and contribute to the selection of varieties suited to their needs. This concept developed out of the recognition that the formal plant breeding programmes conducted by the plant breeders at research stations generally develop varieties for high input agricultural systems. In contrast, most small-scale farmers operate under marginal conditions subject to complex stresses that present high production risks. Therefore, participatory crop improvement programmes are carried out at the farmers' fields. These programmes are of the following two types: (i) participatory varietal selection and (ii) participatory plant breeding. In **participatory varietal selection**, popular local varieties, improved varieties developed by breeders, and advanced generation elite breeding lines are grown in the farmer's fields. The farmers identify and select the varieties/lines most suitable for their needs (Fig. 4.10A). This approach has been successfully used in several developing countries. But in **participatory plant breeding**, farmers participate in the selection process from the segregating materials.

Popular local varieties, improved varieties and advanced breeding lines grown in farmers' fields

↓ Farmers evaluate and select the variety they like

The varieties selected by farmers are made available to them

A. Participatory varietal selection

Crosses made and F_1 generations grown at the experiment station

↓

- F_2-F_6 grown as bulk in the target area, preferably at farmers' fields
- Large number of superior plants selected by farmers
- Seeds from selected plants mixed to raise the next generation as bulk

↓

- In F_6, superior plants selected by farmers, and seeds from the selected plants harvested separately

↓

- In F_7 individual plant progenies are grown, and superior progenies selected with the help of farmers

↓

- Replicated trial with checks may be used to select the best progenies that may be released as new varieties.

B. Participatory plant breeding

Fig. 4.10. A schematic representation of participatory crop improvement. A. Participatory varietal selection. B. Participatory plant breeding.

Thus the plants/lines selected from the segregating generations of crosses will reflect the joint efforts of the breeders and the farmers. The selection may be done in the farmers' fields in the target area and may use selection criteria generally not used by the breeders. One simple and inexpensive scheme for this purpose is to grow the F_2 to F_6 generations as bulks. In these generations, a suitably large number of superior plants may be selected by the farmers, and seeds of the selected plants may be composited to raise the next generation. In the F_6, seeds from the selected plants can be harvested separately and individual plant progenies grown in the next generation. The superior progenies selected and evaluated to identify the purelines to be used as new varieties (Fig. 4.10B). Alternatively, the selected population may be retained as bulk and could be used as a variety developed by mass selection.

4.14. BREEDING OF OFTEN CROSS-POLLINATED CROPS

The **often cross-pollinated** crops show more than 5% to up to 30% cross-pollination. Sorghum, cotton, pigeon pea, safflower are examples of such crop species. Therefore, the genetic constitution of these crop species has some features of both self-pollinated and cross-pollinated species. For example, similar to self-pollinated species; they show little or low inbreeding depression since they are predominantly self-pollinated. As a result, homozygous lines with high performance are easily isolated from these crops and these lines have been used for commercial cultivation. For example, purelines are selected in crops like sorghum or jowar and pigeon pea or arhar and are being used for cultivation. On the other hand, their natural populations retain considerable amounts of heterozygosity, and they show high degree of heterosis; these features are similar to those of cross-pollinated crops. In addition, hybrid seed production is much easier in these crops than in self-pollinated crops.

In view of the above, the breeding methods suitable for both self-pollinated and cross-pollinated crops have been used for the improvement of often self-pollinated crops. A number of pureline varieties have been developed in these crops and used for commercial cultivation. For example, the varieties CSV10 (Coordinated Sorghum Variety-10), CSV11, CSV12, CSV15, SS84, Sel3, HC6 are some of the several pureline varieties of sorghum. The varieties CSV11 and CSV12 yield as much as CSH5 (Coordinated Sorghum Hybrid-5). Similarly, Prabhat, Sagar, Sharda, WB20, etc. are pureline varieties of pigeon pea. Cotton varieties Co4, Virnar, H420, MCU2, Laxmi, etc. are examples of pureline varieties. Hybrid varieties have also been developed in the often cross-pollinated crops, and these have been very popular. Some selected examples are as follows. Cotton hybrids Varalaxmi, Savitri, Jayalaxmi, H2HC, etc. (all *G. hirsutum* x *G. barbadense*), H4, H6, H8, Savita, Fateh, Sugna, Godavary, etc. (all *G. hirsutum*), DH7, DH9 and DDH2 (all *G. arboreum* x *G. herbaceum*); in addition, over 100 *Bt* cotton hybrids are in cultivation and they are very popular. Some examples of sorghum hybrid varieties are CSH1 to CSH11, CSH12R, CSH13, etc.

4.15. BREEDING OF ASEXUALLY PROPAGATED CROPS

Some of our important crops are propagated asexually; these crops are also called **clonal crops**. Some examples of clonal crops are potato, sugarcane, sweet potato, pointed gourd, etc. Clonal crops often show reduced flowering and seed set. They are invariably cross-pollinated, and show severe inbreeding depression. Each variety of these crops is a distinct clone. A **clone** comprises the asexual progeny of a single individual. As a result, all the plants of a clone have the same genotype, which is identical to that of the parent plant of the clone. Therefore, any variation present within a clone does not have genetic basis; this variation is purely due to environmental effects, and is nonheritable. Consequently, selection within a clone is ineffective unless it has become genetically variable. These features of clones are comparable to those of purelines. But purelines are homozygous and are maintained by self-pollination, while clones are highly heterozygous and are maintained by asexual reproduction. Theoretically, clones can be maintained indefinitely, i.e., they are perpetual. But the performance of clones generally declines with time; this is known as **clonal degeneration**. The clonal degeneration is the result of bacterial and/or viral infections and in some cases the consequence of mutations. As is the case with purelines, the clones do become genetically variable with time. The genetic variation in the clones is produced by mutation, mechanical mixture and occasional sexual reproduction. The improvement of clonal crops is done by the following methods: (i) clonal selection, (ii) hybridization, and (iii) mutation.

The **clonal selection** aims to isolate the best clone from a genetically variable clone. A genetically variable clone can be considered as a mixture of several slightly different clones. In the first year of clonal selection, the genetically variable clone is space-planted, and 200 to 2000 superior plants are selected. Asexual progeny of the selected plants are grown in the next year, and several (50-100) superior progenies (=clones) are selected. There is no selection within the clones or progenies, since this would be ineffective. The selected clones are evaluated in a replicated trial, and the clones superior to the checks are promoted to the multi-location trials. A clone outperforming the checks in these trials will be released as a new variety, and its seed will be multiplied for distribution among farmers (Fig. 4.11). Clonal selection utilizes the genetic variation present in the clones, and it does not generate new genetic variation. This method can be used to maintain the genetic purity of clones, and can be combined with hybridization (Fig. 4.12) to improve the clonal crops. Several varieties of clonal crops have been produced by clonal selection, e.g., Kufri Red and Kufri Safed varieties of potato, Bombay Green, Pidi Monthan and High Gate varieties of banana, etc.

Hybridization is the most commonly used method for creation of genetic variation in clonal crops. Selection of parents is the most critical step of the programme, and is generally based on performance of the clones. It is important that the characters that are expected to be present in the new variety should be present in one of the parents. If possible, the estimates of general and specific combining abilities should also be used for selection of the parents. It is desirable to subject the selected parents to inbreeding for 1-3 generations, and to select superior vigorous plants from the inbred progeny for use in hybridization. This step will reduce the occurrence of weak and undesirable plants in the F_1 generation. There will also be inbreeding depression, but the same will be eliminated when the two parents are crossed, and the is produced. The F_1

generation is, in fact, expected to show heterosis. The selected inbred plants from the two parental clones are then crossed to produce the F_1 generation. Generally, single crosses are made by crossing two parents at a time, but complex crosses can also be made. The F_1 generation will be genetically variable since the two parents themselves are heterozygous. About 5,000 to up to 20,000 F_1 plants form a cross are space-planted, and around 500-2000 superior plants are selected (Fig. 4.12). Clonal progenies from the selected plants are grown, and 50-100 superior clones are selected. The selected clones are evaluated in a replicated trial, and the clones superior to the checks are advanced to the multi-location trials. A clone performing better that the checks in these trials will be released as a new variety and its seed will be multiplied.

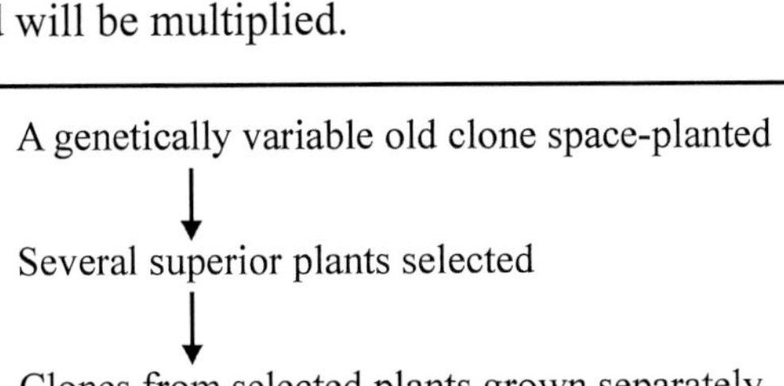

Fig. 4.11. A schematic representation of clonal selection for the improvement of asexually reproducing crops

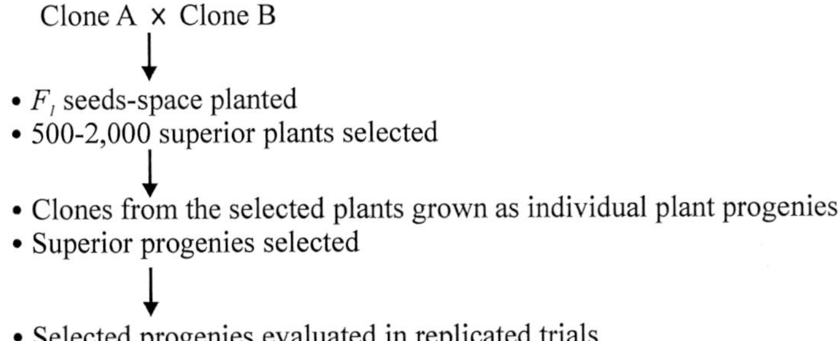

Fig. 4.12. A schematic representation of use of hybridization for the improvement of clonal crops

When several crosses are made, the breeder may like to identify few superior F_1 populations so that he can devote more time to these populations. This can be done by growing about 500 plants from each F_1 population, evaluating these populations visually, and rejecting the inferior populations. Alternatively, a replicated trial may be used to evaluate the F_1 populations, and select the superior performing F_1s. From these F_1s much larger populations are grown for selection of individual plants as described above. In many crops like sugarcane, interspecific hybridization has also been used for variety development. A large number of varieties of clonal crops have been developed by hybridization. Some examples of such varieties are, Kufri Alankar, Kufri Pukhraj, Kufri Surya, Kufri Bahar, Kufri Sadabahar, etc. varieties of potato; and Co1148, Co1158, Co541, CoS109, CoS510, etc. varieties of sugarcane.

4.16. MUTATION BREEDING

Mutation breeding has become a standard technique for the improvement of several clonal crops, e.g., for several ornamental plants in the Netherlands. This method offers the following advantages. (i) A specific trait of an otherwise excellent variety can be improved. (ii) It can be used to improve even such crops that do not reproduce sexually. (iii) Even recessive mutations are useful in clonal crops since these crops are highly heterozygous. Finally, (iv) The time taken for the development of the improved version of a given variety is much smaller than that by hybridization. An excellent example of mutation breeding in a clonal crop is the development of *Verticillium* wilt resistant clone of peppermint (*Mentha piperita*). In USA, the single widely grown clone of peppermint became susceptible to *Verticillium* wilt. A mutation breeding programme was successful in the isolation of a wilt resistant version of this clone and saved the crop from being wiped out. Up to the year 2000, a total of 649 varieties of different clonal crops had been developed through mutagenesis; out of these, 563 varieties belonged to different ornamental crops.

A unique advantage offered by the clonal crops is that any plant with outstanding features, irrespective of whether it is homozygous or heterozygous, can be used to establish a clone that can be used as a new variety. But the breeding of these crops presents the following problems that hamper the progress of breeding programmes. Clonal crops generally show reduced flowering and seed set, and many of them are facultative apomicts. This presents difficulties in the use of hybridization for their improvement. Genetic analysis of these crops is hampered by reduced flowering, sterility and/or perennial life cycle. Finally, the perennial life cycle of many of the clonal crops increases the time required for the development of new varieties.

4.17. CONCLUSION

The natural populations of self-pollinated crops are highly homozygous. Therefore, a genetically variable population of such crops can be regarded as a mixture of several different purelines. These populations can be subjected to **mass selection** or, preferably, **pureline selection** to isolate superior purelines that are used as varieties. Hybridization can be used to create genetic variation and to bring together the desirable alleles present in two or more purelines into a single pureline. The segregating generations from crosses can be handled either by the pedigree, bulk, SSD or the backcross procedure. In the **pedigree method**, individual plants are selected from F_2 and the later generations, and individual plant progenies are grown in the next generation. In F_6, superior progenies are harvested separately in bulk and used for replicated trials. A pedigree record of all the selected plants is maintained, which gives this method its name. In the **bulk method**, the F_2 and later generations are grown at commercial seed rate, seeds are harvested as bulk, and a random sample is used to raise the next generation. The F_7 generation is space-planted, superior plants are selected and individual plant progenies are grown in F_8. Superior progenies are selected and evaluated in replicated yield trials. The bulking period can be extended to F_{20} or F_{30} to allow natural selection to act on the segregating populations. There is considerable evidence that natural selection does increase the performance of the bulk populations. In the single **seed descent method**, one seed from each plant in the F_2 and the later generations is harvested to raise the next generation. In the end (around F_6 generation), superior individual plants are selected and individual plant progenies are grown in the next crop season. Superior progenies are selected and evaluated in replicated yield trials. The pedigree method is the most widely used, while bulk method has been used only to a limited extent. A modification of the bulk method is used by the wheat breeding programme at CIMMYT, Mexico. The SSD procedure is becoming increasingly popular for developing new varieties, and is widely used for creation of mapping populations called **recombinant inbred lines**.

The segregating populations become rapidly homozygous in the pedigree, bulk and SSD procedures, which limits the opportunity for the development of new allelic combinations. This situation can be rectified by intermating among the selected plants in the F_2 and the later few to several generations **(population breeding approach)**. Further, production of **doubled haploids** can be combined with hybridization for rapid isolation of purelines. In addition, **hybrid varieties** are used for commercial cultivation is several self-pollinated crops. When a popular variety has a specific defect, e.g., susceptibility to a disease, **backcross method** can be used to transfer a gene specifying resistance to the given disease. In this method, the F_1 and the later generations are backcrossed to the popular variety (the recurrent parent) so that the backcross progeny rapidly become similar to the RP. Backcross method has been widely used for gene transfers from other varieties and related species of the crop, and for transfer of the cytoplasm to produce CMS lines. It also generates pairs of NILs that are used for linkage mapping of the transferred genes. A comparison among the main breeding methods is given in Table 4.2. Often cross-pollinated species show features of both self- and cross-pollinated species. Therefore, breeding methods used for both self- and cross-pollinated species have been used in breeding of these crops, and both pureline and hybrid varieties have been developed. The clonal crops reproduce asexually, are invariably cross-pollinated and highly heterozygous, and show severe inbreeding depression. Each variety of such a crop is a clone, i.e., asexual progeny of a single plant. Thus, an outstanding plant selected from a genetically variable population can be used to develop a new clone that can be used as a variety (**clonal selection**). The genetically variable population can be an old clone that has become genetically variable or it can be produced by hybridization, including interspecific hybridization. The defects of superior clones with specific weaknesses can be corrected by **mutation breeding**.

Table 4.2: Comparison among the salient features of pedigree, bulk, single seed descent (SSD) and backcross methods of breeding of self-pollinated crops

Feature	Breeding method			
	Pedigree	**Bulk**	**SSD**	**Backcross**
Objective	A new variety superior to the parents	A new variety superior to the parents	A new variety superior to the parents	Removal of a specific weakness of an otherwise excellent variety
Selection of parents	Expected to produce transgressive segregants	Expected to produce transgressive segregants	Expected to produce transgressive segregants	An excellent variety deficient in a specific trait; the other parent has this trait
Mating in F_1 and later generations	Selfing	Selfing	Selfing	Backcross with the RP*
Handling the segregating generations**				
a. Selection	Visual, for observable traits	No selection (mass selection may be done)	No selection	Selection for RP phenotype and the target trait
a. Grown as	Individual plant progenies from selected plants	Bulk of all or selected plants	Bulk of one seed from each plant	Backcross progenies
Improvement in	Both qualitative and quantitative traits	Both qualitative and quantitative traits	Both qualitative and quantitative traits	Generally, qualitative traits
Development of a mapping population[@]	No	No	yes	yes
End product	A new variety superior to both the parents	A new variety superior to both the parents	A new variety superior to both the parents	A new version of RP with improved target trait

* RP, recurrent parent; the parent used for repeated backcrosses, i.e., the excellent variety with a specific defect that is used to produce the F_1.

** F_2, F_3, etc. generations in the pedigree, bulk, and the SSD procedures, and backcross generation in the case of backcross method.

[@] Recombinant inbred lines (RILs) by SSD, and near isogenic lines (NILs) by the backcross method

PRACTICE QUESTIONS

1. Define selection. Briefly describe mass selection and pureline selection methods, and discuss their merits and limitations.
2. List the different breeding methods available for handling the segregating generations from crosses. Describe the pedigree method in some detail, and discuss its merits, limitations and achievements.
3. Describe briefly the bulk and the single seed descent methods of breeding and compare their merits, limitations, and applications.
4. Describe the procedure of backcross method, and discuss its merits limitations, and applications.
5. Briefly discuss the methods of breeding of asexually propagated crops. Discuss the merits, limitations and achievements in the breeding of these crops.
6. Explain the meaning and discuss the usefulness of the following in breeding of self-pollinated crops: (i) hybrid varieties, (ii) mutation breeding, and (iii) participatory plant breeding.
7. Write short notes on: (i) mass selection, (ii) pureline selection, (iii) pedigree method, (iv) bulk method, (v) backcross method, (vi) doubled haploid method, (vii) single seed descent method, (viii) multiline varieties, (ix) mutation breeding, (x) participatory plant breeding, (xii)

RELEVANT LITERATURE

1. Almekinders CJM and Elings A 2001. Collaboration of farmers and plant breeders: participatory crop improvement in perspective. *Euphytica* 122: 424-438.
2. Frey KJ 1981. Plant Breeding II. Iowa State University Press, Ames.
3. Hamblin J 1977. Plant breeding interpretation of the effects of bulk breeding of four populations of bean (*Phaseolus vulgaris*). *Euphytica* 26: 157- 168.
4. Johannsen WL 1926. Elemente der exacten Erblichkeitslehre. Gustav Fischer, Jena.
5. Phillips SL and Wolfe MS 2005. Evolutionary plant breeding for low input systems. *Jour. Agric. Sci*, 143: 245- 254.
6. Suneson CA 1969. Evolutionary plant breeding. *Crop Sci*. 9: 119-121.
7. Van Ginkel M, Trethowan R and Cukadar B 2000. A guide to CIMMYT Bread Wheat Program, WPRS No. 5. CIMMYT, Mexico.

■■■■

5 Breeding of Cross-Pollinated Crops

5.1. INTRODUCTION

The natural populations of cross-pollinated species are highly heterozygous and heterogeneous, i.e., they show variation both at the phenotypic and genotypic levels. In case there is no mutation, migration selection and genetic drift, these populations will be in Hardy-Weinberg equilibrium for the different loci. Selection in such populations tends to change the gene and genotype frequencies in the direction of selection, but it is not likely to either eliminate or fix an allele of a gene. Further, these populations show variable degrees of inbreeding depression that may be rather severe in some species. It is, therefore, necessary to develop such varieties that are heterozygous so that their performance is not reduced due to inbreeding depression. In view of this, the breeding schemes used for the improvement of these crops have been designed to ensure that their end products, (i.e., the populations/lines used as varieties) are heterozygous and free from inbreeding depression. In this chapter, the various schemes used for the improvement of cross-pollinated crops are discussed in some detail. These breeding approaches can be classified into two broad groups, viz., (i) selection or population improvement, and (ii) hybrid and synthetic varieties (Table 5.1). The breeding methods for cross-pollinated species were generally developed for maize, but these methods have subsequently been used for breeding of other cross-pollinated crop species. These breeding schemes have been designed to develop open-pollinated, hybrid, synthetic and composite varieties of cross-pollinated crop species (Table 5.2).

Table 5.1: A summary of the main selection schemes used for the improvement of cross-pollinated crops

Selection scheme	Salient features of the selection procedure	Remarks
Mass selection	Plants with superior phenotype selected, open-pollination, seeds from selected plants harvested in bulk	Simple, one selection cycle in one year, virtually ineffective for traits like yield
Stratified mass	The field divided in small plots or grids, equal number of plants with	Simple, one selection cycle in one year, effective for traits like yield

Selection scheme	Salient features of the selection procedure	Remarks
selection	superior phenotype selected form each plot, open-pollination, seeds from selected plants harvested in bulk	
Ear-to-row method	Plants with superior phenotype selected, open-pollination, seeds from selected plants harvested separately, individual plant progenies grown in next generation, plants with superior phenotype selected form superior progenies, and the process of the second year repeated in the third and later years	Simple, one selection cycle in one year, virtually ineffective for traits like yield
Modified ear-to-row method	Plants with superior phenotype selected, open-pollination, seeds from selected plants harvested separately, individual plant progenies grown in next generation and detasselled, after every 2-3 progenies, one row from bulk seed of all the selected plants grown for pollination, plants with superior phenotype selected form superior progenies, and the process of the second year repeated in the third and the later years	Relatively simple, one selection cycle in one year, effective for traits like yield, perhaps the most successful progeny selection scheme
Simple recurrent selection	Plants with superior phenotype selected and self-pollinated, selfed seed planted in a crossing block in the second year, all possible crosses made and equal seed from all crosses composited to give the selected version of the population, in the next selection cycle, the bulk seed is planted and the processes of years one and two are repeated	Relatively simple, one selection cycle in two years, effective for traits with high heritability; for traits like yield: replicated trial from a part of the selfed seed done in the second year, and superior progenies identified; selfed seeds from the plants producing the superior progenies planted in a crossing block in the third year and all possible crosses made, one selection cycle in three years
Recurrent selection for general combining ability	Plants with superior phenotype selected and self-pollinated, also crossed with a broad genetic base tester; test-cross seed used for replicated yield trial in the second year, superior performing testcross progenies identified; selfed seeds from the plants producing the superior progenies are	One selection cycle in three years, improves the population for GCA and also for performance, the improved population used as a synthetic variety or for isolation of inbreds with superior GCA

Selection scheme	Salient features of the selection procedure	Remarks
	planted in a crossing block in the third year, all possible crosses made and equal seed from all crosses composited to give the selected version of the population; in the next selection cycle, the bulk seed is planted and the processes of years one, two and three are repeated	
Recurrent selection for specific combining ability	Plants with superior phenotype selected and self-pollinated, also crossed with a narrow genetic base tester; test- cross seed used for replicated yield trial in the second year, superior performing testcross progenies identified; selfed seeds from the plants producing the superior progenies are planted in a crossing block in the third year, all possible crosses made and equal seed from all crosses composited to give the selected version of the population; in the next selection cycle, the bulk seed is planted and the processes of years one, two and three are repeated	One selection cycle in three years; improves the population for SCA, also for GCA and performance; the improved population used for isolation of inbreds that combine well with the inbred used as the tester; the inbreds used for hybrid variety development
Reciprocal recurrent selection	Two populations improved for combining ability with each other; plants with superior phenotype are selected from one population and self-pollinated, also test-crossed with a random sample of plants from the second population; test- cross seeds used for replicated yield trial in the second year, and superior performing testcross progenies identified; selfed seeds from the plants producing the superior testcross progenies are planted in a crossing block in the third year, all possible crosses made and equal seed from all crosses composited to give the selected version of the first population; in the next selection cycle, the bulked seed is planted and the processes of years one, two and three are repeated; the same procedure is followed for the second population	One selection cycle in three years; can be used in crops like maize; improves the ability of the two populations to combine well with each other; improved population used for isolation of inbreds that are crossed with each other to produce hybrid varieties or even synthetic varieties

Table 5.2: The different types of varieties in common use for commercial cultivation in cross-pollinated crops

Variety	Description	Produced by	Maintained by	Improved by
Open-pollinated	A population reproducing by natural cross-pollination	A population developed by unintended mass selection	Open-pollination without selection	A selection scheme (population improvement)
Hybrid	The F_1 generation of a cross	Crossing the selected inbreds	Selfing of the parental inbreds	Improving the inbreds or the source populations used for inbred isolation
(a) Single cross	The F_1 generation form a cross between two inbreds	Crossing the two selected inbreds		
(b) Triple cross	The progeny of a cross between an inbred and a single cross	Crossing an inbred with the F_1 generation of a single cross		
(c) Double cross	The progeny from a cross between two double crosses	Crossing F_1 generations of two single crosses		
Synthetic	Open-pollinated generation from a bulk of all possible crosses among several lines with superior GCA	Crossing in all possible combinations a number of lines* with superior GCA and compositing equal amounts of seeds from all crosses	Open-pollination in isolation	A selection scheme (population improvement)
Composite	Open-pollinated generation from a bulk of all possible crosses among several populations with superior performance	Crossing together a number of high performing populations	Open-pollination in isolation	A selection scheme (population improvement)

* The lines may be inbred lines, short-term inbreds, clones, open-pollinated populations or some other populations

5.2. SELECTION IN CROSS-POLLINATED CROPS

The objectives of selection in cross-pollinated crops can be summarized as follows: (i) increasing the frequency of desirable alleles in the selected population, (ii) as a result of the above, increase in the frequency of the desirable genotypes, (iii) improving the characteristics of the population, including its performance, as a consequence of the above changes, and (iv) maintaining the level of inbreeding to the minimum, preferably, zero. (v) Selection in cross-pollinated crops rarely aims for the fixation of desirable alleles, but the alleles governing qualitative traits are usually fixed to maintain homogeneity in the varieties. The main objective of selection methods in cross-pollinated crops is to improve their relevant characteristics, including yield. Therefore, improvement of these crops by selection is often referred to as **population improvement**. A number of selection schemes have been devised for the improvement of cross-pollinated crops. These schemes are grouped into the following two categories: (i) selection without progeny test, and (ii) selection with progeny test. In the first category of selection methods, progeny test of the selected plants is not carried out; mass selection is an example of this group. In contrast, progeny test is an integral part of the selection schemes of the second group; examples of progeny selection are ear-to-row method and recurrent selection (Table 5.1).

5.2.1. Mass Selection

The mass selection and its modifications are the simplest of the selection schemes applicable to both self- and cross-pollinated crop species. In simple terms, the scheme for **mass selection** is as follows (Fig. 5.1A): the population to be improved by mass selection is space-planted, a suitably large number of plants with superior phenotype are selected, and they are allowed to open-pollinate. The seeds from the selected plants are harvested in bulk; this completes the first cycle of selection, and the bulk seed represents the improved population. The bulk seed can be space-planted and the above operations can be repeated one or more times; in such cases, the scheme is often called **phenotypic recurrent selection**. Mass selection is simple and is easily implemented. One selection cycle is completed in one year, is highly effective for traits with high heritability, and some of its modifications are effective in improving even yield. But open-pollination of the selected plants reduces its effectiveness, and inbreeding depression might negate the gains from selection if a sufficiently large number of plants is not selected. Finally, the effectiveness of phenotypic selection for complex traits like yield is limited by low heritability of the traits.

Mass selection is the earliest selection scheme and was extensively used for improvement of cross-pollinated crops. The varieties developed by mass selection are generally referred to as open-pollinated varieties. A large number of varieties were developed by this method in several crops and some of them are still grown in small areas. Mass selection can be used for the improvement of those crops, in which commercial use of hybrid varieties is not feasible due to some reason. In any case, mass selection is used to maintain the purity of open-pollinated, synthetic and composite varieties of crops.

Several modifications of mass selection have been proposed to overcome its limitations. The **stratified** or **grid method of mass selection** scheme was proposed by Gardner in 1961 (Fig. 5.1B). It attempts to minimize the confusing effects of the environment on phenotypes of the plants. In this modification, the field in which selection is to be done is divided into several small plots, each having 40-50 plants. The population is allowed to open-pollinate, an equal number of phenotypically superior plants is selected from each plot, and the seeds from all the selected plants are composited. This completes one cycle of selection; it can be continued for one or more cylses. This method was able to improve the yield of Hay's Golden open-pollinated variety of maize by ~3% per selection cycle or per year.

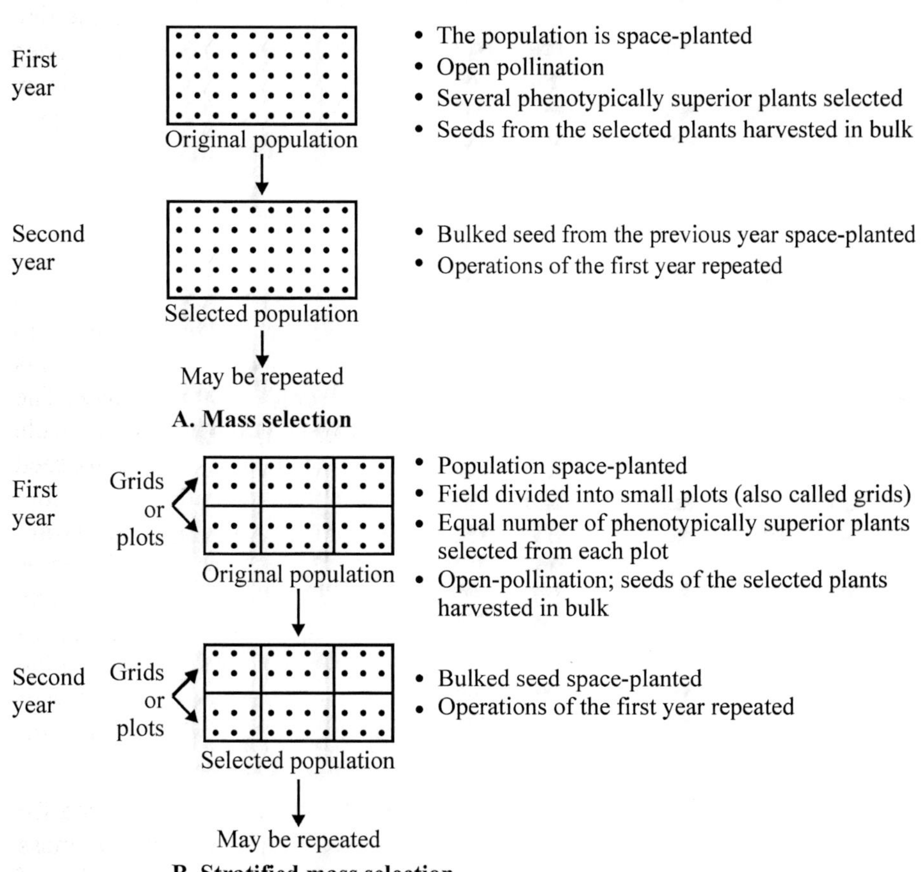

Fig. 5.1. A schematic representation of mass selection in cross-pollinated crops. A. Mass selection. B. Stratified mass selection.

5.2.2. Progeny Selection

In **progeny selection** schemes, the selected plants are subjected to progeny test. The progeny used for evaluation are obtained by open-pollination or selfing of the selected plants or by test-crossing them to specific strains. These progenies are evaluated either visually or in a replicated trial. Selection is then based on the performance of these progenies. The progeny tests can be classified into the following two broad groups: (i) ear-to-row method and its modificaitons, and (ii) recurrent selection schemes. A discussion of these selection schemes is given in the following sections, based mainly on the findings from studies with maize.

5.2.3. Ear-to-Row Method

The ear-to-row method was designed by Hopkins in 1908 for selection in maize populations. In simple terms, the procedure of **ear-to-row method** is as follows (Fig. 5.2). In the first year, the population to be subjected to selection is space-planted, allowed to open-pollinate, and 50-100 plants with superior phenotype are selected. The seeds from the selected plants are harvested separately, and individual plant progenies are grown in the second year. About 10-50 plants are grown in each progeny, the progenies are evaluated visually, and plants with superior phenotype are selected from superior progenies. The plants are allowed to open-pollinate, and seeds from the selected plants are harvested separately. The procedure of the second year can be repeated one or more times. In this scheme, one selection cycle is completed in one year, and the implementation of the scheme is quite simple. But the selected plants are pollinated by both superior and inferior plants present in the population, which reduces the effectiveness of the procedure. In addition, there is likelihood of inbreeding due to selection of plants from a small number of superior progenies. The risk of an increase in inbreeding can be minimized by selecting a sufficiently large number of preferably unrelated plant progenies.

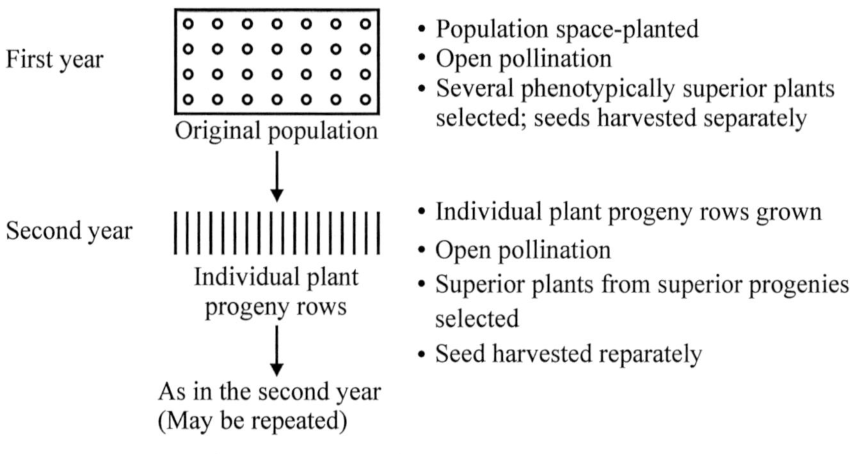

Fig. 5.2. Ear-to-row method of progeny selection proposed by Hopkins in 1908

The ear-to-row method has been modified in several different ways to give us some very useful breeding schemes. One modification, called **modified ear-to-row method**, was proposed by Lonnquist in 1964. In this scheme, phenotypically superior plants are selected in the first year and their seeds are harvested separately. In the second year, the seeds from the selected plants are used for the following three purposes. (i) A part of the seed is used for conducting a replicated trial to assess the performance of the progenies. (ii) A part of the remaining seed is planted in a crossing block as individual plant progenies. (iii) Finally, a bulk is created by mixing equal amounts of seeds from all the selected plants; this seed bulk is planted after every 2-3 progeny rows in the crossing block. The progenies in the crossing block are detasselled. The removal of male inflorescence of maize, i.e., tassel, before it sheds pollen is called **detasselling**. These progeny rows, therefore, are pollinated by the plants in the rows produced from the bulked seed of the selected plants. The superior progenies are identified on the basis of the replicated trial, and the best plants ate selected from the identified progenies. The seeds are harvested separately from each selected plant, and are used for the same activities in the next season. This process can be repeated for one or more cycles. Thus, in this procedure, selection among the plant progenies is based on a replicated trial, which would be more reliable than visual evaluation. The modified ear-to-row method is perhaps the most successful progeny selection scheme.

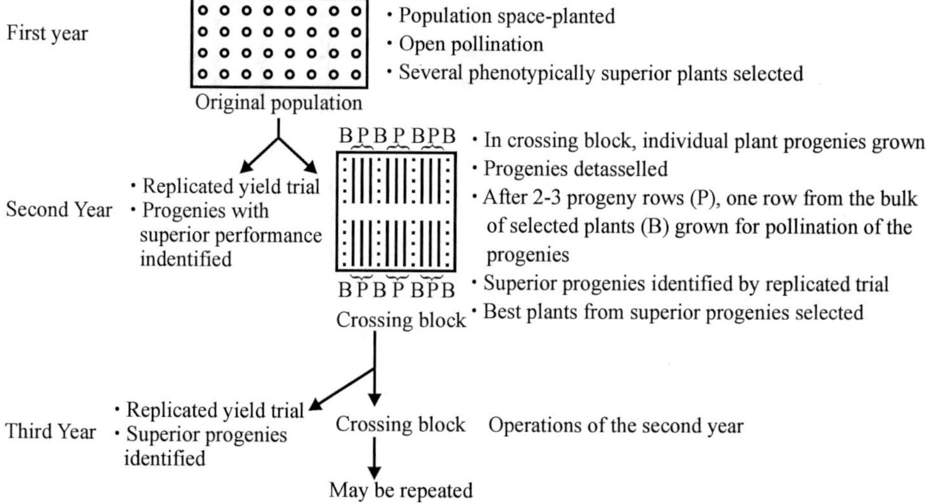

Fig. 5.3. Modified ear-to-row method propose by Lonquist in 1964. P, progeny rows from the selected plants; B, bulk row obtained from the seed bulk created by mixing equal amounts of seeds from all the selected plants.

Some other modifications of the progeny selection scheme are as follows. In one modification, the selected plants may be selfed (in the place of being open-pollinated) to obtain the seed for progeny test. This scheme is the basis of simple recurrent selection, and is also called **S_1 progeny selection**. Another modification involves the production of the seeds for progeny test by crossing the selected plants to a suitable tester. This approach is used in recurrent selection for general combining ability (GCA), recurrent selection for specific combining ability (SCA) and reciprocal recurrent selection.

Progeny selection is used for the development of improved open-pollinated varieties of cross-pollinated crops. Often these schemes are used for purification of the existing open-pollinated, synthetic and composite varieties. The progeny selection is generally more efficient than mass selection, and the selection schemes are still relatively simple and easy to use. The level of inbreeding can be minimized by selection of a sufficiently large number of unrelated progenies. But in most of the schemes, open-pollination among the progenies is allowed; this would reduce the effectiveness of selection. The schemes are more complicated and require more work than mass selection. In most schemes, two years would be needed to complete one selection cycle. But the use of molecular markers would allow the completion of a selection cycle in one year only. Further, if the markers were combined with green- house or off-season nursery facilities, 2-3 selection cycles can be completed in one year (Chapter 14).

5.3. RECURRENT SELECTION

As indicated above, the recurrent selection schemes are variants of progeny selection. Although the idea of recurrent selection was floated around 1920, a breeding scheme was developed much later in 1945 when Hull suggested the method for recurrent selection for specific combining ability (SCA). It is important to note that these selection schemes were developed to facilitate heterosis breeding and not for improving the populations themselves. The logic for these selection schemes is simple and as follows. Inbreds used for the development of hybrid varieties, i.e., heterosis breeding, are isolated from highly heterozygous, genetically variable open-pollinated populations called **source populations**. **Inbreds** are homozygous or almost homozygous lines of cross-pollinated species developed and maintained by continuous close inbreeding, preferably selfing, combined with selection. Recurrent selection in the source populations would enhance their performance, combining ability, or both. In any case, the frequencies of desirable alleles and allele combinations, i.e., genotypes, would increase in the selected populations. Further, the magnitudes of these increases will become larger with the increasing number of selection cycles. Therefore, the chances of isolation of superior inbreds will be much higher in the case of selected source populations than those in the case of original source populations. There are four different types of recurrent selection schemes as follows: (i) simple recurrent selection (SRS), (ii) recurrent selection for SCA (RSSCA), (iii) recurrent selection for GCA (RSGCA), and (iv) reciprocal recurrent selection (RRS). All these selection schemes have the following two common features. (i) The seed used for advancing the generation of the population, i.e., for deriving the selected population, is produced by selfing, followed by intermating among the progenies from the selected plants in all possible combinations. Equal amounts of seed from all the crosses are mixed to create the selected version of the source population. (ii) The selection of plants is based on replicated trial of the progenies of the selected plants. The seed for this trial is obtained by either selfing (in the case of SRS) of the selected plants or by crossing them to a suitable tester (in the cases of RSSCA, RSGCA and RRS).

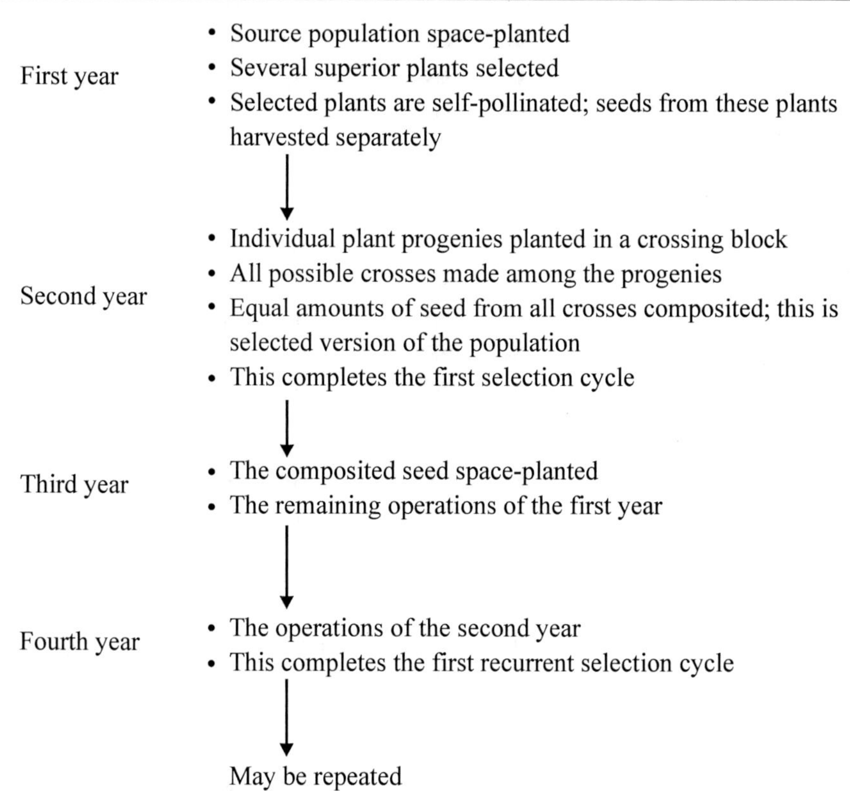

Fig. 5.4. Simple recurrent selection for highly heritable traits

5.3.1. Simple Recurrent Selection

In case of **simple recurrent selection (SRS)**, several superior plants are selected from the original population, i.e., the population to be subjected to selection. The selected plant are self-pollinated, and their seeds are harvested separately. The selfed seeds can be evaluated for seed characteristics and quality traits; in such a case, only the seeds with superior features will be retained. In case the traits under selection have high heritability, the individual plant progenies from the selfed seeds are planted in a crossing block during the second year. All possible crosses are made among the progenies, and equal amounts of seeds from all the crosses are mixed to constitute the selected version of the original population (Fig. 5.4). This completes the first selection cycle; thus, one selection cycle is completed in two years. When this population is subjected to the second selection cycle, the scheme becomes a recurrent selection scheme. Thus, in a recurrent selection scheme, a population is subjected to two or more selection cycles. The SRS scheme differs from the ear-to-row method in the following respects: (i) self-pollination of the selected plants in the place of open-pollination, (ii) intermating of the progenies in the next year and compositing the seeds from all crosses, and (iii) using the composite seed derived population for selection of superior plants. Further, (iv) SRS takes two years to complete one selection cycle in the place on only one year for the ear-to-row method. The use of molecular markers allows the completion of one or more selection cycles every SRS as well as the modified SRSs are described below.

In case of traits like yield that have low heritability, a replicated trial is conducted in the second year using part of the selfed seed from the selected plants. The data from this trial are used for identification of the plants that produced the progenies with superior performance. The remaining selfed seed from these plants are planted in a crossing block in the third year. All possible crosses are made among the progenies and seeds from all the crosses are composited to obtain the selected version of the population. This completes the first selection cycle. The second selection cycle begins by space-planting the composited seed in the fourth year and repeating the operations of the first three years (Fig. 5.5).

The SRS scheme is useful for the improvement of characters having high heritability. The mean of the selected population for the trait under selection shifts in the direction of selection. Generally, there is little decline in the genetic variability present in the selected population. Further, the level of inbreeding remains low; it can be avoided by adopting suitable measures like selection of a sufficiently larger number of plants in each selection cycle. As described above, the SRS scheme can be suitably modified to effectively select for traits having low heritability. The selected populations have improved features and can be used as varieties for commercial cultivation.

5.3.2. Recurrent Selection for General Combining Ability

In 1940, Jenkins proposed a scheme for the development of synthetic varieties from short-term inbreds. A **short-term inbred** is an inbred developed by few (usually, 1-3) generations of inbreeding. The RSGCA scheme is essentially an extension of the above scheme. The main objective of the RSGCA is to improve the GCA of the selected populations. **General combining ability (GCA)** is the ability of a line or plant to produce hybrids with superior average performance when crossed with several different lines, strains or plants, or with a line that has broad genetic base. A line is said to have a **broad genetic base** when it contains plants of several different genotypes, e.g., an open-pollinated, synthetic or composite population a double cross hybrid, or segregating generations of a hybrid. On the other hand, a **narrow genetic base population** comprises plants of a single genotype, e.g., an inbred or a single cross hybrid. The ability of a line or plant to produce superior hybrid progeny when crossed with a specific line or plant is called **specific combining ability (SCA)**. The GCA is believed to be a reflection of the additive gene action, while SCA depends mainly on non-additive gene action. Therefore, GCA is considered to be more relevant in the development of synthetic varieties, while SCA is the basis for heterosis breeding.

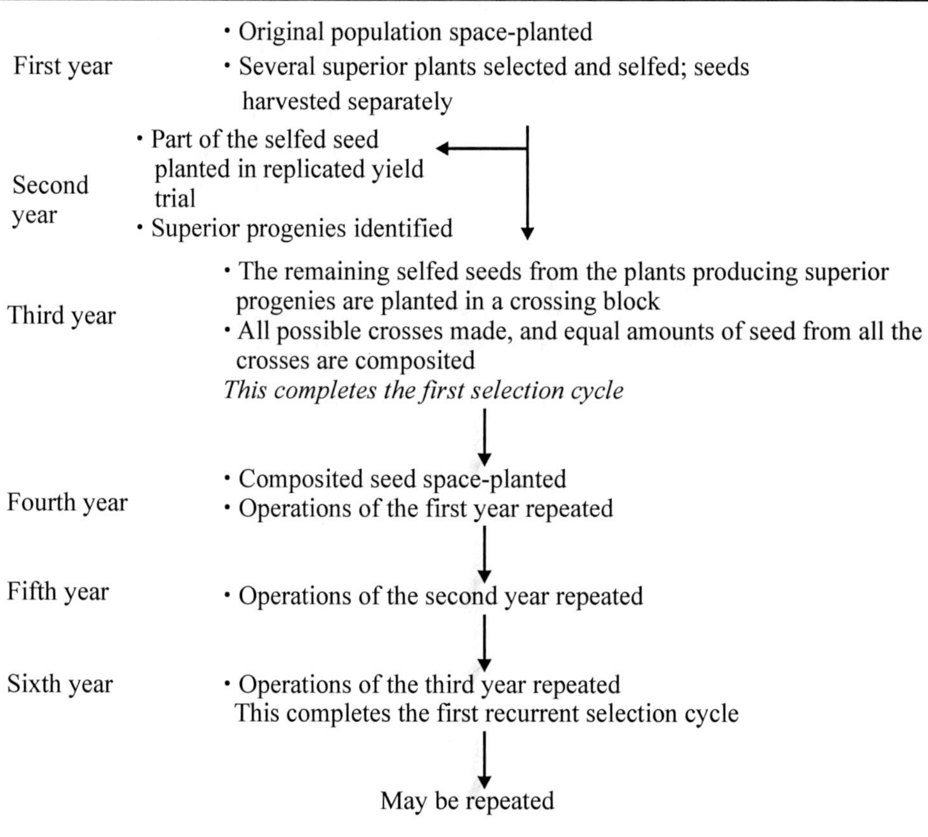

Fig. 5.5. Simple recurrent selection for traits like yield that have low heritability

The RSGCA scheme can be applied to an open-pollinated, synthetic or composite population or to the segregating generations of a hybrid. In this scheme, the seed use for replicated trial is produced by crossing the selected plants to a broad genetic base tester; the seed so produce is termed as **testcross seed**. A **tester** is a line that is crossed with several plants or lines with the objective of estimating their GCA/SCA. The procedure of RSGCA is described here in simple terms (Fig. 5.6). The population to be subjected to selection is space-planted, and several superior plants are selected from this population. Each

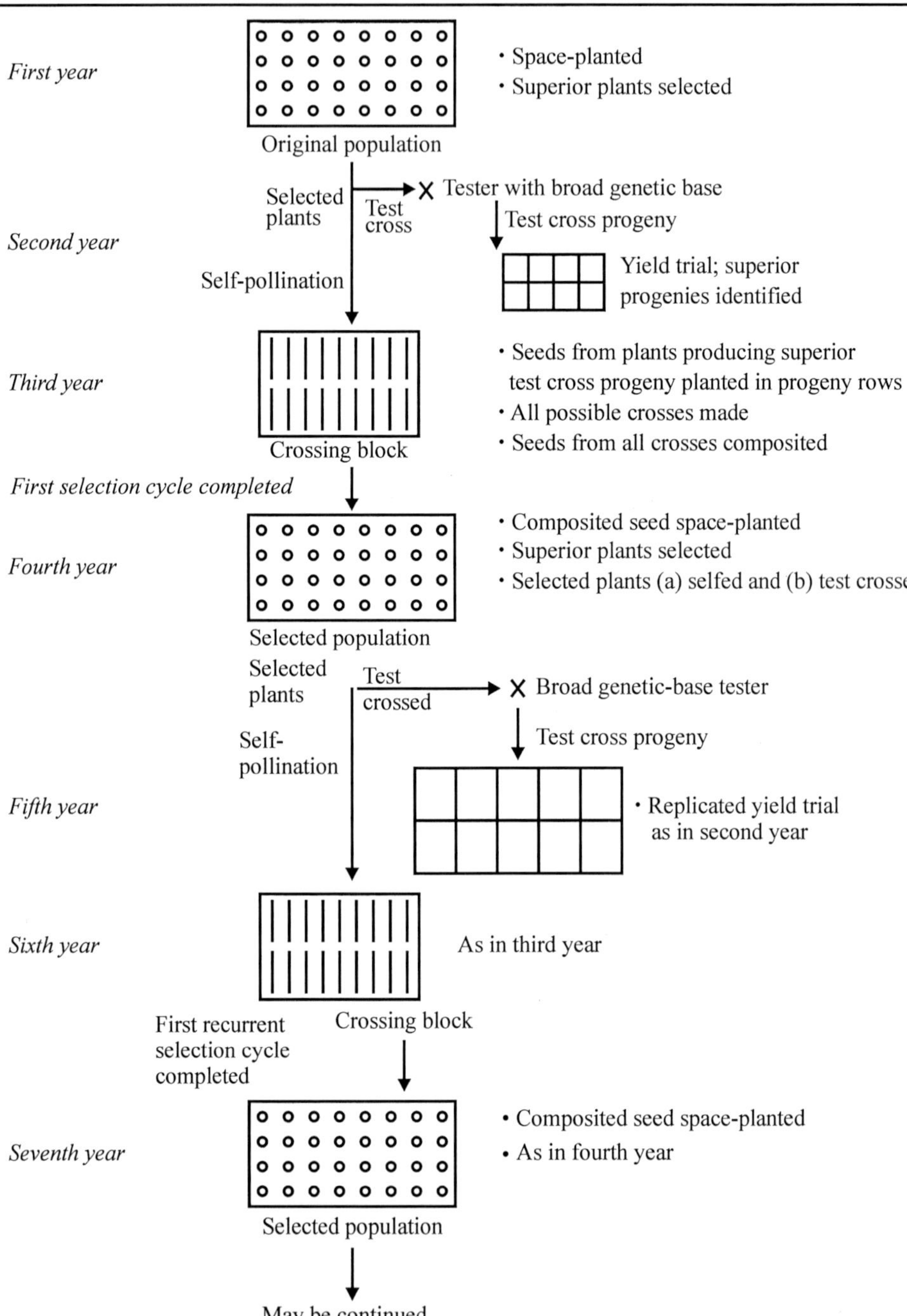

Fig. 5.6. A simplified schematic representation of recurrent selection for general combining ability (RSGCA)

selected plant is self-pollinated as well as crossed with several random plants of the broad-based tester. The test-cross seeds from all the selected plant are harvested separately and used for a replicated yield trial in the second year. The data from yield trial are used to identify the test-cross progeny with superior performance. The selfed seeds from those plants that produced the superior test-cross progenies identified in the second year are planted as individual plant progenies in a crossing block. All possible crosses are made among the individual plant progenies, and equal amounts of seeds from all the crosses are mixed together to constitute the selected version of the population. This completes the first selection cycle. The composited seed is space-planted in the fourth year to initiate the first recurrent selection cycle. In the fourth year the operations of the first year, while in the fifth and sixth years the operations of the second and third years, respectively, are repeated. This completes the first recurrent selection cycles. The selection may be continued for one or more further cycles.

The use of RSGCA scheme improves the general combining ability of the population. In addition, it also improves the yielding ability of the selected population, and this increase may be up to ~10% per selection cycle. Generally, there is some decline in the genetic variability present in the selected populations, particularly for the traits being selected. This decrease is most likely due to an increase in the level of inbreeding, which can be minimized or even avoided by adopting suitable measures. The selected population has improved performance, and is equivalent to a synthetic variety (Section 5.5). Therefore, it can be used as a synthetic variety for commercial cultivation. Alternatively, it can be used for the isolation of inbred lines with superior GCA. These inbreds, in turn, can be used as parents for the development of synthetic and/or hybrid varieties.

5.3.3. Recurrent Selection Specific Combining Ability

The RSSCA was proposed by Hull in 1945 with a view to develop inbreds having superior combining ability with a specific inbred. This objective is of great relevance in the development of hybrid varieties since a large part of heterosis is believed to be due to specific combining ability. The procedure of RSSCA is the same as that of RSGCA, except for the use of a narrow genetic base tester in the place of a broad genetic base tester used in case of RSGCA. The tester used in the case of RSSCA is an outstanding inbred valued as an excellent parent for the development of hybrid varieties. The RSSCA improves the population for combining ability with this inbred. Therefore, the inbreds isolated from the selected population are expected to combine well with this inbred and to produce hybrids with excellent performance.

The procedure of RSSCA is described here in simple terms (Fig. 5.7; the phrase 'broad genetic base' in this figure should be read as **'narrow genetic base'** in the case of RSSCA). The population to be subjected to selection is space-

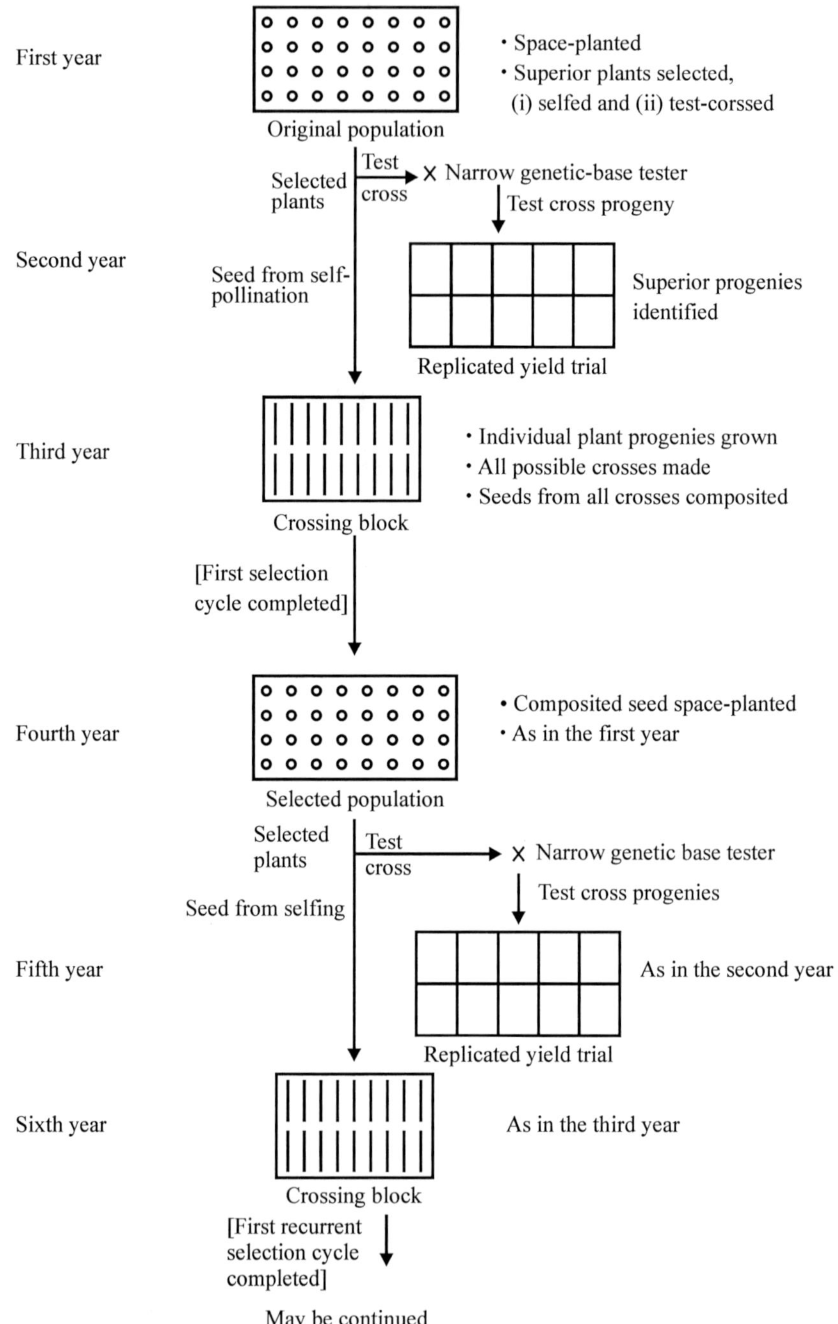

Fig. 5.7. A simplified schematic representation of recurrent selection for specific combining ability (RSSCA).

planted, and several superior plants are selected. The selected plants are self-pollinated and simultaneously test-crossed with the inbred tester. The test-cross seeds are used for a replicated trial in the second year; this allows the identification of test-cross progenies with superior performance. The selfed seeds from the plants producing the superior test-cross progenies are planted in a crossing block, all possible crosses among the progenies are made, and seeds from all the crosses are composited. This completes the first selection cycle. This composite seed represents the improved version of the original population. This seed is space-planted in the fourth year, and the operations of the first year are repeated. In the fifth and the sixth years, the operations of the second and the third years, respectively, are repeated. This completes the first recurrent selection cycle. The selection can be continued for one or more further cycles.

One selection cycle of RSSCA is completed in three years. This scheme improves the population for SCA with the inbred used as the tester. It also improves the population for GCA as well as its performance in terms of yield (about 5% gain per selection cycle) and other traits subjected to selection. The improved population is generally used for isolation of inbreds. These inbreds are expected to combine well with the inbred used as the tester. Further, these inbreds are used for hybrid variety development. Usually, the tester inbred will be used as one of the parents of the hybrid variety.

5.3.4. Reciprocal Recurrent Selection

The RRS scheme, proposed by Comstock and coworkers in 1949, aims to improve two different populations, say, populations A and B, for combining well with each other. The two populations are subjected to selection at the same time. Further, one population, e.g., population A, serves as the tester for the plants selected from the other population, i.e., population B, and vice-versa. As a result, the two populations are improved for combining ability with each other. The procedure for RRS is described here briefly in simple terms and is depicted in Figure 5.8. In the first year, the populations A and B are space-planted in separate fields. Several phenotypically superior plants are selected from population A and self-pollinated. Each of these selected plants is also crossed with several random plants from population B; this constitutes test-cross. The selfed seeds and the testcross seeds from all the selected plants are harvested separately. In the second year, the testcross seed is used for a replicated trial to assess the performance of the testcross progenies. The trial data allow the identification of superior testcross progenies. In the third year, the selfed seeds from the plants producing these superior progenies are planted in individual plant progeny rows in a crossing block, all possible crosses are made among the progenies, and equal amounts of seeds from all the crosses are mixed together. This completes the first selection cycle, and the bulked seed represents the improved version of the original population A, i.e., population A’. This bulked seed is space-planted in the fourth year, and the operations of the first year are repeated. In the fifth and the sixth years, the operations of the second and the third years, respectively, are repeated. This completes the first recurrent selection cycle. The selection can be continued for one or more further cycles.

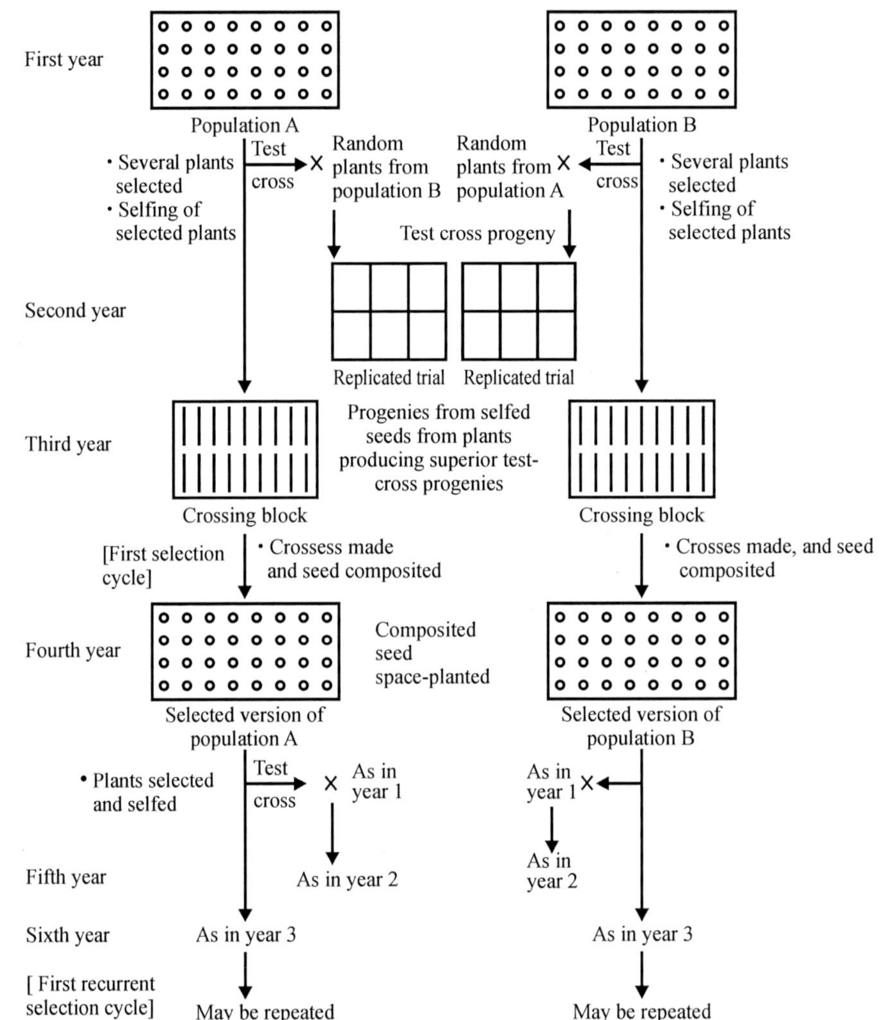

Fig. 5.8. Reciprocal recurrent selection (RS); a simplified schematic representation.

The population B is subjected to the same procedure as described for population A. The only difference from the above procedure is as follows: each plant selected from the population B is crossed with several random plants from the population A, and the testcross progeny produced in this way are used for replicated trial in the second year. The rest of the procedure remains the same as that described for population A. In general, RRS selects mainly for additive and dominance gene effects. The RRS scheme improves the yields of the two populations by approximately 4% per selection cycle. The GCA of the two populations is also improved. Further, the yield of the cross between the populations increases by about 5% for every cycle of selection. The main reason for this yield increase is the improvement in the GCA of the two populations. The improved populations may be crossed together to produce a population cross that can be used as a synthetic variety. Alternatively, inbred lines may be isolated from the two improved populations, and they can be crossed together to develop hybrid varieties.

5.3.5. Comparative Effectiveness of the Different Selection Schemes

The effectiveness of two or more different selection schemes can be compared by applying them to a single population under comparable experimental and environmental conditions. A number of such studies have been carried out in different populations of maize. The findings from these studied allow a number of general conclusions; some of the most relevant conclusions are summarized here.

1. Both RSGCA and RSSCA improve the yields of the selected populations; in some studies, RSGCA was superior to RSSCA in this respect.
2. Both RSGCA and RSSCA improve the GCA of the selected population, and RSSCA seems to be superior to RSGCA in this regard.
3. S_1 family selection seems to be superior to RSGCA in increasing the yielding ability of the selected populations. However, this selection scheme is inferior to RSGCA in improving the GCA of the selected populations.

5.3.6. General Conclusions from Selection Experiments in Maize

1. The major portion of genetic variance in maize is additive.
2. All selection schemes markedly reduce genetic variance in the selected populations. This effect is especially observable during the first few selection cycles. But substantial amount of additive genetic variance remains even after several cycles of selection.
3. All selection schemes mainly select for GCA.
4. The RSGCA, RSSCA and RRS schemes increase the testcross performance of the selected populations most likely due to the improvement in GCA.
5. Generally, partial to complete dominance predominate and there is little evidence for overdominance in maize.

6. The response to selection for yield varies from one population to the other and from one selection cycle to the other. The yield may reach a plateau after a period of selection.
7. Generally, selection for one trait also leads to changes in some such traits, for which selection was not done. This is called **correlated response to selection**, and is most likely due to linkage among the genes governing yield and those controlling these traits.

5.4. HYBRID VARIETIES

A **hybrid variety** is the F_1 generation of a cross between two or more lines that is used for cultivation for commercial crop production. The parental lines of hybrid varieties are generally inbred lines, but they can be short-term inbreds, clones or even open-pollinated populations. An **inbred** is a nearly homozygous line produced by continues inbreeding (selfing if possible) and selection in a cross-pollinated crop, and is maintained by strict inbreeding or selfing, where possible. A hybrid variety is generally a single cross hybrid, double cross hybrid, or triple cross hybrid, but some other types of hybrids have also been used occasionally. A **single cross hybrid** variety is the F_1 generation from a cross between two lines. A **double cross hybrid** is the F_1 progeny from a cross between two single cross hybrids. Similarly, a **triple cross hybrid** is produced by crossing an inbred with a single cross hybrid.

The first hybrid varieties in maize were F_1 generations from crosses between open-pollinated varieties, but they did not become popular. The use of inbreds for hybrid variety development was proposed by Shull in 1909. But this scheme could be used on commercial scale only after the idea of double cross hybrids was put forth by Jones in 1918. The Texas cytoplasmic male sterility (CMS-T) in maize was discovered in 1938; it has been the most widely used male sterility cytoplasm for hybrid seed production in maize. The double cross hybrid scheme of hybrid varieties was quite popular and was extensively used in maize. But with the improvement in the performance of the inbred lines of maize, the use of single cross hybrids for commercial cultivation began in 1960s. The single cross hybrids have now become the most extensively used hybrid varieties in maize. In almost all other crop species, single cross hybrids are exclusively used. In India, the hybrid varieties were first developed in maize; four hybrid varieties were released for commercial cultivation in 1961. This was followed by release of hybrid varieties of sorghum in 1964 and pearl millet in 1965. Subsequently, hybrid varieties were released for commercial cultivation in several other crops species, including some self-pollinated crop species like rice (Table 5.3).

5.4.1. Development of Hybrid Varieties

The commercial use of hybrid varieties can be considered to comprise two separate groups of activities, viz., (i) the development of hybrid varieties, and (ii) the production of commercial quantities of seeds of the hybrid varieties. The development of hybrid varieties is an elaborate and demanding procedure consisting of the following two activities: (i) production of inbred lines, and (ii) evaluation of the inbred lines to identify the outstanding inbreds that can be used as parents of the hybrid varieties. In addition, the features of inbreds have to be continuously improved since the performance of hybrid varieties can be enhanced only by improving the parental inbreds or by isolating new superior inbreds. These procedures are discussed in some detail in the following sections.

Table 5.3: A list of the first hybrid varieties of different crop species to be released for commercial cultivation in India

Crop	Year of release	Hybrid variety	Type of hybrid	Hybrid seed production based on**
Maize	1961	Ganga 101, Ganga 1, Deccan, Ranjit	Double cross*	CGMS
Sorghum	1964	CSH1	Single cross	CGMS
Pearl millet	1965	HB1	Single cross	CGMS
Castor	1968	GCH3	Single cross	Pistillateness
Cotton	1970	H4	Single cross	Manual
Tomato	1973		Single cross	Manual
Sunflower	1980		Single cross	CGMS
Pigeon pea	1991		Single cross	GMS
Rapeseed (*Brassica napus*)	1994		Single cross	CGMS
Rice	1996		Single cross	CGMS

* Single cross hybrids are the currently favored hybrids.

** CGMS, cytoplasmic-genetic male sterility; GMS, genetic male sterility.

5.4.2. Production of Inbred Lines

The inbreds can be derived from open-pollinated, synthetic, composite or hybrid varieties. They can be developed by one of the following methods: (i) continued inbreeding with selection, and (ii) chromosome doubling of haploid plants. The development of inbreds by continued inbreeding should be based on selfing in all such species where it is possible. The procedure for inbred development in maize is briefly outlined here. The genetically variable source population from which the inbreds are isolated is designated as the **S_0 generation**. Several phenotypically superior S_0 plants are selected and self-pollinated. In the S_1 generation, individual plant progenies are grown, and in each progeny 30-40 plants are space-planted. Plants with the best characteristics are selected from the best progenies, and the selected plants are self-pollinated. The steps of the S_1

generation are repeated up to the S_5 generation (Fig. 5.9). With the advancing generation of selfing, the variation within individual plant progenies would rapidly decline. As a result, the individual plant progenies would become increasingly homogeneous, and selection in the S_5 generation would primarily be among the progenies. In the S_6 generation, each individual plant progeny will be an inbred line.

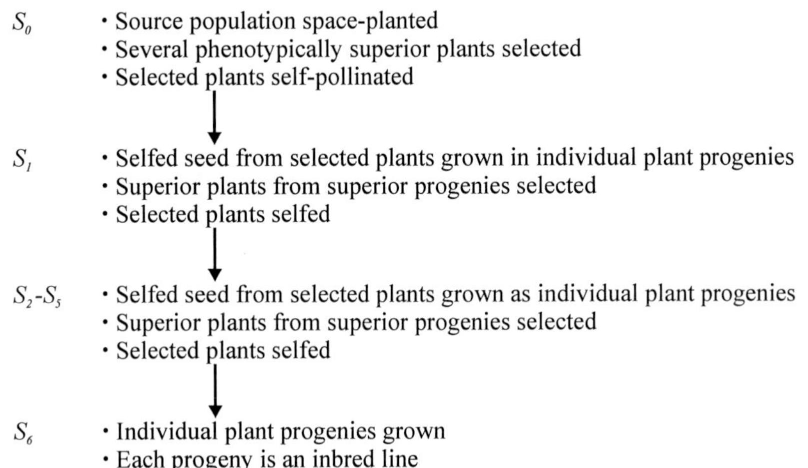

Fig. 5.9. Inbred isolation by continued inbreeding. Selfing should be used wherever it is feasible

A strict selection is done during the inbreeding process; this selection will mainly focus on easily observable traits and would be effective for highly heritable characters. Weak and inferior plants as well as progenies are carefully and strictly removed. It has been suggested that selection for GCA may be carried out especially during the early phase of inbreeding, i.e., during S_0 to S_2 generations (Fig. 5.10). This strategy has been called **early testing**. For early testing, the selected plants in S_0 generation are self-pollinated as well as crossed, as males, with a broad genetic base tester, which is used as the female parent. In the next crop season, the testcross progenies (progenies from the crosses between the selected plants and the broad genetic base tester) are evaluated in a replicated trial, and superior progenies are identified. The selfed seeds from the plants that produced the superior testcross progenies are planted as individual plant progenies. The procedures of the S_0 generation may be repeated in the S_1 and S_2 generations. In the S_3 and the later generations, selfing and selection is done as described in the case of the procedure outlined above without early generation testing. Early testing has been widely used especially in forage and clonal crops for the development of hybrid varieties. Most breeders, however, delay early testing, to most often, the S_3 generation. Early testing allows the breeders to discard the progenies with poor testcross performance early in the inbreeding process. This reduces the workload of the breeders, and permits them to concentrate on superior progenies that are likely to yield superior inbreds. Thus, inbred development by continued selfing takes about seven generations. But in the case of early testing the time taken will be somewhat longer, e.g., eight to 10 generations, depending on the number of cycles of selection for GCA.

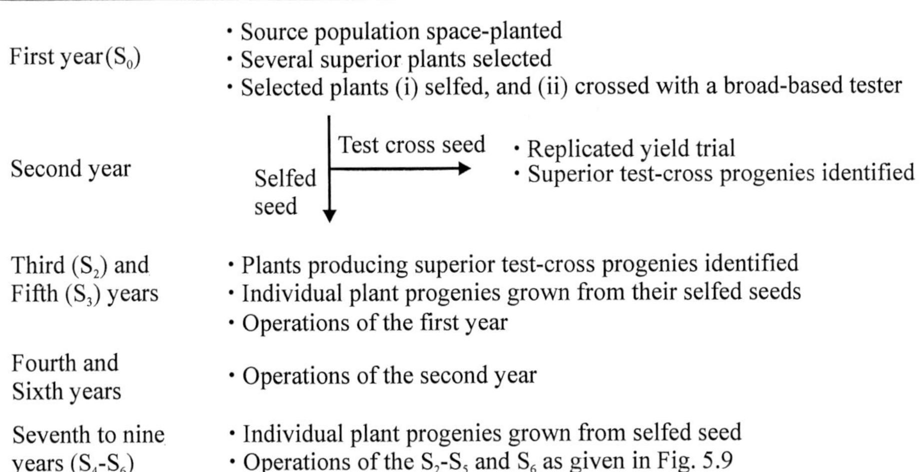

Fig. 5.10. Inbred development by continued inbreeding combined with early testing. The broad genetic base tester selects for GCA

The inbred development can be achieved in merely three generations by producing haploid plants from the heterozygous plants, and doubling the chromosome number of these haploid plants. The haploid plants can be produced by anther culture (Fig. 5.11A), unfertilized ovary culture or pollination with some inducer strains. In case of maize, the inducer strains like RWS and RWK-76 lead to the production of 8-10% haploid plants of the female parent (Fig. 5.11B). Generally, the chromosome number of haploid plants is doubled by the application of a suitable concentration of colchicine. This alkaloid is well known for blocking the spindle apparatus development during mitosis leading to chromosome doubling in the dividing cells. The plants produced from chromosome doubling of haploids are called **doubled haploid (DH) plants**. Seeds from each DH plants are harvested separately and planted as individual plant progenies in the next crop season. Each progeny represents an inbred, and will be completely homogeneous.

5.4.3. Evaluation of Inbred Lines

The inbreds produced by any one of the above methods need to be evaluated for identifying those inbreds that will produce excellent hybrid varieties. It may be pointed out that only about <0.1% of the inbreds give rise to outstanding hybrids that can be profitably used for commercial cultivation. The identification of outstanding inbreds, which constitute a very small proportion of the total inbreds produced in a breeding programme, is the most important and the most expensive operation. In view of the large number of inbreds involved in the evaluation process, the evaluation procedure comprises the following three steps: (i) phenotypic evaluation, (ii) top-cross test, and (iii) single cross test. In addition, when double cross hybrids have to be developed, the single cross performance data are used for the prediction of double cross performance. The salient features of these steps are briefly described in the following paragraphs and are summarized in Table 5.4.

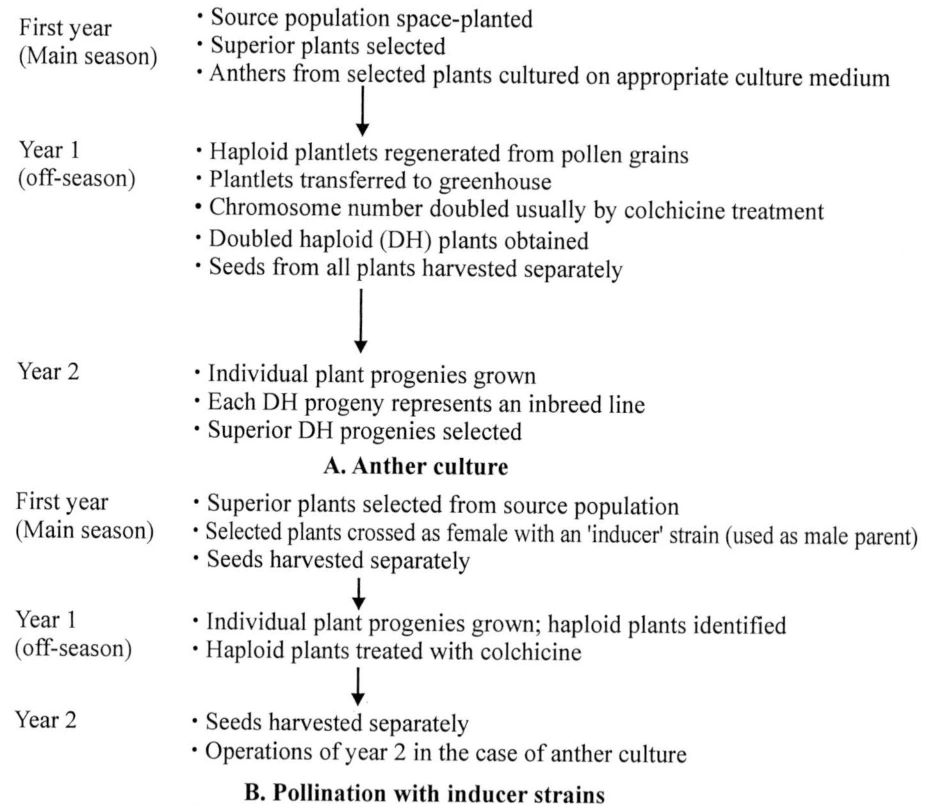

Fig. 5.11. Inbred development by chromosome doubling of haploid plants. A. Haploid plants produced by anther culture. B. 'Inducer' strains are used as male parents in crosses with selected plants to produce the haploid plants.

The first step consists of **phenotypic evaluation** of the inbreds developed in the breeding programme. The inbreds are used for conducting a replicated trial, and their average performance for the relevant traits, including yield is determined. The inbreds with inferior performance are rejected. Phenotypic evaluation is highly effective for traits with high heritability. It is also effective for complex traits like yield to various degrees. For example, selection of the inbreds for yield improves the yields of their hybrids to some extent: the inbred yield shows a small positive correlation ($r = 0.2$) with the yields of hybrids.

In the next step, the inbreds are used for a top-cross test to evaluate them for GCA. A **top-cross** is a cross between an inbred and a broad genetic base tester, e.g., an open-pollinated, a synthetic or a composite variety. The top-cross test provides a reliable estimate of GCA of the inbreds involved in the test. The top-cross progenies produced from all the inbreds are evaluated in a replicated trial; preferably, these trials should be conducted over two or more locations and more than one year. On the basis of the performance of top-cross progenies, about 50% of the inbreds are eliminated. This reduces the number of inbreds that will have to be evaluated in the next step, i.e., the single cross test.

Table 5.4: The salient features of the procedure of evaluation of inbreds for production of hybrid varieties

Evaluation step	Objective	Procedure	Remarks
Phenotypic evaluation	Elimination of inferior performing inbreds so as to reduce their number for top-cross test	The inbreds are evaluated in a replicated trial for the relevant traits including yield	The yield of parental inbreds shows a small positive correlation (r = 0.2) with the yield of their hybrids
Top-cross test	Identification of inbreds with superior GCA	The inbreds are crossed with a broad genetic base tester; testcross progenies evaluated in a replicated trial	Approximately 50% of the inbreds are eliminated; the number of inbreds used for single-cross test is reduced
Single-cross test	Identification of outstanding single cross combinations	Inbreds crossed in all possible single cross combinations; single crosses evaluated in a replicated trial	The final and most demanding step; the outstanding single cross identified and used as a new hybrid variety
Prediction of the best double cross combination	Identification of the best double cross from a set of selected inbred lines	Performance of double cross predicted from those of the nonparental single crosses among the four parental inbreds	The double cross performance is predicted from the data of the previous step; no new data needs to be generated

The single cross evaluation is the final step in the procedure for inbred evaluation, and it allows the estimation of SCA. It is also the most critical step in the development of hybrid varieties since SCA is believed to be the most important component of heterosis. In the **single cross test**, the inbreds remaining after the top-cross test are crossed in all possible combinations to produce single cross progenies from them. The number of all possible single crosses from n inbreds would be n (n -1) in case the reciprocal crosses are included. But if the reciprocal crossed were excluded, the number of crosses will be reduced to half, i.e., $[n(n - 1)] / 2$. The single cross performance is evaluated in a replicated trial that is preferably conducted over locations and years. The best performing single crosses are identified, and they can be released as new single cross hybrid varieties.

The single cross performance can also be used to predict the performance of the double cross hybrids produced from them. The number of double crosses produced from a given number of inbreds is much larger than the number of different single crosses produced from them. For example, the numbers of different single and double cross hybrids from 20 inbred will be merely 190 and 14,535, respectively. Therefore, prediction of double cross performance from single cross data drastically reduces the amount of work needed to obtain these estimates. A set of four inbreds, e.g., A, B, C, and D, will generate six different single crosses, viz., A x B, A x C, A x D, B x C, B x D, and C x D. These four inbreds can be crossed to produce the double cross hybrid (A x B) x (C x D). The yield of this double cross hybrid is predicted to be the average yield of the four single crosses not involved in the production of the double cross hybrid, that is, the single crosses A x C, A x D, B x C, and B x D. Surprisingly, this prediction of double cross performance is quite accurate and agrees well with the actual performance.

5.4.4. Improvement of Inbred Lines

Once outstanding inbreds are identified, they can be improved for various traits so that their hybrids are also improved. The inbred improvement is an important component of hybrid variety development programmes for the following reasons. (i) The isolation and evaluation of inbreds to identify outstanding inbreds is costly and time consuming process. In addition, (ii) the frequency of outstanding inbreds among the newly isolated inbreds is quite low. Therefore, the improvement of various features of inbred lines is highly attractive objective and it can be achieved in the following ways.

1. The backcross procedure may be used to transfer one or more genes from suitable donors into an outstanding inbred to correct some specific weaknesses of this inbred. The weakness may be susceptibility to a disease, an insect pest, etc. It is important that the trait should have high heritability; preferably, it should have oligogenic control. This is the second most popular method of inbred improvement.
2. In addition, genetic engineering can be used to transfer specific transgenes to generate useful features in the inbreds. The hybrids produced from these inbreds will have these useful and often novel features (Chapters 10-11). In some cases, somaclonal variation can be used to improve specific characteristics of inbreds.

3. The pedigree scheme of breeding self-pollinated crops has been used to improve the performance of inbred lines. This approach has been the most widely used in hybrid variety development programmes. In this strategy, two inbreds with desirable and, preferably, complementing features are crossed. The segregating generations of this cross are handled according to the pedigree scheme, and inbred lines with superior features are isolated. The inbreds isolated from open-pollinated populations are called **first cycle inbreds**. In contrast, the inbreds derived from crosses between first cycle inbreds are known as **second cycle inbreds**. Similarly, the inbreds isolated from crosses between second cycle inbreds are called **third cycle inbreds** and so on. In general, second cycle inbreds are superior to first cycle inbreds and so on.
4. The combining ability of inbreds can be improved by a scheme called **gamete selection**. This scheme has not been widely used.
5. Often the CMS (cytoplasmic male sterile) lines used as the female parents of hybrids need to be improved for features specified by either their plasmagenes or nuclear genes.

   5a. In case the weakness of a CMS line is governed by its nuclear genes, one of the following approaches may be used to improve this line. The male sterile cytoplasm may be transferred into a line that lacks this weakness. The weakness can be corrected by transferring the relevant gene from a donor into the maintainer line of the concerned CMS line. Alternatively, the weakness of the maintainer line is corrected by either mutagenesis or isolation of suitable somaclonal variants. Further, a suitable transgene may be introduced by genetic transformation into the maintainer line to remove its defect. The improved maintainer line by any of the above four strategies is then used to maintain the concerned CMS line; this automatically leads to the removal of the defect of CMS line.

   5b. In case the undesirable feature of a CMS line is due to its cytoplasm, i.e., plasmagnes (the genes present in the mitochondrial or chloroplast genomes), somatic hybridization can be used to correct the defects due to plasmagenes. The somatic hybrids having their mitochondria from the CMS line and chloroplasts from the other fusion parent can be selected to remove the defect due to the chloroplast genome. Further, a CMS line could be isolated from among the somatic hybrid plants that has recombinant mitochondrial genome; this line might lack the defect associated with the mitochondrial genome of the CMS line. Often the defect may be associated with the CMS trait, and a more desirable CMS source may have to be discovered. New CMS sources might be obtained by searching the germplasm mutant cytoplasms, induction of plasma gene mutations, and by transferring cytoplasms from related species either by backcross method of breeding or by somatic hybridization.

5.4.5. Production of Hybrid Seed

The seeds of hybrid varieties have to be produced every year by making crosses between their parental inbred lines. For this purpose, two to four rows of the inbred to be used as the female parent are planted in alternation with two rows of the inbred to be used as the male parent. The female parent is emasculated to prevent self-polli-nation. These female parent plants are pollinated by pollen from the male parent. At maturity, seeds from only the female rows are harvested as hybrid seed. Thus, there are two main requirements for the large scale production of hybrid seed, viz., (i) easy, cheap and effective emasculation of the female parents of the hybrids, and (ii) assured and adequate pollination of the flowers of the female parents by the pollen grains produced by the concerned male parents. The pollination efficiency depends mainly on the floral biology of the species, which affects pollen dispersal from the male parent and pollination of the flowers of the female parent. In case of cross-pollinated species, both pollen dispersal and pollination are adequate, but they may present problems in self-pollinated species. Hybrid seed production may be based on one the following strategies: (i) hand emasculation and pollination, (ii) cytoplasmic-genetic male sterility (CGMS), (iii) cytoplasmic male sterility (CMS), (iv) genetic male sterility (GMS). (v) self-incompatibility (SI), and (vi) chemically-induced male sterility.

In the beginning, **hand emasculation** was the only available method of hybrid seed production. The female rows of maize had to be detasselled. Detasselling is the, removal of tassels (male inflorescences) of the plants by hand, before pollen-shed. The detasselled female plants were pollinated by the male inbreds. Hand emasculation was used to produce hybrid seed of cotton in India. In case of cotton, the flowers of the female parent are emasculated by hand and are pollinated by hand by collecting flowers from the male parent plants. The cost of hybrid seed produced by this approach is high, and the amount of hybrid seed produced will be limited. Therefore, this method will be economically viable in such crops, where hybrid crop fetches high economic returns, seed set per fruit is relatively larger, and the need for hybrid seed per unit area is relatively smaller.

The **cytoplasmic-genetic male sterility system** is the most extensively used strategy for hybrid seed production. It is being used in several crops like maize, Pearl millet, sorghum, sunflower, rice, onion, pigeon pea, etc. This system employs three different lines for hybrid seed production. These lines are CMS (male sterile) line or A line (used as the female parent), the **maintainer line** or **B line** (used for propagation of the A line), and the **restorer line** or **R line** (used as the male parent of the hybrid). Therefore, this is often called the **three-line system of hybrid seed production**. This system has been used to produce seeds of both single and double cross hybrid varieties of maize. But in the case of other crops single cross hybrids are produced. For production of **single cross hybrid seed** in maize, four rows of the A line are planted in alternation with two rows of the R line. The plants of A line are pollinated by pollen grains from the plants of R line. Therefore, the seed produced on the A line plants is hybrid seed. The hybrid plants will be male fertile since they will be heterozygous for the restorer gene (Fig. 5.12).

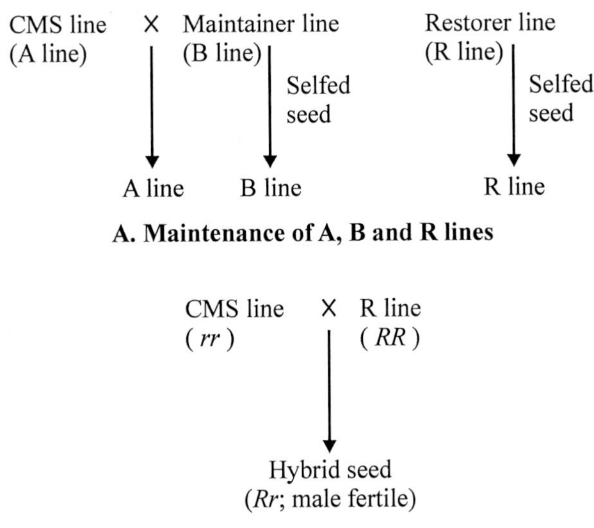

Fig. 5.12. Production of seed of a single cross maize hybrid using eytoplasmic male sterility. CMS, cytoplasmic male sterile.

The scheme for the production of the double cross hybrid seed is presented in Figure 5.13. It involves the production of seeds of the two single crosses as the first step. These single crosses are then crossed to produce the double cross hybrid seed. For example, let us suppose that the inbreds 1, 2, 3, and 4 combine well with each other to produce the outstanding double cross (1 x 2) x (3 x 4). The inbred 1 may be used as a CMS line, and it is planted in isolation with the inbred 2, which is a non-restorer line (genotype, *rr*). The seed produced on inbred 1 will be the seed of single cross 1 x 2, and this single cross will be male sterile (Fig. 5.13). Similarly, the inbred 3 may be used as a CMS line, and it is planted in isolation with the inbred 4, which is a restorer line (genotype, *RR*). The seed produced on inbred 3 will be the single cross 3 x 4; this single cross will be male fertile. In the next crop season, the two single crosses 1 x 2 and 3 x 4 are planted in isolation in the ratio 6 rows to 2 rows, respectively. The seed produced on the single cross (1 x 2) will be the double cross seed. However, half of the plants of the double cross hybrid will be male sterile (Fig. 5.13); this may cause some problem under certain environments. This can be avoided by using the inbreds 3 and 4 as restorer inbreds (genotype, *RR*). The inbred 3, in this scheme, will have to be manually detasselled for producing the single cross 3 x 4 seed. But all the plants of the double cross hybrid will be male fertile.

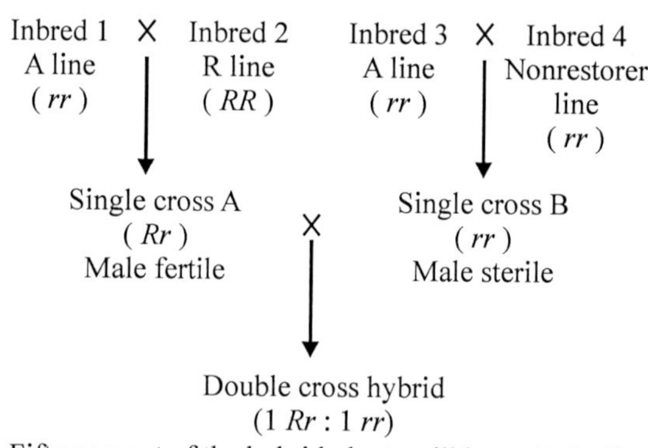

Fig. 5.13. A schematic representation of the method for seed production of a double cross hybrid. The inbreds 1, 2, 3 and 4 combine well with each other to produce an outstanding double cross hybrid.

In some cases, a restorer system for the CMS line may not be available; this is known a **cytoplasmic male sterility**. The CMS line can be crossed with any male fertile line that combines well with the CMS line to produce a single cross hybrid seed in the same way as depicted in Fig. 5.12B. Since the hybrid plants will be sterile, this system cannot be used in grain crops. In the case of **genetic male sterility (GMS) system**, the hybrid seed is produced by crossing the male sterile line with any male fertile line that combines will with the GMS line. This hybrid will be fully male fertile. In this system, multiplication of the GMS line is based on crossing it with heterozygous (*Ms ms*) male fertile plants. Therefore, 50% of the progeny plants in each generation are male fertile (*Ms ms*). These male fertile plants have to be identified and eliminated before they shed pollen. Ingenious strategies had been developed to eliminate the male fertile plants. But this system has been used only in few crops, e.g., castor in USA and pigeon pea in India.

In some species, the GMS may be either photoperiod sensitive (PGMS) or temperature sensitive (TGMS). In such a system, the male sterile line is fully male fertile under specific environments, while it is completely male sterile under another set of environmental conditions. For example, a rice TGMS line is male fertile when the temperature is less than 28°C, but it is completely male sterile when the temperature is above 30°C (Fig. 5.14). This TGMS line is multiplied in a location where the temperatures remain below 28°C during the critical period of the growth. The hybrid seed, on the other hand, is produced in an area where the temperatures are above 30°C during the critical growth period. The TGMS line will be male sterile in this area, and it can be pollinated by any male fertile line that combines well this line. This system of hybrid seed production is the most efficient of all the systems, and requires only two lines, viz., the TGMS line and the male fertile line, for hybrid seed production. Therefore, this system is referred to as the **two-line system of hybrid seed production**. In contrast, the CGMS system requires three lines (the A, B, and R lines), and is called the **three-line system of hybrid seed production.**

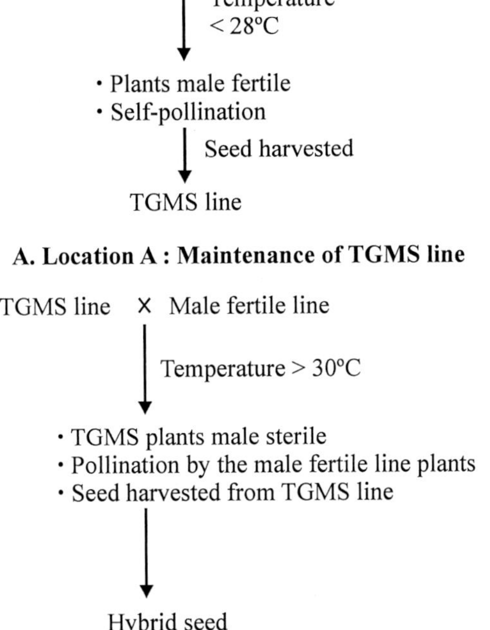

Fig. 5.14. The use of TGMS lines of rice (e.g., line 54605) for hybrid seed production. TGMS line is multiplied at a location where this line is male fertile. Hybrid seed is produced at a different location, at which the plants of this line are male sterile due to the prevailing temperature. TGMS, temperature-sensitive genetic male sterile.

Self-incompatibility is used for hybrid seed production in some Brassica crops like Brussel's sprouts and cabbage in Japan and Europe. In one approach, two self-incompatible lines that are cross-compatible with each other are planted in alternate rows. In this case, seeds produced in both the lines will be hybrid seed. In another scheme, a self-compatible line and a self-compatible line may be planted in alternate rows. In this case, the seed produced only in the self-incompatible line will be hybrid seed. The use of self-incompatibility for hybrid seed production presents some problems, including the production and maintenance of the self-incompatible inbred lines.

Male sterility is induced by several chemicals when they are applied to the plants at specific developmental stages; these chemicals are **chemical hybridizing agents (CHAs)**. When plants are treated appropriately with a suitable CHA, their pollen grains become nonfunctional; this constitutes **chemical emasculation**. These plants are essentially male sterile, and this condition is called **chemically induced male sterility**. The CHA treated line serves as the female parent of the hybrid (Fig. 5.15). Some of the CHAs show high effectiveness, and they are being used for hybrid seed production in crops like rice and maize. Some examples of CHAs are zinc methyl arsenate (MG1), sodium methyl arsenate (MG2), ethephon or ethrel, RH0007 (Hybrex) and WL84811 (Shell).

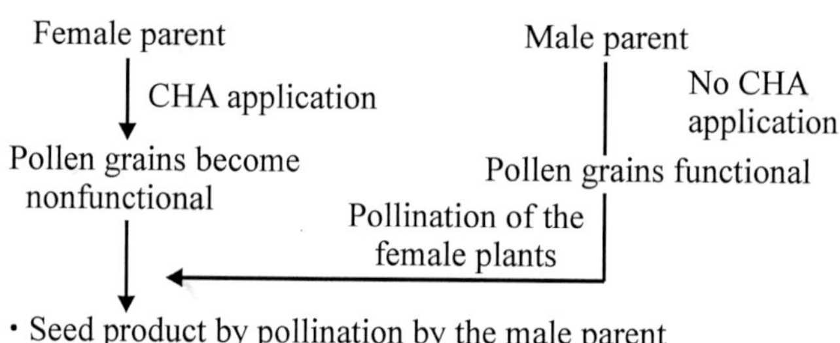

Fig. 5.15 Schematic representation of hybrid seed production by application of CHAs on the plants of the female parent. The seed produced by the CHA-treated female parent plants will be hybrid seed. CHA, chemical hybridizing agent.

5.4.6. Maintenance of Hybrid Varieties

The hybrid varieties are, as a rule, maintained in the form of their parental inbreds. The inbreds themselves are maintained by strict self-pollination or close inbreeding (in case self-pollination cannot be done). These varieties cannot be maintained by using either the open-pollinated of selfed seed produced from them. This is because the plants of these varieties are highly heterozygous. As a result, there will be extensive segregation and recombination leading to formation of a large number of different genotypes in the progeny. Therefore, the genotypes of progeny from hybrid varieties will be considerably different from those of concerned hybrid varieties. For this reason, these varieties are maintained as their parental inbreds. Since the inbreds are almost homozygous, their selfed progeny will have the same genotype as the concerned inbreds generation after generation. As a result, the hybrid varieties produced by crossing them will have the same genotype as the original hybrid variety.

5.4.7. Merits of Hybrid Varieties

1. Hybrid varieties make the best use of heterosis as they utilize both GCA and SCA components of heterosis.
2. Hybrid varieties are the only means of utilization of heterosis in self-pollinated crops where the use of synthetic, composite or open-pollinated varieties is not possible.
3. Since the hybrid varieties are maintained in the form of parental inbreds, their genetic composition does not change due to selection, inbreeding, random drift, etc. from one generation to the next.
4. The hybrid varieties of self-pollinated crops give up to 30% higher yields than the pureline varieties. Therefore, these varieties are becoming popular among farmers even when the cost of their seed is much higher than that of the latter.
5. The produce from hybrid varieties is much more uniform than those of the other types of varieties of cross-pollinated crops. This characteristic is particularly relevant in the case of single cross hybrids.

5.4.8. Limitations of Hybrid Varieties

1. The farmers have to buy fresh hybrid seed every year. This is because the seed produced by hybrid varieties shows genetic variation due to segregation and recombination. As a result, this seed cannot be used for raising the next crop since the performance of this crop will be inferior to that of the hybrid variety.
2. The production of hybrid seed is demanding and requires considerable technical skill. As a result, the cost of hybrid seed is much higher that that of other types of varieties.
3. The large scale production of hybrid seed requires a suitable and feasible means of pollination control (Section 5.4.5). Such pollination control methods may not be available in many crop species. This makes commercial utilization of heterosis in these crops quite difficult.
4. The exploitation of full potential of hybrid varieties requires adequate resources like irrigation water, fertilizer, efficient disease and insect pest management, etc. Many farmers, especially those with small holdings may not be able to provide for these resources.
5. In case the two parents of a hybrid have considerably different grain quality, the produce from the hybrid variety will have variable grain

quality. This problem was encountered in the case of initial hybrid varieties of rice developed in India. However, this problem has been resolved in the present-day rice hybrids.

5.4.9. Achievements with Hybrid Varieties

Hybrid varieties have been developed and used in a number of crop species throughout the world. These varieties were first developed in maize, and subsequently used in many other crop species. In India, hybrid varieties are being used in maize, sorghum, Pearl millet, cotton, pigeon pea, rice, castor, etc. In case of maize, four double cross hybrids, namely, Ganga 1, Ganga 101, Ranjit, and Deccan, were the first to be released in 1961. These hybrids yielded 25-40% higher than the open-pollinated varieties, and became very popular. Maize improvement has focused mainly on hybrid variety development, and by the year 2009 72 hybrid varieties had been released for commercial cultivation (Table 5.5). It would be seen that the bulk of these hybrid varieties are single cross hybrids, although the initial hybrid varieties were double cross hybrids. Eight of the hybrid varieties have the *opaque2* gene, as a result of which they have improved lysine content in their seed proteins. These quality protein hybrids are HQPM7, Vivek-9QPM, HQPM5, Shaktiman 3, Shaktiman 4, HQPM1, Shaktiman 2, and Shaktiman 1.

Table 5.5: The total number of hybrid varieties released up to 2003 for commercial cultivation in India in some selected crop species

Crop	Number of hybrid varieties	Remarks
Pearl millet	107	CMS-based
Sorghum	53	Includes 3 hybrids for rabi season, and 38 private sector hybrids
Maize	72 (44 single cross, 17 double cross, 11 triple cross)	Up to 2009; 8 hybrids are *opaque2*-based quality protein hybrids
Cotton	31	Tetraploid cotton intraspecific hybrids
	13	Tetraploid cotton interspecific hybrids
	9	Diploid cotton (both inter- and inter-specific hybrids)
Rice	15	-
Castor	10	-
Pigeon pea	6	GMS-based; since 2005, several CMS-based hybrids released
Safflower	3	GMS-based; includes one private sector hybrid

The success of initial maize double cross hybrids prompted the initiation of hybrid development programmes in sorghum and Pearl millet as well. HB1 was the first Pearl millet hybrid to be released in 1965; subsequently, HB3 and HB5 were released. These hybrids had Tift23A, a CMS line, as their female parent. This CMS line was highly susceptible to downy mildew and ergot, as a result of which the hybrids were also highly susceptible to these diseases. The disease susceptibility of the Tift23A line was governed by its nuclear genes. Backcross breeding and mutation breeding were used to correct these defects of this CMS line, and these improved CMS lines have been used to develop downy mildew and ergot resistant hybrids like PHB10, PHB11, BJ104, BK560, etc. The largest number of hybrid varieties has been released in Pearl millet. The public sector hybrids HB3, BJ104, BK560, MH179, Pusa-23, HHB50 and HHB67 are very popular among the farmers.

The first hybrid variety of sorghum, CSH1, was also released in 1965. Up to 2005, over 50 hybrid varieties had been released for commercial cultivation (Table 5.5). Most of these hybrids are recommended for cultivation during the kharif season, but three of the hybrids (CSH12R, CSH13R, and CSH15R) have developed for the rabi season. The new hybrids like CSH10 and CSH11 can give yields of up to 50 q/ha. Cotton is the only crop species in which interspecific hybrids have been used for commercial cultivation in India. The interspecific hybrids have been produced both in the tetraploid (*G. hirsutum* x *G. barbadense*; Varalaxmi, DCH32, Savitri, Jayalaxmi etc.) as well as diploid (*G. arboreum* x *G. herbaceum*; DH7, DH9, DDH2) levels. In addition, several intraspecific hybrids have also been produced, e.g., H4, H6, H8, H10, Savita, Fateh, Raj HH16, Sugna, etc. hybrid varieties of *G. hirsutum*. Hybrid varieties have been produced in several other crops, including rice, castor, safflower (Table 5.5), coconut, tomato, chillies and several other vegetable crops.

5.5. SYNTHETIC VARIETIES

A **synthetic variety** is produced by crossing in all possible combinations a number of inbred lines, clones, open-pollinated varieties, short-term inbred lines or some other populations, and is maintained by open-pollination in isolation. Thus, the development of synthetic varieties has three steps as follows: (i) evaluation of a large number of different lines for GCA, (ii) construction of the synthetic varieties from several lines having high GCA and (iii) multiplication of the synthetic varieties so constructed. These steps are briefly discussed in the following paragraphs and are depicted in Figure 5.16.

1. The first step in the development of synthetic varieties is to evaluate large number of lines to be used as parental lines for their GCA. The GCA is generally evaluated either by topcross test or polycross test. In a polycross test, the testcross progeny are produced by pollinating each line included in the test with a random sample of pollen from the lines included in the polycross test. In general, the polycross progeny are produced by planting the lines included in the test in isolation and

allowing open-pollination among them. The seed produced in each line would represent polycross seed as it will be produced by natural pollination with the lines included in the test. The performance of the testcross or the polycross progenies is evaluated in a replicated trial. The trial data are used to estimate the GCA of the test lines, and the lines with high GCA are selected for use as parents of the new synthetic varieties. The selected parental lines constitute the Syn_0 generation of the synthetic varieties.

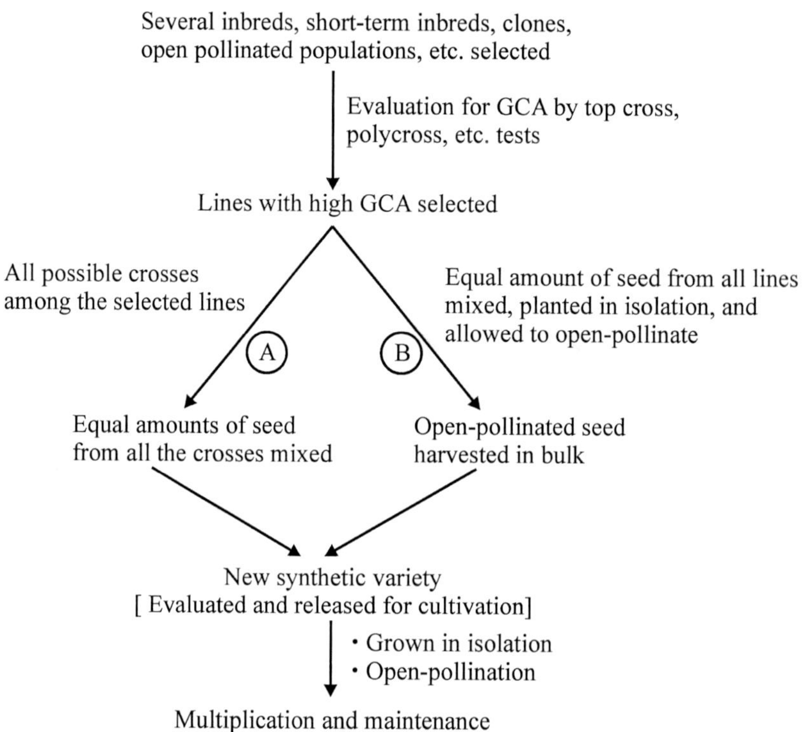

Fig. 5.16. Development and maintenance of synthetic varieties. The schemes A and B yield comparable results. Farmers can save a random sample of the open-pollinated seed produced from synthetic varieties for planting the crop in the next year

2. The selected parental lines are now used to construct a new synthetic variety in one of the following two ways. A simple approach involving little effort is to mix equal amounts of seeds from all the selected parental lines. This seed bulk is planted in isolation, and the plants are allowed to open-pollinate. The seeds from all the plants in the population are harvested in bulk; this seed constitutes the Syn_1 generation. An alternative method that involves considerably more work is as follows.

The parental lines are planted in isolation, and they are crossed in all possible combinations by hand pollination. Equal amounts of seeds from all the crosses are composited to give the Syn_1 generation of the new synthetic variety. The experimental evidence shows that the synthetic varieties created by either of the above two methods are comparable in their performance.

3. The synthetic varieties are maintained and multiplied in open-pollination. The open-pollinated progeny of the Syn_1 **generation** is called Syn_2 **generation** and so on. In general, the Syn_2 or a later generation seeds are used for commercial cultivation. But in case of some crops like sugar beet the Syn_1 generation seed itself is distributed to farmers for commercial cultivation.

5.5.1. The Performance of a Synthetic Variety

In fact, the performance of Syn_1 generation is higher than those of the Syn_2 and the subsequent generations. The yield of Syn_2 generation can be predicted from the performance of the parental lines and that of the Syn_1 generation itself by the following simple formula.

$$\overline{Syn_2} = \overline{Syn_1} - \left[(\overline{Syn_1} - \overline{P}) / N\right]$$

Where, $\overline{Syn_2}$ is the performance of the Syn_2 generation of the synthetic variety, $\overline{Syn_1}$ is the performance of the Syn_1 generation of the synthetic variety, $\overline{P}$ is the average performance of the parental lines of the synthetic variety, and N is the number of parental lines involved. This yield estimate is generally good reflection of the actual yield of the synthetic varieties. Thus the yield of Syn_2 generation of a synthetic variety is lower than that of the Syn_1 generation. But there is no further decline in the performance of the synthetic variety in the subsequent generations. The yield reduction in the Syn_2 generation occurs due to the following reasons: (i) the level of heterozygosity declines in Syn_2 as compared to that in Syn_1 due to random mating, and (ii) segregation and recombination lead to the production of many new gene combinations that may not be as good as those present in the Syn_1 generation.

5.5.2. Improving the Performance of a Synthetic Variety

The above formula (Section 5.5.1) suggests the following approaches for enhancing the yield of a new synthetic variety that is being developed. (i) As the number of parental lines of a synthetic increases, the level of yield reduction in Syn_2 as compared to that of Syn_1 decreases. Therefore, the yield of Syn_2 generation can be improved by increasing the number of parental lines. But the number of lines with good GCA is limited in any given crop species. Therefore, the number of parental lines cannot be increased beyond a point. In any case, the number of parental lines should not be less than four; a smaller number of parental lines will adversely affect the performance of the Syn_2 generation. (ii) The reduction in performance of the Syn_2 generation is lower when the performance of the parental lines themselves is higher. The performance of parental lines can be increased by using short-term inbreds (lines isolated after only one to three selfed generations), second or higher cycle inbreds (inbreds improved for one or more rounds by, usually, the pedigree method), or inbreds isolated from source populations that have been subjected to population improvement programme (several cycles of a suitable selection scheme). In some species, even plants from open-pollinated populations or clones are used as parents of the synthetic varieties. Finally, (iii) an increase in the performance of the Syn_1 generation will lead to improved Syn_2 performance. The Syn_1 performance, however, would depend on the GCA of the parental lines. Since the number of lines with high GCA is likely to be limited, a compromise has to be made between the number of parental lines and the magnitudes of their GCA.

The synthetic varieties contain considerable amounts of genetic variability that can be utilized by a suitable selection scheme to improve the performance of the variety. A recurrent selection scheme is likely to improve the synthetic variety for several generations. Care should, however, be taken to avoid or minimize inbreeding.

5.5.3. Maintenance of a Synthetic Variety

Synthetic varieties are multiplied and maintained by growing them in isolation and allowing the plants to open-pollinate. A random sample of the seed produced by this population is used for cultivation in the next crop season. Theoretically, the genetic constitution of the variety should not change if the population were large, there was random mating and selection were absent. But the available evidence indicates that genetic constitution of the synthetic varieties changes with time. This change most likely occurs due to natural selection, inbreeding, and/or random genetic drift. It is, therefore, desirable to reconstruct the synthetic varieties at regular intervals from their parental lines. In case the parental lines of a synthetic variety are inbred lines or clones, the synthetic variety can be exactly reconstituted whenever desired. This is because the inbred lines and clones can be maintained without any change in their genetic constitution by close inbreeding or selfing and asexual reproduction, respectively. But if the parental lines are open-pollinated populations or short-term inbreds, their genetic constitutions are likely to change with time. In such cases, the synthetic varieties have to be maintained by the seed produced by their own populations grown in isolation and allowed to open-pollinate. Alternatively, the synthetic varieties have to be reconstituted from the parental lines that are available at the given time. It may be pointed out that the parental lines themselves might have undergone changes in their genetic constitution, but this is the best that can be done under the given circumstances.

5.5.4. Composite Varieties

A **composite variety** is produced by mixing together the seeds of several lines with outstanding performance, raising a population from this bulked seed, and allowing the plants to open-pollinate. The open-pollinated seed from this population constitutes the seeds of the new composite variety. The lines used for the production of composite varieties are rarely, if ever, tested for their GCA. Often the parental lines of composites are themselves open-pollinated varieties. The number of parental lines of composites varies considerably, and some composites may have as few as two parents. The composite varieties are maintained by open-pollination in isolation.

5.5.5. Merits of Synthetic and Composite Varieties

1. The synthetic and composite varieties are the only feasible means of utilization of heterosis in such crop species where controlled pollination is not practicable. The controlled pollination is necessary for hybrid variety development as well as hybrid seed production.
2. The farmers do not have to purchase new seed every year as is the case with hybrid varieties. They can save the seeds from their own crops and use them for raising the crop during the next crop season.
3. These varieties are expected to perform better than hybrid varieties under environments that are variable.
4. Their seed production is simpler and less technically demanding than that of hybrid varieties. Therefore, their seed is relatively much cheaper than that of the latter.
5. These varieties have considerable genetic variability. This variability can be exploited by selection to bring about further improvement in the performance of these varieties.

5.5.6. Limitations of Synthetic and Composite Varieties

1. The synthetic and composite varieties utilize only the GCA component of heterosis. Therefore, the performance of these varieties is generally lower than that of hybrid varieties.
2. In case of synthetic varieties, often lines with inferior GCA have to be included as parental lines. This has to be done to keep the number of parental lines reasonably high so that the performance of the Syn_2 generation is not unduly reduced. This, however, reduces the performance of Syn_1 generation, leading to a lower performance of the Syn_2 generation.
3. These varieties cannot be used in self-pollinated crop species.

5.7. ACHIEVEMENTS WITH SYNTHETIC AND COMPOSITE VARIETIES

Synthetic varieties are in common use in such crop species as sugar beet, clovers and other forage crops. In India, several maize composite varieties have been released for cultivation. The first composite varieties to be released in 1968 were Amber, Jawahar, Kisan, Vikram, Sona and Vijay. Subsequently, several other maize composites were released; some of these varieties are, Moti, Chandan 1, Chandan 3, Agaiti 76, Co1, NLD, Renuka, Kanchan, Diara 3, etc. Many of the composite varieties yield up to 90% of the best commercial hybrids. The composite varieties like Shakti, Ratan and Protina contain the *opaque2* gene; as a result, the lysine contents of the seed proteins of these composite varieties are almost twice as much as those of the normal coposites/hybrids. In addition, one composite variety of toria (*Brassica campestris*) has been released. This variety was developed by mixing the seeds of 10 superior parental lines and harvesting the open-pollinated seed from the population raised from this bulked seed.

5.8. CONCLUSION

The performance of populations of cross-pollinated crops can be improved by selection, which is quite effective for traits with high heritability. The mass selection and progeny selection schemes have been suitably modified to effectively improve even complex traits like yield. The recurrent selection schemes are the most elaborate of the progeny selection schemes and are widely used to improve the performance of cross-pollinated crops. The improved populations can be used either as open-pollinated varieties or as source populations for the isolation of inbred lines. The inbred lines, in turn, can be used either to produce hybrid varieties or synthetic varieties. Hybrid varieties are the best means of utilizing heterosis, and are preferred wherever possible. The main limitation of hybrid varieties is the need for seed replacement every year. As a result of this, large quantities of hybrid seed have to be produced every year. This requires a dependable means of easy emasculation of the female parent of the hybrid. CGMS has been the most widely used system of genetic emasculation of the female parents of hybrid varieties, but this system is based on three different lines. In contrast, the TGMS and PGMS systems are based on only two lines and are technically less demanding. But in these systems, two separate locations with the specified temperature and/or photoperiod regimes are required. A more convenient system of genetic emasculation has been created by producing a suitable transgenic line that is normally male fertile. But this line becomes male sterile when it is sprayed with a specific chemical, e.g., a growth regulator. This male sterility system also requires only two lines, needs only a chemical spray for effective emasculation of the female line, and any male line can be used as the male parent of the hybrid. But in many species, controlled pollination may not be practicable. In such cases, synthetic and composite varieties are used to utilize the portion of heterosis that is due to GCA.

QUESTIONS

1. Define cross-pollinated crops. How do breeding strategies for these crops differ from those for self-pollinated crops?
2. Define progeny selection. Briefly describe the ear-to-row method and its modifications, and discuss their merits and limitation.
3. Define recurrent selection. Briefly describe the scheme for recurrent selection for GCA (or SCA) or reciprocal recurrent selection and discuss its merits and limitations.
4. What is the most effective approach for utilization of heterosis? Briefly describe the development of hybrid varieties.
5. Describe the various approaches for hybrid seed production and discuss their chief advantages and limitations.
6. Define synthetic variety. Briefly describe the development of synthetic varieties and discuss their advantages and limitations.
7. Briefly describe the development of inbred lines and evaluation of these lines for hybrid variety development.
8. Write short notes on the following: (i) population improvement, (ii) progeny selection schemes, (iii) recurrent selection, (iv) recurrent selection for GCA, (v) recurrent selection for SCA, (vi) reciprocal recurrent selection, (vii) hybrid varieties, (viii) synthetic varieties, (ix) inbred lines, (x) production of inbred lines, (xi) evaluation of inbred lines, (xii) improvement of inbred lines.

RELEVANT LITERATURE

1. Stalker HT and Murphy JP (eds) 1991. Plant Breeding in the 1990s. C.A.B. International, Wallingford, UK.
2. Banga SS and Banga SK 1998, Hybrid cultivar development. Narosa Publishing House, New Delhi.
3. Hayes HK 1963. A professors Story of Hybrid Corn. Burges, Minneapolis, USA.
4. Kalloo G, Rai M, Singh M and Kumar S (eds) 2006. Heterosis in Crop Plants. Researchco Book Centre, New Delhi.
5. Kaul J, Dass S, Dhekhar JC and Bharadwaj M 2009. Maize Hybrid and Composite Varieties Released in India, Vol. 2. Directorate of Maize Research (ICAR) New Delhi.

■■■■

6 Breeding for Biotic Stress Resistance

6.1. INTRODUCTION

It is well recognized that the performance of a line depends on its genotype (*G*), the prevailing environment (*E*) and the interaction between these two factors (*G* x *E* interaction). In simple terms, the **environment** for a given plant or organism would include all the factors other than the concerned plant or organism. Thus the environment would include both biological (biotic) and non-biological (abiotic) factors. The full expression of the genotypic potential of a plant or line would require the optimum environmental conditions. The **optimum environment** would be the one in which none of the biotic and abiotic factors have an adverse effect on the performance of the given plant or line. When an environmental factor has an adverse influence on the performance of a plant or line, this situation is called **stress**. A stress caused by biological environmental factor(s) is known as **biotic stress**. In contrast, the stress imposed by non-biological factor(s) of the environment is termed as **abiotic stress**. The two main biotic stresses are generated by the various plant diseases, and the insect pests. There are several different abiotic stresses, of which drought, temperature, and mineral stresses are the most important. In this chapter, we shall examine the nature of chief biotic stresses and discuss the breeding strategies devised so far for their management and alleviation. The major abiotic stresses and the breeding strategies for developing varieties resistant to them are discussed in the next chapter (Chapter 7).

6.2. BREEDING FOR DISEASE RESISTANCE

In simple words, **disease** may be defined as an abnormal condition in plants generated by some other organisms (called pathogens) or some environmental factor(s). The definition of disease by Agrios (1970) is more inclusive and is as follows: **Disease** is a series of invisible and visible responses of plant cells and tissues to a pathogenic microorganism or an environmental factor that result in adverse changes in form, function or integrity of the plant and may lead to partial impairment or death of the plant or its parts. Thus abnormal states in plants resulting from both biological and non-biological factors are regarded as disease.

However, the term disease is used in a rather restricted sense in this chapter and indeed throughout this book: the term **disease** is limited to the abnormal conditions produced in plants by microorganism, such as fungi, bacteria, viruses, mycoplasma/phytoplasma, viroids and nematodes. The microorganism that produces a disease is called **pathogen**, while the affected plant is termed as **host**. The discussion in this chapter is based mainly on the knowledge gained from diseases caused by fungal/bacterial pathogens. This is because these pathogens cause the greatest damage to the crop plants, followed by bacteria, viruses and nematodes in the given order. Plant diseases reduce the total biomass production, production of the economically important part of the crop plant, and the quality of produce and may even make the produce hazardous to human health.

Some general features of plant diseases are as follows. (i) The extent of damage caused by the different pathogens varies from one crop to the other. (ii) Even in a single crop, the importance of a given disease changes with time mainly due to the breeding activities. (iii) Many excellent varieties of crops had to be withdrawn because they became susceptible to some disease. (iv) In fact, most disease resistant varieties become susceptible to the disease after some time. For example, the wheat varieties resistant to rust become susceptible in, on an average, about 4-5 years. (v) But some varieties resistant to certain diseases have not become susceptible to the disease even after a long period of time. Therefore, the durability of resistance seems to be remarkably different for the different host-pathogen systems.

6.2.1. Disease Development

A disease progresses from the landing of the pathogen onto the host tissue (**contact**), through entry of the pathogen into the host tissue (**infection**), proliferation of the pathogen (**establishment**) to the appearance of the typical symptoms of the disease (**development**). The amount of disease produced in the host depends on the following three factors: (i) the pathogen (representing the sum total of virulence genes, abundance of inoculum, etc.), (ii) the host plant (sum total of resistance genes and other factors favoring susceptibility), and (iii) the environment (sum total of the environmental factors affecting disease development) (Fig. 6.1). The host genotype has a strong influence on the establishment and the disease development phases: some cases of host resistance prevent establishment, while others interfere with symptom development and spore production by the fungal pathogens. In contrast, the pathogen genotype determines whether it will be able to produce disease symptoms in the given strain of the host plant species. The environmental factors also affect disease development especially during the infection stage. Many disease susceptible plants may not develop disease simply because the pathogen inoculum (spores, cells, etc.) could not reach them or the infection was prevented by adverse environmental conditions. Such disease susceptible plants that appear disease resistant are called **escapes**. The escapes are a nuisance in programmes designed to breed for disease resistance. Therefore, every effort is made to minimize their occurrence in the breeding populations.

6.2.2. Physiological Races and Pathotypes

The ability of a pathogen to produce a disease is called **virulence**. In general, virulence is governed by several genes, which are known as **avirulence** (*avr*) genes. The dominant allele at an avirulence locus determines avirulence, i.e., inability of the pathogen to produce the disease, while the recessive allele specifies virulence. The discovery and identification of virulence genes depends on the availability of host lines having a number of different resistance genes for the concerned pathogen. The host lines used for the identification of avirulence genes present in a pathogen strain are called **host testers** or **differential hosts**. Often the lines making up a set of differential hosts may have one or more genes for resistance to the concerned disease. When such a set of differential hosts is used to classify the strains of a pathogen into distinct groups, these groups are called physiological races. The **physiological races** of a pathogen differ from each other in their ability to attack the host lines making up the set of differential hosts. It is important that often the identity of the avirulence genes present in the physiological races are not known. Physiological race classification is used in diseases like wheat stem, leaf and yellow rusts, and some other diseases.

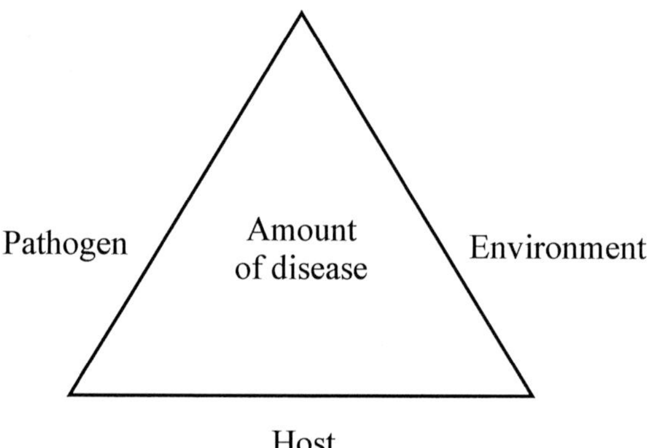

Fig. 6.1. The amount of disease produced in a host plant depends not only on the host itself, but also on the pathogen and the prevailing environment.

In the case of **ideal differential hosts**, each line of the set has a single distinct gene for resistance to the given disease. The pathogen strains classified on the basis of their virulence against the ideal differential hosts are known as **pathotypes**. Since the resistance gene present in each of the ideal differential hosts of the set is known, the avirulence genes present in the pathotypes are also known. Pathotype classification is commonly used in the case of late blight of potato caused by *Phytophthora infestans.* Pathotype differentiation is common in all pathogens as well as insect pests. Pathogens have a variety of efficient mechanisms for generation of new pathotypes. These mechanisms include mutation, segregation and recombination, heterokaryosis, and parasexual reproduction. **Mutation** is the ultimate source of all the genetic variation present in the biological world. However, the frequency of mutations is very low and assumes significance only over a long period of time. **Segregation and recombination** are integral to sexual reproduction, and are the basis of the bulk of pathotype variability. Heterokaryosis is the presence of two genetically different nuclei in a single cell. This phenomenon occurs in some fungi, and produced new races of fungal pathogens. In parasexual reproduction, the two nuclei of a heterokaryotic cell fuse, and this is followed by production of haploid nuclei due to mitotic irregularities. This may lead to the formation of new gene combinations and the generation of new races of fungal pathogens.

6.2.3. Disease Resistance

Disease resistance is the ability of a host line to prevent a pathogen from producing disease. This ability of hosts was clearly shown by Biffen in 1905 with reference to the yellow rust of wheat; he also showed that the wheat resistance to yellow rust was due to a single recessive gene. Barrus in 1911 demonstrated that different isolates of a pathogen differ in their ability to attack different varieties of the host crop. This finding is the basis for identification of physiological races of the pathogens (Section 6.2.2). The extensive studies on the responses of plants against pathogens have established that plants possess an elaborate defense system against the various types of pathogens. The defense systems of plants can be classified into two main groups as follows: (i) basal defense mechanisms and (ii) defense based on specific resistance (*R*) genes. The **basal defense** system is the first line of defense against a wide range of pathogens. This system of defense is activated by a group of compounds derived from plant cell walls. These compounds are generated by the hydrolytic enzymes produced by the pathogens infecting the host plants. In addition, some highly conserved molecules of microbial origin, *e.g.* chitins, glucans, lipopolysaccharides and flagellins, also activate the basal defense system. These molecules are generally referred to as pathogen-associated molecular patterns (PAMPs)/microbe-associated molecular patterns (MAMPs)/danger-associated molecular patterns (DAMPs). It is noteworthy that all microbes possess PAMPs. Therefore, both pathogenic and nonpathogenic microbes are capable of activating the basal resistance response in plants. The pathogenic microbes however, have also developed a group of specific proteins designed to suppress the basal defense responses of their respective host plants. These proteins are referred to as **effector proteins**. Similarly, in response to this strategy developed by the pathogenic microbes, the host plants have evolved a group of proteins that function as specific receptors for the pathogenic effector proteins. These receptor proteins of host origin are known as **resistance proteins** (R proteins). The R proteins interact with the effector proteins in a highly specific manner. This interaction initiates a series of responses that ultimately lead to the hypersensitive response. This type of host response to pathogenic microorganisms is called ***R*-gene-mediated defense**. In this chapter, the discussion focuses primarily on the *R*-gene mediated disease resistance.

It is now well established that the *R*-gene-mediated host response to a pathogenic microbe can be grouped into four different categories as follows: (i) susceptible, (ii) immune, (iii) resistant, and (iv) tolerant responses. In case of **susceptible response**, the host genotype and physiological conditions do not interfere with the disease development and pathogen reproduction. As a result, there is profuse development of disease symptoms and spore production (in the case of fungal pathogens). **Immune response** is the other extreme of host reaction to a pathogen: in this case, the host plants do not show any symptom of the disease. Generally immune response is the consequence of **hypersensitive reaction**, in which a group of cells surrounding the point of infection by the pathogen dies in response to the infection. This type of reaction is found in the case of biotrophic pathogen, which survive only in the living tissues of their hosts. As a result, the pathogen infecting the host tissue also dies along with the host cells that die in the wake of hypersensitive response.

In cases of **resistance response**, the pathogen infects the host, establishes within the host tissues, produces disease symptoms, and is able to reproduce. But the host genotype and the resultant physiological conditions within the host interfere with both disease symptom development and reproduction of the pathogen. As a result, both the level of disease symptom and the rate of reproduction are lower in magnitude than those in the case of susceptible response. In the case of **tolerance response**, both the magnitude of disease symptom and the rate of pathogen reproduction are comparable to those in the susceptible lines of the host. However, the extent of reduction in yield or biomass in this type of response is much lower than that in the susceptible response. Tolerance response is known to occur in certain cases, but it is very difficult to determine or demonstrate. *Plant breeders, however, use the term tolerance to describe moderate levels of disease resistance.* Further, whenever yield is used as the basis for selection for disease resistance, tolerance will be automatically selected for.

The mechanism of diseased resistance is well known in some cases, and it can be classified into the following three groups: (i) mechanical barrier, (ii) hypersensitive reaction, and (iii) nutrient restriction. In some cases, the resistance to a pathogen is the result of certain anatomical or mechanical features of the host plants. For example, some varieties of wheat and barley have closed flowering habit, i.e., their flowers do not open. In such varieties the spores of ovary infecting fungi like smuts cannot reach the stigma. As a result, these varieties are resistant to these pathogens. The resistance against obligate parasites is mediated by hypersensitive response; the molecular basis of hypersensitive response is considered in some detail in Section 6.2.5 (gene-for-gene relationship). It has been proposed that the resistance response to pathogens might involve a restriction on the availability of nutrition required for optimum proliferation and reproduction of the pathogen.

6.2.4. Genetics of Disease Resistance

The inheritance of disease resistance may be governed by oligogenes (or major genes), polygenes or plasmagenes (or the cytoplasm). In many cases, the disease resistance is governed by one or many major genes; this type of resistance is called **oligogenic resistance**. Generally, resistance is dominant over susceptible response, but recessive gene governed resistance is also known in several cases. The disease response of plants can be readily classified into clear-cut resistant and susceptible groups. This type of resistance is race-specific and is often termed as **vertical resistance**. When a variety shows resistance to some races of the given pathogen, but is susceptible to some other races, it is known as **race-specific resistance**. The race-specific resistance response is due to a gene-for-gene relationship between the host plant resistance (*R*) genes and the concerned *avr* genes of the pathogen (Section 6.2.5). In addition, oligogenic resistance is associated with hypersensitive response. As a result, the disease reaction of the plants is of only two distinct types: (i) disease symptoms present (susceptible response), and (ii) disease symptoms absent (resistant response). Thus, the inheritance of this type of disease resistance is qualitative in nature. The various salient features of the two chief types of resistance are listed in Table 6.1.

The resistance response to a disease may be governed by several polygenes, each of which has a small effect and the effects of all the genes controlling the resistance are cumulative. The polygenes generally show additive gene action, but non-additive gene effects are also common. In addition, the disease response is markedly affected by the prevailing environment. As a result, this type of resistance does not allow the plants to be classified into clear-cut resistant and susceptible groups. In contrast, the disease response of the plants or lines forms a continuous curve as in the cases of other polygenically governed traits. This type of resistance is associated with slow growth to the pathogen within the host tissues, and reduced rate of spore production by the pathogen. There is little evidence for race-specific resistance response, i.e., the resistance is race-nonspecific. In case of race-nonspecific resistance, a given host variety shows comparable levels of resistance to all the races of the pathogen. Polygenic resistance is often referred to as **horizontal resistance**. The concepts of vertical and horizontal resistance were proposed by Van der Plank. Polygenic resistance may be assessed as percent of the area exhibiting disease symptoms, percent of the area covered by the pathogen, and severity of the disease symptoms. In almost all cases of host-pathogen systems, there is evidence for the presence of polygenic resistance.

Table 6.1. The salient features of vertical (= oligogenic) and horizontal (= polygenic) disease resistance

Feature	Vertical resistance	Horizontal resistance
Genetic control	Oligogenic	Polygenic
Inheritance pattern	Qualitative	Quantitative
Host reaction	Generally, hypersensitive	As a rule, resistance response
Race-specificity	Race-specific	Race-nonspecific
Expressed during	Seedling stage to maturity	The level of expression increases with plant age
Scoring of disease reaction	Easy	Difficult
Selection	Easy	Difficult
Durability of resistance	Rarely durable	Durable
Risk of disease epidemics	Always present	Essentially absent
Useful for the hosts that are	Annuals	Both annuals and perennials
Most useful in cases of pathogens	Pathogens that are immobile, e.g., soil pathogens	All types of pathogens
Specific deployment (in time and/or space) of resistant varieties	Critical for successful use in case of mobile pathogens like air-born fungi	Not required
Specific breeding strategy to increase usefulness	Gene pyramiding*	QTL pyramiding*
Other control measures	Needed	Might be needed in few cases
Efficiency of resistance	Highly efficient against specific races of the pathogen, but ineffective against the other races	Effective against all the races, but the level of effectiveness varies; generally, lower than that for vertical resistance

* Different resistance genes for the same disease are introgressed into a single variety to increase the durability of resistance. Similarly, different QTLs specifying to the same disease are introgressed into a single variety to increase the level resistance.

The relevant features of the oligogenic and polygenic resistance are summarized in Table 6.1. The vertical resistance is easy to score and, as a result, easy to select for in the breeding programmes. In contrast, horizontal resistance is difficult to score as well select for in the segregating generations of crosses. The vertical resistance is race-specific so that it is effective only so long as the race virulent against the concerned *R* gene has not evolved. Once this race develops, the resistance is broken down and the susceptible response would be produced. The horizontal resistance, on the other hand, is not race-specific and remains effective against all the races of the pathogen so that it is highly durable. The relative importance of horizontal and vertical disease resistance depends mainly on the host-pathogen systems. For example, breeders exclusively used vertical resistance against wheat stem rust, but were forced to deploy horizontal resistance against potato late blight. However, the wheat breeders at CIMMYT, Mexico have now initiated a huge genomic selection programme for the exploitation of horizontal resistance to manage the stem rust of wheat. Thus the interest in the deployment of horizontal disease resistance is increasing in view of its durability and broad range protection. This is also aided by the development of molecular marker technology (Chapters 13-15), which allows relatively easy selection for polygenic traits (in fact, selection for quantitative trait loci, QTLs).

There are a small number of cases, where disease resistance shows cytoplasmic inheritance. In such cases, the resistance is governed by plasmagenes that are located in the organellar genomes. A good example of cytoplasmic inheritance of disease resistance is provided by the resistance of maize to *Helminthosporium/Bipolaris* leaf blight or southern leaf blight. The gene governing resistance/susceptibility in this case is located in the mitochondrial genome. Further, the susceptible reaction to the disease is associated with CMS-T, the Texas male sterility cytoplasm of maize.

6.2.5. Gene-for- Gene Relationship

The gene-for-gene hypothesis was proposed by Flor in 1955 based on his findings on rust resistance in flax (*Linum usitatissimum*). This hypothesis can be explained, in simple terms, as follows. A host plant has *R* genes that specify resistance to a given pathogen that produces a disease in this host. Generally, the dominant alleles of these genes specify resistance to the pathogen, while their recessive alleles lead to susceptible response. The pathogen, in turn, has similar number of avirulence (*avr*) genes; the recessive alleles of these genes specify virulence, while their dominant alleles generate avirulence. For each *R* gene of the host, the pathogen has a corresponding *avr* gene. A host *R* gene, say, R_1, produces resistance to a pathogen race that has the dominant allele of the corresponding *avr* gene, e.g., Avr_1 in this case. But a pathogen race having the recessive allele of this *avr* gene, viz., avr_1, will overcome the resistance due to R_1 gene and will lead to susceptible response. In addition, a race with the Avr_1 allele will produce susceptible response in a host line homozygous for the r_1 allele, i.e., has the $r_1 r_1$ genotype (Fig. 6.2). The gene-for-gene relationship has been found to be valid for almost all host-pathogen systems that have been investigated in detail.

		Host	
		R_1R_1	r_1r_1
Pathogen*	*Avr*[1]	Resistant	Susceptible
	avr[1]	Susceptible	Susceptible

Fig. 6.2. The gene-for-gene relationship for *R* genes of host and *avr* genes of pathogen. The relationship between a single *R* gene of host and the corresponding *avr* gene of the pathogen is depicted here. Resistance is dominant over susceptibility, while avirulence is dominant over virulence. The pathogen in generally haploid; as a result, only one allele of *avr* gene is present.

When a host line has two or more different genes specifying resistance to a given disease, the pathogen race must have the virulence alleles of all the corresponding *avr* genes to be able to produce disease in this host line (Fig. 6.3). As a result, the chances of evolution of a race capable of attacking a host line having multiple *R* genes is much lower than that of a race virulent to a single *R* gene. Therefore, two or more genes may be introgressed into a single variety of the host crop; this is called **gene pyramiding**. Gene pyramiding is generally achieved by the backcross method of breeding. Further, it is greatly facilitated by the use of molecular makers linked to the different resistance genes. The chief objective of gene pyramiding is to create simultaneous resistance to several races of the pathogen. This also reduces the risk of epidemics caused by the concerned pathogen. Thus gene pyramiding in case of oligogenic resistance tends to mimic some of the properties of horizontal resistance.

Another feature of vertical resistance is that some resistance genes are strong while others are weak resistance genes. The strong resistance genes tend to remain effective for a longer period of time than do weak resistance genes. Therefore, **strong resistance genes** are useful in crop improvement, while **weak resistance genes** are of little use. For example, the gene *Sr6* for resistance to wheat stem rust is a strong resistance gene. It may be pointed out that a resistance gene is strong or weak is not determined by its own intrinsic property. In fact, it depends on the competitive ability of the pathogen race virulent to a given resistance gene in the absence of this resistance gene from the host population. The pathogen race virulent to a strong resistance gene is a poor competitor against the other races of the pathogen when this strong resistance gene is not present in the host population. In contrast, the pathogen race virulent to a weak resistance gene is a strong competitor against the other races of the pathogen when this strong resistance gene is absent from the host population.

Pathogen race	Host line: $R_1R_1\,R_2R_2$	$R_1R_1\,r_2r_2$	$r_1r_1\,R_2R_2$	$r_1r_1\,r_2r_2$
$Avr_1\,Avr_2$	R*	R	R	S
$Avr_1\,avr_2$	R	R	S	S
$avr_1\,Avr_2$	R	S	R	S
$avr_1\,avr_2$	S	S	S	S

Fig. 6.3. Gene-for-gene relationship between a host plant and its pathogen. Two *R* genes (*R1* and *R2*) and the corresponding *avr* genes (*avr1* and *avr2*) are shown here. In this case, 7 of the 16 different host line pathogen race combinations show resistance response, while only one combination out of four combinations was resistant when only one *R* gene was considered (Fig. 6.2). Thus, pyramiding of *R* genes increases the range of pathogen races, to which the host variety will be resistant. More importantly, the host line with both *R1* and *R2* genes is resistant to three out of the four races of the pathogen (with respect to the two *Avr* genes). In contrast, the host line with one *R* gene was susceptible to one out of two races with respect to the corresponding *Avr* gene (Fig. 6.2).

There are some notable deviations from the ideal gene-for-gene relationship in case of some host-pathogen systems. For example, two different genes for virulence are reported to be specific to a single host gene for resistance in the cases of flax rust and barley smut. On the other hand, a single *R* gene may specify resistance to two different pathogens. For example, wheat gene *Sr20* (for stem rust resistance) is, as far as is known, identical to the wheat gene *Lr15* (for leaf rust resistance). Similarly, wheat gene *Sr24* seems to be the same as wheat gene *Lr19.* The basis for this type of situation might be that the same or similar metabolic products confer resistance to the two pathogens. As a result, the product generated by *Sr20* serves the same purpose as that of the gene *Lr15.* But this situation could also be the result of a very tight linkage between the genes *Sr20* and *Lr15*, and between *Sr24* and *Lr19.* In several cases of resistance, multiple alleles of *R* genes have been reported. But in some such cases, tight linkage between two or more *R* genes was shown to give rise to the multiple alleles, i.e., the multiple alleles were, in fact, pseudoalleles.

6.2.6. Molecular Basis for Gene-for Gene Relationship

The gene-for-gene relationship involves specific interaction between the products of the host plant *R* genes and the pathogen *avr* genes. In general, the protein encoded by an *R* gene recognizes and interacts with the product of the corresponding Avr gene. This interaction is highly specific and leads to the activation of an array of genes involved in the hypersensitive response (Fig. 6.4A). Thus, an interaction between the *R* gene product and the corresponding *Avr* gene product leads to resistant reaction in the host to the concerned pathogen. It starts as a signaling cascade leading to the production of reactive oxygen species. It also activated defense genes, which results in phytoalexin biosynthesis, reinforcement of host cell walls, and liberation of hydrolytic enzymes and/or other inhibitory proteins by the host cells. These changes ultimately lead to rapid cell death. **Phytoalexins** are low molecular weight compounds that are either polyphenolic or terpenoid in nature and are produced by plant cells. In view of the above, hypersensitive response is often regarded as **programmed cell death** or **apoptosis**. But when the R protein is unable to recognize the Avr protein, the hypersensitive response is not initiated, and the host would show susceptible response to the pathogen. This would happen when the host has the recessive *r* allele of the resistance gene, the pathogen has the *avr* allele of the virulence gene, or both host and the pathogen have the mutant alleles of the resistance and virulence genes, respectively (Fig. 6.1). The tomato-*Cladosporium fulvum* host-pathogen system provides an excellent example of this type of interaction. The Avr9 protein is encoded by the pathogen gene Avr9. Protein Avr9 is recognized by the *Cf-9* R protein specified by the tomato gene *Cf*-9; the two proteins interact in a highly specific manner and generate resistance response.

But is some cases, the R protein has not been found to directly interact with the Avr protein. The Avr protein, however, does interact with another protein (other than the R protein) produced by the host plant; this host protein has been termed as **guardee protein**. This interaction between the Avr protein and the guardee protein somehow leads to the activation of hypersensitive response. This observation has led to the proposal of the **guard hypothesis**. According to this hypothesis, the R protein functions as guard of the host protein (the guardee protein) that is targeted by the Avr protein. When the Avr protein interacts with the guardee protein, it either degrades it or brings about a change in the conformation of the guardee protein. These changes in the guardee protein are detected by the R protein, which then initiates the hypersensitive response (Fig. 6.4B). The *Arabidopsis thaliana* and the *Pseudomonas syringae* pv. Tomato host-pathogen system serves as good example for the guard hypothesis. Two *Arabidopsis* genes, *RPM1* and *RPS2*, together confer resistance to this bacterial pathogen. The Avr protein produced by this pathogen does not interact with either the RPM1 or the RPS2 protein. It, however, acts on another host protein RIN4, and induces changes in the conformation of RIN4. These changes are detected by the R proteins RPM1 and RPS2, and a hypersensitive response is initiated. In this example, the RIN4 is the guardee protein, while RPM1 and RPS2 function as the guard proteins. These two types of resistance mechanisms operate in the cases of obligate parasites or biotrophic pathogens.

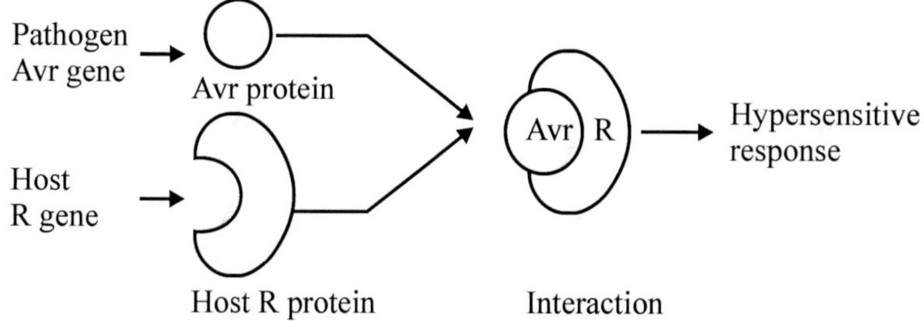

A. Interaction between Avr and R proteins

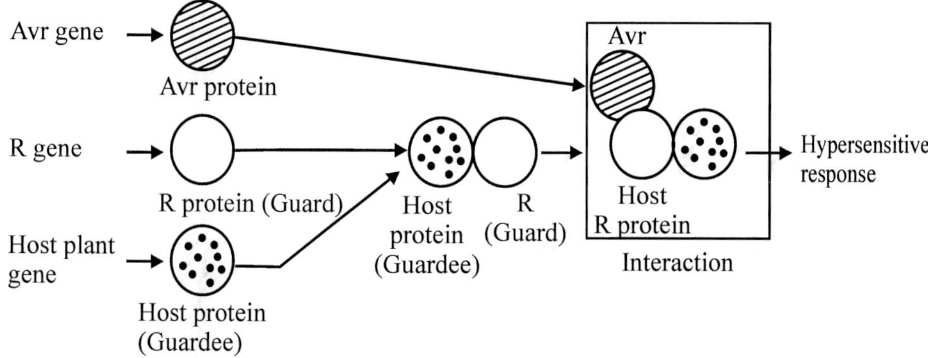

B. Interaction between Avr protein and a host protein other than the R protein

Fig. 6.4. The molecular basis of hypersensitive response and gene-for-gene relationship. (A) R protein interacts directly with the Avr protein. This interaction initiates hypersensitive response. (B) The Avr protein does not interact with the R protein directly. But it interacts with another host protein called the guardee protein. This interaction changes the guardee protein. The R protein detects this change and initiates hypersensitive response.

6.2.7. The Boom and Bust Cycle

A newly released crop variety, say, variety A, possessing vertical or oligogenic resistance to a disease like rust is expected to be virtually free from the disease. Therefore, the variety A will rapidly become popular among the farmers, and there would be rapid increase in the area on which this variety is cultivated. This situation represents the **boom phase** of the variety A as it quickly gains popularity with the associated increase in its area of cultivation. The biological systems of the pathogens allow the evolution of new races that are virulent against different resistant varieties of the host species. This would also lead to the evolution of the pathogen race virulent against this new variety A as well. Initially, this virulent race will be in very low frequency. But the virulent race will easily infect the plants of variety A and multiply and reproduce freely to increase in frequency. Since the variety A is being cultivated on a large area, the inoculum of the pathogen race virulent on the variety A will rapidly increase, and would ultimately cause an epidemic. **Epidemic** may be simply defined as a severe outbreak of a disease starting with a low level of initial infection. As a consequence, the variety A that was popular so far would no longer be favored by the farmers for cultivation. This will lead to a sharp decline in the area on which the variety A is cultivated, and ultimately variety A would be withdrawn from cultivation. This represents the **bust phase** for variety A (Fig. 6.5). Thus, a variety with vertical resistance for a disease invariably experiences the boom phase, followed by the bust phase. For example, the wheat variety Kalyan Sona had vertical resistance for leaf rust and became very popular with the farmers. But after several years, it became susceptible to the disease and had to be replaced with other rust resistant varieties.

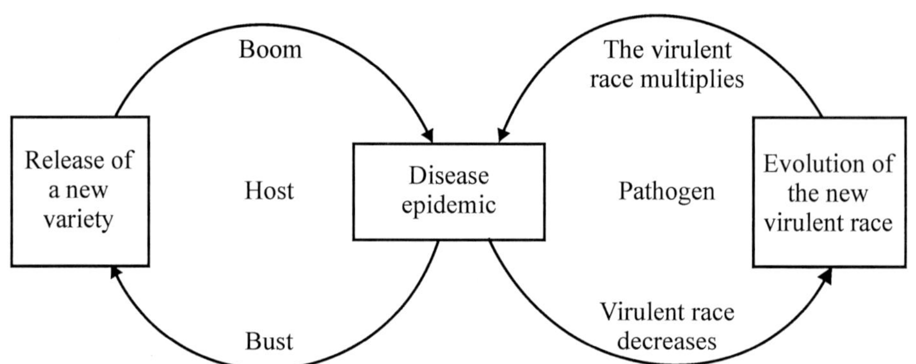

Fig. 6.5. The boom and bust cycle. When a new host variety with vertical resistance is released, its area increases rapidly. This is called boom phase. Soon the pathogen race virulent on this variety evolves and multiplies. Ultinately, it produces disease epidemic in the new variety. As a result, area under this variety declines rapidly (this is the bust phase).

After a variety (the variety A above) with vertical resistance becomes susceptible to the disease, a new variety with vertical resistance (say, variety B) would be released. The variety B will also experience a boom phase, followed by the bust phase. The boom and bust cycles are a common feature of the vertical disease resistance. This cycle is the consequence of the nature of vertical resistance coupled with the genetic systems of the pathogens. The vertical resistance is effective so long the *R* allele of the resistance gene present in the host is not matched by the pathogen *avr* allele of the avirulence gene. Once the pathogen race with the corresponding *avr* gene is evolved, the resistance due to the *R* gene is broken down. In addition, when breeding programme is designed to select for vertical resistance, the genes conferring horizontal component of resistance to the host cannot be selected for. As a result, the level of horizontal resistance inevitably declines in the varieties developed by the breeding programme. Therefore, once the virulent race is produced, it multiplies on the widely grown variety that has now become susceptible to this race. The virulent race might already be present in very low frequency in the pathogen population. This low frequency is the result of the selection pressure exerted by the host varieties being widely cultivated before the new variety became popular. The resistance genes present in those varieties will allow multiplication of only those pathogen races that have appropriate virulence genes. Other races of the pathogen will survive on other varieties of the host, which can be infected by them. The increased inoculum of this virulent race leads to disease epidemic, and the popular variety rapidly phases out of cultivation.

6.2.8. Systemic Acquired Resistance

When a plant is infected by a pathogen, it shows increased resistance to the subsequent pathogen infections; this is called **systemic acquired resistance** (*SAR*). The SAR involves the spread of defense response, initiated by the pathogen infection in the affected tissue, from this tissue to the other parts of the plants. The SAR is generally induced by those pathogens that produce necrotic or hypersensitive response. It is usually long lasting, and is often effective against viral bacterial and fungal pathogens. SAR is associated with increased levels of salicylic acid, which may be involved in signal transduction leading to *SAR*. The pathogen infection activates the expression of a set of plant genes, which are often termed as *SAR* genes. It appears that different sets of genes comprise the SAR genes in monocot and dicot crops. In general, the SAR genes are a group of proteins that produce pathogenesis related (PR) proteins. The PR proteins play active role in disease resistance. SAR can be activated by certain chemicals like salicylic acid, INA (2,6-dichloroisonicotinic acid), BTH [benzo (1,2,3)-thiadiazole-7-carbothiolic acid-S-methyl ester], etc. These chemicals are effective in inducing SAR in both dicot and monocot crops. BTH is so far the most promising SAR inducing chemical. For example, a single spray of 30g/ha BTH is able to protect a wheat crop from powdery mildew for the entire crop season. Similarly, the protein 'Harpin' produced by the bacterium *Erwinia amylovora* generates SAR response in plants. The 'Harpin' induced SAR is effective against bacteria, fungi, viruses, insects, mites and nematodes. The Harpin-based biopesticide is being marketed as **Messenger**. Harpin is being produced on commercial scale by expressing the Harpin encoding gene from *E. amylovora* in *Escherichia coli.*

6.2.9. Breeding Strategies for Disease Resistance

The breeding strategy for the development of disease resistant varieties depends primarily on the source of disease resistance, and the genetic control of disease resistance, i.e., oligogenic, polygenic or cytoplasmic genetic control. The desired resistance may be available in one or more of the following: (i) an improved variety, (ii) a germplasm accession of the concerned crop species, (iii) a related species, (iv) mutagenesis, (v) somaclonal variation, and (vi) an unrelated organism. The oligogenic resistance is easily scored and readily used in breeding programmes for the development of new varieties. The resistance present in an improved variety is the most easily utilized. In case such a variety is available in another country, it could be introduced and released for commercial cultivation. This variety will be evaluated in the area where it is to be released, and it may also be subjected to selection, if so required (plant introduction). There are many successful examples of introduction of disease resistant varieties in India, e.g., rust resistant wheat varieties Kalyan Sona, Sonalika, etc. from CIMMYT, Mexico. Sometimes selection within cultivated varieties may be able to identify disease resistant genotypes. This approach would be useful in such varieties of self-pollinated and clonal crops that have become genetically variable. It would also be useful in the varieties of cross-pollinated crops. Some successful examples of selection of disease resistant varieties from cultivated varieties are as follows: potato variety Kufri Red from the cultivated variety Darjeeling Red Round, Pusa Sawani okra from a germplasm collection from farmer's field, etc.

But often the resistance would be present in an inferior variety, unadapted germplasm line of the crop species or in a related species of the crop. In such cases, the resistant line has to be used in a hybridization programme with the elite germplasm of the crop in order to develop a superior improved disease resistant variety. **Hybridization** is the most important method of breeding used for the development of disease resistant varieties. One of the parent's used for hybridization is a popular improved variety of the concerned region, and the other parent has high degree of resistance to the disease. In case the parent used as the source of resistance is an improved variety that might also contribute to some other desirable features, the segregating generations from the cross are handled according to the **pedigree method** of breeding. The scheme of pedigree method remains essentially the same as described in Section 4.4.1, except for the following activity. The segregating generations are subjected to the appropriate disease tests to enable a reliable selection of the disease resistant plant/lines. It is expected that the new disease resistant variety will inherit from the resistant parent some desirable features in addition to disease resistance.

Often the disease resistance may be present in an agronomically inferior line or germplasm accession, or in a related species. In such cases, the **backcross method of breeding** (Section 4.8) is used for the development of a new disease resistant variety. In the backcross programme, the disease resistant parent is used as the nonrecurrent parent, while a popular improved variety is used as the recurrent parent. The new disease resistant variety developed by this programme will be the same as the recurrent parent, but with the improved feature of disease resistance. This method is the only approach available for transfer of disease resistance from the related species of the crops.

When a source of resistance to the disease is either not known or is not available to the breeder, **mutagenesis** may be used to recover disease resistant mutant lines. This approach would be especially useful in the case of clonal crops, where hybridization may present problems or may not be desirable in view of the highly heterozygous nature of the popular clones. There are several successful examples of the isolation of useful disease resistant mutants. The occurrence of genetic variation in the plants regenerated from tissue culture (**somaclonal variation**) also offers opportunities to isolate disease resistant somaclonal variants. There are some successful examples, e.g., Fiji disease resistant somaclone of sugarcane, bacterial wilt resistant somaclone of tomato, etc. However, this approach has not been widely used.

The development of **recombinant DNA technology** (Chapter 9) has made it possible to create transgenic plants with novel features as specified by the researcher (Chapters 10, 11). Appropriate genes from other organisms, including the plant pathogens themselves, have been transferred and expressed in crop plants to create transgenic disease resistant plants (Chapter 11). For example, the coat protein of virus pathogens has been transferred into crop plants to produce virus resistant varieties that are already in commercial cultivation. This approach has been used to develop insect resistant and drought resistant varieties of certain crop species. This seems to be emerging as the technology of choice for achieving these objectives.

Polygenic resistance presents problems both in scoring and selection since it is affected to a considerable degree by the environment. This type of resistance is quantitative in nature and is governed by **quantitative trait loci** (QTLs). Polygenic resistance can be measured only in the absence of oligogenic resistance. Therefore, the parents for the hybridization programmes are selected on the basis of the levels of their polygenic resistance. This can be assessed by the levels of their resistance to such races of the pathogen, which are able to overcome their oligogenic resistance. Further, in the segregating generations of the cross between of the selected parents, all the plants with oligogenic resistance are removed and the remaining plants are evaluated for polygenic resistance. It may be pointed out that this selection will be more effective when the disease resistance is evaluated on the basis of a replicated trial. Therefore, generally the resistance response of individual plant progenies is used as the basis of selection. Polygenic resistance is being used for the development of disease resistant varieties in the case of several host-pathogen systems like potato-late blight pathogen, wheat-yellow or stripe rust pathogen and so on. But the development of super-races is forcing the breeders to breed for polygenic resistance even in host-pathogen systems like wheat-stem rust pathogen. In these cases, oligogenic resistance was regarded as the resistance of choice so far. Breeding for polygenic resistance is greatly facilitated by the use of molecular markers.

6.2.10. Disease Tests

An effective breeding for disease resistance requires the plants in the segregating generations to be reliably assessed for their disease responses; this procedure is called **disease test**. This would allow the identification of such plants that are disease resistant and distinguish them from those that are disease susceptible. An essential requirement of the evaluation for disease response is to minimize or even eliminate the risk of presence of disease escape plants in the population. Therefore, the plant population is exposed to a suitable level of the pathogen inoculum under appropriate environmental conditions. In addition, the plants should be in the optimum developmental stage for the disease development. Specific procedures for disease test have been developed for all the important diseases of the crop plants. The procedures of disease test depend mainly on the nature of the pathogen, and the mode of its transmission to new host plants. The general outline of disease tests for some selected pathogens is briefly described below.

1. Many plant diseases like wilts, root rots, seedling rots, etc. are **soil-born**, i.e., the concerned pathogens are present in the soil from where they infect the seedlings emerging from the seeds planted in these soils. In case of such diseases, sick plots are created. A **sick plot** is a field in which a high level of inoculum of the concerned pathogen is present in the soil. As a result, all the seedlings/plants grown in such plots are uniformly infected by the pathogen.
2. In case of **air-borne** diseases like rusts, smuts, etc. the pathogens spread from plant-to-plant by means of spores dispersed by wind. The disease test for diseases might use infector rows for spreading the pathogen spores. An infector is such a variety of the crop that is highly susceptible to the pathogen so that the pathogen multiplies in the plants of this variety without any restriction. The infector is planted after every few rows of the test plants. Alternatively, the test plants may be inoculated with a suspension of pathogen mycelial fragments or spores.
3. Some diseases are, however, **seed-born**, i.e., the pathogen inoculum is present on/in the seeds of the crops. Examples of such diseases are wheat bunt, some smuts, etc. in the cases of such diseases; the seeds to be used for disease test are inoculated with the pathogen before they are planted in the field.
4. Most plant viruses are **insect-transmitted**. The insects that serve as vehicles for the transmission of virus pathogens are called **insect vectors**. The important insect vectors for viruses infecting crop plants are aphids, white fly, leafhoppers, etc. One commonly used method of infecting test plants with an insect-transmitted virus is as follows. The insect vectors of the virus are raised on plants infected by the virus.

These insects are collected and placed onto the test plants, which become infected by the virus.

6.2.11. Management of Disease Epidemics

A **disease epidemic** is a severe outbreak of a disease starting with a low level of initial infection. Epidemics are rare in natural populations of plants. But they are often encountered in case of cultivated species. The epidemics are generally associated with air-born fungal pathogens. Soil-born fungal pathogens may also produce epidemics when they are introduced in a new area because most of the cultivated varieties of the area are likely to be susceptible to the pathogen. The main reasons for disease epidemics are as follows: (i) narrow genetic base of the cultivated varieties of the crop, (ii) large continuous or almost continuous areas planted to a single variety with vertical resistance, (iii) failure of the vertical resistance of this variety due to the evolution of a new race of the pathogen that is virulent against the vertical resistance of this variety, (iv) reduced level of horizontal resistance in the cultivated varieties of the crop, and (v) unintended introduction of a new pathogen or a new race of an existing pathogen in the area.

The management of disease epidemics is a continuing challenge for the plant breeders. Several strategies based on vertical resistance for the management of epidemics have been used/proposed; some of the important strategies are summarized here. (i) Varieties with different strong genes for vertical resistance may be planted in successive years. (ii) In some cases like wheat-stem rusts in India, the pathogen inoculum moves along a specific path from one region to another. Varieties with different strong vertical resistance genes should be planted in different regions that run perpendicular to the path of pathogen movement. (iii) The varieties grown in the neighboring fields should have different resistance genes. (iv) Multiline varieties may be developed and used since they tend to mimic the horizontal resistance. Finally, (v) Multiple resistance genes may be pyramided into a single variety to reduce the chances of resistance to be overcome by the pathogen.

Many breeders feel that all the measures based on vertical resistance are likely to be successfully defeated by the air-born fungal pathogens. In their opinion, the only durable strategy for an effective prevention of epidemics is to deploy adequate levels of horizontal resistance. The use of molecular markers enables easy and effective selection for QTLs for horizontal resistance even in the presence vertical resistance genes. Therefore, it is now feasible to develop a variety with vertical resistance along with high levels of horizontal resistance. Such a variety will produce clean disease-free crop as long as vertical resistance is effective. But when the vertical resistance is broken down by the pathogen, there will not be an epidemic caused by this pathogen.

6.2.12. Merits of Breeding for Disease Resistance

1. The disease resistant varieties are an effective means of disease management. They are generally able to prevent the occurrence of frequent disease outbreaks in epidemic proportions.
2. Disease resistant varieties are the cheapest means of disease control, and there is no recurring expenditure. The only expenditure is for the purchase of the seed of the disease resistant variety.
3. For many diseases, e.g., wheat rusts, resistant varieties are the only practicable means of disease control.
4. The environmental conditions usually do not markedly affect the effectiveness of disease resistant varieties. Disease resistant varieties help in reduction of the use of fungicides for disease control. This reduces the damage to the environment and minimizes the risk of presence of harmful pesticide residues in the crop produce.
5. Deployment of disease resistant varieties has a stabilizing effect on crop yields. This is because these varieties are able to reduce the variable losses in crop yields due to variation in the intensity of the diseases from one year to other.
6. The disease resistant varieties can be effectively combined with other methods of disease management for an effective control of the disease.

6.2.13. Limitations of Breeding for Disease Resistance

1. Generally, disease resistance breeding programmes are based on oligogenic resistance since this is easily utilized for the development of disease resistant varieties. But the duration of effectiveness of oligogenic resistance is relatively short, and boom and bust cycles are a common feature.
2. Breeding for oligogenic resistance does not allow selection for polygenic resistance. As a result, most varieties with oligogenic resistance have only low levels of horizontal resistance. Therefore, when their vertical resistance is overcome by the race virulent to the oligogene, there is severe disease outbreak leading to the vertifolia effect.
3. The polygenic disease resistance is generally highly durable. However, breeding for this type of resistance is quite demanding. But the use of molecular markers linked to the QTLs governing the horizontal resistance greatly facilitates the breeding effort.
4. A useful crop variety has to be resistant to more than one, often several, different diseases. The evolution of a variety with resistance to multiple diseases is a demanding activity.

5. Many resistance genes impose a yield penalty on the variety carrying them. This negative effect of the *R* genes may be either due to linkage or pleiotropy. In either case, considerable effort would be required to minimize such detrimental effects of the *R* genes.

6.3. BREEDING FOR INSECT RESISTANCE

Insects attack all crop plants to varying degrees. But the degree of damage caused by insect pests varies to a great deal from one crop to the other. For example, cotton is attacked by over 160 species of insects and nearly 60% of all the insecticides used in agriculture are applied to cotton. In comparison, crops like wheat and barley are infested by insect pests and none of these insects causes serious damage to these crops. The type of damage caused by the insect pests to the crops is of the following two types. (i) Many insect pests suck the sap from the plant tissues, e.g., aphids; they are called **sucking pests.** (ii) Many other insects like pod borers, shoot borers, etc. feed on the tissues of the plant parts like leaves, fruits, etc.; these pests are known as **tissue feeders**. The insect pests cause both direct and indirect damage to the plants. The direct damage is the result of the tissue eaten or the sap consumed by the insect. The indirect damage is caused by transmission of viruses and by creating conditions suitable for fungal growth and/or infection. Many sucking insects build up population on young leaves of host plants and reduce their photosynthetic abilities, e.g., aphids of cluster bean (guar) and safed aak (*Calotropis* sp.). Insect pests are estimated to cause a loss of about 14% in yield averaged over all the crops and the countries. Therefore, the development and use of insect resistant crop varieties is an important component of insect pest management. There are many common features between breeding for disease resistance and those of insect resistance breeding. Therefore, only the salient features of insect resistance breeding are described here.

6.3.1. Insect Biotype Differentiation

As in the cases of disease causing pathogens, the different strains of an insect pest differ in their ability to attack different varieties of the host crop. Such strains of an insect pest are called **biotypes,** and this situation is called **biotype differentiation**. However, biotype differentiation in insect pests appears to be much less frequent that the differentiation of physiological races in the case of pathogens, especially fungal pathogens. This situation may be the result of the following factors. (i) An insect female produces only a small number of progeny in one generation as compared to the very large numbers produced by pathogens, especially fungal pathogens. (ii) The insect resistance may be based on some morphological/anatomical feature or nonspecialized biochemical or physiological factor. It may be difficult for the insect pest to overcome these mechanisms of resistance. (iii) A sufficient number of lines of the host differing in resistance may not be available, so that insect biotypes, even when present, may not be identified. (iv) The cultivation of insect resistant varieties is generally on a limited scale, so that the opportunities for biotype differentiation are rather small. Finally, (v) The genetics of insect resistance has not been investigated as extensively as that of disease resistance.

A biotype of an insect pest that has adaptation specifically for the mechanism of resistance present in a host variety is called **true resistance-breaking biotype**. Such a biotype will not be able to attack other varieties of the host that have different resistance mechanisms. In contrast, a **highly vigorous biotype** of an insect pest attacks all the varieties of the host crop more vigorously than the other biotypes of the pest. In this discussion, only the true resistance-breaking biotypes of pests are considered. It may be pointed out that the emergence of true resistance-breaking biotypes has rarely resulted in insect epidemics leading to the withdrawal of the resistant variety that has become susceptible to the new biotype. This situation is in contrast to that with the diseases, especially those caused by air-borne fungi, where evolution of virulent races leads to the withdrawal of the concerned disease resistant varieties (Section 6.2.7).

6.3.2. Levels and the Mechanisms of Insect Resistance

Insect resistance of plant a line or variety of crop can be measured only in relation to the level of insect infestation of known susceptible line or variety of the crop. When a large number of different lines/varieties of the crop are evaluated for their resistance to a given insect pest, they can be divided into the following five groups: (i) immune, (ii) highly resistant, (iii) moderately resistant, (iv) susceptible, and (v) highly susceptible. An **immune variety** does not suffer any damage due to the insect pest under any condition. On the other hand, a highly susceptible variety suffers much greater damage due to the insect than the average damage sustained by a number of different lines or varieties of the crop.

The mechanism of insect resistance varies considerably, and is classified into the following four groups: (i) nonpreference, (ii) antibiosis, (iii) tolerance and (iv) avoidance. The resistance of given line or variety may involve a single mechanism or it may be the result of two or more of the above mechanisms. For example, the resistance of maize varieties to European corn borer is due to antibiosis, while in the case of corn leaf aphid both antibiosis and nonpreference contribute to resistance (Table 6.2). In cases of nonpreference, the resistant variety is not liked by the insect for colonization, oviposition or both. Generally, the insect prefers to infest other varieties of the host crop in comparison to the resistant variety. But in extreme cases, the insect would not invade the resistant variety even when no other host plants are available. The nonpreference may be due to either morphological or biochemical features of the host plant. This type of resistance is generally highly effective and is likely to be durable. Some examples of nonpreference are aphid resistance in sugar beet, brown plant hopper resistance in rice, corn earworm resistance in maize, etc. (Table 6.2).

Table 6.2: The salient features of resistance to selected insect pests in some crop plants

Crop	Insect pest	Host trait involved in resistance	Mechanism of resistance
Wheat	Stem-sawfly	Solid stem	Nonpreferance and antibiosis
	Greenbugs	High benzyl alcohol content	Antibiosis
Barley	Cereal leaf beetle	Hairiness and waxiness of leaves	Tolerance and antibiosis
Rice	Rice stem borer	High silica content	Antibiosis
		Lignified stem	Nonpreferance
		Greater amount of sclerenchymatous tissue in the stem	Tolerance
	Brown planthopper	Low asparagine content in leaves	Nonpreferance
Maize	European corn borer	High DIMBOA content in leaves	Antibiosis
	Maize stem borer	High aspartic acid, and low nitrogen and sugar contents	Antibiosis
	Corn ear worm	Toughness of husk	Nonpreferance
	Corn leaf aphid	High DIMBOA content in leaves	Nonpreferance and antibiosis
Cotton	Bollworm	High gossypol content	Antibiosis
		Smooth leaves	Nonpreferance
		Nectariless	Nonpreferance
		Earliness	Avoidance (escape)
	Jassids	Hairiness of leaves	Nonpreferance
Sugar beet	Aphids	Low free sugar	Nonpreferance and antibiosis
Alfalfa	Spotted alfalfa aphid, pea aphid	High saponin content	Antibiosis
Brassica spp.	Cabbage aphid	Low sinigrin content	Nonpreferance and antibiosis
Potato	Aphid	Sticky trichome exudate	antibiosis

In case of **antibiosis**, the insect pests invade the resistant varieties, but their growth and reproduction are adversely affected as compared to the insects feeding on the susceptible lines or varieties of the host. In severe cases of antibiosis, the insects feeding on the resistant lines may even die. This type of resistance may be due to morphological (Table 6.3), biochemical or physiological features of the host plants (Table 6.2). In case of **tolerance**, a host line or variety shows a significantly lower level of yield reduction than a susceptible line/variety when both of them have comparable levels of insect infestation. The tolerance response of a line may reflect a lower damage caused by the insect or a faster recovery from the damage as compared to the susceptible line. Insect pest **avoidance** is the same as disease escape, and is not a case of true insect resistance. However, in the cases of some insect pests avoidance may be quite effective in minimizing losses due to the insect attack. The avoidance may be the result of some feature of the host plant. For example, early varieties suffer much less damage since they reach a developmental stage that is much less prone to damage at the time the insect population is at its peak. Avoidance may also be the result of some environmental factors.

Table 6.3: Some examples of host plant morphological features involved in insect resistance

Morphological trait	Host	Insect pest	The host trait affects
Color	Cotton	Bollworm	Landing on the plant and colonization
Thicker cell walls and tougher plant tissues	Cotton	Bollworm	Feeding and oviposition by the insect
Solid stem and other features of stem	Wheat	Stem sawfly	Feeding, oviposition
Trichomes	*Solanum* spp.	Aphids	Feeding, digestion, oviposition, movement on the plant surface, etc., including antibiosis
	Cotton	Jassids	
	Sorghum	Shoot fly	
Husk toughness	Sweet corn	Corn ear worm	Various effects

6.3.3. Features of the Host Plant Responsible for Insect Resistance

Different cases of insect resistance may involve divergent features of the host plants. The resistance may be based on (i) morphological traits, (ii) physiological characteristics or (iii) biochemical features of the plants. In several cases, insect resistance is produced by the **morphological traits** of the host plants; some well-known examples are listed in Table 6.3. For example, hairiness of leaves is associated with resistance to jassids in cotton, with resistance to leaf beetle in cereals, etc. Similarly, color of the plant also contributes to insect resistance, e.g., certain Lepidopteran insect pests do not prefer colored cabbage for oviposition. In many cases, certain **physiological factors** of the host plant, e.g., high osmotic concentration of the cell sap, presence of various types of exudates, etc., may lead to insect resistance. For example, some *Solanum* species have hairy leaves and these hairs secrete a gummy exudate. Aphids become trapped in these gummy exudates, as a result of which they fail to feed and reproduce. This phenomenon leads to the resistance of these plants from aphid infestation.

There are many well-known cases of insect resistance based on biochemical features of the host plants. In general, these factors generate resistance due to non-preference and antibiosis. In fact, biochemical factors are more important in non-preference and antibiosis than are morphological and physiological features. Some examples of insect resistance based on biochemical factors are as follows: rice resistance to shoot borer due to high silica content in stem, resistance of wheat and barley to green bugs produced by high benzyl alcohol content, resistance to spotted alfalfa aphid and pea aphid generated by high saponin content, low sinigrin content of *Brassica* species associated with resistance to cabbage aphid, etc.

6.3.4. Genetics of Insect Virulence and Host Plant Resistance

Virulence is the ability of an insect biotype to invade and survive on an insect resistant line or variety of the host crop species, and such a biotype is called **virulent**. This can be explained by the following example. The wheat variety Monon is resistant to the Hessian fly race GP, which cannot successfully attack Monon. But the race E of hessian fly is able to invade and survive on Monon. Thus, race E is virulent on Monon, while race GP is avirulent. A cross between race GP and race E revealed the virulence of race E to be governed by a single gene and that virulence was recessive to avirulence. In general, insect virulence seems to be under monogenic control. But in some cases, e.g., in brown plant hopper (*Nilaparvata lugens*), the virulence is under polygenic control.

The resistance of host plants to insect pests is also governed primarily by one or few major genes, but in some cases it is polygenically controlled. In addition, some cases of insect resistance are specified by the cytoplasm (Table 6.4). In case of some host plant–insect pest systems, the resistance shows more than one type of genetic control. For example in the case of maize-European corn borer system, oligogenic, polygenic as well as cytoplasmic controls of resistances have been documented. In case of **oligogenic resistance**, one or few major genes or oligogenes of the host plant determine resistance to the insect pest. In addition, the difference between resistant and susceptible plants is large, and there is little risk of misclassification. Many cases of oligogenic resistance are governed by single major genes, i.e., they are cases of **monogenic resistance**. Monogenic resistance shows simple inheritance pattern, and is easy to select for and use in breeding programmes. The resistance involves only one trait of the host plant, and the insect pest is often able to overcome the host plant resistance. But in some cases, monogenic insect resistances are quite durable. When the resistance is governed by more than one oligogene, all the genes may affect a single trait of the host plant, or they may govern different traits. The latter type of insect resistance is generally more durable that the former. In addition, in some cases the multiple oligogenes might show gene interaction.

Table 6.4: The genetic control of resistance to selected insect pests of some crop species

Crop	Insect pest	Genes specifying resistance	Biotype differentiation	Remarks
Oligogenic resistance				
Wheat	Greenbug (*Schizaphis graminum*)	5D + 1d*	Yes	Modifying genes known
	Hessian fly (*Myetiola destructor*)	19D + 1d	Yes	-
Barley	Corn leaf aphid (*Rhopalosiphum maidis*)	2d	Yes	Complementary gene action
	Hessian fly (*M. destructor*)	3D	Yes	Complementary gene action
Cotton	Jassids (*Empoasca* spp.)	3D	Yes	Modifying genes
	Thrips (*Thripps* spp.)	3D	Yes	-
	Tobacco budworm (*Heliothis virescens*)	2D	Yes	Additive interaction
Maize	European corn borer (*Ostrinia nubilalis*)	3 genes	Yes	Polygenic control and cytoplasmic effects also
Muskmelon	Melon aphid (*Aphis gossypii*)	4D	Yes	2D for antibiosis and tolerance, 2D for nonpreference
Rice	Brown planthopper (*Nilaparvata lugens*)	21 genes (D and d)	Yes	Two genes show inhibitory interaction
	Gallmidge (*Orseolia oryzae*)	3d	Yes	Complementary gene action
	Striped stem borer (*Chilo suppressalis*)	Single gene	-	-
Polygenic insect resistance				
Alfalfa	Spotted aphid (*Therioaphis maculata*)	Polygenic control	-	-
Cotton	Pink bollworm (*Pectinophora gossypiella*)	Polygenic control	-	Additive action
Maize	Corn ear worm (*Heliothis zea*)	Polygenic control	-	Additive action
	European corn borer (*O. nubilalis*)	Polygenic control	-	Oligogenic and cytoplasmic control also

Crop	Insect pest	Genes specifying resistance	Biotype differentiation	Remarks
Rice	Stem borer	Polygenic control	-	-
	Striped stem borer	Polygenic control	-	Oligogenic control also
Sorghum	Shoot fly (*Atherigona socata*)	Polygenic control	-	
Soybean	Mexican bean beetle (*Epilachna varivastis*)	Polygenic control	-	Oligogenic control also suggested
Wheat	Cereal leaf beetle (*Oulema melanopus*)	Polygenic control	-	Additive action
Cytoplasmic insect resistance				
Cotton	Boll weevil (*Anthonomus grandis*)	Cytoplasmic control		Oligogenic control also
	Tobacco budworm (*H. virescens*)	Cytoplasmic control		Oligogenic control also
Lettuce	Root aphid (*Pemphigus bursarious*)	Cytoplasmic control		Oligogenic and polygenic control also
Maize	European corn borer (*O. nubilalis*)	Cytoplasmic control		

* D, dominant alleles of the genes specify resistance; d, recessive alleles of the genes confer resistance.

Some cases of insect resistance are known to be under polygenic control (Table 6.4). In such cases, the resistance is governed by several polygenes, each of which produces a small but cumulative effect on resistance. This situation is similar to horizontal disease resistance. The polygenic insect resistance involves more than one trait of the host plant. The resistance shows continuous variation, and is affected by the environment. As a result, selection for resistance is difficult, and its transfer into a variety is far more problematic than that of oligogenic resistance. However, polygenic insect resistance is generally much more durable than oligogenic resistance.

In at least four cases, the insect resistance is known to be governed by the plasmagenes, i.e., genes located in the cytoplasm. The examples of **cytoplasmic insect resistance** are, maize resistance to European corn borer, resistance to root aphid in lettuce, and resistance of tobacco to boll weevil and tobacco budworm.

6.3.5. Durability of Insect Resistance

The **durability of resistance** is the time taken for the appearance of a true resistance-breaking biotype of the insect pest from the time a variety with this resistance was used for commercial scale cultivation. The evolution of true resistance-breaking biotype is affected by the following factors: (i) insect species, (ii) the genetics of resistance, and (iii) the host characters associated with the resistance. The appearance of new biotypes is more frequent in case of some insect pests than for some others insects. For example, new biotypes appear more frequently in case of Hemiptera, particularly aphids, than in Lepidoptera (moths and butterflies), and Coleoptera (weevils and beetles). The genetic basis of resistance is associated with the durability of resistance. In general, insect resistance governed by single major genes is less durable than that specified by two or more oligogenes. In addition, polygenic insect resistance is much more durable than oligogenic resistance. But some cases of monogenic resistance are very durable, while in few cases polygenic resistance is no more durable than oligogenic resistance. For example, polygenic resistance of rape varieties to cabbage aphid is overcome by true resistance-breaking biotypes of the cabbage aphid. In general, the durable cases of resistance involve more than one feature of the crop plants.

The nature of host plant feature associated with resistance has considerable influence on the durability of resistance. In general, resistance based on morphological and/or anatomical features of host plants is much more durable than that associated with other traits. For example, resistance to some insect pests is based on hairiness of plant surfaces like leaves. Hairiness interferes with insect oviposition and impedes the movement of insect larvae leading to their retarded growth and development. Clearly, the changes in the insect oviposition and movement of the larvae will be required to overcome this resistance. Such gross changes in the insect pest may not be easily accomplished. However, some cases of insect resistance based on biochemical features of the host plant are also durable. For example, resistance of rice to stem borer is associated with high silica content of the stem; this resistance is highly durable. Other such cases of durable resistance are resistance of alfalfa to alfalfa aphid, resistance of maize to European corn borer, etc.

6.3.6. Breeding Strategies for Insect Resistance

The general, issues in breeding for insect resistance are similar to those in breeding for disease resistance. The specific breeding strategy will largely depend on the source from which insect resistance is to be used in the breeding programme. Insect resistance may be found in the following types of materials: (i) a cultivated variety, (ii) a germplasm accession of the crop species, (iii) a related wild species, and (iv) an unrelated organism. The resistance to an insect may be found in a **cultivated variety**. For example, cotton (*G. hirsutum*) varieties Khandwa 2 and SRT1 are excellent sources of resistance to jassids. The resistance to many insect pests may not be found in cultivated varieties; in such cases, **germplasm accessions** are screened for resistance. Germplasm collections are good sources of insect resistance. For example, many cotton germplasm accessions with good levels of jassid resistance are known. The resistance present in germplasm accessions is relatively more problematic to use in breeding programmes than that found in cultivated varieties. But sometimes, the desired resistance may be available only in a wild relative of the crop species. This source of resistance is the most difficult to use in breeding programmes, and should be the last choice for the breeder. Often, useful resistance to an insect pest may not be available in the germplasm, including the wild relatives, of a crop species. In such cases, a suitable gene from an **unrelated organism** may be introduced into the crop by transgenic technology or genetic engineering (Chapters 9-12) to create a transgenic plant with insect resistance. For example, the *cry* gene from *Bacillus thuringiensis* has been transferred into crops like cotton, maize, etc. to produce insect resistant crop varieties.

The resistance available in cultivated varieties is the easiest to use in breeding programmes. Therefore, cultivated varieties should be the first to be searched for resistance to the concerned insect pest. If required and whenever feasible, a resistant variety may be introduced from another country, and either directly released for cultivation or used in breeding programmes for developing resistant varieties. Plant introduction of good insect resistant varieties is perhaps the simplest and the quickest breeding method for the purpose. There are several good examples of introduction of insect resistant materials. For example, the introduction of grape phylloxera (*Phylloxera vitifoliae*) resistant rootstocks from USA saved the grape industry in France. In cases of cross-pollinated crops, open-pollinated varieties contain large amounts of genetic variation for almost all the traits. Therefore, **selection** in such varieties for insect resistance is likely to be successful. For example, selection for resistance to potato leafhopper and spotted alfalfa aphid in alfalfa was quite effective. Thus, in such cases, even when the cultivated variety is not resistant to the concerned insect pest, a resistant variety may be developed by selection for resistance in this variety.

In many cases, the line with insect resistance may itself not be suitable for direct release as a new variety. In such cases, the resistant line is crossed with a popular high yielding variety of the area to evolve a new improved insect resistant variety. Hybridization is the most extensively used method of breeding for insect resistant varieties. When the line with insect resistance also has some other desirable features, the segregating generations from the cross are handled according to the **pedigree method**. The insect resistant variety developed by this method is expected to combine the good agronomic features from both the in addition to the insect resistance. The pedigree method has to be used when insect resistance is under polygenic control. In such cases, the selection for insect resistance is delayed till F_3. and a large number of plants is selected in F_2, and the insect resistance is generally evaluated on the basis of progeny test. But when the insect resistance is present in an agronomically inferior germplasm accession or in a related species, the **backcross method** of breeding has to be used to transfer the resistance into an insect susceptible, but a highly popular variety of the crop. The insect resistant line is used as the donor parent, and the popular variety is used as the recurrent parent. The new variety developed by this method will be exactly like the recurrent parent, except for the insect resistance trait transferred from the donor parent. In such cases, often there may be linkage drag, i.e., some undesirable genes linked to the insect resistance gene may also be transferred from the donor parent. Whenever such problems are encountered, efforts are made to eliminate the undesirable effects associated with the insect resistance by raising large segregating populations and selecting for insect resistant plants without the undesirable features. Molecular markers can also be used to eliminate linkage drag by implementing recombinant selection (Section 15.9.3).

It may be pointed out that insect pests themselves are generally attacked by other insects; these insects (as well as fungi, bacteria and viruses) are called **natural enemies**. The natural enemies of insect pests contribute to the control of the insect pests. Further, certain features of the host plant may affect the activity and the effectiveness of the natural enemies. Such plant traits may be morphological or biochemical in nature. These features may interfere with the mobility of the natural enemies, may affect their survival, or may restrict the ability of the natural enemies to find and attack the insect pests. Plant breeders could attempt to modify such features of the host plants in such a way that the activity and the effectiveness of the natural enemies are enhanced. There is some evidence that efforts to modify the crop plant characteristics to enhance the effectiveness of natural enemies of insect pests can serve a useful purpose in control of the insect pest. This approach of breeding may be called **breeding for enhanced biological control of insect pests**.

6.3.7. Tests for Insect Resistance

An effective breeding for insect resistance is dependent on accurate and reliable identification of resistant plants from the susceptible ones. Reliable tests for insect resistance often require considerable technical skill, and are costly. The tests for insect resistance can be done in a greenhouse or in the field. In case of **greenhouse tests**, the test plants are grown in a greenhouse under controlled environmental conditions. Each test plant is inoculated with equal numbers of eggs or larvae of the insect pest. Thus, inoculation of each plant is ensured and the risk of escapes is greatly minimized. The results from greenhouse insect tests are more reliable than those from field tests. The chief limitation of this approach is that only a limited number of plants can be evaluated for insect resistance.

In case of **field tests for insect resistance**, the test plants are grown in a field. One of the following approaches is used to ensure that all the test plants are uniformly invaded by the insect pest. (i) After every few rows of the test plants, one row of a known highly susceptible line of the concerned host is planted. The insect pest multiplies on the susceptible line and invades the test plants. (ii) The tests plants may be evaluated in an area where the concerned insect pest is known to occur in high frequency so that the insect pest is expected to invade every test plant. (iii) It is desirable to transfer equal numbers of eggs or larvae of the insect pest onto each test plant. This approach should be used wherever it is practicable. (iv) The insect test may be carried out in a season, in which the concerned insect is known to cause heavy infestation. For example, rice shoot borer is known to cause more serious infestation during the off-season than during the main season. Finally, (v) in case of soil inhabiting insects, the test plants are grown in a field that is heavily infested by the concerned insect pest.

The field insect tests offer the following advantages: (i) a large number of plants can be evaluated for insect resistance, (ii) the insect tests require less investment, are cheaper, and less labour intensive, (iii) they allows automatic evaluation of the test plants for resistance to other insect pests of the crop that are prevalent in the area, and (iv) provide the response of the plants to the insect pest under the environmental conditions in which they will be grown if released as commercial varieties. In general, the results from insect tests in greenhouse are more reliable than those from the field tests. Therefore, generally the plants are first evaluated in the field, resistant plants or lines are identified, and then their resistance is confirmed by a greenhouse test. Further, the resistance of plants identified by a greenhouse test has to be evaluated in the field to determine if it is expressed and effective under the field conditions.

6.3.8. Management of Insect Outbreaks

An **insect outbreak** is a severe infestation by an insect pest over a large area in which the host crop is cultivated. Insect outbreaks can be managed by applying one or more of following approaches. (i) Two or more genes for resistance to the same insect pest may be introgressed into a single variety; this is called **gene pyramiding**. This greatly reduces the chances of evolution of an insect biotype capable of overcoming all the resistance genes. This will be particularly so if each resistance gene specified resistance based on a different mechanism of resistance, e.g., nonpreference in combination with antibiosis. (ii) Wherever feasible, polygenic resistance should be used in preference to oligogenic resistance. (iii) If required, multiline varieties may be used to minimize the risk of insect outbreaks as in the case of plant diseases. (iv) An effective strategy to prevent insect outbreaks is to grow different variety every year from among a set of insect resistant varieties. Each variety of this set has a different gene for insect resistance. Finally, (v) molecular markers can be used to combine oligogenic resistance with adequate levels of polygenic resistance in such cases where the both the types of resistance are available. This approach is likely to give a highly effective insect control and a resistance that is quite durable.

6.3.9. Problems in Breeding for Insect Resistance

1. In some cases, breeding for resistance to one insect pest of a crop may increase susceptibility to another insect pest. This may, in some cases, be due to the fact that the same trait leads to resistance to one insect pest but increases susceptibility to another insect pest. For example, the hairless trait of cotton strains increases their resistance to bollworm, but increases their susceptibility to jassids.
2. An insect resistant variety may have inferior produce quality. As a result, it may not be liked by the farmers and the consumers. For example, cotton lines with increased resistance to bollworm often have high gossypol content. But this feature makes their oilcake unsuitable for use as feed.
3. Resistance to many insect pests may be available in a wild relative of the crop species. The utilization of such resistance is not easy and generally plagued by many problems.
4. The selection for insect resistance is based on insect tests. Generally, evaluation of insect resistance is cumbersome, costly and labor intensive. The use of molecular markers in close linkage with the insect resistance gene would greatly facilitate selection for the gene without any need for insect resistance test.
5. Breeding for insect resistance is a long-term activity. Therefore, it requires continued financial support on a long-term basis, which may often be difficult to ensure.

6.3.10. Achievements of Breeding for Disease and Insect Resistance

Breeding for disease resistance is an integral component of breeding programmes of virtually every crop species. The improved varieties released for cultivation in an area are evaluated for their reaction to the diseases important in the area, and only those varieties that are resistant to these diseases are released. For example, the wheat varieties must have specified minimum levels of resistance for the three rusts, viz., stem, leaf and stripe rusts, for being released for cultivation. In addition, wheat varieties should be resistant to Karnal bunt for release in those areas where this disease is important.

Similarly, insect resistant varieties of many different crops have been developed in several countries. For example, over two dozen wheat varieties resistant to Hessian fly have been developed in USA. In addition, wheat varieties like Rescue that are resistant to stem saw fly (*Cephuscinctus*) have been developed in USA. The barley variety Will is resistant to greenbugs, while Cody, Moapa and Zia varieties are resistant to spotted alfalfa aphid. In India, several insect resistant cotton varieties have been developed. For example, B1007, SRT1, Khandwa 2, DHY286, PKV081, PKVHy2, etc. varieties of cotton are resistant to jassids. The G27, Lohit, DS1, Sharda, Abadhita, etc. cotton varieties show tolerance to bollworms. Similarly, cotton varieties Kanchana, Supriya, LK861, etc. exhibit tolerance to white fly.

6.3.11. Adverse Effects of Breeding for Disease and Insect-Resistance

Resistance breeding unquestionably offers considerable to complete protection to the crops from the damages caused by diseases and/or insect pests. However, this protection may impose yield penalties in at least some of the cases. For example, barley genes *Mla9* and *Mlk1* specify powdery mildew resistance. It is noteworthy that two resistance genes do not have any adverse effect on the performance of the resistant varieties. But the recessive gene *mlo* confers broad resistance to all the races of the pathogen, and reduces the yields of resistant varieties by about 4%. It also leads to reduction in the grain test weights of the resistant varieties. The undesirable effects of the resistance genes could be due the following two reasons: (i) close linkage with one or more undesirable genes (**linkage drag**), and (ii) pleiotropic effects of the resistance gene. In case the undesirable effects of a resistance gene are due to a close linkage with one or more undesirable genes, breeding strategies, e.g., recombinant selection (Section 15.9.3) can be used to break this linkage. This would allow the recovery of resistant plants that do not show the undesirable effects. But in cases of pleiotropy, the undesirable effects are produced by the resistance genes themselves. Therefore, it is not possible to develop resistant lines without the undesirable effects associated with these genes. One possibility of minimizing the undesirable pleiotropic effected by transferring the resistance gene in a suitable genetic background. For this, the resistance gene will have to be transferred into several different genetic backgrounds, and lines with the minimum undesirable effects will have to be identified. It is generally not easy to distinguish between linkage drag and pleiotropy. One approach to resolve this issue is to clone the resistance genes, and transfer them into the concerned crops. But care has to be taken in such studies because the performance of transgenic lines is likely to be affected by several factors other than the transgene itself.

In addition, insect resistant varieties may sometimes generate toxic effects in humans as well. This is because many compounds like alkaloids, phytoalexins, etc. that are involved in insect resistance are also known to produce adverse effects in humans (Table 6.5). Breeding for enhanced levels of resistance to an insect pest or even a disease may sometimes lead to an increase in the concentration of one of these substances in the crop produce. In some cases, the concentration of the toxic substance may increase beyond the safe level. In such cases, consumption of the produce from the resistant varieties may produce toxicity. For example, an insect resistant potato variety was developed from an interspecific cross involving *Solanum chacoense*. This variety produced acute toxicity in humans and had to be withdrawn from cultivation. The toxic effects were generated by a much higher level of glycoalkaoids in this variety as compared to those in the normal varieties. Similarly, an insect resistant variety of celery had 8-times higher concentration of a phytoalexin, psoralen, which is known to generate toxic, mutagenic and carcinogenic effects. In view of the above, this variety of celery was not used for cultivation.

Table 6.5: Some of the substances involved in insect and/or disease resistance in plants and their adverse effects on humans

Compound	Role in plant defense	Adverse effects in humans
Alkaloids	Insect resistance	Toxic effects
Cry protein	Insect resistance (transgenic)	No adverse effects reported
Lectins	Insect resistance, fungal resistance	Allergenic effects; weight loss and hyper-growth of small intestine in rats
Phytoalexins	Fungus and insect resistance	Toxic effects
Proteinase inhibitors	Insect resistance	Allergenic effects; pancreas enlargement and some other adverse effects
Thionins	Antifungal action	Toxic effects on intravenous injection, seem nontoxic on oral intake
Virus coat protein	Virus resistance (transgenic)	No adverse effects reported

6.4. CONCLUSIONS

Crop plants are attacked by several pathogens (fungi, bacteria, viruses, and nematodes) and insect pests. In general, the economic losses caused by these biological agents are in the order fungi > bacteria > viruses > nematodes = insects. Further, the relative importance of different pathogens and insects varies greatly depending on the crop. Each pathogen/insect has both virulent and avirulent strains; this situation is called **physiological race differentiation** (in case of pathogens) or **biotype differentiation** (in case of insects). A virulent strain is capable of attacking a host variety, while an avirulent strain is unable to do so. Virulence is governed by specific genes and avirulence is generally dominant over virulence. Similarly, the varieties of the host crop differ in their ability to resist invasion by the pathogens/insects; this is called **resistance**. Resistance is also governed by specific resistance genes, and generally resistance is dominant over susceptibility. The pathogen virulence and host resistance genes show **gene-for-gene relationship**. This relationship specifies that for every resistance gene in the host, the pathogen has a specific virulence gene. A pathogen race having the virulence allele corresponding to a resistance gene will overcome the resistance generated by this gene and invade the host line. The above type of resistance is governed by oligogenes, is called vertical resistance, and is associated with disease epidemics and boom-and-bust cycles. Host resistance is also polygenically controlled, and in some cases the inheritance of resistance is cytoplasmic.

The resistance may be available in a cultivated variety, germplasm accession, related variety or an unrelated organism. The breeding method used for the development of a resistant variety will depend on the source of resistance. The different methods of breeding used for this purpose are introduction (source, cultivated variety), selection (cross-pollinated crops), hybridization (pedigree and backcross methods), mutagenesis/somaclonal variation (source of resistance not available), and genetic engineering (genes from unrelated organisms also used). Recently developed tools of editing plant genomes (Section 2.11) using site directed nucleases like CRISPR/Cas9 system are quite promising for the development of disease resistant genotypes. Efficient breeding for resistance requires cheap, quick but reliable disease tests/insect tests. Several strategies have been suggested to minimize the risk of epidemics; the most useful approach seems to be the deployment of horizontal or polygenic resistance. Molecular markers can be used for effective selection for polygenic resistance and even for combining oligogenic and polygenic resistances. Resistance breeding has been extremely successful, and has prevented the elimination of some of our valuable crop species from commercial cultivation. However, some disease resistance genes might impose a yield penalty, and in few cases the resistant varieties might cause toxicity/allergenicity problems.

In addition, some crops like chickpea, sorghum and pearl millet are attacked by parasitic weeds, e.g., *Striga* and *Orobanche*. Efforts for identification of lines resistant to these parasitic weeds have been successful. In several crop species, resistance sources have been identified and high yielding lines with resistance to the parasitic weeds have been released for commercial cultivation by farmers.

Breeding for disease and insect resistance is critically important for ecological and genetic safety as applications of fungicides and insecticides cause pollution and health hazards. They also harm soil, water, useful insects (like pollinators, etc.) and reduce air quality and allow breeding of insecticide-resistant insects. However, some insects are famous for their catastrophic invasion, which has been mentioned in Bible as the eighth plague. The desert locust (*Schistocerca gregaria*) has been a serious threat to agriculture in Africa, Middle East and Asia up to India for centuries. The locusts come as clouds of insects spread over several squares of kilometers; they move rapidly and eat plant products voraciously. It is difficult to manage such pests through breeding and wide spread application of insecticides from the air seems to be the only feasible strategy for their control. The insecticide application programmes, of necessity, have to be managed through international agencies and inter-governmental cooperation.

PRACTICE QUESTIONS

1. Define vertical resistance. Explain the concept of gene-for-gene relationship and discuss its molecular basis.
2. Define horizontal resistance. Discuss the usefulness of horizontal resistance in breeding of disease resistant varieties.
3. Define physiological races, pathotypes and biotypes. Discuss the relevance of physiological races and biotypes in the development of disease/insect resistant varieties.
4. Define insect resistance. Briefly describe the various levels, types and the mechanisms of insect resistance.
5. There are many similarities as well differences between breeding for disease and insect resistant varieties. Discuss this statement with the help of suitable examples.
6. Briefly discuss the sources and the methods of breeding for insect or disease resistance.
7. Write short noted on: (i) biotype differentiation (or physiological race differentiation), (ii) disease epidemic (or insect outbreak), (iii) pathogenicity, (iv) gene-for-gene relationship, (v) vertical resistance, (vi) horizontal resistance, (vii) negative effects of resistance breeding, (viii) boom-and-bust cycle, (ix) disease test, (x) insect test, (xi) systemic acquired resistance.

RELEVANT LITERATURE

1. Agrios, GN 1997. Plant Pathology, 4th edn. Harcourt Asia PTE Ltd., Singapore.
2. Allard, RW 1960. Principles of Plant Breeding. John Wiley and Sons, Inc., New York.
3. Brown, JKM 2002. Yield penalties of disease resistance in crops. *Curr. Opin. Biol.* 5: 239-344.
4. Day PR 1974. Genetics of Host-Parasite Interaction. Freeman, San Francisco.
5. Khush, GS 1977. Disease and insect resistance in rice. *Adv. Agron.* 29: 265-341.
6. Painter, RH 1951. Insect Resistance in Crop Plants. McMillan Co., NY.
7. Russel, GE 1978. Plant Breeding of Pest and Disease Resistance. Butterworths, London.
8. Sharma, HC and Ortiz, R 2000. Transgenics, pest management, and the environment. *Current Sci.* 79: 421-437.

9. Singh, BD 2015. Plant Breeding, 10th edn. Kalyani Publishers, New Delhi.
10. Singh, DP 1991. Breeding for Resistance to Diseases and Pests. Springer-Verlag, Berlin.
11. van der Plank, JE 1968. Disease Resistance in Plants. Academic Press, NY, London.

■■■■

7 Breeding for Resistance to Abiotic Stresses

7.1. INTRODUCTION

The expression of the full genotypic potential of a crop variety can be achieved only when the optimum environment is provided. The **optimum environment** for a given variety is such that the performance of the variety is not adversely affected by any factor of the environment. Such an environment is also called **stress-free environment**. **Stress** is an environmental factor that has an adverse effect on the performance of a crop variety. When the stress is due to a non-biological environmental factor, it is called **abiotic stress**. Of the several different abiotic stresses, drought, temperature, and mineral stresses are the main abiotic stresses. It is estimated that approximately 90% of the world's arable area is subject to one or the other abiotic stress. Further, a significant proportion of this area is subject to more than one abiotic stress. The abiotic stresses together are estimated to be responsible for about 71% of the total reduction in crop yields due to the various factors including abiotic stresses. In this chapter, we shall examine the nature of these stresses and discuss the breeding strategies devised so far for their management and alleviation.

7.2. BREEDING FOR DROUGHT RESISTANCE

Drought is the most important of the abiotic stresses; it affects about 26% of the arable land area. **Drought** may be defined as inadequate availability of water during the life cycle of a crop so that the full genetic potential of the crop is not expressed. The inadequacy of water supply may be in terms of quantity of the available water or in distribution of the available water over time. The inadequate water availability may be the result of low rainfall, poor water storage capacity of the soil or high evapotranspiration rates. The occurrence of drought is unpredictable, and it varies during the crop season. The drought environment is quite variable, but it can be grouped into the following three main categories: (i) stored moisture environment, (ii) variable moisture environment, and (iii) optimal moisture environment. In the case of **stored moisture environment**, the crop is grown on the moisture stored in the soil, and the crops are exposed to drought during the terminal stages of growth and development. The **variable**

moisture drought environment, on the other hand, has alternate wet and dry periods of varying lengths of time due mainly to variation in the rainfall. Finally, in the **optimal moisture environment**, the crop is grown under optimal moisture conditions during most of the period of time. Drought occurs for variable durations of time and the occurrence is unpredictable. It is generally the result of low rainfall, but may also result from higher rates of evapotranspiration than the rate of water uptake by the roots.

7.2.1. Drought Resistance

Drought resistance is the mechanism that minimizes the loss in yield under a drought environment as compared to that in a stress-free environment. Thus, drought resistance of a variety is measured by comparing the yield of the variety under a drought environment with its yield in the stress-free environment. The drought resistance mechanisms can be classified into the following three groups: (i) drought escape, (ii) dehydration avoidance, and (iii) dehydration tolerance. When a drought susceptible variety avoids the period of drought, it is termed as **drought escape**. For example, early varieties often mature before the onset of drought in the terminal phase of crop growth. In the case of **drought avoidance**, the plants retain a relatively higher level of hydration under drought stress than the susceptible plants. The following characters contribute to dehydration avoidance: reduced transpiration, osmotic adjustment of the cells and tissues, cuticular wax, abscisic acid accumulation, leaf characters like pubescence, leaf rolling, etc. **Dehydration tolerance**, on the other hand, is the ability of plants to suffer less damage due to water stress in the cells than the susceptible plants. The characters that may contribute to dehydration tolerance are as follows: maintenance of membrane integrity, plant growth characteristics like large number of awns, and proline accumulation (Table 7.1).

Table 7.1: The various plant characteristics contributing to drought resistance and their utility as selection criteria for developing drought resistant varieties

Plant character	Ease in measurement	Association with yield under drought	Useful as selection criterion
Drought escape			
Earliness	Easy	Positive	Quite useful
Dehydration avoidance			
1. Reduced transpiration (water saver plants)			
Stomatal sensitivity	Difficult	Positive (?)	May be useful
Osmotic adjustment	Difficult	Positive	Useful
Cuticular wax	Easy	Positive (?)	Possibly useful
Abscisic acid content	Difficult	Positive (?)	Questionable

Plant character	Ease in measurement	Association with yield under drought	Useful as selection criterion
Leaf pubescence (hairiness)	Easy	Positive (?)	Questionable
Leaf angle and leaf movement	Easy	-	-
Leaf rolling*	Easy	Negative	Quite useful in rice
2. Increased water uptake (water spender plants)			
Root traits	Very difficult	Variable	Limited
(a). Deep root system	Very difficult	Positive (?)	Useful[@]
(b). Root-length density	Very difficult	Questionable	Useful[#]
(c). Root hydraulic resistance	-	Positive (?)	Questionable
Dehydration tolerance			
Membrane integrity[$]	Easy	Doubtful	Doubtful
Seedling survival[$]	Easy	Doubtful	For seedling drought tolerance
Seedling rooting pattern[$]	Easy	Doubtful	Doubtful
Seed germination[$]	Easy	Nil	Nil
Translocation of stem reserves	Easy	Positive	Useful
Proline accumulation	Difficult	Positive	Useful (?)

* Visual score, 1-5; selection for delayed leaf rolling.
[@] When the deeper soil layers have water reserve.
[#]The deeper soil layers do not have water reserve.
[$] Under water stress conditions

7.2.2. Genetics of Drought Resistance

The existence of genetic variation for drought resistance has been shown in several crops, including wheat, rice, maize, barley, sorghum, soybean, cotton, common bean, *Brassica* spp., etc. The drought resistance was estimated as yield stability, leaf water potential, leaf rolling, root growth, root xylem diameter, osmotic adjustment, seedling establishment and regrowth, etc. The genetics of these traits varies from oligogenic to polygenic, depending on the trait in question (Table 7.2). For example, leaf characters like glabrous leaves, glossy trait, waxy bloom, etc. show oligogenic control. Oligogenic control has also been found for some other traits like ABA accumulation in wheat, proline accumulation in a barley mutant, resistance to flower abscission in common bean. Most of the traits associated with drought resistance are under polygenic control. The mode of gene action involved is both of additive and dominance type. The heritability estimates vary from low to high depending mainly on the trait under consideration. The various traits for drought resistance show variable and usually low association with yield under drought stress.

Table 7.2: The mode of inheritance for some of the traits used for evaluation of drought resistance

Character	Crop	Inheritance	Correlation with yield/remarks
Dehydration avoidance			
Root traits*	Rice	Polygenic	Positive correlation with shoot; moderate heritability
Stomatal conductance	Cotton	Polygenic	Negative correlation
Osmoregulation	Wheat	Simple (oligogenic)	Positive correlation with yield
	Sorghum	?	Positive, with yield; may reduce growth
ABA accumulation	Wheat	Simple	Positive, with yield; selection effective
Waxy bloom	Sorghum	Oligogenic	Negative
Epicuticular wax load	Sorghum	Oligogenic	Negative
	Barley	Oligogenic (56 genes)	Negative
Glossy trait	Sorghum	Oligogenic	Negative
Glaucousness	Wheat	Oligogenic	Modified composition of cuticular wax
Glabrous leaves	Soybean	Single gene	-
Leaf water retention	Wheat	Polygenic	-
Dehydration tolerance			
Seedling recovery	Maize	Polygenic	-
Seedling growth	Maize	?	Selection effective
Pod formation and no flower abscission	Common bean	Oligogenic	-
Proline accumulation	*Brassica* spp.	Polygenic	Selection effective
	Barley	Oligogenic	Constitutive accumulation
Delayed leaf rolling	Rice	-	Negative with drought resistance; selection effective
Leaf water potential	Several crops	-	Positive with yield
Canopy temperature	Maize	-	Negative with yield

* Long root, high root number, high root/shoot ratio, root length density, root dry weight.

7.2.3. Sources of Drought Resistance

The drought resistance traits may be found in the following types of materials: (i) elite germplasm and cultivated varieties, (ii) land races, (iii) wild relatives, and (iv) unrelated organisms. Generally, ample genetic variation for the traits contributing to drought resistance is present the **elite germplasm** of most of the crops. In case of some crops even the cultivated varieties may contain usable genetic variation for these traits. This variation is the easiest to use, and its utilization should be the first priority of the plant breeders. The **land races** of the concerned crops are rich storehouses of genetic variation for various traits, including that for resistance to abiotic stresses. This is particularly true for the germplasm accessions collected from those areas that are known to be regularly affected by the concerned abiotic stress. For example, germplasm collected from drought prone areas would be valuable source for drought resistance traits. Often these accessions would need to be evaluated for their drought resistance and to characterize the traits contributing to the resistance. The use of these materials would be more difficult than that of elite germplasm mainly due to linkage drag. Therefore, these accessions should be used only when the elite materials have been exhausted as the source for drought resistance.

Table 7.3: Some selected examples of wild relatives of crops having resistance to abiotic stresses

Crop	Related wild species	Resistant to abiotic stress
Wheat (*Tritium aestivum*)	*Aegilops variabilis*	Drought
	Ae. speltiodes	Drought
	Ae. umbellulata	Drought
	Ae. squarrosa	Drought
	Agropyron pontica	Salinity
	A. cliac	Salinity
Sugarcane (*Saccharum officinarum*)	*Saccharum spontaneum*	Drought and salinity
Tomato (*Lycopersicon esculentum*)	*Lycopersicon cheesmanii*	Salinity

The **related wild species** are valuable sources of useful traits for crop improvement. The wild relatives adapted to arid environments are known to be excellent sources of drought resistance traits (Table 7.3). In many cases, the characteristics of these species that confer drought resistance may not be usable in the concerned crop species. In any case, their utilization may present problems in hybridization and linkage drag might reduce the performance of the lines derived from their crosses. They should be used in breeding programmes only when sources of drought resistance are not available even in the land races of the crop. Finally, appropriate genes from other organisms may be isolated and cloned; these genes can be used to produce transgenic plants resistant to different stresses, including drought. A number of genes potentially useful for creation of abiotic stress resistant lines have been cloned and evaluated. Some of these transgenic lines have been released for commercial cultivation (Chapter 10).

7.2.4. Selection for Drought Resistance

In breeding programmes, germplasm lines and the segregating generations from crosses have to be evaluated for the features of the crop species being improved. The selection for drought resistance would be based on one or few traits that confer drought resistance. In case the drought avoidance feature is to be improved, the selection can be based on delayed leaf rolling, lower canopy temperature, a combination of delayed leaf rolling and leaf firing (drying of leaves), leaf attributes (dense pubescence, heavy glaucousness, etc.), root traits (deep root system, root-length density, etc.), and leaf water retention. Out of these, leaf rolling (in rice at IRRI, Philippines), canopy temperature (in case of wheat at ICARDA, Syria), and leaf attributes are easily scored on a large scale and are being used in different crops. The root traits, on the other hand, are very difficult to evaluate, and are not useful as selection criteria for segregating populations.

Several different measurement criteria are available for dehydration tolerance. It can be assayed by seedling growth under PEG (polyethylene glycol) stress, growth under drought stress in the field, changes in plant phenology (e.g., delayed or accelerated flowering), grain filling by translocated stem reserve, cellular membrane stability under drought stress, seed germination in an osmoticum, the $^{13}C/^{12}C$ ratio of the biomass, and selection of cell lines for resistance to osmotic stress created by PEG. Out of these selection criteria, seedling growth under PEG stress, changes in plant phenology, and grain filling by translocated stem reserve are relatively easy to assay and are being used to some extent is certain crop species.

A good selection criterion should be (i) easy to score, (ii) have reasonably high heritability, (iii) be significantly correlated with drought resistance, and (iv) exhibit positive association with yield under drought. In addition, (v) there should be adequate genetic variability for the trait in the concerned crop species. It may be that none of the traits for drought resistance can serve as a satisfactory criterion of selection for drought resistance. Therefore, the breeding programmes generally use a combination of few to several traits as selection criterion for drought resistance. These traits are generally used to construct a selection index that is used for selection of the plants/lines. In general terms, a **selection index** for a plant/line is a numerical value derived from the performance of this plant/line in terms of the different traits comprising the selection index; the values for the different traits are adjusted according to the specific importance assigned to the individual traits. For example, a selection index for drought resistance in wheat is based on canopy temperature, chemical desiccation, and the presence of large awns; this index is used by the wheat breeders in Israel.

7.2.5. Creation of Drought Environment

The success of a breeding programme for drought resistance depends to a large extent on creation of the correct drought environment for evaluation of the plants/ lines/populations. The creation of the correct drought environment will not only involve the control of the water regime, but also of the mineral nutrition, weeds, diseases, etc. Drought conditions can be created in a more controlled manner under greenhouse conditions. But the number of plants/lines that can be evaluated in a greenhouse is limited, and these conditions may not accurately represent the conditions prevailing in the field to which the selected materials will be exposed. The drought conditions can be created under the field conditions in one of the following three ways.

1. The field may be located in a low rainfall area, and the level of water stress can be regulated by providing controlled irrigation water.
2. Alternatively, a line-source gradient can be generated as follows: the irrigation water is provided by either a row of sprinklers or an irrigation channel. The test lines are grown in lines on either side of the channel/ sprinklers, and these lines run perpendicular to the channel/sprinkler row (Fig. 7.1). The moisture availability declines as one move away from the irrigation channel/sprinklers, and the degree of drought stress increases. Thus, a line of the crop is simultaneously evaluated under a variety of drought stress regimes.

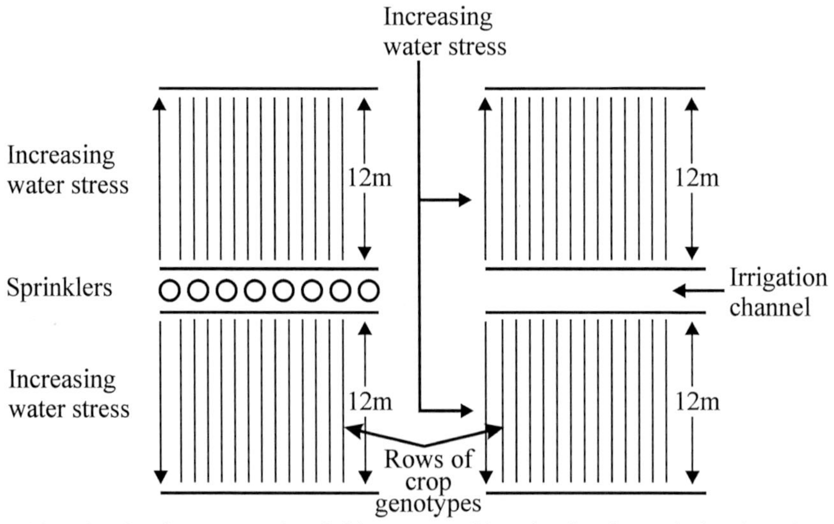

Fig. 7.1. Creation of drought environment by the line-source gradient strategy. (a) Irrigation by a row of sprinklers. (b) Irrigation by an irrigation channel.

3. Finally, two sets of plots may be created: in one set, there is no water stress; while the other set is subjected to water stress. A given material is evaluated under both the conditions, and the difference in the performance under the two conditions indicates the level of drought resistance of the material. In general, the larger is this difference; the lower is the drought resistance of the material.

7.2.6. Breeding Strategies for Drought Resistance

The breeding objective should ideally be to combine drought resistance with high yield potential. This presents a problem: the yield potential of a genotype can be evaluated only under an optimal or non-stress environment, while drought resistance has to be assessed under a drought environment. Attempts have been made to devise breeding strategies that facilitate achieving of both the above objectives. One of these strategies is to develop varieties with enhanced adaptation to a specific environment. In this approach, the breeding materials are evaluated and selections are carried out under the drought environment. This approach is used for those regions where the crop is grown on moisture stored in the soil. It suffers from several limitations like the need for extreme care in preparation of the fields, reduced effectiveness of selection, etc. In any case, the varieties developed by this approach would not be able to take advantage of favorable moisture regimes that may occur during some of the years.

The other approach has been designed to develop varieties for variable drought environments. In this scheme, the crosses are made and the generation is advanced to F_3 under optimal environment. The F_3 individual plant progenies are harvested separately and the F_4 progenies are evaluated under a range of environments ranging from optimal to drought conditions. The progenies with high mean yield over the environments are selected, and their F_5 progenies are grown under optimal conditions. Individual plants are selected for yield and yield traits and their F_6 progenies are evaluated as in F_4. The F_7 progenies are handled as in F_5, and the best progenies are selected and evaluated for release as new varieties.

In another approach, the yield potential and the traits contributing to drought resistance, e.g., cuticular wax, are used as the bases of selection. The cross in made and the generation is advanced to F_4 under optimal moisture conditions. In F_4, a large number of plants are harvested, their F_5 progenies are grown under moisture stress and selected for cuticular wax. In F_6, the selected progenies are grown under optimal moisture, and selection is based on yield, yield traits and quality. The procedure of F_5 may be repeated in F_7, and that of F_6 in F_8. The superior progenies selected from this scheme are evaluated for release as new varieties.

The development of molecular markers has greatly facilitated the improvement of complex traits like resistance to abiotic stresses. Markers are used to identify and map the quantitative trait loci (QTLs) involved in the control of the trait of interest, say, drought resistance. The markers flanking the QTL regions are used to select for the target trait, drought resistance in this case. For examples, QTLs for drought resistance have been identified and mapped in maize, and markers have been used to select for these QTLs. Similarly, markers were used to identify and map the QTLs governing the stay-green trait in sorghum, and to transfer these QTLs in the desired genotypes. Stay-green trait is the ability of some genotypes to remain green much longer than other genotypes of the same crop when subjected to drought stress, i.e., it reflects drought resistance. The use of molecular markers for selection of the QTLs involved in abiotic stress resistance avoids the use of other selection criteria that may present one or the other problem. It also makes the selection independent of the prevailing environment, and thereby eliminates the need for creation of the stress environment each time the selection is to be done. The lines selected in this way are rigorously evaluated for their abiotic stress resistance before they are used for cultivation/in breeding programmes. Molecular markers can also be used in the breeding for resistance to other abiotic stresses.

7.2.7. Difficulties in Breeding for Drought Resistance

1. It is very difficult to precisely define the drought environment of the region for which the variety is to be developed. This is because there is considerable year-to-year variation in the moisture stress prevailing in the region.
2. The creation of the drought environment for selection is generally quite problematic.
3. The results from greenhouse have to be confirmed in the field.
4. Selection for drought resistance cannot be based on a single trait, and an index of few to several traits has to be used. The application of a selection index is quite a demanding activity.
5. The assays for many drought resistance traits are cumbersome and tedious.
6. Many traits associated with drought resistance tend to reduce yield, e.g., earliness.
7. Generally, the breeding schemes for drought resistance are more elaborate, and require considerable resources and additional work.
8. The use of wild relatives in breeding for drought resistance is difficult and demanding.

7.2.8. Drought Hardening

Drought hardening is the increase in resistance of a crop genotype/line to drought stress as a result of certain seed/seedling treatment(s). The drought hardening treatments comprise pre-sowing and post-sowing treatments. The **pre-sowing treatments** are applied to seeds prior to their planting in the field. An example of pre-sowing treatments involves soaking of the seeds in water for 24 h, followed by sun-drying, after which the seeds are sown in the field. This simple seed treatment has been found to improve drought resistance in several crops like barley, sunflower, etc. The **post-sowing treatments**, in contrast, are applied to seedlings. For example, the young seedlings are subjected to mild moisture stress for a relatively brief period. This exposure to mild moisture stress is reported to improve the drought resistance of wheat, oats, etc. plants in the later stages of growth. Different varieties of the crops species differ in their drought hardening response. Thus, the level of drought hardening seems to have a genetic basis, and it might be possible to improve it by suitable breeding approaches.

7.3. BREEDING FOR RESISTANCE TO SALINITY STRESS

The mineral salts present in soil often generate stress in the crop plants grown in these soils. The stress due to soil minerals is the second most important abiotic stress after the drought stress. These stresses are primarily of the following three types: (i) deficiency of a specific mineral salt (mineral deficiency), (ii) adverse effects generated by an excess amount of a specific mineral salt (mineral toxicity), and (iii) stress caused by an excess accumulation of soluble salts in the root zones of the plants (salinity and alkalinity). The various aspects of mineral deficiency and mineral toxicity are discussed in Sections 7.4 and 7.5. In this section, the nature of salinity resistance and the breeding strategies for this trait are described.

7.3.1. Salt-Affected Soils

A soil having an excess accumulation of soluble salts in the root zones of the plants is called **salt-affected soil**. In India, nearly 5% of the cultivated land is salt-affected. Salt-affected soils are mainly of two types, viz., saline and alkaline soils. In simple terms, the electrical conductivity of the saturated soil paste of saline soils is more than 4 dS/m (decisiemens per meter). The saline soils are classified into the following four categories: (i) slightly saline (electrical conductivity, 2-4dS/m), (ii) moderately saline (4-8dS/m), strongly saline (8-16 dS/m), and (iv) very strongly saline (>16 dS/m). In case of slightly saline soils, the yields of only sensitive crops are affected, while in very strongly saline soils only very tolerant crops would give satisfactory yields. The alkali of sodic soils, on the other hand, have excess, usually >15%, of exchangeable sodium. In addition, their pH is always more than 8.2. Some other types of salt-affected soils are also known, e.g., acid sulphate soils, degraded alkali soils, etc.

The agricultural usefulness of salt-affected soils can be improved by the following two approaches: (i) reclamation, and (ii) management of these soils. **Reclamation** comprises a suitable treatment that effectively removes the excess salts from the root zone of the plants. The management of salt-affected soils comprises the following: appropriate cultural practices, including method and time of planting, irrigation and other cultural practices; selection of tolerant crops for cultivation (Table 7.4); and development of salt tolerant varieties. The plants are divided into two categories, namely, glycophytes and halophytes, on the basis of their adaptation to saline conditions. The glycophytes are not adapted to saline conditions and are represented by the crop plants. But the halophytes are adapted to grow in saline soils and are represented by wild species of plants.

Table 7.4: A list of some selected crop species with different levels of salinity tolerance

Category of salinity tolerance	Crop
Sensitive	Common bean, carrot, okra, onion
Moderately sensitive	Cowpea, cabbage, maize, peanut, potato, rice, sugarcane, sweet potato, tomato
Moderately tolerant	Barley (forage), safflower, sorghum, soybean, wheat
Tolerant	Barley (grain), cotton, sugar beet

7.3.2. Salinity Stress and Salinity Resistance

The plants growing in saline soils are subjected to the following three types of stresses. (i) These plants are subjected to water stress generated by the salt solution in the root zone. The water stress is considered to be an important component of the salt injury to plants. (ii) However, excess salt accumulation leads to salt toxicity. For example, sodium chloride adversely affects the enzymes of glycolytic pathway. Salinity has adverse effects on energy generation by oxidative phosphorylation, interferes with protein synthesis, etc. In short, ion toxicity remains the central feature of salinity stress. finally, (iii) The excess salts present in the root zone may disturb the balanced mineral nutrition of the plants. The different crop species differ considerably in their tolerance to salinity stress (Table 7.4).

Salinity resistance is the ability of some genotypes of a given species to perform better than the other genotypes under salinity stress. Salinity resistance may arise due to the following three features of the crop genotypes. (i) The resistance to the water stress produced due to salinity is the major mechanism that minimizes the adverse osmotic effects of salinity. The resistant genotypes crop plants show superior osmoregulation to that of sensitive genotypes. The osmoregulation may be achieved by organic compounds like sugars, organic acids, proline, etc. or by accumulation of K^+ ions in the vacuole (in case of sorghum). The organic compounds involved in osmoregulation may also protect the proteins and enzymes, and cellular membranes. (ii) Resistance to salt toxicity effects of salinity involves that maintains the salt level in the cell cytoplasm to nontoxic levels. There are two main mechanisms for achieving this, viz., salt exclusion and cellular compartmentation. **Salt exclusion** is the ability of some genotyped to take up smaller quantities of the ions like Na^+ and Cl^- than the other genotypes. Salt resistant lines of pearl millet, rice, soybean, tomato, etc. are known to exhibit salt exclusion. In some cases, salt tolerance is presumed to be based on cellular compartmentation of the injurious ions. Finally, (iii) a combination of both the above features may be the basis of salt resistance in some cases.

7.4. BREEDING FOR SALINITY STRESS RESISTANCE

There are three issues relevant to the breeding for salinity resistance, viz., (i) creation of saline environment, (ii) measurement of salinity resistance, and (iii) the genetics of the salinity resistance. The saline environment can be created in one of the following three different ways as discussed below.

1. The saline condition is created in a nonsaline field with sand-loam soil and good drainage. The field is first irrigated with excess amounts of water of known salinity levels to remove the accumulated salts and to establish known levels of salinity. The test population of the concerned crop is grown in this field, and excess irrigation is applied as required.
2. Controlled levels of salinity can be created by constructing especially designed micorplots.
3. The greenhouse environment can be used to generate environments of specified salinity levels using Petri plates with filter papers or antibiotic supplemented agar medium, pots filled saline soil/sand/gravel, or hydroponic culture system. The salt solution used for creating the saline environment and in evaluation should represent the salt composition present in the soil at the target site. The salts that are common in saline soils are NaCl, Na_2SO_4, KCl, $MgCl_2$, and $CaCl_2$.

The salinity resistance of a plant genotype or line can be assessed by several different tests, some of which are as follows.

1. Cell survival is regarded as a better criterion than dry weight or yield. This is measured by immersing tissue sections in a saline solution (generally, a range of salinity levels is used) followed by determining the proportion of plasmolysed cells in response to a hypertonic glucose solution.
2. Seed germination in a saline medium may be a good selection criterion in those species that are more sensitive to salinity during germination.
3. The dry weights of seedlings or plants is a good measure of salinity resistance. However, the dry weights of the test genotypes should be determined both under saline and nonsaline conditions.

4. The level of osmoregulation can be estimated in terms of proline or carbohydrate accumulation is response to salinity. This assay is important in such species, in which osmoregulation is a significant contributor to salinity resistance.
5. Leaf death in response to salinity is scored visually and may be really useful in the selection of potentially resistant plants from segregating populations; the resistant plants would show much lower leaf death than the susceptible ones.
6. Yield is an important selection criterion for salinity resistance. This criterion ensures that the selected lines are resistant to salinity during the grain filling stage as well.
7. Some other measures of salinity resistance are leaf ion content, leaf necrosis, and root growth.

Salinity resistance can be found in a cultivated variety, germplasm collection, related species, or it may be produced using suitable transgene(s). Sometimes, evaluation of the cultivated varieties may identify a variety resistant to salinity, e.g., in muskmelon. Generally, the **germplasm accessions** collected from salinity affected areas provide excellent sources of resistance, e.g., Kharchia wheat collected from Kharchi region of Rajasthan, many rice lines collected from salt affected areas, etc. Many wild relatives of crop plants are good sources of salinity resistance, e.g., some *Elytriga* (Syn., *Agropyron*) species like *E. pontica*, *E. elongata*, etc., and some wild relatives of tomato like *L. peruvianum*, *L. pennellii*, etc. In addition, cultured plant cells can be selected for resistance to elevated levels of salt with a view to isolate somaclonal variants resistant to salinity. Many studies on this aspect have been carried out with variable levels of success and at least one salt tolerant somaclone has been released as a variety ('Andro' in flax). Some transgenes like *ADC, coda, mtl1D*, etc. involved in osmoregulation are also useful in generating salinity resistance (Chapter 10). The resistance present in cultivated variety is the easiest to utilize, while those present in a land races and related species present several problems including linkage drag. The transgenes can be used only by genetic transformation (discussed in some detail in Chapter 9).

The breeding methods used for breeding for salinity resistance are selection, hybridization, cell selection, and genetic engineering. Selection in the germplasm collected from the salt affected areas has been used successfully to develop several salt resistant varieties, e.g., in rice (Table 7.5). Hybridization has been widely used to develop salt resistant varieties with superior yield potential. When the salt resistant parent has some useful agronomic features as well, the pedigree method of breeding would be used. But if the salt resistant parent is agronomically inferior, the backcross method of breeding should be used. This is particularly true for transfer of resistance from the related species. Wild relatives of some crop species possess salinity resistance and some of them have been used in breeding programmes. For example, salinity resistance from *Lycopersicon cheesmanii* ssp. *Minor* has been transferred into tomato and some salinity resistant lines have been developed. Similarly, attempts have been made to transfer salinity resistance from *Elytriga* spp. into wheat. Interspecific gene transfers generally present several problems, and often the usefulness of these transfers may not be as expected.

Table 7.5: A list of some selected salt resistant rice varieties developed by selection in materials collected from salt affected areas.

Salt resistant selection	Local variety, from which selected
Arya 33	Arya
BR4-10	Bhura Ratta
MCM2	Budda Molagolukulu
Vytila 1	Choottupokkali
CSR1	Damodar
CSR2	Dasal
CSR3	Getu
Jhona 341	Jhona
KR1-24	Kala Ratta
SR26B	Kalambank
Hamilton	Nona Bokra
SCR6	Nonasail

Breeding for salinity resistance presents the following problems. (i) The creation of controlled saline environment suitable for selection work is demanding and difficult. (ii) The scoring of salinity resistance is not easy and dependable selection criteria are not available. (iii) The genetic control of salinity resistance is complex, which makes the breeding work quite difficult. Finally, (iv) the mechanisms involved in salinity resistance are poorly understood, which makes the genetic analysis of the trait quite difficult.

7.5. BREEDING FOR RESISTANCE TO MINERAL STRESSES

Mineral stress may be defined as adverse effects of either an excess of minerals or a deficiency of an essential mineral nutrient in the soil on the performance of a crop. The adverse effects on crops of the excess amounts of minerals in root zones of plants are called **mineral toxicity**. In contrast, the adverse effects on crops of low amount of a specific mineral nutrient in the soil are termed as **mineral deficiency**.

7.5.1. Breeding for Resistance to Mineral deficiency Stresses

The ability of some genotypes of a crop species to suffer less adverse effects due to mineral deficiency stress than the other genotypes of the same crop is called **mineral deficiency resistance**. Further, such genotypes are referred to as **mineral deficiency resistant genotypes** or **mineral efficient** genotypes. The mineral efficient genotypes usually absorb the deficient mineral in higher quantities from the soil than the other genotypes. Alternatively, they are able to transport greater quantities of the deficient mineral to the appropriate plant parts. These strategies are called **avoidance of mineral deficiency** since the relevant tissues/organs of the plants do not experience deficiency of the concerned mineral(s). On the other hand, some genotypes might suffer less damage to the essential life processes than other genotypes of a crop; this is known as **tolerance to mineral deficiency**. Mineral deficiency tolerance seems to be a rare phenomenon.

The performance of mineral-deficiency resistant genotypes under mineral deficiency conditions generally comparable to their performance under conditions without mineral deficiency. These genotypes resist mineral deficiency due to one or more of the following: (i) mineral redistribution within the plants, (ii) efficient uptake of minerals from the soil, (iii) increased growth of roots as compared to that of shoots, and (iv) increased transport of minerals to the shoots from the roots. Redistribution of N, P and K is well known, but that of Mg is also reported in some cases. The roots of mineral-efficient genotypes release higher amounts of organic acids, which enhance the acidification of the soils in the root zone. This increases mineral availability and their uptake by the plants. Increased transport of Fe from root to shoot is known to be involved in the cases of crop varieties that are resistant to iron-chlorosis when grown in calcareous soils, e.g., maize variety WF9. Finally, several genotypes of crops show increased root/shoot ration under conditions of mineral deficiency. Further, genotypes with higher root/shoot ratio seem to be superior in their ability to avoid mineral deficiency. But this feature is likely to restrict their yields under mineral deficiency.

There is ample genetic variation within species for mineral deficiency resistance in various crops. In many crops, mineral deficiency resistance is reported to be under polygenic control. But in many other cases, mineral-efficiency is governed by single major genes. For example, the tomato gene *btl* is considered to reduce boron transport from the roots. Similarly, iron-efficiency in soybean and maize is governed by single genes, but is also affected by several minor genes (Table 7.6). The sources of mineral deficiency resistance are found in the following materials: (i) cultivated varieties, (ii) land races or local varieties, (iii) mutant stocks, and (iv) related wild species. Resistance to mineral deficiency may be found in the cultivated varieties of crops. This is particularly true when varieties of a crop have been developed in an area that has deficiency of the given mineral. For example, several of the wheat varieties developed in

Australia were found to be resistant to Mn-deficiency since Australian soils are generally Mn-deficient. Mineral deficiency resistance is often found in the land races or local varieties grown in and adapted to the regions having deficiency of a given mineral. Cultivated varieties are the easiest to use in breeding programmes, while the use of land races may present some problems. The related wild species are the most difficult to be used in breeding programmes. Therefore, they should be the last option for the development of mineral deficiency resistant varieties.

Table 7.6: Some selected examples of mineral deficiency resistance in crop plants

Crop	Genetic control	Mineral efficiency/ inefficiency	Remarks
Tomato	Single gene (*btl*)	B-inefficiency	Causes brittle stem
Maize	Single gene (*ysl*)	Fe-efficiency	The mutant is not able to use Fe^{3+} ions
Soybean	Single gene (*fe*)	Fe-efficiency	Minor genes also involved
Tomato	Single gene (*Fer*)	Fe-efficiency	Minor genes also involved
Wheat	Single gene (*Ku*)	K-efficiency	Increased uptake from sodic soils
Barley	Polygenic (?)	Mn-inefficiency	Introduced from Egypt
Several crops	Polygenic	P-efficiency	Likely to be based on root traits
Soybean	Single gene (*np*)	P-sensitivity	The mutant strain *np np* is highly sensitive to P-deficiency

Selection for mineral deficiency resistance may be based on the the following selection criteria. (i) The selection could be on the basis of visible symptoms of mineral deficiency since the deficiencies for different elements produce easily recognizable symptoms. This strategy is simple, fast and cheap. It is generally used for selection of resistance to iron deficiency, and has been used for selection for zinc deficiency resistance as well. (ii) The extent of deficiency for certain elements can be determined by assaying the activities of certain enzymes and accumulation of certain biochemicals. For example, peroxidase activity is positively correlated with iron content of tissues in oats, tomato and citrus. Similarly, ascorbic acid oxidase is positively correlated with tissue copper content in several crop species. These assays are highly specific and often quite sensitive, but they are costly and time taking. (iii) The yield of the crop under mineral deficiency could be used as the basis of selection. But the heritability of yield declines markedly under the conditions of stress. Therefore, one strategy is to determine the yield of different genotypes both under mineral deficiency stress and nonstress conditions. The selection should be based on the mean yields of the genotypes under stress condition, and the ratio of yield under stress to that under nonstress condition. Alternatively, the different genotypes are evaluated for yield only under the stress condition, and a check variety with known level of mineral deficiency resistance is planted after every few test genotypes. The yields of the test lines are compared with that of the check variety, and superior performing lines are selected. Thus, the yield cannot be used as a criterion for the selection of individual plants from segregating populations. (iv) The mineral content of plant tissues has a limited usefulness in selection for resistance to mineral deficiency for a variety of reasons.

The breeding schemes used for the development of mineral deficiency resistant genotypes are similar to those for other abiotic stresses. If the resistance is present in a cultivated variety or elite germplasm, the resistant parent can contribute to some useful agronomic traits as well. In this case, the segregating generations are handled according to the pedigree scheme. But if the mineral deficiency resistance is present in a land race or, especially, in a related species, the backcross method of breeding has to be used to transfer the genes for resistance into a high yielding but mineral deficiency sensitive variety. The segregating generations have to be evaluated under conditions of deficiency of the concerned element(s). The mineral deficiency condition can be created in one of the following ways: (i) growing the population in a field that has been determined to show a defined level of homogeneous mineral deficiency, (ii) growing the plants in pots filled with soil having the desired level of mineral deficiency, or (iii) the plants may be grown in a hydroponics system. The hydroponics system provides a good control on mineral deficiency level, but it requires some degree of expertise and the necessary infrastructure. In addition, only a limited number of plants can be evaluated at a time in hydroponics. Therefore, it is generally used for evaluation of the parental lines and the selected lines.

7.5.2. Breeding for Resistance to Mineral Toxicity Stresses

As noted above, **mineral toxicity stress** is generated by the adverse effects of high levels of available minerals in the root zones of plants. About 40% of the global arable lands are acid soils. Such soils often contain toxic levels of aluminum (Al) and manganese (Mn). **Mineral toxicity resistance** is the ability of some genotypes of a crop species to suffer less amount of damage due to mineral toxicity than other genotypes of the same crop. The mechanism of mineral toxicity resistance may differ from one mineral to the other. For example, Al toxicity is generally observed at a pH below 5.5. Al forms complexes with various molecules, including proteins, which are considered to lead to the adverse effects of Al. Al toxicity also induces deficiency for P, Ca or Fe, so that the deficiency symptoms of these minerals may also develop. Al resistance in some crop plants like rice, peas, and barley, wheat and maize results from their ability to raise the pH of the rhizosphere. In some cases, the uptake of Al by roots is greatly reduced. In some other cases, Al is excluded from shoots and is stored in the roots, it may be accumulated in older leaves, or it may be safely stored in a cellular compartment.

Similarly, Mn toxicity occurs in soils with pH below 5.5, and is the second most common mineral toxicity. Mn toxicity reduces activities of many enzymes and affects metabolisms of Ca and Fe. The chief strategy of resistance to Mn toxicity is avoidance, but in some cases tolerance may be involved. In some crops like maize, there is reduced transport of Mn from roots to shoots. It has been suggested that Mn^{2+} is oxidized to Mn^{4+} in the roots. The rate of this oxidation is presumably higher in the resistant than in the sensitive genotypes. Finally, tolerance to toxic levels of Mn has been implicated in several cases, but the evidence supporting this inference is inconclusive.

Resistance to mineral toxicity is governed by a single gene in the case of some crops. For example, resistance to Al toxicity in barley (gene *Alp*) and maize (single gene with multiple alleles) are governed by single major genes. But in many cases, resistance to mineral toxicity is polygenically inherited. Some examples of polygenic inheritance are Al toxicity resistance in sorghum and alfalfa, and Mn toxicity resistance in soybean and alfalfa. In the case of Mn toxicity resistance in soybean, maternal effects have also been recorded. The important sources of mineral toxicity resistance are as follows: (i) old/local varieties and germplasm accessions adapted to acid soils with known mineral toxicities, and (ii) improved cultivated varieties developed for cultivation or being cultivated in acid soils. For example, the wheat varieties developed in Brazil for cultivation in acid soils are excellent sources of Al resistance.

Selection for mineral toxicity resistance would require the creation of a reliable mineral toxicity environment with known/specified mineral toxicity. Such an environment can be created by the following approaches: (i) a field with known mineral toxicity problem, (ii) pots filled with soil with the specified mineral toxicity problem, (iii) a sand culture system based on a nutrient solution with the specified mineral toxicity problem, and (iv) a hydroponics system based on such a nutrient solution. A field with mineral toxicity problem is generally used to carry out selection since it allows evaluation of a large number of plants at a very low cost. Further, the plants are evaluated under conditions in which they will be grown, and the yields of the selected plants/lines can also be determined. However, selection for individual mineral toxicities is not possible in the field since most acid soils show both Al and Mn toxicities. The pot sand and hydroponics cultures are maintained under controlled conditions and should be used to supplement the field data. The sand cultures are cheaper and less technically demanding than hydroponics systems, but both these systems allows selection for individual mineral toxicities.

The breeding strategies for the development of mineral toxicity resistant varieties are similar to those for production of mineral deficiency resistant genotypes. If the source of toxicity resistance is a cultivated variety, the resistant parent may provide some genes governing useful agronomic traits in addition to the mineral toxicity resistance. Therefore, the segregating generations form crosses involving such parents are handled according to the pedigree method of breeding. But if the mineral toxicity resistance is present in a land race, the backcross method of breeding has to be used to transfer the concerned genes for resistance into a high yielding but mineral toxicity sensitive variety. In case of polygenic resistance, the pedigree method is desirable so long as the toxicity resistant parent is not an agronomically inferior line.

The criteria of selection would depend on the specific mineral toxicity being selected for. In case Al toxicity, shoot dry mass (in pot and nutrient cultures), root length, root weight, and root discoloration and deformity (in hydroponic cultures) are evaluated. But in those cases, where resistance is based on exclusion of Al from roots, the seedling roots are stained with hematoxylin, and seedlings with little or no stain are selected. Yield is the selection criterion used for identification of resistant lines in field evaluations. It is desirable that the yield of each line is determined both under toxicity stress and nonstress conditions. The lines with high mean yield and low yield reduction in response to stress are selected. In case of Mn toxicity resistance, shoot growth estimated as shoot dry matter or by some other suitable method is an excellent selection criterion. The shoot growth should be measured both under Mn toxicity stress and without the Mn toxicity stress. The lines with high mean shoot growth and low reduction in shoot growth response to Mn toxicity stress are selected as Mn toxicity resistant lines/genotypes.

7.5.3. Problems in Breeding for Mineral Deficiency/Toxicity Resistance

1. The characteristics of the mineral toxicity/deficiency in the soil vary with the location of the problem soil. As a result, varieties developed for mineral toxicity/deficiency resistance in one area may not be useful in the other areas with problem soils.
2. Most acid soils show both Al and Mn toxicities. Therefore, breeding should address both these problems.
3. Both Al and Mn interact with other mineral nutrients. This requires careful selection and evaluation protocols for effective development of resistant varieties.
4. Root studies require hydroponics systems, which are expensive and technically demanding.
5. Chemical and biochemical analyses would be needed when the mineral concentration in the soil not low enough to cause deficiency symptoms. These tests are costly and require considerable expertise.
6. Finally, the selected materials have to be evaluated both under stress and nonstress conditions. This doubles the amount of work needed for the evaluation.

7.6. BREEDING FOR RESISTANCE TO TEMPERATURE STRESS

Each crop species has an **optimum range of temperatures** within which it shows normal growth and development, and its performance is not adversely affected. The actual range of the optimum temperature depends on the crop species as well as the developmental stage of the plants. For example, the stage between initiation of spike and flowering in wheat is highly sensitive to heat stress. When the prevailing temperature during the crop growth and development is outside the optimum range, it adversely affects the performance of the crop; this is referred to as **temperature stress**. The temperature stress is classified into the following three different categories: (i) heat stress, (ii) chilling stress, and (iii) freezing stress.

7.6.1. Breeding for Resistance to Heat Stress

Heat stress is the adverse effect on the performance of a crop by a temperature higher than the optimal temperature range for the concerned crop. Higher temperatures affect the following features of a crop: various physiological processes, growth and development, and ultimately survival. The nature and degree of the adverse effects of a given temperature depend mainly on the crop species and the process of the crop being monitored. When the temperatures are as high as 48°C, cell and tissue death is common in organs like fruits, twigs, trunks, etc. since they lack transpirational cooling. Heat causes flower abscission, pollen death and poor fruit set. The heat effect on growth and development depends on the temperature and the duration for which the plants are exposed to this temperature. In general, heat accelerates developmental processes and, consequently, reduces yield; sometimes, it may cause premature senescence.

Temperature affects a variety of physiological processes, of which the following are the most relevant: respiration, photosynthesis, photosynthate translocation, protein denaturation, membrane composition and stability, and heat shock proteins. Heat increases respiration, suppresses photosynthesis, and reduces photosynthetic translocation to the grain, especially in cereals. Photosynthesis is far more sensitive to heat than respiration, and photosystem II is much more sensitive than photosystem I. Heat might cause conformational changes in proteins particularly near the lethal temperatures. The fatty acid composition of the membranes changes with temperature, and the fluidity of the membrane lipids declines in plants exposed to higher temperatures. Heat exposure of plants accelerates the synthesis of a group of proteins called **heat shock proteins** (Hsps). Hsps are a group of diverse proteins that are produced at all the temperatures, and some of them function as chaperonins (they minimize protein denaturation). Some Hsps may condition heat tolerance in crop plants.

The ability of a line/variety of a crop to perform better than other lines/varieties of the crop when subjected to the same level of heat stress is known as **heat stress resistance**. There are two basic mechanisms of heat resistance, viz., heat avoidance and heat tolerance; the factors that contribute to these phenomena are summarized in Table 7.7. **Heat avoidance** comprises such features that either reduce the light energy absorbed by plants (leaf pubescence and glaucousness) or dissipate the light energy absorbed by the plants (transpiration). However, cooling by transpiration increases the water loss from plants. Therefore, the effects of heat avoidance are closely associated with dehydration avoidance particularly in conditions of water stress. In contrast, **heat tolerance** is the ability of some genotypes of crops plants to perform better than other genotypes of the same crop when their internal temperatures are comparable and higher than the optimum temperature range for the concerned crop. Heat tolerance is expressed following heat hardening. **Heat hardening** is the improved ability of a genotype to withstand a period of heat stress as a result of an earlier exposure to high temperature for a period of time (generally, for <1 hr). But when heat hardened plants are exposed to lower temperatures, they become as sensitive to heat as the non-hardened plants; this is called **dehardening**. Heat tolerance involves the following features: (i) increase in membrane thermostability (reflected by reduced membrane fluidity), (ii) increased thermostability of photosystem II (assayed as variable chlorophyll fluorescence at 685nm), (iii) greater photosynthate translocation, (iv) better mobilization of stem reserves particularly in cereals, (iv) improved osmoregulation due to proline, glycine-betaine, etc. accumulation.

Table 7.7: The features/processes that contribute to heat avoidance and heat tolerance in plants

Mechanism of heat resistance	Feature/process	Effect/remarks
Heat avoidance	Transpiration	Cooling of cells and tissues
	Leaf reflectance generated by:	The light received by leaves is greatly reduced
	1. Pubescence	
	2. Glaucousness	
	Insulation by bark	Heating of the shoot is reduced
Heat tolerance	Membrane stability	These processes are associated with heat hardening and, possibly, Hsp synthesis
	Photosystem II stability	
	Photosynthate translocation	
	Stem-reserve mobilization	
	Osmoregulation	

Genetic variation for heat resistance is present in most crop species, e.g., in maize, potato, tomato, soybean, common bean, etc. Generally heat resistance is under polygenic control. In most cases, both additive and dominance gene effects are involved, but in most cases, additive gene effects appear to be more important than dominance effects. But in some cases, e.g., in tomato, the dominance effects are more important than additive effects. The heritability estimates vary from moderate to high depending on the crop and the genotypes used in the study. However, there is at least one case of oligogenic control of heat stress; in snap bean (cultivars of the common bean, *Phaseolus vulgaris*, grown for green pods) one or two major genes are reported to control heat resistance measured as number of pods/plant under heat stress.

In most crop species, sources of heat resistance can be found in the breeding materials provided they are carefully evaluated for the heat resistance traits. In case, the source for heat resistance has to be searched in materials other than the breeding germplasm, the accessions collected from areas that are exposed to heat stress should be evaluated. The use of breeding materials is the simplest and would involve the pedigree scheme of selection in the segregating generations. But the use of resistance present in land races and other unadapted germplasm would be based on backcross breeding. The selection for heat resistance can be done in a suitable environment having the appropriate temperature regime. It can be done in a **natural field environment**; this is the simplest approach, but the temperature regime is difficult to control. The selection can be carried out in the field environment at a location exposed to heat stress, or in the off-season that allows evaluation for heat stress resistance. **Controlled temperature regimes** of greenhouses and growth chambers can also be used to select for heat stress resistance. It is highly desirable to avoid water stress during evaluation for heat stress. Finally, certain assays for heat tolerance, e.g., membrane thermostability, can be done in the **laboratory** in the test tubes.

The following criteria can be used for selection for heat resistance. In general, (i) growth and (ii) yield under stress are used as a selection criteria. (iii) The reproduction related traits like flower, fruit, seed, etc. formation, pollen fertility etc. should be used for selection when heat stress is likely to occur during the reproductive stage of the crop. In some situations, heat stress may occur during germination; in such cases, (iv) germination under heat stress should be used as a criterion for selection. In addition, (v) recovery after heat stress, (vi) sensitivity of the photosynthetic process, and (vii) membrane stability following heat shock can also be used as selection criteria. Of these, membrane stability following heat shock is a promising criterion, and it shows high positive correlation with yield under heat stress at least in crops like soybean.

7.6.2. Breeding for Resistance to Chilling Stress

Chilling is the exposure of plants to temperatures above the freezing point, i.e., >0°C. Chilling sensitive plant species are those that are adapted to tropical environments. Chilling injury can be described both at the plant/organ level and at subcellular level. All these effects are mediated by the effects of chilling on the biological membranes. At the plant level, Chilling stress reduces germination (imbibition being the most sensitive stage) and seedling growth, and increases the risk of attack by soil pathogens. The low temperature reduces root growth, and delays the onset of photosynthesis in seedlings; severe chilling may irreversibly damage the roots. As a result of chilling, the stomata become locked open. Chilling also leads to the formation of abnormal flowers and fruits, reduce pollen fertility, and seed and fruit set. In general, flower development is more sensitive to chilling than fruit development.

At the subcellular level, chilling reduces membrane stability, chlorophyll synthesis, photosynthesis and respiration, and may generate H_2O_2. Chilling is proposed to induce a solid-gel state in the membranes in the place of the normal liquid-crystalline state. This affects the membrane function and the functions of the membrane-bound enzymes. The effect on membranes is regarded as the primary effect of chilling. Low temperatures interfere with chlorophyll synthesis and function, and reduce photosynthesis as well respiration. Chilling affected plants are unable to degrade H_2O_2, which tends to accumulate and cause injury. The photosynthates tend to accumulate in the chloroplasts of chill-affected plants and this may totally inhibit photosynthesis. In any case, respiration is much less sensitive to chilling than photosynthesis.

Some genotypes of crop plants survive and/or perform better than others under chilling stress; this is called **chilling tolerance**. In general, chilling tolerance is expressed in response to chill hardening. **Chill hardening** is the improved chilling tolerance of a genotype as a result of its earlier exposure to a near chilling temperature for the appropriate duration. Chilling tolerance may involve one or more of the following: (i) increased membrane lipid unsaturation, (ii) reduced sensitivity of photosynthesis to chilling, (iii) increased chlorophyll accumulation, (iv) decreased sensitivity of germination to low temperature, (v) improved fruit and seed set, and (vi) increased pollen fertility at low temperatures. Chill hardening increases membrane lipid unsaturation and chill tolerant varieties of crops like maize, beans, apple, etc. show higher degree of membrane lipid unsaturation. In some cases like maize, chill tolerant lines show reduced sensitivity of photosynthesis.

The studies on the inheritance of traits involved in chilling stress reveal that in most of the cases they are under polygenic control, and the additive gene effects seem to be predominant. But in some cases, these traits are governed by major effect genes. For example, the oligogene *Cts1* is involved in the reduced discoloration of rice in response to chilling stress (Table 7.8). In few cases, maternal effects are also implicated. The sources of chilling tolerance are as follows: (i) the elite germplasm or the breeding populations (at least in some crops) (ii) germplasm accessions, especially those collected from cooler climatic regions, (iii) induced mutations, (iv) somaclonal variants, (v) related wild species, and (vi) unrelated organisms. There are some instances of isolation of low temperature tolerant mutants in few crops like chrysanthemums. Cold tolerant somaclonal variants have been isolated by cell selection for survival at low temperatures. Low temperature tolerance has also been transferred from related wild species, e.g., from *Lycopersicon hirsutum* strains adapted to high altitudes into tomato. Finally, genes conferring low temperature tolerance can be transferred form unrelated species and expressed in the desired crop species to create chilling tolerant transgenic lines. For example, a chilling tolerant variety of strawberry has been produced by transferring the antifreeze protein encoding gene isolated from winter flat-fish.

Table 7.8: The inheritance pattern of some selected traits involved in chilling stress tolerance in some selected crops

Crop	Trait involved in chilling tolerance	Inheritance	Remarks
Tomato	Electrolyte leakage	Polygenic	GCA more important than SCA
	Germination	Polygenic	Additive gene effects; maternal effects; high heritability (~70%)
	Plant growth	At least three genes involved	-
Maize	Germination	Polygenic	Additive gene effects; maternal effects; high heritability
	Leaf discoloration	Simple Mendelian	-
Cucumber	Germination	Polygenic	Additive gene effects predominant
Rice	Germination	Polygenic	Three QTLs* identified
	Floret sterility	Polygenic	Additive gene effects predominant; heritability low to high (22-88%)
	Leaf discoloration	Oligogenic	Gene *Cts1*
	Seedling vigor	Four to five genes involved	

*QTLs, quantitative trait loci

The chilling resistance present in the elite germplasm is readily used in breeding programmes, and generally the pedigree scheme of selection is used to handle the segregating generations. The use of chilling resistance present in the inferior germplasm accessions and particularly in related species have to be used by backcross breeding. Genetic transformation technique (Chapter 9) has to be used for the transfer of genes from unrelated organisms. The segregating generations from crosses have to be evaluated in a suitable low temperature environment. Such an environment can be created by one of the following approaches: (i) growing the plants in the field at a suitable site that experiences low temperatures, and (ii) the use of controlled temperature environments like growth chambers, germination cabinets, thermo-gradient incubators, etc. Field environment is the most widely used, and selects for cold tolerance during germination and seedling stages as well. It is the simplest and the easiest to use, but there is little control on the temperature regime. The creation of controlled temperature environment is expensive and technically demanding, but allows evaluation at several different temperature regimes. The different selection criteria used are as follows: germination, growth, chlorophyll loss, membrane stability, photosynthesis, seedling mortality, seed/fruit set, and pollen sterility (all evaluated under chilling stress). Each selection criterion has some merits and some limitations.

7.6.3. Breeding for Resistance to Freezing Stress

Freezing stress comprises the detrimental effects of below freezing (<0°C) temperatures on the plants. Freezing stress is the most extensively studied stress in plants. Further, the breeding efforts for alleviating freezing stress are perhaps the most extensive of those directed at any abiotic stress management. The survival after freezing depends on the interactions among the crop plants, the temperature and the pathogens. When the temperature of water in plants falls below 0°C, it may either freeze and form ice or it may supercool to subzero temperatures without ice formation. Ice formation needs a nucleation source around which ice crystals develop; the nucleation sources may be present inside the plants or outside them. Plant cells usually are free from ice nucleators. The nucleation sources are generally present outside the plants. These nucleation sources include dust particles, particles of organic matter, some strains of bacteria like *Pseudomonas syringae*, *P. fluorescence*, and *Erwinia herbicola* or even gas bubbles. Some of these nucleation sources can lead to ice formation even at temperatures as high as -1°C.

The initiation ice formation at the surface of plants is enough to induce extracellular (in the intercellular spaces, xylem vessels, etc.) ice formation within the plants. Extracellular ice formation causes loss of water from the plant cells, and thus creates water stress in the cells and tissues. Intercellular ice formation is generally lethal most likely due to physical damage to the subcellular structures caused by the ice crystals. Damage to the plasma lemma may lead to loss of solutes from the cells. Thus, intracellular ice formation is the chief and the terminal freezing stress. Further, some factors external to the plants like ice sheet formation above and/below ground, pathogen attacks on the plants surviving freezing, and toxicity caused by dead cells to the surviving cells/tissues impose additional stress on the plants. In case of supercooling, the water cools below 0°C without ice formation. In case of plants, the internal water may cool down to up to -15°C in most herbaceous species and to up to -45°C in winter -hardy trees without formation of ice. It is important that these species lack any internal ice nucleators. Supercooling is considered as an important mechanism of freezing avoidance.

Freezing resistance is the ability of a genotype to survive the freezing stress, and to regrow after thawing. Freezing resistance may involve freezing avoidance or freezing tolerance. **Freezing avoidance** is the ability of tissues and organs of plants (but not whole plants) to avoid ice formation at below zero temperatures.

Freezing avoidance is based on supercooling, which is common in the meristematic tissues of flower buds of hardy plants and in some other tissues. It is favored by lack of internal ice nucleators, small cell size, little or no intercellular space, low moisture content, etc. But if the temperatures fall below -41°C, the supercooled cells would freeze and this will be lethal to the cells/tissues/plants. The **freezing tolerance**, on the other hand, is the ability of plants to survive freezing stress produced by extracellular ice formation, and to recover and regrow after thawing. Freezing tolerance is considerably affected by freeze-hardening. The enhanced freezing tolerance of plants as a result of their exposure to low temperatures for an appropriate duration at an earlier occasion is called **freeze-hardening**. In case of winter cereals, variation in freezing tolerance is largely due to the variation in the water contents of their crowns. Hardening is at least partly responsible for the initiation of the dormant state or the shoot flower buds. Freezing tolerance has the following components: (i) osmotic adjustment, (ii) the amount of bound water, (iii) stability of the plasma membrane, and (iv) production of cold-response proteins. The osmotic adjustment has but limited effectiveness, while plasma lemma stability and the extent of cellular water binding are much more effective in contributing to freeze tolerance.

The different cereal crops are winter hardy in the following order: rye > wheat > barley > oats. Further, breeding has not been able to enhance the level of winter tolerance in the cultivated varieties of these crops beyond the level of the varieties available around 1950s. Winter hardiness traits are generally under polygenic control and the additive effects predominate. However, the heritability of these traits is low to moderate primarily because of the high G x E interaction effects. The freezing resistance traits may be found in (i) cultivated varieties, (ii) germplasm accessions, (iii) induced mutations, (iv) related wild species, and (v) unrelated organisms. Many varieties of cereals developed for cold regions are excellent source of winter hardiness. In addition, germplasm collected from cold regions experiencing freezing may serve as good sources of freezing resistance. Winter hardy mutations have been isolated in some cases, e.g., in Bermuda grass. In addition, related wild species of some crops are winter hardy but their use in breeding may present problems. Finally, some transgenes may prove useful in the development of freeze tolerant transgenic lines of crop plants.

The use of freezing resistance present in cultivated varieties is the easiest and is based on pedigree method of breeding. The backcross method is used to transfer freezing resistance from inferior germplasm lines and especially from related wild species. Chemical or physical mutagenesis may be used to isolate winter hardy mutants from freezing sensitive cultivars. The selection for freezing tolerance may be carried out under the field environment, or under a controlled freezing environment created for the purpose. It may, however, be pointed out that survival under controlled freezing is not a good indicator of the survival under field conditions. Therefore, the results from controlled environments are generally supplemented with field evaluation. The basis of selection is generally survival during freezing, and recovery and regrowth after thawing. Hardiness of plants can also be assayed by subjecting them or their parts to freezing in the laboratory, and monitoring their viability, degree of browning, electrolyte leakage, etc. after freezing. These tests are generally used with cereals, but they can be applied to only a limited number of plants. Some other criteria used for selection for freezing resistance are freezing of isolated crowns of cereals, and measuring the degree of osmoregulation following freeze hardening.

The problems encountered in breeding for freezing tolerance are as follows: (i) problems posed by the complex inheritance of freezing resistance, (ii) difficulties in reliable assessment of the trait especially under field conditions, (iii) the field evaluations are complicated by pathogen attacks after thawing, (iv) the large G x E interaction effect exhibited by freezing tolerance, and (v) the difficulties in the use of laboratory tests on large breeding populations.

7.7. ACHIEVEMENTS OF BREEDING FOR ABIOTIC STRESS RESISTANCE

A number of varieties resistant to different abiotic stresses have been developed in several countries. For example, a drought tolerant rice variety called Sahbhagidhan was released for cultivation in 2009; it yields up to 35 q/ha under drought situations. The rice variety Swarna Sub-1 was developed by marker-assisted selection. It is submergence tolerant (it can tolerate up to 10 days complete submergence) and can yield up to 65 q/ha.

7.8. CONCLUSIONS

An abiotic stress is a nonliving environmental factor that has an adverse influence on the performance of the crop in question. About 90% of the global arable land area is affected by one or the other abiotic stress and drought is the most important of these stresses. Drought resistance reduces the adverse effects of drought, and may be due to (i) drought escape, (ii) dehydration avoidance, and (iii) dehydration tolerance. Salinity stress is generated by the presence of high soluble salt concentrations in the root zones of plants. Salinity resistance may be the result of water stress resistance, salt toxicity resistance or both. Mineral stress comprises stress due to mineral deficiency and that due to mineral toxicity. Mineral deficiency resistance may involve (i) mineral redistribution within plants, (ii) efficient uptake of minerals from the soil, (iii) greater growth of roots than shoots, and (v) increased transport of minerals to the shoots from the roots. Mineral toxicity occurs in acid soils and is mainly due to Al, Mn or both. Resistance to mineral toxicity is generally due to reduced uptake of the concerned mineral, but in some cases cellular compartmentation and even tolerance may be involved. Temperature stress may be generated by heat (temperatures above the optimum range for the crop), chilling (temperatures close to 0°C) or freezing (temperatures below 0°C). There are two basic mechanisms of heat resistance, viz., heat avoidance and heat tolerance, and several features of the plants contribute to these phenomena. Chilling tolerance is expressed in response to chill hardening. Chilling tolerance may involve, (i) increased membrane lipid unsaturation, (ii) reduced sensitivity of photosynthesis, (iii) increased chlorophyll accumulation, (iv) decreased sensitivity of germination, (v) improved fruit and seed set, and (vi) increased pollen fertility at low temperatures. Freezing resistance may involve freezing avoidance or freezing tolerance.

Resistance to the various abiotic stresses may be found in (i) cultivated varieties, (ii) elite germplasm or breeding materials, (iii) land races, (iv) wild relatives, and (v) unrelated organisms. The resistance present in the cultivated varieties and breeding materials is the easiest to use. The segregating generations from these crosses are handled according to the pedigree scheme. The resistance present in inferior germplasm accessions and in related species has to be transferred in agronomically superior lines/varieties using the backcross method. The use of these sources in breeding programmes is likely to present problems. Therefore, it should be used only when resistance is not found in the elite germplasm. Appropriate transgenes from other organisms can be transferred into crop plants and transgenic lines with abiotic stress resistance can be created.

Selection for abiotic stress resistance is based on traits contributing to resistance; these traits are different for different abiotic stresses. In addition, a suitable stress environment has to be created to select for the resistance. In most cases, selections are carried out under field conditions, and fields and locations are carefully selected for this purpose. However, molecular markers allow the identification and mapping of the QTLs governing abiotic stress resistance. The markers flanking these QTLs can be used to select for the QTLs under optimal conditions. Thus markers can be used to select for abiotic stress resistance independent of the environment and without the evaluation of any other selection criteria. The lines selected in this way are rigorously evaluated for their abiotic stress resistance before they are used for cultivation/in breeding programmes.

PRACTICE QUESTIONS

1. Explain the meaning of abiotic stress, briefly describe the different types of abiotic stresses, and discuss their relevance to plant breeding.
2. Briefly discuss the breeding for drought resistance under the following heads: genetics and sources of resistance, selection criteria, and breeding strategies.
3. Define salinity stress. Explain the mechanism of salinity resistance, and discuss the various aspects of breeding for salinity resistance, including inheritance, sources and selection for salinity resistance.

4. Explain the meaning of mineral stress. Discuss the mechanism of and breeding for mineral deficiency stress, including the sources and inheritance pattern of, and selection for this characteristic.
5. Explain the meaning of mineral toxicity stress. Discuss the mechanism of and breeding for mineral toxicity stress, including the sources and genetic basis of, and selection for this characteristic.
6. Briefly explain the nature of heat stress and discuss the mechanism of resistance to heat stress. Discuss the breeding for improvement of resistance to heat stress.
7. Briefly explain the nature of freezing stress, and discuss the mechanism of resistance to this stress. Discuss the breeding for improvement of resistance to freezing stress.
8. Briefly explain the nature of chilling stress and discuss the mechanism of resistance to this stress. Discuss the breeding for improvement of chilling stress.
9. Discuss the various opportunities and limitations of breeding for resistance to various abiotic stresses.
10. Write short notes on the following: (i) drought environment, (ii) drought resistance, (iii) drought hardening, (iv) salinity stress, (v) salt-affected soils, (vi) salinity resistance, (vii) mineral stress, (viii) mineral deficiency stress, (ix) mineral toxicity stress, (x) supercooling, (xi) freeze hardening, (xii) heat stress, (xiii) chilling stress, (xiv) freezing stress
11. Differentiate between the following: (i) drought avoidance and dehydration tolerance, (ii) avoidance of mineral deficiency and mineral deficiency tolerance, (iii) mineral deficiency and mineral toxicity, (iv) Al and Mn toxicities, (v) chilling and freezing stresses.

RELEVANT LITERATURE

1. Blum A 1988. Plant Breeding for Stress Environments. CRC Press, Inc., Boca Raton, Florida.
2. Chopra VL and Paroda RS (eds.) 1986. Approaches for Incorporating Drought and Salinity Resistance in Crop Plants. Oxford and IBH Publishing Co. Pvt Ltd., New Delhi.
3. Christiansen MN and Lewis CF (eds.) 1982. Breeding Plants for Less Favorable Environments. John Wiley and Sons, Inc., New York.
4. Fujino K and Iwata N 2011. Selection for low temperature germinability on the short arm of chromosome 3 in rice cultivars adapted to Hokkaido, Japan. *Theor. Appl. Genet.* 123: 1089-1097.

5. Grover A et al. 1998. Engineering crops for tolerance against abiotic stresses through gene manipulation. *Curr. Sci.* 75: 689-696.
6. Ishitani M et al. 2004. Integration of genomic approach with traditional breeding towards improving abiotic stress adaptation: drought and aluminum toxicity as case studies. *Field Crops Res.* 90: 35-45.

■■■■■

8 Breeding for Quality Traits

8.1. INTRODUCTION

The produce from every crop species has more than one use. For example, wheat grains are used for making bread, biscuits, pastry, chapatti, etc. For each use, the crop produce should possess a set of characteristics; these features of the crop produce are termed as **quality traits**. The specific traits that determine the quality of crop produce, therefore, will vary with the crop and the specific uses of the produce (Table 8.1). In simple terms, therefore, the term **quality** would mean the degree of suitability or desirability of an object for a specific use or purpose. In the case of crop produce, therefore, the term **quality** would signify the suitability of the produce for a particular use. For example, the suitability of tomato for fresh use and for processing is determined by several different features of the tomato fruit (Table 8.1). The various quality traits can be grouped into the following categories: (i) morphological, (ii) organoleptic, (iii) nutritional, (iv) biological and, (v) other quality traits. **Morphological quality traits** are related to the appearance of the crop produce, e.g., size, color, etc., of the produce. These traits are easy to observe and are very important for consumer acceptance.

The **organoleptic quality traits** affect the palatability of the crop produce, e.g., softness, juiciness, aroma, taste, etc. Although they are not visible to the eye, they can be readily assessed by the consumer. The **nutritional quality traits** determine the value of the crop produce in human and animal nutrition, and include protein content and quality, oil content and quality, vitamin content, mineral content, etc. However, these traits are not easily determined, and they are not appreciated either by the consumers or the farmers. The **biological quality traits** reveal the actual usefulness of a crop produce when consumed by humans and animals. Examples of these traits are biological value, body weight gain, etc.; they are the most relevant to the consumers, most difficult to determine, and least appreciated by the consumers and the farmers. The group **other quality traits** comprises all such trait are not included in the above categories. Examples of such traits are cooking quality of rice, keeping quality of fruits and vegetables, etc. In this chapter, the discussion will focus on nutritional quality and its improvement by breeding.

Table 8.1: A list of some important quality traits of some selected crops

Crop	Quality trait	Preferred phenotype	Remarks
Wheat	Grain size	Bold and plump	-
	Grain color	Amber to white	Amber is the most common grain color
	Grain texture	Hard, vitreous, shiny	-
	Protein content	High	Few varieties like Atlas 66 have up to 19% protein
	Gluten content	High	The water insoluble flour protein
	Polshenke value	Strong (bread making) Medium (chapati making) Weak (biscuit and confectionary making)	Measures gluten strength
	Sedimentation value	20-40	The lower the value, the weaker is the gluten
	Alveogram	Strong, medium and weak	Measures gluten strength
	Mixogram	-	Indicates baking quality and gluten strength
Tomato	Used as fresh fruit		
	Fruit appearance:		
	1. Size	Large	Polygenic control
	2. Shape	Round	Oligogenic control; allele *O* specifies round fruits
	3. Color	Red	Oligogenic control; allele og^c increases red color due to lycopene
	Texture and firmness	Firm for transportation; soft and juicy for fresh consumption	-
	Flavor	Flavor preferred	Flavor due mainly to acids, sugars and volatile compounds
	Nutritional value	Vitamins A	Affected by the alleles og^c and *hp*
		Vitamins C	Varieties with up to 50mg/ 100g fresh fruit developed
	Processing quality		
	Color	Red	Measured by colorimeter
	Fruit pH	Acidic (<4.5)	Important for sterilization of the fruit products
	Titrable acidity	High	Polygenic control
	Total soluble solids	High	Mainly due to sugars
	Viscosity	High	Estimated by a viscometer or as insoluble solids

8.2. BREEDING FOR NUTRITIONAL QUALITY

Improvement in the quality traits of a crop constitutes **breeding for quality**. It is needless to emphasize that breeding for quality is an integral part of every crop improvement since the produce of a given crop must be suitable for the various purposes, for which it is used. This is necessary for the produce of a given crop variety to be accepted by the consumers. The morphological quality traits are visible to the naked eye, and are easily selected for during the segregating generations of crosses. In fact, these traits are always selected for by the breeder because they are the basic quality traits that determine the consumer acceptance of the produce. The plant products are the primary source of nutrition for humans as they provide energy or calories, proteins, fats, vitamins, minerals, etc. Therefore, continued consumption of a crop produce affects the human health, which constitutes **nutritional quality** of the crop produce. The nutritional quality of a crop produce depends mainly on the following components of the food: (i) protein content, (ii) protein quality (the amino acid balance of the proteins), (iii) digestibility of the proteins, (iv) oil content, (v) oil quality (fatty acid composition), (vi) vitamin content, (vii) contents of various minerals, and (viii) the absence of anti-nutritional compounds.

It may be pointed out that evaluation of these traits requires biochemical tests and even studies with animals. Therefore, selection for these quality traits is generally delayed till the preliminary yield trials are conducted. At this stage, the number of lines to be evaluated for quality traits is reduced to relatively smaller numbers and enough seeds/grains would be available for each line to support the various quality tests. For each crop, the newly evolved lines have to satisfy a specific set of nutritional and other important quality traits before they can be recommended for release as new varieties. Breeding efforts have been directed at vitamin content, protein content and quality, oil content and quality, and elimination of anti-nutritional factors. The breeding for these aspects is briefly considered in this chapter in the following sections. There has been some recent breeding activity aimed at the modification of the contents of certain minerals in the produce of few crops.

8.3. INHERITANCE OF NUTRITIONAL QUALITY TRAITS

The various quality traits, similar to the other characters of crop plants, are governed by either oligogenes or polygenes. In some cases, the involvement of modifying genes has also been documented, and in some other cases maternal effects and cytoplasmic inheritance have been reported. **Oligogenic inheritance** has been reported for several quality traits, including nutritional quality traits. Some examples of quality traits showing oligogenic inheritance are as follows: (i) β-carotene content in tomato fruits (gene *hp* causes increases; gene og^c leads to decrease), trypsin inhibitor content in soybean seeds (gene *ti* leads to reduced content of the inhibitor), lysine content in the seeds of barley, maize and sorghum

(Section 8.6.3), erucic acid contents in the oils of *Brassica* spp., etc. The oligogenic quality traits are the easiest to select for in the breeding populations and they are readily transferred from one line into another. The expression of many oligogenes governing quality traits is affected by **modifying factors**. For example, the undesirable effects on kernel characteristics produced by some oligogenes for high lysine content in maize and sorghum are eliminated by the accumulation of favorable modifying genes (Section 8.6.3).

Polygenic control of quality traits is quite common and is found to be applicable in the cases of most of the quality traits. Some selected examples of quantitatively inherited quality traits are as follows: (i) protein content, (ii) oil content, (iii) carotene content in carrot, (iv) mineral content, (v) grain size, (vi) fruit size, (vii) fiber length and other fiber quality traits in cotton, etc. The gene action is mainly additive, but in some cases the dominance effects may be predominant. The improvement in quantitatively inherited traits is much more demanding than that in oligogenic traits, and depends mainly on heritability of the trait.

The expression of some quality traits is markedly affected by the genotype of the female parent, i.e., these traits show **maternal effects**. Generally, such traits are concerned with the features of seeds or grains, e.g., seed size, seed protein content, seed oil content, etc. For example, chickpea seed protein content shows marked effect of the maternal genotype. Similarly, oil contents in maize and *Brassica napus* seeds show considerable maternal effects. The glucosinolate contents in the seed cakes of *Brassica* spp. are remarkable as they are determined completely by the genotype of the plants producing the seeds. In contrast, the fatty acid composition of seed oil is governed by the genotype of the seed embryos. Maternal effects, like other environmental factors, reduce the effectiveness of selection for the concerned quantitative traits. Finally, in *Brassica juncea*, the glucosinolate content is also affected by plasmagenes (**cytoplasmic inheritance**).

8.4. ELIMINATION OF ANTI-NUTRITIONAL FACTORS

An **anti-nutritional factor** is a component of crop produce that has adverse effect on the health of those humans/animals that consume this produce. Examples of anti-nutritional factors are trypsin inhibitors, cyanogens, tannins, erucic acid, glucosinolates, phytic acid, etc. (Table 8.2). The seeds and *dal* of khesari (*Lathyrus sativus)* normally contain ~0.55% of the anti-nutritional β-OAA (β-N-oxalyl-α, β-diaminopropionic acid). β-OAA is involved in paralysis of lower limbs caused by prolonged consumptions of khesari seeds/*dal*. Breeding efforts have been successful in reducing this content to 0.05% or even less. *Brassica* spp. seeds contain glucosinolates, which are hydrolyzed in the human body to some compounds that cause goiter. ***Goiter*** is the condition in which the thyroid gland is enlarged, and is generally produced by a deficiency of iodine.

Efforts to develop low glucosinote varieties of *Brassica* spp. have been quite successful. Goiter is also caused by a phenolic glycoside present in the seed skin of groundnut. This glycoside is inactivated by heat treatment during roasting or cooking. Therefore, raw seeds of groundnut should not be consumed. Another example of an anti-nutritional factor is phytic acid, which is found in for bajra or pearl millet seed coat and germ/embryo. Phytic acid serves as storage compound for phosphorus, but it forms complexes with minerals like iron present in the food and reduces their availability to humans/ animals.

Table 8.2: Some examples of anti-nutritional factors present in the produce of our important crop plants

Anti-nutritional factor	Crop produce	Effect on human health
βOAA	Seeds and *dal of* khesari (*Lathyrus sativus*)	Paralysis of lower limbs (lathyrism)
Erucic acid	Oil from *Brassica* spp.	Reduction in body weight gain
Glucosinolates	Oil and oil cake from *Brassica* spp.	In extreme case, goitre
Gossypol	Oil and oil cake from cotton	Toxic effects
Hemagglutinin	*Vicia faba* (broad bean) and other legume pods/seeds	Favism (caused by consumption of raw beans)
Phenolic glycoside	Seed skin of groundnut	Goiter (caused by consumption of raw seeds with skin)
Phytic acid	Pearl millet seeds	Reduces availability of minerals leading to mineral deficiency

8.5. BREEDING FOR ENHANCED VITAMIN CONTENT

Vitamins are essential for good health. We obtain most of our vitamin requirements from our diets. This is particularly correct for the vitamins A, C and B_6. Vitamin A is derived from β-carotene, which is present in ample quantities in carrots, tomatoes, maize and sweet potatoes. Similarly, vitamin C is found in good quantities in oranges, tomatoes, cabbage and potatoes. Although vitamins are very valuable for human health, little effort has been made to enhance vitamin contents in our food crops. For example, efforts were made to increase the β-carotene contents of tomato and carrot using suitable breeding strategies. The red fruit color of tomato is due to lycopene, which is now regarded as an important nutritional component. β-carotene is present in tomato fruits at a concentration of around 5% of that of lycopene. The high β-carotene trait was transferred into tomato from the wild tomato species *Solanum habrochaites* (Syn. *Lycopersicon hirsutum*). The high β-carotene content in the lines derived from this programme was governed by two major genes. In addition, modifying genes from the cultivated tomato, *Solanum lycopersicum* L. (Syn., *Lycopersicon esculentum*), also affect the β-carotene content. These high β-carotene lines were used in hybridization programmes to develop the variety Caro Red. The fruits of this variety had 10-times more β-carotene than those of normal tomato varieties. But this variety was not liked by the consumers due to the fruit color being different from the preferred red color.

In case of carrots, consumers generally prefer orange colored roots. The intensity of orange color is positively associated with the total carotene content in the roots. Therefore, total carotene content in carrot roots can be selected visually on the basis of root color. The root carotene content in carrots shows a complex inheritance. But breeding has been successful in the development of lines with as low as 1 μg/g root fresh weight to as high as 400 μg/g root fresh weight. However, the commercial varieties of carrot have 100-120 μg/g root fresh weight carotene. The main components of the carotene are α-carotene and β-carotene. Some commercial hybrid varieties of carrot have in their roots carotenes between 140 and 160 μg/g root fresh weight. It is likely that hybrids with over 180μg/g root fresh weight of carotenes would be developed for commercial cultivation.

Some other examples of improvement in vitamin/mineral contents of food crops through breeding are as follows. Some relatives of tomato, e.g., *Solanum arcanum* (Syn. *Lycopersicon peruvianum*) have much higher vitamin C contents in their than tomato. Some tomato lines derived from crosses with these species also show improved vitamin C content. For example, tomato variety Pusa Red Plum was derived from a cross of tomato with *Solanum pimpinellifolium* (Syn. *Lycopersicon pimpinellifolium*), and has high vitamin C content. More recently, marker-assisted selection has been used to develop some varieties of crops like peal millet that accumulate higher amounts of specific minerals in their grains than do conventional varieties of the same crops.

8.6. PROTEIN CONTENT AND QUALITY

Proteins are a group of macromolecules composed of 20 different amino acids. Humans obtain their protein nutrition from their diets. The proteins are the source of amino acids that are used to synthesize human proteins that serve a variety of functions, including building up of body tissues. The structures and cellular functions of proteins are determined by their amino acid sequences, which are themselves specified by the base sequences of the gene encoding them. Proteins are an essential component of human diet and a minimum amount of proteins has to be consumed every day for good health. The total protein requirement varies with the age: it is the highest for infants (1.8g/kg body weight/day), intermediate for preschool children (0.9g/kg body weight/day), and the lowest for adults (0.55 g/kg body weight/day). Protein malnutrition is known to cause specific health problems like kwashiorkor and marasmus. Similarly, protein-energy imbalance results in the utilization of proteins for energy production in place of building of body tissues.

The nutritional usefulness of proteins is generally assayed in terms of digestibility, protein efficiency ratio, and biological value. **Digestibility** of a protein is the proportion of ingested protein digested and absorbed by the test animal; it ranges from 70 to 90% for the proteins from different sources. The **protein efficiency ratio** of a protein is the increase in body weight (in grams) of the test animal on consumption of 1g of the protein. Similarly, the proportion (in percent) of nitrogen retained in the body of the test animal out of the total nitrogen absorbed from diet is known as the **biological value** of the protein. It ranges from 50 to 70% for legume proteins.

8.6.1. Breeding for Protein Content

It has been estimated that the human protein needs are provided primarily form plant sources: plants provide ~90% and ~70% of the protein requirements in the developing and the developed countries, respectively. Further, cereals are the chief source of plant proteins as they supply ~70% and ~50% of the human protein needs in the developing and the developed countries, respectively. In contrast, legumes provide ~20% of the protein requirement in both the developing and the developed countries. **Protein content** of a crop is estimated as the amount of protein present in a unit weight of its produce and is generally expressed as percent of the produce dry weight. The protein content of the crop produce shows polygenic inheritance, and in several cases maternal effects have also been reported. In addition, protein content is markedly affected by the prevailing environment, especially the factors that affect grain filling.

The protein contents of different crops show considerable variation. However, at least in the case of cereals, protein content shows a negative correlation with grain yield; this has hampered the progress of breeding programmes directed at improving their protein contents. In a few cases, however, breeding efforts have been able to overcome this negative association between yield and protein content. For example, genes for high protein content were transferred from the wheat variety Atlas 66 into Lancota. The improved version of Lancota had 1-2% more protein in its grains and there was no reduction in its yield. In contrast, most of the improved varieties of rice have about 7% protein, while the range for this trait in rice is from 7 to 14%. Breeding efforts to develop improved varieties of rice with higher (say, 8-9%) protein content have not been successful. It has been suggested that this failure is most likely due to the low heritability of the trait. Similarly, efforts to isolate mutants with higher protein content without any reduction in yield have been largely unsuccessful. It may be pointed out that an increase in protein content of a crop would necessitate an increased supply of nitrogen nutrition to the crop. It has been estimated that an increase of 1% in the protein content of the crop produce would require 6-11% more nitrogen input for the crop.

8.6.2. Protein Quality

The nutritional quality of proteins depends on their amino acid composition and not their amino acid sequence. The term **protein quality** may be defined in simple terms as follows: it is the degree of similarity of a given protein with the ideal protein in terms of the amounts of the eight essential amino acids. An **ideal protein** is defined as a protein that has the amino acid composition that enables it to provide the necessary amounts of the essential amino acids when consumed in the recommended doses. The amounts of the essential amino acids needed by humans vary with the age: it is relatively higher for pre-school children than for infants, and is the lowest for adults. An **essential amino acid** is one that is not synthesized by human body. Humans are not able to synthesize the following eight amino acids: (i) isoleucine (Ile), (ii) leucine (Leu), (iii) lysine (Lys), (iv) methionine (Met), (v) phenylalanine (Phe), (vi) threonine (Thr), (vii) tryptophan, and (viii) valine (Val). It may be pointed out that a protein that has the essential amino acid composition almost comparable to that of the ideal protein for infants (Table 8.3) is said to have an **excellent amino acid balance**. Such proteins are suitable for consumption by all age groups of humans.

Table 8.3: The chief sources of proteins in vegetarian diets, and the status of these proteins in terms of essential amino acids

Food source	Protein deficient in	Remarks
Cereals *	Lysine, tryptophan, threonine	Deficiency due to the prolamin** fraction of the seed storage proteins; rice and oat seeds have low prolamin contents so that the amino acid balance of their proteins is excellent
Legumes *	Methionine, tryptophan	Deficiency due to the globulin** fraction of the seed storage proteins
Nuts and oilseeds	Lysine	
Leafy vegetables	Methionine	

* A 3:1 combination of cereals and legumes in the diet provides a nearly excellent balance of amino acids.

** The seed storage protein fractions albumins and glutelins have balanced amino acid profiles.

In general, few of the crops produce foods that have excellent amino acid balance. The cereal proteins are deficient in lysine, tryptophan and/or threonine. But rice and oat seeds have low prolamin contents so that the amino acid balance of their proteins is excellent. On the other hand, the legume proteins are deficient in tryptophan and sulphur containing amino acids like methionine (Table 8.3). The proteins from nuts and oilseeds are poor in lysine, while those from leafy vegetables are deficient in methionine. Thus a combination of cereals and pulses and/or leafy vegetables would provide proteins with nearly excellent amino acid balance. It has been estimate that a 3:1 combination of cereals and pulses would provide and excellent amino acid balance.

8.6.3 Breeding for protein quality

Several mutants, both spontaneous and induced, with improved protein quality in their seed storage proteins have been successfully isolated (Table 8.4). But in general, the grains of the mutant lines are poorly filled. It is encouraging that intensive breeding efforts have been successful in improving the grain filling of the protein quality mutants. Mutant lines with well-filled plump grains have been developed by concerted breeding efforts in crops like wheat, maize and barley. In case of high lysine mutants of cereals, generally the prolamin content is reduced. But in the case of *Hiproly* barley mutant, the prolamin content is not affected; instead the albumin and globulin fractions of seed storage proteins are increased, and there is an increase in the lysine content of the glutelin fraction. The high lysine trait of the barley mutants is specified by a single recessive gene; this has been transferred in different genetic backgrounds by backcross breeding. In case of sorghum, the mutant *P721 opaque* was induced by a chemical mutagen treatment. The grains of this mutant have 60% more lysine than the normal sorghum lines, but its grains are opaque. The mutant gene has been transferred in different lines, and lines with vitreous grains have been isolated. In addition, the *P721 opaque* mutant was subjected to chemical mutagenesis, and mutant lines with high lysine and vitreous grains were isolated.

Table 8.4: Some selected examples of cereal mutants with improved protein quality

Crop	Mutant	Improvement in protein quality	Remarks
Barley	Hiproly	High protein and lysine	Increased lysine due to: (i) increase in albumin and globulin fraction, and (ii) increased lysine content of glutelins; single recessive gene; spontaneous mutant
	Notch-2	High protein and lysine	Increased lysine due to reduced prolamin; single recessive gene; chemically induced mutant
Maize	Opaque-2	High lysine and tryptophan	Increased lysine due to reduced prolamin; single recessive gene; spontaneous mutant; modifiers eliminate the undesirable kernel features; commercial hybrids released
	Fluory-2	High lysine and sulphur containing amino acids	Increased lysine due to reduced prolamin; single recessive gene; spontaneous mutant
Sorghum	P721 opaque	High lysine	Increased lysine due to reduced prolamin; opaque grains; the grains become vitreous due to modifying genes; single partially dominant gene; chemically induced
	P721 vitreous	High lysine	Chemically induced mutant from P721 opaque; has vitreous grains

High lysine mutants have been isolated in maize as well. Some of these mutants are *opaque-2, opaque-7, opaque-16*, and *fluory-2*. In the grains of all these mutants, the prolamin content is greatly reduced, and the kernels have undesirable features like opaque appearance and soft endosperm. Out of these mutants, *opaque-2* has been extensively used in breeding programmes. The *opaque-2* mutant grains have high lysine and tryptophan contents as compared to those of the nonmutant normal maize lines. It has been demonstrated that the undesirable features like opaque appearance and the soft endosperm features can be eliminated by accumulation of modifying genes. This approach has been used to develop several varieties and hybrids of maize that have high protein quality along with normal kernel characteristics. For example, the maize hybrid Vivek QPM 9 is based on *opaque-2*, and has the yield and grain characteristics comparable to the parental hybrid Vivek Hybrid Maize 9. Vivek QPM 9 was developed by transferring the *opaque-2* gene into the parental inbreds of the single cross maize hybrid,Vivek Hybrid Maize 9.

The legume seeds generally contain between 18 and 28% protein. Soybean seeds are an exception in that they contain as high as 43% protein. The legume seed proteins are overall deficient in sulphur-containing amino acids. There are mainly two types of seed storage proteins in legumes, viz., albumins and globulins. The albumins have balanced amino acid composition, while the globulins are deficient in sulphur-containing amino acids. There has been little breeding effort directed at the improvement of the amino acid profile of legumes. This situation may be mainly due to the fact that in the case of most legumes, the breeding programmes are directed mainly at the improvement and stabilization of their yielding ability. In addition, suitable genetic stocks with improved quality are not available. In any case, legumes are generally consumed in combination with cereals, and they are seldom consumed alone. Therefore, the deficiency of sulphur containing amino acids in the grain legume seed proteins may not be that critical a nutritional issue as that of the cereal proteins as cereals are often consumed without pulses.

8.6.4. Transgenics with Improved Protein Quality

A gene introduced into a plant by means of genetic engineering is called **transgene**. A **transgenic plant** contains in its genome a transgene and expresses the same. The transgenic plants can be created by transferring into them and expressing appropriate transgenes to create novel trait phenotypes. For example, a transgene may be introduced into a plant with a view to produce proteins with improved amino acid balance. The following approaches have been used in an effort to improve protein quality of different crops:

(i) Introduction of a natural gene,

(ii) Expression of a natural gene with modified amino acid content,

(iii) Transfer of a completely synthetic gene,

(iv) Overexpression of an endogenous gene, and

(v) Suppression of an endogenous gene. It may be emphasized that these transgenes are driven by such seed-specific promoters that are highly efficient. Therefore, the transgene encoded proteins are produced in relatively high quantities in the seeds of the transgenic lines.

In the **first** approach, a natural gene encoding a protein that has high amounts of the desired amino acid is introduced and expressed in the concerned crop. For example, pea seed storage proteins are deficient in sulphur containing amino acids. The sunflower albumin 8 (SFA 8) protein is rich in sulphur containing amino acids (it has 23% methionine plus cysteine). The gene encoding SFA 8 has been transferred in alfalfa and subterranean clover to improve their forage quality. The **second** strategy consists of modifying the base sequence of an endogenous seed storage protein gene in such a way that it now encodes a protein that has adequate amounts of the deficient amino acids. For example, the zein proteins stored in maize seed are deficient in lysine and tryptophan. The base sequence of the gene encoding zein was modified to encode a protein rich in lysine, tryptophan, or methionine. These genes have been transferred into maize rice and wheat. In the **third** approach, a completely new gene is chemically synthesized. This gene encodes a protein that has high concentrations of the essential amino acids that are deficient in the dietary proteins. DuPont, USA has synthesized a gene that encodes a protein, designated as CP3-5, which has 35% lysine and 22% methionine. This gene has been transferred and expressed in plant cells and lines.

The **fourth** method involves overexpression of an endogenous gene that encodes a seed storage protein having the desired amino acid balance. This can be done by introducing additional copies of the gene that are driven by a strong promoter. For example, the gene encoding high molecular weight glutelin has been overexpressed in wheat in order to improve the quality of dough. The **fifth** and the **final** strategy consist of suppressing the expression of genes encoding seed proteins deficient in one or more essential amino acids. The techniques used for this purpose are (i) antisense RNA, (ii) ribozyme, (iii) cosuppression, (iv) gene disruption, and (v) RNA interference (RNAi). The phenomenon of RNA interference has been described in Section 2.9, and antisense RNA is discussed Section 10.6.2. The antisense and RNAi technologies have been used to suppress the expression of zein genes in maize. This resulted in improved amino acid balance of the maize seed proteins.

8.7. OIL CONTENT AND QUALITY

Oils and **fats** are esters of glycerol (CH_2OH-$CHOH$-CH_2OH) and straight chain fatty acids. Generally, the different fatty acids have 12, 14, 16, 18, 20, 22, or 24 carbons in their molecules. A fatty acid without a double bond in the carbon chain is called **saturated fatty acid**, while that with a double bond is known as

unsaturated fatty acid. The –COOH groups of the fatty acids react with the –OH groups of the glycerol molecules to produce esters. Ordinarily, each glycerol molecule forms esters with three fatty acid molecules; therefore, the oils are often called **triglycerides**. The physical properties of a given oil and its usefulness for various purposes depend mainly on the types and the proportions of the fatty acids present in it. Oils constitute an important component of human diet; they are rich in energy, source of some vitamins, and are important constituents of cell structures like membranes. Oils and fats have many important industrial applications, e.g., manufacture of soaps, detergents, lubricants, paints, varnishes, and biofuels. In any case, oils are the cooking medium of a great variety of dishes, especially the Indian culinary preparations. The recommended daily consumption of oils is 30g/day/person, but the availability of oils to the Indian population is much lower (around 11g/day/person).

Table 8.5: The salient features and uses of oils from some important oilseed crops

Oilseed crop	Oil content (%)	Main fatty acid	Common uses	Remarks
Soybean	20	Linoleic acid (18:2; 49%)**	Cooking and salad oil	Reduction in linolenic acid content is an important objective
Oil palm	56*	Palmitic acid (16:0; 44%)	Industrial and edible uses	Efforts being made to increase 18:1 and reduce 16:0
Rapeseed and mustard	28	Erucic acid (22:1; up to >50%)	Industrial and edible	Low erucic acid and low glucosinolate genotypes preferred for human consumption
Sunflower	29	Linoleic acid (18:2; 67%)	Edible	Best oil for human consumption
Groundnut	45	Oleic acid (18:1; 71%)	Edible	Becomes rancid due to linoleic acid
Cottonseed	25	Linoleic acid (18:2; 54%)	Edible	Removal of gossypol is essential for human use
Coconut	-	Lauric acid (12:0; 50%)	Industrial and edible uses	
Linseed	38	Linolenic acid (18:2; 47%)	Making of paint	

* Oil content of oil palm fruit mesocarp

** The figures 18:2, 16:0, etc. within parenthesis represent the number of carbon atoms (the figure before the colon) in the fatty acid chain and the number of double bonds (the figure after the colon) in the chain; and the figures 49%, 44%, etc. denote the proportion of the given fatty acid commonly found in the oils of the concerned species

Vegetable oils and fats are obtained from seeds of some plant species, which are called **oilseed plants**. The following seven crops, listed in the order of their importance, are the most important sources of vegetable oils: (i) soybean, (ii) palms, (iii) rapeseed and mustard, (iv) sunflower, (v) cotton seed, (vi) groundnut, and (vii) coconut. In addition, there are several other minor oilseed crops like sesame, safflower, linseed, castor, maize, etc. Further, some wild plants are also important sources of oils and fats, e.g., *Madhuca*, sal, neem, etc. The oils from different oilseed plants are generally suitable for different purposes (Table 8.5).

Table 8.6: Salient features of some selected fatty acids found in vegetable oils

Fatty acid	Relevance to human health	Remarks
Saturated fatty acids	Not desirable for human health	Many industrial uses
Oleic acid (18:1)		High oleic acid oils used for cooking and as salad oil
Linoleic acid (18:2)	Lowers blood cholesterol, helps prevent atheroschlerosis, constituent of cell membranes, precursor of some hormones; edible oils should have at least 35% linoleic acid	
Linolenic acid (18:3)	Important for making paint	Reduces keeping quality of oil by becoming rancid

8.7.1. Breeding for Oil Content

Oil content shows polygenic inheritance, and in some crops like maize and *B. napus* it shows considerable maternal effects. In general, gene action is mainly additive, but in some cases dominant effects are also important. In case of sesame, partial dominance of low oil content has been reported. Breeding programmes for improved oil content have been remarkably successful in the cases of different oil seed crops. For example, the oil content in sunflower was around 30% during 1920s. A recurrent selection scheme was used to improve oil content. As a result, the oil content of this crop increased to 54% by the 1950s. It is noteworthy that this increase in oil content was accompanied with increased seed yield. As a consequence, sunflower has become the most profitable oilseed crop in the world. In case of maize, the germ oil content is generally less than 5%. But maize hybrids with about 8% germ oil content are now in cultivation. Similarly, oil content of safflower seeds has been increased from 37% to 50% by breeding for lowered hull content of seeds.

8.7.2. Breeding for Oil Quality

Oil quality, in simple terms, means the types of fatty acids and their proportions in the given oil. Since oils are used for a variety of purposes, and the oils from different crops are primarily used for some specific purposes, the desired oil quality would largely depend on the oilseed crop and the intended use of the oil from the variety being developed. For example, oil seed crops like maize, soybean, groundnut, sunflower, safflower, cotton seed, etc. are generally termed as oleic-linoleic acid crops since these fatty acids constitute up to 70% of the fatty acids in their oils. In case of soybean, reduction in the linolenic acid content from about 8% to around 1% is an important objective. Commercial soybean varieties having high percentage of oleic acid and low proportion of polyunsaturated fatty acids in their oils have been developed in USA.

The sunflower oil has 67% linoleic acid and 19% oleic acid. Genotypes with over 72% linoleic acid have been identified. Similarly, it has been possible to develop by mutagenesis such lines that have up to 86% oleic acid in their oils. Breeding efforts to develop high linoleic acid and low oleic acid lines, and those with high oleic acid and low linoleic acid have been successful. The sesame oil has around 47% oleic acid and about 39% linoleic acid, and it does not contain linolenic acid. The proportions of oleic acid and linoleic acid are affected by a single gene: the dominant allele of this gene specifies high oleic acid (52%) and low (32%) linoleic acid in oil, while its recessive allele leads to low oleic acid (39%) and high linoleic acid (47%) in the oil. Sesame oil has the antioxidant sesamol that is responsible for its long shelf life.

The erucic acid group comprises 10 species of the family Cruciferae; of these, three species (*Brassica campestris*, *B. napus* and *B. juncea*) are the most important. The oils from these crops have high erucic acid (22:1) content, which may be as high as 60% or even more. In addition, the oilseed cakes from these crops contain glucosinolates (content ranges between 70 and 130μmol g^{-1} of seed cake). Both erucic acid and glucosinolates produce anti-nutritional effects. Therefore, reduction in the amounts of these two compounds is an important breeding objective in these crops. It is desirable to reduce the erucic acid content to <3% of the total fatty acids in the oil; lines are generally called **zero erucic acid lines** or simply as **zero erucic acid lines**. The erucic acid content in these crops is governed by one or two major genes, and each gene is proposed to have three alleles (Table 8.7). Further, the gene action governing the trait is additive. Zero erucic acid lines have been isolated in all the three *Brassica* spp., and commercial varieties have been developed in *B. campestris* and *B. napus*.

Table 8.7: The proposed genetic control of erucic acid content in the oils of important oil producing *Brassica* spp.

Brassica spp.	Genetic control
Brassica campestris	A single gene with three alleles; the gene action is additive
Brassica napus	Two genes with three alleles; the gene action is additive
Brassica juncea	Two genes with three alleles; the gene action is additive

It is desirable to reduce the glucosinolate content to <30 μmol g^{-1} of seed cake; such lines are called **zero glucosinolate lines**. The glucosinolate content is proposed to be governed by 6-8 genes having multiple alleles. In addition, it is specified by the maternal plant genotype, and is also affected by plasmagenes. It has, however, been possible to isolate zero glucosinolate lines in *B. campestris* and *B. napus*. Further, commercial zero glucosinolate varieties have been developed in both *B. campestris* and *B. napus*. The zero erucic acid and zero glucosinolate traits have been combined to develop **double zero** or **Canola quality** varieties of *B. napus*.

8.7.3. Transgenics with Improved Oil Quality

The biosynthesis of fatty acids occurs in plastids of plants and involves 7-8 distinct enzymes. These enzyme activities can be specifically altered to achieve a modification in the fatty acid profile of the oils produced by the transgenic lines. Several studies have attempted to develop transgenic lines producing oils with modified quality. One such effort has resulted in the commercial release of a transgenic variety, called Laurical, of *B. napus*. This variety was developed by modifying the activity of the enzyme lauroyl-ACP thioesterase. The gene encoding the enzyme lauroyl-ACP thioesterase was isolated from the plant California bayleaf and transferred into *B. napus*. Some of the transgenic line of *B. napus* accumulated up to 60% lauric acid. One of these transgenic lines was released as Laurical for commercial cultivation. Genetic engineering has been used to develop soybean lines with high (90%) oleic acid content, rapeseed/ Canola lines with high (89%) oleic acid content, and cotton lines with 78% oleic acid content

8.8. SOURCES FOR QUALITY TRAITS

The improvement in a quality trait depends on the availability of the desired phenotype fir the trait in some line that can be used in breeding programmes. The desired quality traits can be found in the following types of materials: (i) a cultivated variety, (ii) a germplasm accession or line, (iii) a mutant line (either spontaneous or produced by a mutagen treatment), (iv) a somaclone of the crop species, or (v) a related species. When the desired trait is not available in any of the above, (vi) a suitable transgene may be introduced into the crop species genome. The expression of this transgene is expected to produce the desired phenotype of the concerned quality trait. The desired quality trait may sometimes be present in a **cultivated variety** of the crop. This source is the most easily used in breeding programmes and, for this reason, is the most preferred by plant breeders. For example, wheat varieties Atlas 66 and Naphal have been widely used as sources of high protein content. When the desired quality trait is not known to occur in a cultivated variety, the **germplasm collection** of the species should be screened. This search might identify a germplam line that can be used as a source of the quality trait. There are several examples of this type of search. For example, the search among Ethiopean sorghum lines allowed the identification of some lines with high lysine content. These sources of quality traits are more difficult to use in breeding programmes because such lines may have poor agronomic characteristics. In many cases, the desired quality trait has been found some spontaneous or induced **mutants**. Some examples of such mutants are the high lysine mutants of barley, maize and sorghum (Table 8.4), mutants for oil quality traits, etc. Some **somaclonal variants** with the desired quality traits have been isolated in sweet potato (somaclone Scarlet with deep stable root color), khesari (somaclone Ratan with reduced βOAA content), *Citronella java* (somaclone Bio-13 with increased oil content), etc. The plants regenerated from plant cells cultured in vitro often show heritable genetic variation. A plant showing a heritable change in one or more of its traits is called a **somaclonal variant**.

In several cases, the desired quality trait has been transferred from a **related species**. For example, genes for increase vitamin C content have been transferred in tomato from its wild relative *S. pimpinellifolium*. Similarly, tomato lines with enhanced red fruit color have been derived from its cross with the wild tomato species *Solanum habrochaites*. In such cases, where the desired quality trait is either not available or usable even in a related species, genetic engineering may be used to produce **transgenic lines** that express the concerned quality trait. Such transgenic lines have been developed for protein quality and oil quality traits. These examples are discussed in a greater detail in Sections 8.6.4 and 8.10..

8.9. BREEDING APPROACHES FOR QUALITY TRAITS

Several breeding approaches have been used for the improvement of the quality traits, which are as follows: (i) domestication, (ii) screening of germplasm accessions; (iii) mutation induction, including isolation of somaclonal variants, (iv) hybridization, including interspecific hybridization, and (v) genetic engineering to produce transgenic plants. Sometimes, the produce from cultivated plants may not be suitable for some specific use. In such cases, the option of **domestication** of wild species should be explored if such species were available. An example of domestication of wild plant species for a specific purpose is that of jatropha (*Jatropha curcas*) for the production of biodiesel from its oils. Similarly, efforts are being made to domesticate *Cuphea* spp. for obtaining oils suited for specific applications. The **screening of germplasm accessions** often leads to the discovery of lines having the desirable quality traits. For example, screening of the Ethiopian germplasm of sorghum allowed the identification of lines having high lysine in their seed proteins. The use of desirable traits present in germplasm accessions would require some to considerable breeding effort depending on the agronomic features of the lines possessing these traits.

Sometimes, **spontaneous mutants** having the desired quality trait may be recovered from the breeding materials. Spontaneous mutants have been isolated both for protein quality and oil quality. For example, the opaque and floury mutants of maize, and Hiproly mutant of barley have high protein quality. Similarly, low erucic acid lines were isolated from breeding materials of *B. napus* and *B. campestris*. In addition, a mutagen treatment may be used to isolate **induced mutants** possessing the desired quality trait. For example, high lysine mutants have been isolated in barley and sorghum. Similarly, several mutants affecting oil quality have been isolated in soybean, sunflower, *Brassica* spp. and linseed. Often the mutant may have some undesirable characteristics, which have to be minimized/ eliminated by suitable breeding approaches. There are two basic approaches for achieving this objective: (i) accumulation of modifying genes and (ii) mutation induction in the mutant line (Table 8.4). In some cases, **somaclonal variants** have been isolated with improvements in quality traits; some of relevant examples are given in Section 8.9.

Hybridization is the most widely used breeding approach to develop a high yielding variety with superior quality. When the desired quality trait is present in a line that has good agronomic features, the segregating generations would be handled according to the pedigree scheme of breeding. But if the quality trait is present in an agronomically inferior line, a backcross programme would be the preferred breeding scheme. Often only 2-3 backcrosses may be made to the recurrent parent when the line with the quality trait possesses some other desirable agronomic features. In addition, recurrent selection schemes have been used to successfully improve the seed protein contents in soybean and French bean (rajma). However, when the quality trait is to be transferred from a related species, a backcross programme is the only suitable approach. Generally, the transfer of quality traits has two stages: (i) isolation from the interspecific cross such lines of the crop that have the quality trait being transferred, and (ii) use of these lines in breeding programmes to develop high yielding varieties with the improved quality trait. In some cases, the lines extracted from interspecific crosses show improvement in such quality traits that were not expected on the basis of the traits present in the parents. For example, tomato lines derived from the cross tomato x *S. habrochaites* showed improved red color of their fruits, while the fruit color of *S. habrochaites* is green.

Many quality traits may not be available in the germplasm of the concerned crop species. In all such cases, **genetic engineering** can be used to advantage and such transgenic plants can be created that have the desired quality trait. The use of this approach would require information about the biochemical basis of the desired quality trait, identification of the genes that would give rise to this trait, cloning of these genes and their transfer into the crop species where these genes would be expressed and are likely to generate the desired phenotype of the concerned quality trait. The methods of production of transgenic plants and their various uses are discussed in some detail in Chapters 9 to 12. The use of genetic engineering for improving protein and oil quality are described in section 8.7.2, respectively.

8.10. LIMITATIONS OF BREEDING FOR QUALITY

1. Most of the quality traits are quantitative in nature, and they often show low heritability; this makes their improvement difficult.
2. Estimation of many quantitative traits is difficult and time consuming.
3. Since the quality of a crop produce is determined by several component traits, the various traits have to be evaluated and selected for during the segregating generations. This greatly increases the time, effort and resources required for the breeding programme.
4. In many cases, the desirable quality trait may be associated with some undesirable feature, e.g., the opaque and soft kernel in the case of high lysine mutants of maize and sorghum. In such cases, considerable breeding effort would be required to eliminate the undesirable features.
5. The superior nutritional quality of the produce may require additional inputs, e.g., in the case of high protein content varieties. However, the superior quality produce does not fetch a better price in the market.

PRACTICE QUESTIONS

1. Define crop produce quality. Briefly describe the different categories of quality traits, and discuss the relevance and limitations of breeding for quality traits.
2. Nutritional quality of crop produce is very important for human health. Discuss the breeding efforts directed at the improvement of nutritional quality of crops.
3. Explain the meaning of protein quality. Discuss the various approaches for improving the protein quality of some important crops.
4. Explain the meaning of oil quality. Discuss the various approaches for improving the oil quality of some important crops.
5. Briefly describe the various sources for quality traits, and discuss the various breeding approaches for the improvement of these traits.
6. The term quality of crop produce does not refer to a single trait but it signifies a group traits that show varied modes on inheritance. Comment on this statement with the help of suitable examples.
7. Write short notes on the following: (i) Canola quality mustard oil, (ii) oil quality, (iii) protein quality, (iv) essential amino acids, (v) nutritional quality, (vi) anti-nutritional factors, (vii) amino acid balance of proteins, (viii) breeding for vitamin content.

RELEVANT LITERATURE

1. Khanna KR and Singh SP 1991. Genetics of fatty acid composition and breeding for modified seed oils. In KR Khanna (ed) Biochemical Aspects of Crop Improvement, pp. 255-283. CRC Press, Boston.
2. Chopra VL 1989. Plant Breeding. Oxford and IBH Publishing Co. Pvt. Ltd., New Delhi.
3. Chopra VL 2001. Breeding Field Crops. Oxford and IBH Publishing Co. Pvt. Ltd., New Delhi.
4. Ram HH 1998. Vegetable Breeding, Principles and Practices. Kalyani Publishers, New Delhi.

■■■■

Part B.
Transgenic technology

9 The Basic Techniques of Transgenic Technology

9.1. INTRODUCTION

Plant breeding primarily relies on genetic variation created by hybridization followed by segregation and recombination. Further gene transfers are based exclusively on mating between the selected individuals/strains. This limits the utilization of only such genes that are present in lines and species, which are sexually compatible with the concerned crop species. Interspecific gene transfers present many problems, including the loss of F_1 hybrids during embryo development due to abortion of the endosperm in the developing hybrid seeds. The technique of **embryo culture** was developed to rescue such interspecific hybrid embryos, and it has been widely used for this purpose. Embryo culture consists of isolation of young embryos from the developing seeds and culturing them in vitro under conditions that enable the recovery of complete plantlets from them. Embryo rescue has allowed the development of haploid plants from certain interspecific crosses, and the production of a large number of interspecific hybrids. It may be noted that embryo rescue can be used only in such cases where sexual hybridization is successful in producing zygotes that begin to develop into embryos.

In addition, the technique of **somatic hybridization** was developed to enable gene transfers even from such species that do not produce zygotes on hybridization. In this method, the cell wall is removed by enzymatic digestion from the cells of the two selected species to produce protoplasts. The protoplasts of the two species are then fused by a suitable treatment, and ultimately plants are regenerated from the hybrid cells resulting from the protoplast fusion. These somatic hybrids may receive the complete genomes of the two fusion parents, or they may receive only few chromosomes or even merely the cytoplasm from one of the parents. This method has been extensively used and some useful lines have been derived. But this method is limited to utilization of genes present in plant species. Further, the results from somatic hybridization are remarkably unpredictable since many somatic hybrids fail to regenerate complete plants and the transfer of specific genes or groups of genes depends entirely on chance.

The development of recombinant DNA technology or, in popular terms, genetic engineering, has permitted the transfers of genes from any organism into any other organism. In addition, even chemically synthesized genes can be transferred and integrated into genomes of the various organisms to enable the expression of these genes. A gene transferred into an organism by genetic engineering is called **transgene**, and the individuals carrying such genes are known as **transgenic organisms**. Therefore, a **transgenic plant** can be simply defined as a plant that contains in its genome a transgene and expresses this transgene. The production of transgenic plants involves identification and isolation of the gene to be transferred, production of a large number of copies of this gene, introduction of the gene into plant cells, and regeneration of complete plants from the cells having the transgene in their genomes. Thus plant tissue culture and recombinant DNA technology are the basic requirements for the production of transgenic plants. The basics of these techniques are described in simple terms in this chapter. The next two chapters (Chapters 10 and 11) are devoted to the description of various types of useful transgenic plants. The last chapter of this section of the book, i.e., Chapter 12, examines the biosafety issues relevant to transgenic plants, and the agricultural and socioeconomic impacts of their adoption for commercial cultivation. It may be pointed out that these chapters are designed to introduce the basics of the transgenic technology. Confusing technical details are avoided to facilitate comprehension, but enough information is provided so that the students are prepared for pursuing the subject in greater detail from appropriate sources, and are able to readily understand and assimilate the same.

9.2. PLANT TISSUE CULTURE

The term **plant tissue culture** is generally used to describe cultivation of plant cells, tissues and organs in vitro, i.e., in culture vessels on suitable nutrient media and under appropriate conditions. Plant tissue culture began in 1902, when Haberlandt attempted, with limited success, to grow isolated single leaf cells in a simple medium. Research efforts by subsequent workers have enabled rapid developments in this field so that the tissue culture technology is now a highly developed field with a variety of applications. Not only the plant cells and organs can be grown and maintained virtually indefinitely, complete plants can be regenerated from single cells/protoplasts. The ability of a cell to produce a complete plant is called **totipotency**; this ability is of great significance in many applications of plant tissue culture technology for crop improvement.

The plant tissue cultures can be grouped into several categories mainly on the basis of the explants used to establish the culture and/or the type of growth produced in the culture. **Explant** is the plant part used to initiate the tissue culture. Some of the common types of plant tissue cultures are briefly described here. The term **cell culture** is used to describe growth of plant cells and tissues in an unorganized state; cell cultures are called **callus cultures** when they are maintained on agar medium and **suspension cultures** when they are grown in liquid medium. An **organ culture** is initiated by using an organized plant part, e.g., shoot tip, as an explant and maintaining the organized state, i.e., growth as shoots in this case, during the in vitro culture phase as well. **Embryo culture** involves culture of immature embryos and recovery of complete plants from them; it is also an example of organ culture. **Ovary culture** involves culture of ovaries generally to produce haploid plants, while **anther culture** comprises culture of the anthers generally for obtaining haploid plants. A **meristem culture** generally consists of the cultivation of shoot-tip meristems to obtain multiple shoots and, ultimately, complete plants.

9.2.1. General technique of plant tissue culture

The in vitro culture of the plants and tissues must ensure the following: (i) the cultures must be free from microbes, and (ii) the proper nutritional and the environmental conditions should be provided to obtain the type of growth and development desired from the cultures. The infection of a plant tissue culture by bacteria and/or fungi is called **contamination**. Contamination is prevented by effective sterilization of all the materials, including the explants, the instruments, and the working area used for various operations during the culture. **Sterilization** is the inactivation of microbes using a suitable treatment e.g., dry heat, flame, autoclaving, 70% ethanol, a chemical sterilant, or filtration (Table 9.1). Even the hands of the worker handling the cultures are surface-sterilized by wiping them

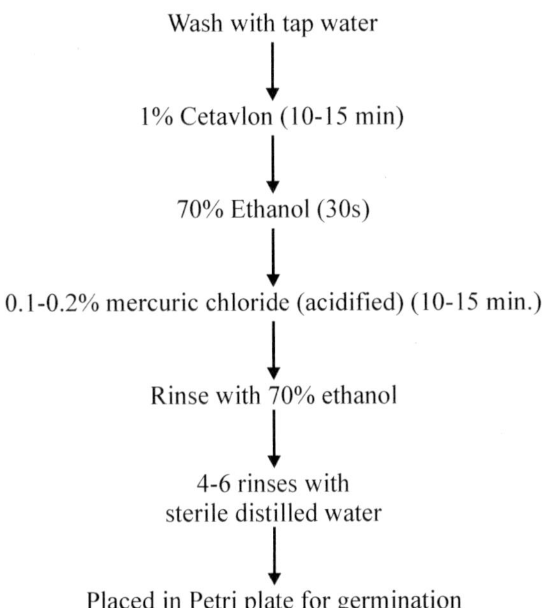

Fig. 9.1. A simple procedure for surface sterilization of seeds for obtaining explants for culture initiation.

with 70% ethanol. The media and all such vessels and instruments that can be autoclaved are sterilized by autoclaving at 1.06kg/cm^2 (15 psi, pounds per square inch) for 15 to 40 min. The air in the laminar airflow cabinets is sterilized by passing it through the high efficiency particulate air (HEPA) filters. The explants are decontaminated by treatment with a suitable chemical like calcium hypochlorite, mercuric chloride etc. according to a suitable procedure (Fig. 9.1). These procedures aim to inactivate the microbes and their spores present on the surfaces of the explants; therefore, this process is called **surface sterilization**.

Table 9.1. A summary of the methods used for the of sterilization of various materials used for plant tissue culture work as well as the persons handling the cultures

Material	Sterilization technique	Remarks
Empty glassware, certain plastic ware, scalpels, forceps, needles, etc.	Dry heat: 160-180°C for 3 hr	In an oven
Media, glassware, several plastic ware, contaminated cultures	Autoclaving	In an autoclave
Scalpels, forceps, etc., mouths of flasks	Flame	A suitable burner used
Enzymes, GA3, ABA and other heat labile compounds	Membrane filter; pore size 0.45μm or less	Used with a filter assembly
The air blown though laminar airflow cabinets	HEPA filter	
Hands of operators, and surfaces of the working area, culture vessels, etc.	70% ethanol	The hands and surfaces are wiped with alcohol
Explants	One of several chemical sterilizing agents	Surface sterilization

Plant cells and organs are cultured on a completely synthetic medium, and several standard media formulations have been developed. The MS (Murashige and Skoog) and the B5 media are two of the most widely used tissue culture media; their compositions are given in Table 9.2. The media formulations provide six macronutrients (N, P, K, Ca, S, Mg), six micronutrients (Mn, Zn, Fe, Cu, B, Mo), some vitamins (inositol, nicotinic acid, pyridoxine, thiamine), an energy source (generally sucrose), and the desired growth regulators. Some media formulations like MS and several other media may contain some amino acids and/or other complex organic additives like yeast extract, potato extract, etc. Growth regulators are used in low concentrations (generally, 0.1 to 5.0 μg ml^{-1}) and are critical in determining the type of growth and development obtained in the cultures (Table 9.3). All callus and suspension cultures require an auxin, but callus cultures of dicot species also need a cytokinin. Shoot regeneration generally requires a cytokinin, while somatic embryogenesis and rooting in shoots are induced by an auxin. Other growth regulators (GRs) are needed in some cases and under certain conditions. The pH of the medium is adjusted to

5.8. When semisolid media are to be prepared, 6-8g l^{-1} agar is added to the medium. The media are sterilized by autoclaving.

Table 9.2: The composition of MS (Murashege and Skoog) and B_5 media, the two of the most widely used plant tissue culture media

Compound	Concentration (mg/l)	
	MS[1]	B5[3]
Inorganic		
NH_4NO_3	1650	—
KNO_3	1900	2527.5
$MgSO_4.7H_2O$	370	246.5
KH_2PO_4	170	—
$(NH_4)_2SO_4$	—	134
$Ca(NO_3)_2.4H_2O$	—	—
Na_2SO_4	—	—
$NaH_2PO_4.H_2O$	—	150
KCl	—	—
$CaCl_2.2H_2O$	440	150
KI	0.83	0.75
H_3BO_3	6.2	3
$MnSO_4.4H_2O$	22.3	—
$MnSO_4.H_2O$	—	10
$ZnSO_4.7H_2O$	8.6	2
$Na_2MoO_4.2H_2O$	0.25	0.25
MoO_3	—	—
$CuSO_4.5H_2O$	0.025	0.025
$CoCl_2.6H_2O$	0.025	0.025
$Fe_2(SO_4)_3$	—	—
$FeSO_4.7H_2O$	27.8	—
$Na_2.EDTA.2H_2O$	37.3	—
Sequestrene 330 Fe[3]	—	28
Organic		
(a) *Vitamins*		
Inositol	100	100
Nicotinic acid	0.5	1
Pyridoxine HCl	0.5	1
Thiamine HCl	0.1	10
(b) *Amino acid*		
Glycine	2	—

Compound	Concentration (mg/l)	
	MS[1]	B5[3]
(c) *Carbon source*		
Sucrose (g/l)	30	20
(d) *Growth regulators*	(Section 6.3.5.4)	
Gelling agent (for callus cultures)		
Agar	Generally 6-8 g/l of the medium *Medium pH generally adjusted to 5.8 with* 1N HCl or 1N NaOH	

1. Murshage, T. and Skoog, F. 1962. Physiol. Plant. 15:473-497
2. Gamborg, O.L., Miller, R.A. and Ojima, K. 1968. Exp. Cell Res. 50: 151-158
3. Fe.Na.EDTA may be used when this compound is not available.

Table 9.3: The growth regulators commonly used in tissue culture media

Class of growth regulator	Examples	Generally used for				
		Callus culture	Suspension culture	Shoot regeneration	Somatic embryogenesis	Rooting of shoots
Auxin	2,4-D, NAA, IAA, IBA	Yes	Yes	In some cases	Yes	Yes
Cytokinin	BAP, kinetin	In dicot species	No	Yes	For somatic embryo (SE) maturation	No
Gibberellins	GA_3	No	No	In few species	For SE germination	No
-	Abscisic acid	No	No	Promotory effect	For SE maturation	No

The surface sterilized explants are transferred into culture vessels containing the desired media. The explants are cultured in Petri plates, culture tubes, conical flasks, especially designed culture flasks, bottles or some other vessels suitable for culturing plant cells and tissues. The culture vessels are generally made of glass or plastic, but vessels made up of other materials like teflon are also used. The mouths of culture tubes, culture flasks, etc. can be plugged with nonabsorbent cotton or closed with suitably designed caps. The cultures stored in a culture room with the temperature around 25 ± 2°C and illumination at 1000 lux or lower intensity of light from cool fluorescent tubes.

9.2.2. Callus and Suspension Cultures

The living cells of the explants are induced by the growth regulators present in the culture medium to undergo cell division. The explants cultured on semisolid agar media provide a mass of unorganized cells called **callus** (Fig. 9.2A, 9.3).

But the explants cultured in liquid media produce a mass of single cells and cell aggregates of various sizes that are suspended in the medium; these cultures are called **suspension cultures** (Fig. 9.2B). The suspension cultures have to be continuously shaken to ensure the supply of adequate amounts of oxygen essential for survival of the cultured cells. Generally, the suspension cultures are continuously agitated at 100-250 rpm (revolutions per minute) with the help of a gyratory shaker. The agitation of suspension cultures, in addition, helps breaking up of the larger cell masses into smaller ones and ensures continuous mixing of the medium. The latter enables a more efficient and uniform availability of the nutrients to the cultured cells.

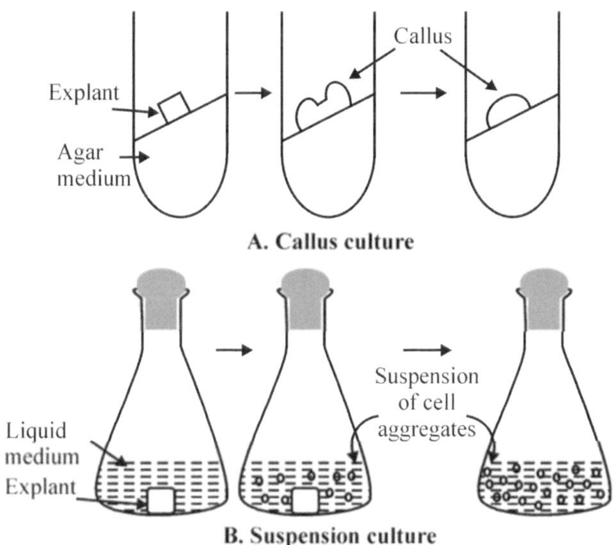

Fig. 9.2. Initiation of callus and suspension cultures from plants. A. Callus culture, B. Suspension culture

The growing plant cells continuously utilize the nutrients present in the culture media. As a result, the nutrients present in the media become progressively depleted. In addition, there is continued moisture loss from the media, and an increase in the cellular biomass and the wastes liberated from the cultured cells. Therefore, it becomes necessary to transfer the cells and tissues into new culture vessels containing fresh culture media. Generally, a small portion of the callus/suspension cultures is taken and transferred to fresh media; this is called **subculture**. In case of suspension cultures, generally 20-25% of the culture volume is used as inoculum during subculture. The suspension cultures need to be subcultured every 3-10 days, while callus cultures are generally subcultured every 4-6 weeks. Theoretically, with regular subcultures and proper care to avoid contamination, the callus and suspension cultures can be maintained indefinitely.

Fig. 9.3. Callus culture maintained on MS medium. MS, Murashige and Skoog.

9.3. REGENERATION OF COMPLETE PLANTLETS

In general, the callus and suspension cultures do not support the development of organized structures like shoots, roots, etc. But under suitable conditions, these cultures can be induced to produce shoots, roots, somatic embryos (SEs), etc.; the production of these structures is called **regeneration**. The most critical factor that affects regeneration of specific structures is the combination growth regulators present in the media. Regeneration of complete plantlets is essential for application of the plant tissue culture technology for crop improvement. Complete plantlets can be regenerated from cultured cells and tissues in the following two ways: (i) regeneration of shoots from the cultured cells and then rooting of these shoots (Section 9.3.1), and (ii) regeneration of somatic embryos (SEs) from the tissue cultures and then germination of the SEs to obtain complete plantlets (Section 9.3.2). These plantlets are subjected to a hardening procedure to make them capable of surviving out of the culture vessels, and the hardened plantlets are ultimately taken to the field.

9.3.1. Shoot Regeneration and Rooting of Shoots

The *de novo* development of shoot buds from unorganized cultures is known as **shoot regeneration**. Some explants like shoot-tips and nodal segments contain pre-existing shoot buds; in such cases, shoots are generally produced by proliferation of the shoot buds, but some new shoot buds may also develop. When unorganized cultures are transferred onto a suitable medium, several groups of meristematic cells develop in the cultures. This is followed by differentiation of vascular bundles in these cell clusters. Ultimately, leaf primordia and the shoot apical meristem develop from many of these clusters and give rise to shoot buds. The shoot buds continue to grow to give rise to shoots (Fig. 9.4). In general shoot regeneration is promoted by cytokinins like kinetin and benzyl aminopurine (BAP), while auxins have an inhibitory effect. But in some species, shoots regenerate on growth regulator (GR)-free medium, while in some others auxin has a promotory effect on shoot regeneration.

Fig. 9.4. Regeneration of multiple shoots from callus cultures of plant species

In any case, GRs appear to be the most important factor of the culture system that determines shoot regeneration. In addition, shoot regeneration is affected by explant genotype, physical condition (agar or liquid) of the medium, temperature and light regimes, and some other factors as well. The genotype of explant has a remarkable effect on shoot regeneration. As a result, the frequency of shoot regeneration varies considerably among the varieties of the same crop, and some varieties may not show any shoot regeneration. The regeneration ability is under genetic control and can be improved by a suitable breeding scheme.

In order to obtain complete plantlets, roots have to be regenerated from the lower ends of the shoots produced from the culture (Fig. 9.5). For this, the shoots are excised and transferred to a rooting agar medium. Generally, the rooting medium has an auxin like IBA (indole butyric acid) in suitably low concentration, but in several plant species rooting may occur on a GR-free medium. In addition, the rooting medium may have low sugar concentration (generally $1 g\ l^{-1}$) and/or reduced salt concentration (1/2 or even ¼ salts of the MS medium). The recent trend is to root the shoots directly in the potting mix: the cut ends of shoots are treated with an auxin and transplanted into pots where they regenerate roots. In this approach, rooting step is combined with hardening of plantlets and their establishment in the soil (Section 9.3.3).

Fig. 9.5. Regeneration of roots in shoots produced in tissue culture. The shoots were kept on MS medium supplemented with a suitable concentration of IBA (indole butyric acid).

9.3.2. Somatic Embryogenesis

In **somatic embryogenesis**, a somatic cell divides and becomes organized into a somatic embryo (SE). **Somatic embryos** (SEs) are similar to zygotic embryos, except for the fact that they originate from somatic cells. Generally, SEs originate from single cells, which divide to form meristematic clusters of cells. The outer walls of the cells on the periphery of each cell cluster generally become cutinized. As a result, each cell cluster becomes separated from the cell mass surrounding the culture. The cells in each cluster continue to divide and develop into a somatic embryo. The somatic embryo develops through the following successive stages: globular, heart-shaped, torpedo-shaped and cotyledonary stages (Fig. 9.6 and 9.7). The SEs have both plumule and radicle so that they germinate to give rise to complete plantlets. The initiation of somatic embryogenesis generally requires a high concentration of a strong auxin like 2, 4-Dichlorophenooxyacetic acid (2,4-D). But once a culture acquires the ability for embryogenesis, it can be maintained in the embryogenic state by culturing it on a medium with low auxin concentration.

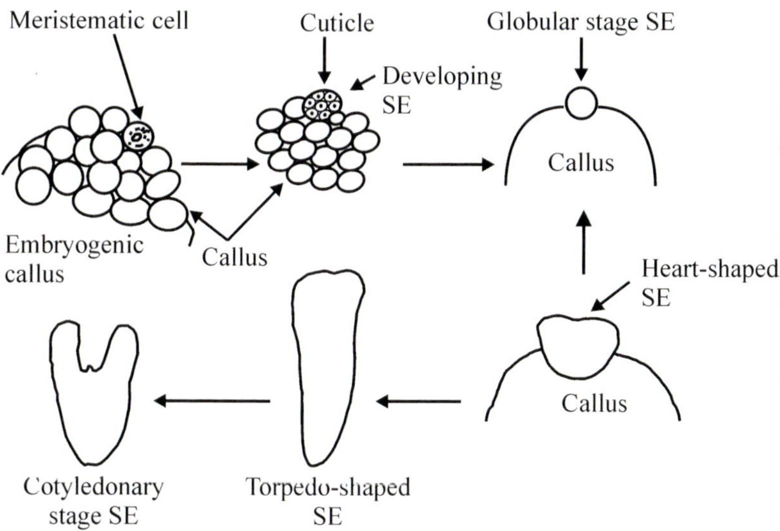

Fig. 9.6. The development of a somatic embryo (SE). The stages of SE development are comparable to those of zygotic embryos. The torpedo-shaped and cotyledonary stage embryos are shown here without the callus tissue supporting them, but they do remain associated with the cell mass from which they originated.

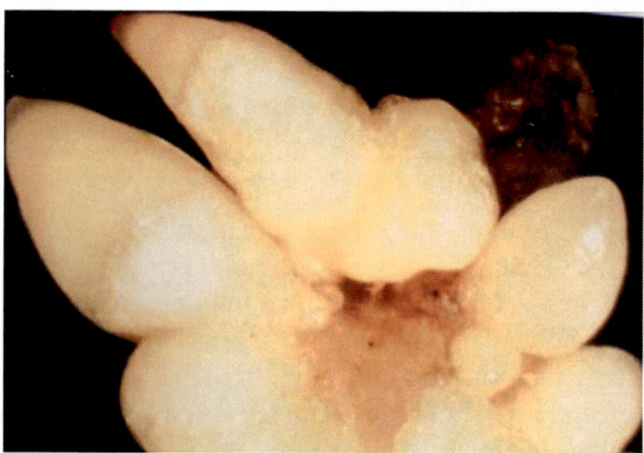

Fig. 9.7. Torpedo-shaped somatic embryos regenerated from cell/callus cultures of *Crocus sativa* cultured on MS medium supplemented with 0.5 mgL^{-1} NAA and additives (50.0 mg L^{-1} of ascorbic acid and 25.0 mg L^{-1} each of adenine sulphate, argine and citric acid)

The differentiation of cultures beyond the globular stage generally occurs on a medium having no auxin, and in some cases a low level of cytokinin may be useful. After the embryos are fully developed, they undergo a process of maturation, which improves germination of the SEs. SE maturation is supported by ABA, partial desiccation of the SEs, and culturing the SEs on a high sucrose medium. Thus, GRs are the critical factor influencing somatic embryogenesis. In addition, the genotype of the explant has a marked effect on somatic embryogenesis so that cell cultures of some varieties of a given species may fail to regenerate SEs. As in the case of shoot regeneration, the ability for somatic embryogenesis is genetically controlled, and in some cases **plasmagenes** are also known to be involved. SE regeneration is promoted by NH_4^+ nitrogen, and the cultures seem to require a minimum level of endogenous NH_4^+ for SE regeneration. Some other factors like K^+ concentration, maltose as carbon and energy source, and low levels of dissolved oxygen might promote somatic embryogenesis in some cases.

9.3.3. Hardening of the Plantlets and Their establishment in the Field

In the culture vessels, the shoots and plantlets are maintained in conditions of high humidity and low light intensity. As a result, they do not have well developed vascular bundles and cuticle, and are not suitable for growing under the relatively harsh conditions prevailing in greenhouses and, especially, fields. In addition, they are maintained on media having sucrose, as a consequence of which their photosynthetic apparatus is not adequate to support their survival outside the culture vessels. Therefore, these plantlets have to be subjected to suitable procedures designed to increase their hardiness and the ability to survive on transfer to pots and fields. These procedures are termed as **hardening procedures** and the process is called **hardening**. The hardening procedure may be carried out as the terminal stage of their in vitro culture. This procedure might involve culturing the plantlets on a suitably modified medium and/or culture conditions. Alternatively, they may be cultured in the presence of specific chemicals added into the medium or in association with appropriate microorganisms.

More generally, the plantlets are hardened simultaneous with their establishment in the soil. After rooting, the plantlets are removed from the culture vessels, and their roots are washed free of the medium. They are then transplanted into small pots containing a suitable potting mix, and kept in an environment having low intensity of light and high humidity. Over the next 7-10 days, the light intensity is gradually increased and the humidity level is gradually reduced to the level of normal greenhouse conditions. The pots are later shifted to a greenhouse, then to a net house and finally transplanted into the field. There will be some plant loss during this procedure, but this loss can be minimized by careful designing and handling of the hardening procedure.

9.4. RECOMBINANT DNA TECHNOLOGY

A **recombinant DNA** comprises a vector molecule into which a DNA segment has been integrated. A **vector** is a DNA molecule that is capable of independent replication in the cells of a given species. The DNA fragment integrated into the vector molecule is termed as **DNA insert**. The recombinant DNAs are constructed with the help of restriction enzymes that cut the DNA into pieces, and DNA ligases, which join the desired DNA pieces together. Once the recombinant DNA has been produced, it is generally multiplied in *E. coli*; this is called **gene cloning** or **DNA cloning**. Thus **recombinant DNA technology** consists of construction and propagation of recombinant DNA. It also involves the use of recombinant DNAs to obtain the protein products encoded by the DNA inserts, and to generate transgenic plants and animals. Thus, the entire procedure of recombinant DNA technology consists of the following steps: (i) isolation and fragmentation of DNA from the selected organism, (ii) integration of the DNA fragments into a suitable vector, (iii) introduction of the recombinant DNA into a suitable host, e.g., generally, *E. coli*, (iv) isolation of the host cells/colonies having the desired DNA fragment and isolation of multiple copies of this fragment, and (v) use of this fragment for the desired purpose.

There are several different standard procedures for the isolation of DNA from different organisms. In case of plants, the cetyltrimethyl ammonium bromide (CTAB) method of DNA isolation is the gold standard and yields DNA of very high quality. There are several modifications of the original CTAB procedure, and many other procedures of DNA isolation suited for different applications. A brief discussion of these procedures is given in Singh and Singh (2015). The other steps of the recombinant DNA technology are briefly described in the following sections.

9.4.1. Restriction Enzymes and DNA Ligase

The isolated DNA is fragmented in one of the following three ways. The DNA can be fragmented either by mechanical shearing (called **nebulization**) or by sonication; both these methods produce blunt-ended fragments. Alternatively, the DNA may be partially digested with a suitable restriction enzyme; this approach would yield fragments with either blunt or protruding ends depending on the enzyme used for digestion of the DNA. The DNA is digested only partially so that fragments of size suitable for cloning are obtained since complete digestion may generate fragments too small for cloning. A **restriction enzyme** or **restriction endonuclease** cleaves DNA molecules at specific sites, called **recognition sites**. The recognition sites are palindromes with rotational symmetry so that the base sequence of the first half one strand is the mirror image of the second half of the other strand (Fig. 9.8A). A large number of restriction enzymes are known, and most of them have distinct recognition sequences (Table 9.4). Most restriction enzymes cleave the two DNA strands at different points so that the cut ends have protruding single-stranded regions; such cuts are known as **staggered cuts**. The single-stranded regions of the two strands are complementary to each other so that they pair together (Fig. 9.8B). This property is of great advantage since the fragments generated by a single enzyme from two different DNA preparations will pair with each other at their ends (Fig. 9.9). The nicks present at the paired ends of these fragments are sealed by the enzyme DNA ligase.

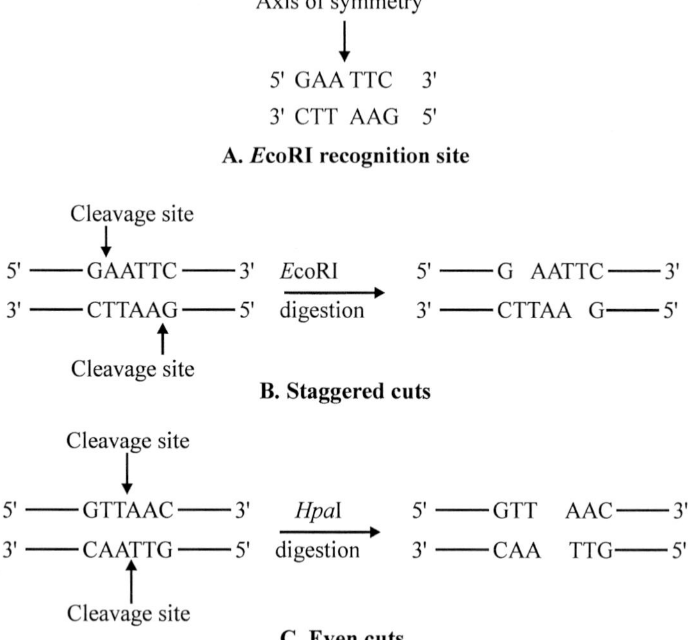

Fig. 9.8. The types of cuts produced by restriction enzymes. A. The recognition site of *EcoRI* restriction enzyme. The arrow indicates the axis of symmetry. B. Staggered cuts generated by *EcoRI*; the arrows indicate the sites of cleavage within the recognition site. C. Even cuts induced by *HpaI*; the arrows indicate the locations of cleavage.

Table 9.4: A list of some selected restriction enzymes and their recognition sequences

Restriction enzyme	Recognition sequence*
*Alu*I	AC/GT
	TG/CA
*Bam*HI	G/GATCC
	CCTAG/G
*Eco*RI	G/AATTC
	CTTAA/G
*Hin*dIII	A/AGCTT
	TCCGA/A
*Hpa*I	GTT/AAC
	CAA/TTG
*Pst*I	CTGCA/G
	G/ACGTC
*Taq*I	T/CGA
	AGC/T

* The / indicates the site of cleavage.

Some other restriction enzymes, however, cut the two strands at the same point to generate blunt ends; such cuts are called **even cuts** (Fig. 9.8C). The fragments with blunt ends can be converted into fragments with protruding ends or they can be joined as such by using a suitable DNA ligase. The enzyme **DNA ligase** seals the nicks that remain in the recombinant DNA molecules during their construction. A **nick** signifies that the phosphodiester bond between two adjacent nucleotides of a single DNA strand is broken. The ligase enzyme catalyzes the formation of phosphodiester bond between the two adjacent nucleotides and thereby seals the nick present in the strand. This restores the integrity of the DNA strand. For this reason, the DNA ligase if often referred to as **molecular glue**. The DNA ligase enzyme uses energy from ATP for the ligation reaction.

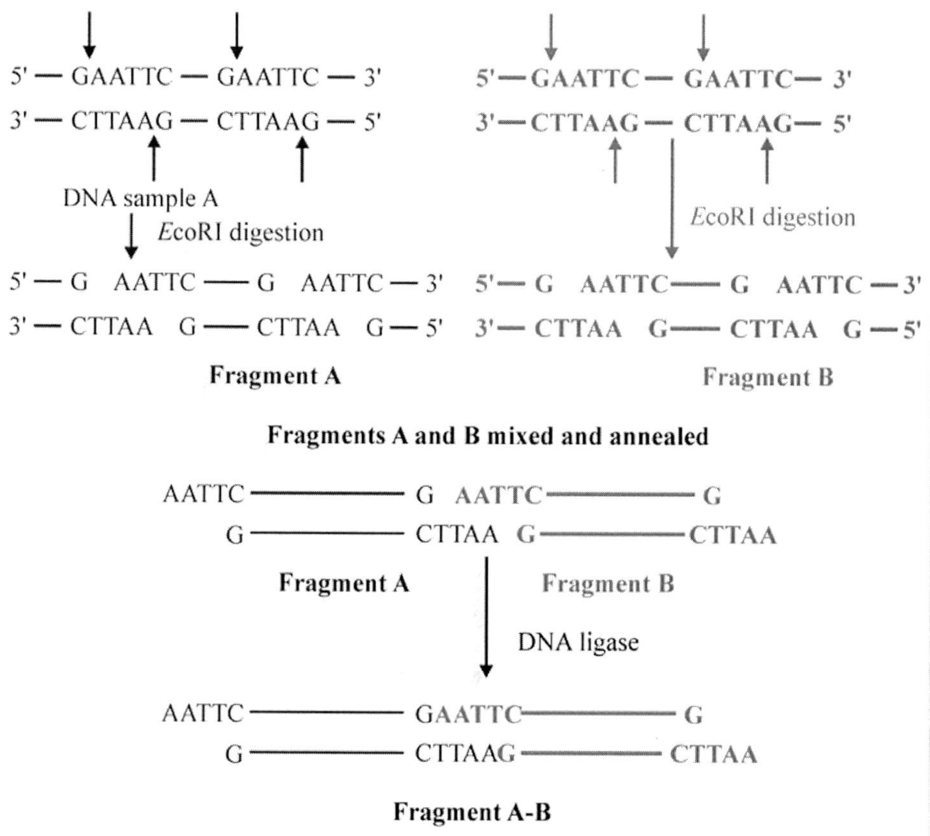

Fig. 9.9. Digestion of two different DNA samples (samples A and B) with the same restriction enzyme produces fragments that have complementary protruding ends. When these fragments are mixed, their protruding ends base-pair or anneal to yield a DNA molecule composed of both the fragments A and B. DNA ligase is used to seal the nicks present at the points where the ends of the two fragments meet.

9.4.2. Production of Recombinant DNA

A **recombinant DNA** comprises a vector and a DNA insert integrated into the vector. A **DNA insert** is a DNA fragment that is integrated into a vector for either cloning or expression. A **vector** is a DNA molecule that is capable of independent replication in a suitable host. **Host** is an organism into which vector and recombinant DNA are multiplied or cloned. In general, all cloning is initially done in the bacterium *E. coli.* The vectors are constructed from natural DNA molecules like plasmids and phage genomes that are capable of independent replication. The present day vectors are derived from these DNA molecules by deleting their unnecessary sequences and adding useful sequences from a variety of sources. The sequences included in a vector perform a number of different useful functions, including replication in the cells of the given host. Vectors are generally grouped into two categories, viz., cloning vectors and expression vectors. The **cloning vectors** are used for production of large number of copies of the DNA insert. Therefore, these vectors contain such origins of DNA replication that allow production of multiple copies of the vector in each host cell. The **expression vectors**, on the other hand, are used for expression of the DNA insert so that it is transcribed and even translated in the host cells. Therefore, these vectors also contain the regulatory sequences, including promoters, that are required for transcription and translation of the DNA insert.

The vector pUC18/19 (p, plasmid; UC, University of California) is an example of a good vector for the bacterium *E. coli* (Fig. 9.10). This vector contains several different functional sequences, of which the following three sequences are the most relevant for our discussion. (i) It has an origin of replication (*Ori*) for multiplication in *E. coli* cells. (ii) It has the *bla* gene for ampicillin resistance; this allows the elimination of cells lacking the vector pUC18/19 or the recombinant DNA. (iii) It also has the gene *lacZ-α* that encodes the α- fragment of β-galactosidase enzyme. The β-galactosidase hydrolyzes lactose as well as X-gal. The hydrolysis of X-gal produces a blue colored dye. The hosts used for cloning of the pUC18/19 recombinant DNA are specific strains of *E. coli*, e.g., JM103 and JM109. In these strains, the sequence of *lacZ* gene that encodes the α- fragment of β-galactosidase is deleted. Therefore, cells of these strains are unable to hydrolyze X-gal. But when the vector pUC18/19 is present in these cells, it provides the missing *lacZ-α* fragment. As a result, these cells are able to hydrolyze X-gal, and they yield blue colored colonies when they are grown on a medium containing X-gal. The *lacZ-α* gene of pUC18/19 has a multiple cloning site (MCS) that has recognition sequences for several restriction enzymes. The DNA insert is integrated into the MCS by using a suitable restriction for opening the vector molecule followed by integration of the DNA insert (Fig. 9.11). The integration of the DNA insert into the MCS disrupts the *lacZ-α* gene sequence so that the α fragment of the β-galactosidase enzyme cannot be produced. Therefore, the host cells carrying the pUC18/19 recombinant DNA are unable to hydrolyze X-gal so that they produce white colonies. This allows easy selection of the colonies that carry the recombinant DNA and to eliminate the colonies possessing the unchanged vector pUC18/19.

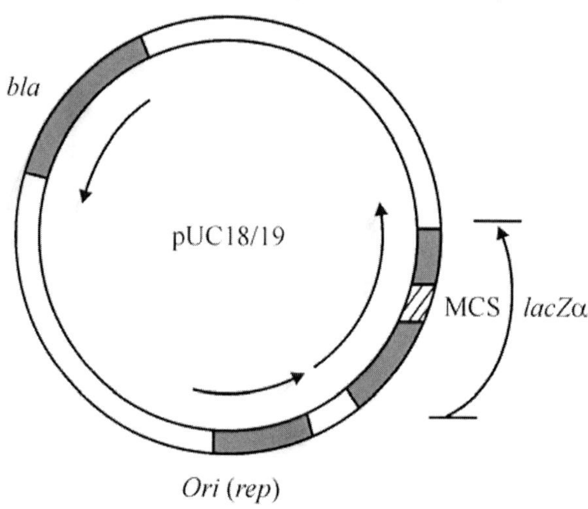

Fig. 9.10. A highly simplified and schematic representation of the *E. coli* vector pUC18/19. *bla,* gene specifying ampicillin resistance; *Ori (rep),* origin of replication for replication in *E. coli; lacZα,* sequence encoding the α fragment (the N-terminal fragment) of β-galactosidase enzyme. This enzyme hydrolyzes lactose; it also hydrolyzes X-gal to generate a blue dye.

The production of recombinant DNA involves opening the vector with the same restriction enzyme that was used to generate the DNA insert. The opened vector and the DNA insert are mixed and allowed to anneal; the nicks remaining in the annealed molecules are sealed by DNA ligase. This generates a mixture of the following three types of DNA molecules. (i) The vector molecules themselves become circularize and yield unchanged vector molecules (Fig. 9.11A). (ii) The DNA insert becomes integrated into the vector to generate recombinant DNA molecules (Fig. 9.11B). Finally, (iii) the DNA insert itself may circularize and give rise to circular DNA insert (Fig. 9.11C). This mixture of the three types of DNA molecules is introduced into *E. coli* cells; this is called **transformation**. Incubation of the bacterial cells in a solution of 50 mM $CaCl_2$ increases the frequency of transformed cells. These cells are plated on a medium supplemented with ampicillin and X-gal. Bacterial cells that are either not transformed or are transformed with the circularized DNA insert will be killed by ampicillin. The surviving cells will produce blue colonies in case they are transformed by the unchanged vector. Those bacterial cells that are transformed by recombinant DNA will produce white colonies. Therefore, the white colonies are identified and isolated to obtain multiple copies of the recombinant DNA/DNA insert.

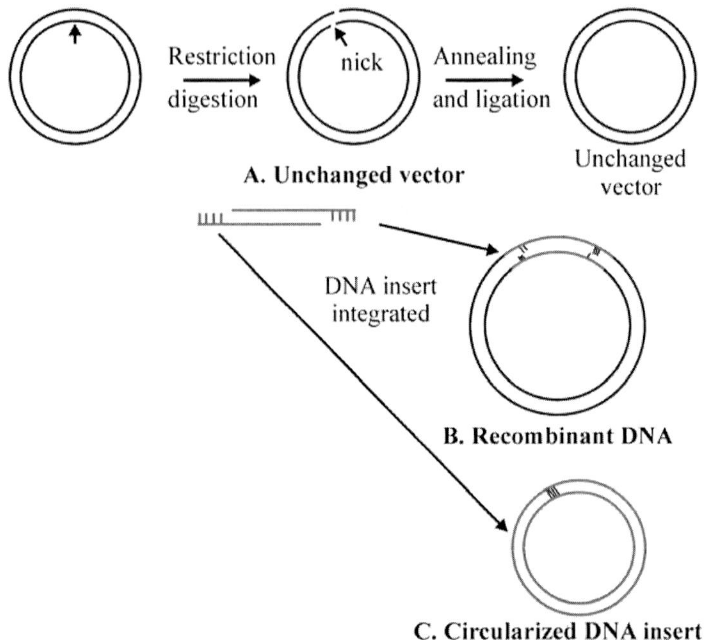

Fig. 9.11. The vector is cut open with the same enzyme that was used to generate the DNA insert. The opened vector and DNA insert are mixed, allowed to anneal and then the nicks are sealed with DNA ligase. Three types of DNA molecules will be obtained: (A) unchanged vector (vector without DNA insert), (B) recombinant DNA (vector with DNA insert integrated into it), and (C) circularized DNA insert.

9.4.3. Genomic and cDNA Libraries

Copies of a desired gene can be obtained by one of the following strategies. The **first method** comprises construction of a genomic or a cDNA library of the concerned organism and then screening the library to identify and isolate the desired gene. The **second strategy** uses polymerase chain reaction (PCR) to amplify the desired gene from the genome of the concerned organism. The **third approach** is to chemically synthesize the complete gene. In the first approach, the base sequence of the gene or the amino acid sequence of the protein encoded by it is not required. But in the case of the second method, the sequences at the two ends of the gene to be amplified are needed to design the primers used for PCR. In the third strategy, however, the base sequence of the entire gene or the amino acid sequence of the complete protein encoded by it must be known to allow chemical synthesis of the desired gene. In this section, the gene isolation from genomic and cDNA libraries is described, while PCR is considered in Section 9.4.4 and chemical synthesis of gene is briefly narrated in Section 9.4.5.

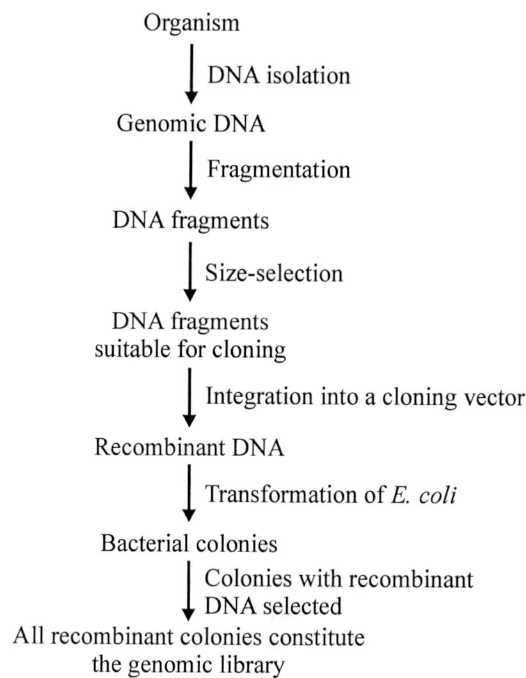

Fig. 9.12. A simplified schematic representation of construction of a genomic library. Ideally, all the DNA inserts present in the library together represent the complete genome of the organism.

1. Genomic library: It is a collection of recombinant bacterial clones or phage lysates containing DNA inserts from the genomic sequences of the concerned organism. Ideally, the sum total of the DNA inserts in all the recombinant DNAs included in the library represents the entire genome of the concerned organism. But this ideal is seldom achieved. For construction of a genomic library, the genomic DNA from the concerned organism is isolated (Fig. 9.12). The DNA is then fragmented by a suitable method; generally, a restriction enzyme with four-base recognition sequence is used for partial digestion of the genomic DNA. Fragments of the suitable size are isolated either by agarose gel electrophoresis or sucrose gradient centrifugation. The size-selected fragments are then integrated into the selected vector; usually, cosmid vectors and vectors derived from λ phage genome have been used. The appropriate strain of *E. coli* is transformed and recombinant clones are isolated; these clones together constitute the genomic library of the concerned organism.

2. cDNA library. DNA is transcribed by RNA polymerase to produces RNA copies of DNA. On the other hand, RNA sequences are copied into DNA sequences by the enzyme reverse transcriptase (Fig. 9.13); this is called **reverse transcription**. The DNA produced by reverse transcription is known as ***cDNA*** (**complementary DNA** or **copy DNA**). Therefore, **cDNA library** is a collection of recombinant bacterial clones or phage lysates containing DNA inserts from the cDNA prepared from the given tissue of the concerned organism. The cDNA library would contain the cDNAs from all the mRNAs encoded by those genes that are expressed in the concerned tissue at the specific developmental stage and under the environmental conditions prevailing at the time of sampling the tissue for mRNA preparation. The cDNAs are prepared from RNA isolated from the selected tissue of the concerned organism with the help of reverse transcriptase enzyme. In case of eukaryotic mRNA, a poly-T oligonucleotide is used as primer for the reverse transcription reaction. A DNA.RNA hybrid molecule is produced at the end of the reverse transcription step (Fig. 9.13).

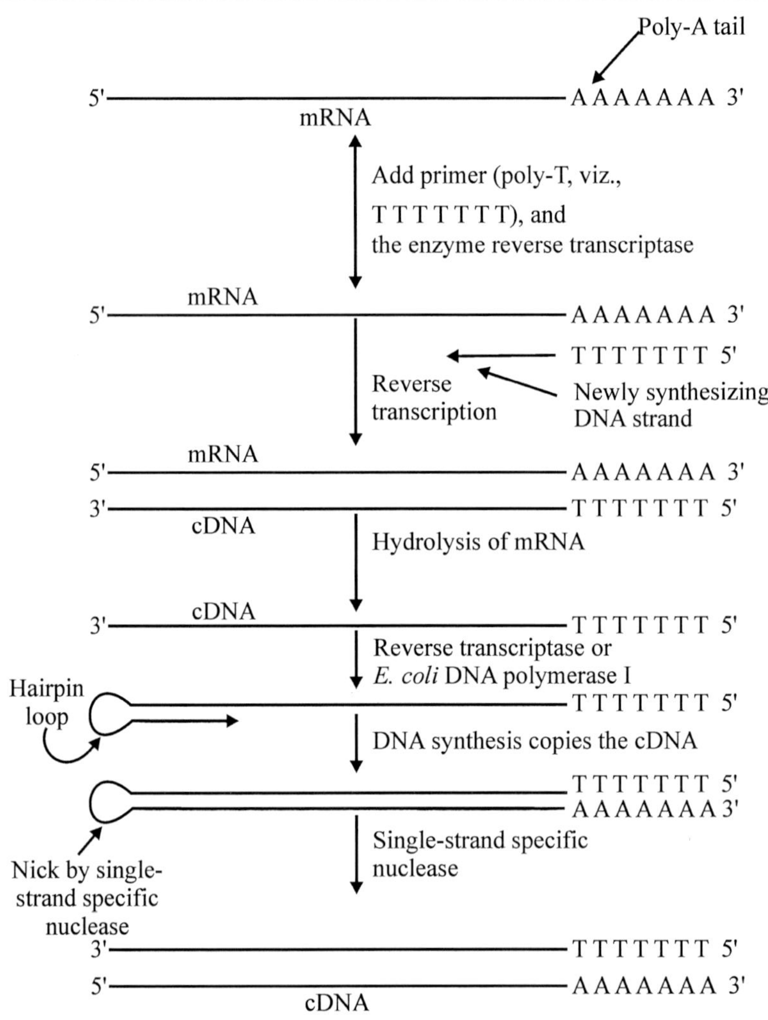

Fig. 9.13. A schematic and simplified representation of production of cDNA from eukaryotic mRNA. cDNA, complementary or copy DNA.

The RNA strand of this hybrid molecule is digested by either RNase H or alkaline hydrolysis, leaving behind the single-stranded cDNA produced by reverse transcription. This cDNA single-strand is made double-stranded by replication catalyzed by either reverse transcriptase itself or by *E. coli* DNA polymerase I. These enzymes use the 3' end of cDNA single- strand as primer and synthesize the complementary strand of the cDNA, making the cDNA double-stranded. This generates a hair-pin loop in the cDNA, which is cut open by a single- strand-specific nuclease. The cDNA preparation is used for production of recombinant DNA that is cloned in a host like *E. coli.* The recombinant bacterial colonies or phage lysates together comprise the cDNA library. It may be pointed out that the composition of a given cDNA library of an organism will depend on the tissue from which the mRNA was isolated, and the stage of development and the environmental conditions prevailing at the time of sampling.

3. Comparison between genomic and cDNA libraries. The genomic and cDNA libraries of an organism differ with respect to several features, the most relevant of which are listed in Table 9.5. When prokaryotic or eukaryotic genes are to be expressed in eukaryotes, genes isolated from genomic libraries can be used to produce transgenics. But when eukaryotic genes are to be expressed in prokaryotes, it is absolutely essential to use cDNAs of the concerned genes for production of transgenics. This is because eukaryotic genes contain introns that are removed by splicing from the RNA transcripts of the genes to yield mature mRNA molecules. Since prokaryotes do not have the capability for intron splicing, the eukaryotic genes cannot yield functional mRNAs in prokaryotic cells.

Table 9.5: The salient features of genomic and cDNA libraries

Feature	Genomic library	cDNA library
DNA inserts represent	Ideally, all genomic sequences	Transcribed sequences only
The composition of library affected by:		
i. Cell or tissue type	No	Yes
ii. Stage of development	No	Yes
iii. Environmental conditions	No	Yes
iv. Frequency of the given mRNA in the cell	No	Yes
Presence of introns in the DNA inserts*	Yes	No
Parts of a single gene may be present in	One or more clones	A single clone
Regulatory sequences of genes	Present	Absent

* This is critical when a eukaryotic gene is to be expressed in a prokaryotic organism.

4. Screening of genomic and cDNA libraries. The bacterial clones/phage lysates of the concerned genomic/cDNA library are analyzed using highly specific and sensitive techniques to identify and isolate the clone(s)/lysate(s) having the desired DNA insert or gene. This analysis is called library screening. The various techniques of library screening are grouped into two categories, viz., sequence-based approaches and protein product-based methods. Each of these groups comprises several highly specific and efficient screening methods. The protein product-based techniques are used for screening of libraries prepared in expression vectors so that the DNA inserts are able to express and encode the proteins specified by them. It may be pointed out that the expression libraries of eukaryotes are, of necessity, cDNA libraries. On the other hand, the sequence-based methods can be used for screening of the libraries prepared in cloning vectors as well as expression vectors.

In the **sequence-based screening techniques**, nucleic acid hybridization or PCR are used to identify clones having the desired DNA inserts. The various techniques included in this group are colony hybridization, chromosome walking/ jumping, PCR, differential screening, etc. In simple terms, the technique of **colony hybridization** is as follows. The bacterial cells of the library are grown in a Petri plate in such a way that each cell forms a distinct colony. The colonies of this plate are replica-plated on a nitrocellulose filter. **Replica-plating** involves lowering a sterilized velvet cloth-covered cork block into the Petri plate till the velvet cloth touches the bacterial colonies. The cork block is of size that it covers the whole inner area of the plate. The bacterial cells from the colonies stick to the velvet cloth. Now the cork is lifted out and lowered onto a nitrocellulose filter kept onto the medium in another Petri plate. The bacterial cells are transferred onto the filter, and grow to form colonies. The filter is removed from the plate, the bacterial cells are lysed, and the DNA is denatured and fixed onto the filter. Now a labeled probe is used for hybridization with the DNAs fixed onto the filter; probes not hybridized to the DNA fixed onto the filter are washed, and the colonies, DNAs from which hybridize with the probe are identified by autoradiography. The **probe** is generally a single-stranded DNA fragment of 25-30 nucleotides and is radioactively labeled with ^{32}P. The probe is complementary to a part of the gene or DNA fragment that is being searched. Therefore, the base sequence of at least a part of the gene/DNA fragment must be known so that specific probes can be designed for colony hybridization.

The group **protein product-based library screening methods** includes complementation, colony/plaque screening with antibodies, fluorescence activated cell sorting, north-western screening, south-western screening, etc. The method for rapid screening of the colonies/plaques using antibodies is as follows. The bacterial cells carrying recombinant DNAs are grown in a Petri plate. These colonies are replica-plated, and the cells in the replica plate are lysed. The antibody specific to the protein encoded by the desired gene is spread uniformly over a plastic or paper disc. This disc is lowered onto the plate till it comes in contact with the lysed cells. If the cells of a colony had the desired gene, they would have produced the concerned protein. These protein molecules will bind to the antibody present in the plastic/paper disc. The disc is removed and treated with another antibody that interacts with a different region the same protein. This second antibody is labeled with radioactivity. Thus, the second antibody will bind only at such sites in the disc that contained the concerned protein. The positions of these sites are determined by autoradiography. These positions represent the sites of such bacterial colonies in the master plate (the plate with recombinant colonies that was used for replica-plating) that have the desired gene as insert. These colonies are isolated and the DNA insert present in them is the desired gene. It would be seen that this method is similar in approach to the colony hybridization, except for the use of antibodies in the place of DNA probes for identification of the desired bacterial colonies. It should be noted that for application of this technique, the protein product of the desired gene should have been purified. In addition, two different antibodies specific to distinct regions of this protein should be available.

9.4.4. Polymerase Chain Reaction

In case the base sequences of at least the two end regions of the desired gene are known, the gene can be isolated by PCR. The sequences of the end regions are used to design a primer specific to each of these regions. A **primer** is a short (around 18 nucleotides long) oligonucleotide that is used to initiate replication of a DNA strand. The pair of primers so designed is used to enable repeated replication of the DNA fragment located between these primer sites. This yields millions of copies of this fragment in a matter of few hours; this process is called **PCR amplification**. The amplification occurs through the following three distinct steps; (i) denaturation, (ii) annealing, and (iii) primer extension; these three steps together constitute the **polymerase chain reaction** (**PCR**). PCR is carried out in small PCR tubes using automated machines called thermal cyclers or PCR machines. Each PCR tube contains a reaction mixture that includes all the reagents necessary for amplification. These reagents include the DNA sample from which the desired fragment is to be amplified, an excess of the two primers, the four deoxyriboside triphosphates, and a heat stable DNA polymerase like Taq polymerase.

The PCR reaction mixture initially heated, generally, to 94°C. This causes separation of the two strands of the sample DNA duplex, and constitutes the **denaturation step** (Fig. 9.14). The mixture is now cooled to between 40 and 60°C. This allows the two primers to pair with the complementary sequences present in the DNA sample; this is the **annealing step**. The temperature of the mixture is then raised to around 72°C when Taq polymerase is used. The DNA polymerase sequentially adds nucleotides to the 3' ends of the two primers using the sample DNA strands as template. This constitutes **primer extension**. In this way, the DNA polymerase carries out replication of the DNA segment lying

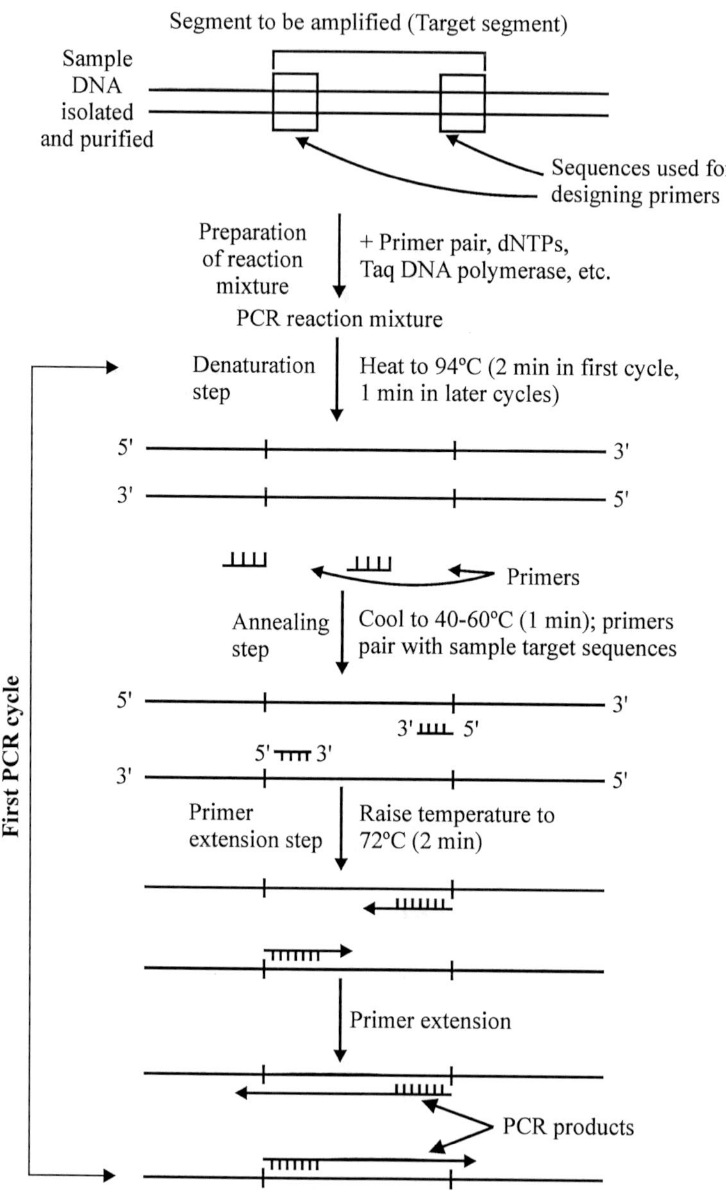

- Two copies of the target segment
- The number of copies doubles after every PCR cycle

Fig. 9.14. A simplified schematic procedure for PCR. The target segment is the segment of sample DNA that is to be amplified. The PCR reaction mixture contains, in addition, PCR buffer and reagents needed for successful PCR. In the second and subsequent cycles, the newly produced copies of the target sequence also are used as template for amplification.

between the binding sites of the two primers used for PCR. The completion of primer extension steps completes one PCR cycle. The DNA fragment produced by the PCR cycle is called **PCR product**. In the next PCR cycle, both the DNA sample and the PCR products will serve as templates for PCR amplification. Generally, one PCR cycle is completed in 4-5 minutes. The PCR is generally carried out for about 40 cycles. This would take less than four hours and would produce several million copies of a single DNA fragment.

PCR generally amplifies fragments of 2 kb (kilo base pairs) or less although longer fragments can be amplified by providing suitable conditions. Since **Taq polymerase** lacks proof reading activity, it cannot correct the errors it commits during DNA replication. As a result, the PCR products might contain some changes in their base sequences as compared to the sequence of the amplified segment of the sample DNA. Therefore, the PCR products are generally cloned and sequenced to enable the isolation of that PCR product, which has the same sequence as the sample DNA fragment.

9.4.5. Chemical Synthesis of Genes

The desired gene can now be chemically synthesized provided its base sequence or the amino acid sequence of its protein product is known. The first notable success in chemical synthesis of DNA was achieved by the phosphodiester method. In 1970, Har Govind Khorana synthesized a complete gene using this approach. This was replaced by the relatively more convenient method, the phosphotriester approach. This method could synthesize 10-20 nucleotide long oligonucleotides in a few days. It was used for oligonucleotide synthesis using automated machines, called **oligonucleotide synthesizers** or **gene machines**. But the most successful method of chemical synthesis of DNA is called the **phosphite triester method**. This method can synthesize oligonucleotides as long as 50 nucleotides. The present-day gene machines are based on this approach.

The phosphite triester method uses deoxynucleosides with protected $-NH_2$ groups of the bases A, G and C (T does not have a free $-NH_2$ group), and the 5' –OH group of deoxyribose molecules. The $-NH_2$ and –OH groups are protected by linking with them suitable nonreactive groups that can be specifically removed as per need. The protection prevents these groups from participating in any reaction. When the protecting group is removed by the appropriate treatment, the –OH group becomes capable of taking part in chemical reactions. The free 3' –OH group of the first nucleoside is fixed to a suitable solid support. The 5' –OH of this nucleoside is now freed by the appropriate treatment. The second and the subsequent nucleosides are used as 3' phosphoramidites. After the second nucleoside is added, it is suitably treated to convert the phosphoramidite to phosphate. During this process, the second nucleoside becomes linked to the 5' –OH of the first nucleoside. Now the 5' –OH of the second nucleoside is freed, the third nucleoside 3' phosphoramidite is added, and the rest of the process is carried out as for the second nucleoside.

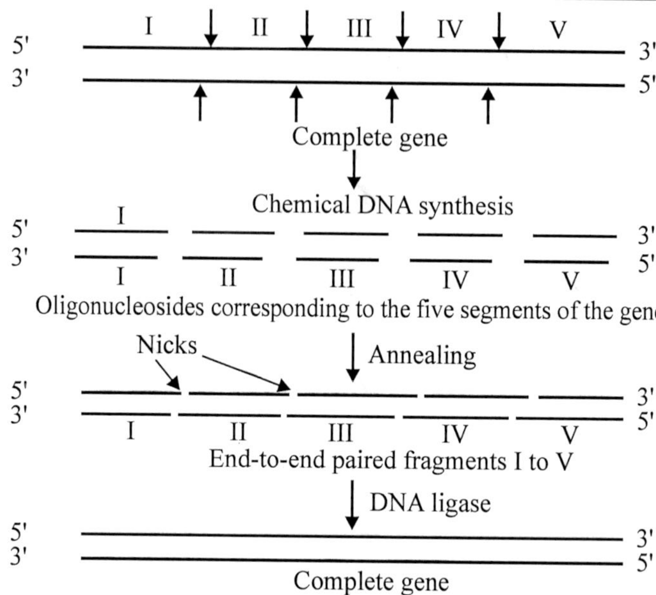

Fig. 9.15. Simplified schematic representation of the chemical synthesis of a complete gene. The gene sequence is divided into segments of suitable size; these segments have protruding ends (indicated by arrows). Oligonucleotides representing these segments are synthesized, mixed and annealed. The protruding ends facilitate correct end-to-end pairing of the fragments, which are joined by DNA ligase

In this manner, the nucleosides are sequentially added to the growing oligonucleotide chain. It may be pointed out that after linking of every nucleoside, the reagents have to be removed by washing before the new reagents are added. It takes about 15 minutes for the addition of one nucleoside to the growing oligonucleotide. The complete gene can be synthesized as follows. The whole gene is divided into several segments of size suitable for chemical synthesis. The two strands of each segment are synthesized in such a way that they pair to yield DNA duplexes with protruding ends (9.15). The protruding ends facilitate the sequential end-to-end pairing of the different fragments so that the complete gene becomes assembled. The nicks remaining after the annealing of the fragments are sealed by DNA ligase.

9.5. TRANSGENIC PLANTS

Transgenic plants are produced by **stable genetic transformation**, which signifies that the transgene is integrated into the genomes of the plant cells. Transgenic plants have been produced for a variety of different purposes, some of which are listed below.

1. In many cases, they produced to facilitate basic studies in molecular biology, including investigation of regulation of gene expression, identification of promoter sequences, isolation of the desired genes, etc.
2. They have been used to produce specific biochemicals, including pharmaceutically valuable proteins, research reagents, and pathogen proteins used for vaccines, etc.
3. Transgenes have been introduced into plant genomes to produce plants with novel and useful traits like herbicide resistance, insect resistance, virus resistance, etc.
4. Transgenic technology has been used to create crop varieties that yield produce with improved quality. Some successful examples include traits like keeping quality and oil quality.
5. Transgenic plants have been developed with enhanced tolerance to abiotic stresses like drought, and even with improved yield potential.
6. Transgenes have been introduced into plants in order to suppress the expression of some endogenous gene, e.g., genes involved in ethylene biosynthesis.

Thus, the transgenic plants with practical utility are produced by transferring transgenes that are expressed or transcribed in these plants. Therefore, the transgenes should have the regulatory sequences required for their transcription and translation of the mRNA encoded by them. The following sections outline the design of gene constructs, the marker genes, the vectors for plant transformation, etc.

9.6. GENE CONSTRUCTS FOR PLANTS

A typical plant gene comprises several functional regions (Fig. 9.16). The 5' region of a gene always has a promoter, and it also may have an enhancer and/or silencer. The 5' and the 3' regions represent the regions located towards the 5' and 3' ends, respectively, of the mRNA encoded by the concerned gene. An **enhancer** promotes the transcription of the gene, while a **silencer** suppresses its transcription. The enhancer and silencer sequences make the gene expression tissue-specific, development stage-specific, or responsive to specific stimuli. This region is followed by the **cap site** or the transcription startsite at which transcription of the gene begins. But the translation initiation codon usually occurs some distance downstream of the cap site; the region before the initiation codon (AUG in mRNA) encodes the 5' leader sequence of mRNA. The sequence encoding a signal peptide is included beyond the initiation codon; the amino acid sequence of peptide determines the specific cellular compartment to which the protein will be transported. The region beyond the signal peptide codes the protein specified by the transgene, which has at its 3' end the translation termination codon (generally, UAG in mRNA). Some distance beyond the termination codon lies the poly(A) site where transcription termination occurs. The expression vectors used for gene transfers generally have the regulatory sequences, including the translation initiation and termination codons, and only the gene to be transferred is added at the correct sites in the vectors. This assembly of the various sequences is referred to as the **gene construct**.

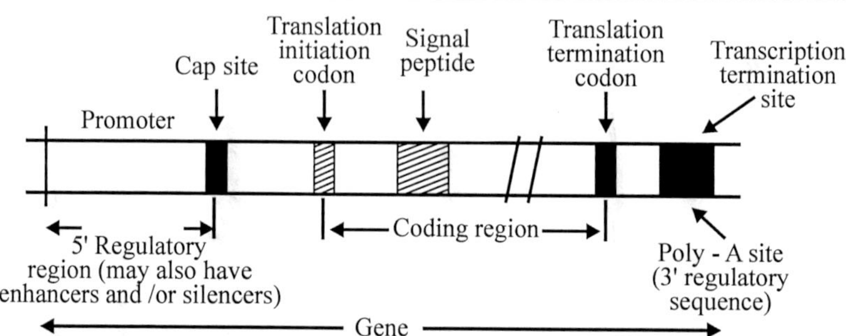

Fig. 9.16. Organization of a typical plant gene. The promoter and other 5' regulatory sequences are responsible for initiation and regulation of transcription. The 3' regulatory region is involved in transcription termination. The cap site is the transcription startsite; it represents the 5' terminus of mRNA, where cap is formed. The appropriate signal peptide is included when the protein is to be transported to a specific cellular compartment.

Promoter is the critical regulatory region of the gene construct. The promoter determines the level of gene expression, the tissue in which the gene will be expressed and the stimuli that will cause or suppress gene expression. It may be pointed out that the promoters from dicots are less efficient in monocots and vice-versa. The promoters are generally classified into the following four types: (i) constitutive, (ii) tissue-specific, (iii) inducible, and (iv) synthetic promoters. **Constitutive promoters** are always active; they support transcription in all the tissues in every developmental stage and under all the environmental conditions. The *35S* promoter from cauliflower mosaic virus (CaMV) is a good constitutive promoter for dicots and has been widely used. Similarly, the *Ubi* promoter from the ubiquitin encoding gene is a good constitutive promoter for monocots (Table 9.6). Both these are naturally occurring promoters, and they have been modified in various ways to enhance their activities. In addition, many promoters are active only in a specific tissue; they are known as **tissue-specific promoters** since they allow tissue-specific expression of the concerned gene. For example, the *vicilin* promoter from pea is active only in developing seeds, and is therefore seed-specific. Some other promoters, on the other hand, become active only when they receive a specific stimulus; these are called **inducible promoters**. For example, the *Adh-1* promoter of the maize gene encoding alcohol dehydrogenase is active under anaerobic conditions. The **synthetic promoters** are constructed by combining sequences from different sources to create a functional promoter.

Such promoters may involve some relatively simple modification of a natural promoter or their construction may use sequences from different organisms including bacteria.

Table 9.6: A list of some selected promoters used for production of transgenic plants

Promoter	Description
35S	Promoter of CaMV* *35S RNA* gene; high constitutive activity in dicots; widely used
35S + sh1-I1	*35S* promoter modified by inclusion of the first intron from the *sh1* (*shrunken 1*) gene of maize; enhanced promoter activity; constitutive gene expression
Ubi1 + Ubi1-I1	Promoter of the *ubiquitin* gene of maize modified by inclusion of the first intron of the same gene; high activity in cereals; constitutive gene expression
Adh1	Promoter of the *alcohol dehydrogenase 1* gene of maize; moderate activity in cereals; gene expressed under anaerobic conditions
Hsp	Promoter of the *heat shock* protein gene of maize/soybean; gene expression in response to heat shock
Hordein	Promoter of the *hordein* seeed storage protein gene of barley; seed specific gene expression in cereals
Vicilin	Promoter of the *vicilin* seed storage protein gene of pea; seed specific gene expression in dicots

* Cauliflower Mosaic Virus

The base sequence of the coding region of the transgene also affects the level of protein produced in the transgenic plants. In the cases of transgenes from prokaryotes, the mRNA might contain sequences that reduce mRNA stability and/or translation. In addition, the codons present in the transgenes may not be those that the host plant normally uses in its genes. It is now well known that different groups of organisms may prefer different codons in the case of few or more of those amino acids that are encoded by more than one codon; this is called **codon bias**. For example, *E. coli* prefers the codon CUG for leucine, while yeast cells prefer the UUG codon. Even the monocots and dicots differ in their codon bias for some of the amino acids. The codon bias would reduce the level of translation of the transgene mRNA in the host plants. The difficulties posed by the base sequence of the transgene have been minimized/eliminated by the following approach. The base sequence of the transgene is changed in such a way that the problems of codon bias and other interfering sequences in the mRNA are reduced/removed, while the amino acid sequence of the encoded protein remains unchanged.

9.7. MARKER OR REPORTER GENES

When a transgene is introduced into plant cells, only a small percentage of the cells are genetically transformed. The transformed plant cells have to be specifically selected, multiplied and plants have to be regenerated from them.

Some transgenes can be easily selected for on the basis of the character they produce in the transformed cells. For example, when a transgene specifying herbicide resistance in transferred, the transformed plant cells express this gene and, as a result, become resistant to this herbicide. But the vast majority of transgenes do not generate such a trait at the cell level that can be used to either identify or select for the cells expressing these transgenes. In all such cases, some other genes are also transferred along with the desired transgenes; these serve as reporter or marker genes. A **reporter** or **marker gene** is a gene that generates a trait at the cell level, which allows either the identification of or the selection for the cells that contain and express this gene. The marker genes are generally classified into two groups, viz., scorable markers and selectable markers. **Scorable markers** are such marker genes that generate a phenotype, which enables easy and rapid identification of the cells that express them. However, these markers do not permit the selection for such cells. Scorable markers are used many basic investigations like studies on the functions of promoters and other regulatory sequences, effectiveness of the transformation techniques, etc. In contrast, the expression of **selectable markers** allows easy and efficient selection for the cells expressing these markers. Several scorable and selectable markers are available for use in plants; some of the commonly used markers are listed in Table 9.7, and are briefly described below.

Table 9.7: A list of some important reporter/marker genes used in plants

Marker gene	Description
Scorable marker genes	
uidA (β-glucuronidase, GUS)	The gene is from *E. coli*; enzyme β-glucuronidase hydrolyzes X-gluc; X-gluc is colourless, but the hydrolysis product has blue color
GFP (green fluorescent protein)	The gene is from jelly fish; the GFP protein yields green fluorescence under UV or blue light; mutant proteins fluoresce red or yellow
Selectable marker genes:	
nptII (neomycin phosphotransferase)	The gene is from Tn5 transposon; the enzyme makes plant cells resistant to kanamycin; used in dicots
hptIV (hygromycin phosphotransferase)	The gene is from *E. coli* or *Streptomyces hygroscopicus*; the enzyme produces resistance to hygromycin B
bar (phosphinthricin acetyl trasferase)	The gene is from *S. hygroscopicus*; it confers on plant cells resistance to phosphinothricin or bialaphos
ALS (mutant acetolactate synthase)	The gene is from *Arabidopsis thaliana*, tobacco, *E. coli*, etc.; it generates resistance to sulfonyl ureas and imidazolinones
Marker genes other than antibiotic/herbicide resistance	
Xylose isomerase gene	Isomerises xylose to xylulose; xylose is used as carbon source in the medium; plant cells cannot utilize xylose but they can use xylulose

An ideal reporter gene should encode a protein that has relatively short half-life in plant cells, and that generates an easily and efficiently detectable phenotype. In addition, a similar protein should not be produced by plant cells. But none of the available markers is an ideal marker. The GUS (β-glucuronisase) marker has been the most widely used scorable marker. The GUS enzyme hydrolyzes the substrate X-gluc and the product of this reaction is a blue colored precipitate. Thus the presence of this enzyme can be visually detected and even it can be scored in individual cells. But the GUS substrate is costly, the enzyme is highly stable and several plant species show variable amounts of Gus activity. Green fluorescent protein (GFP) is rapidly replacing the GUS assay since its use does not require a substrate and can be done with the living unprocessed tissues. In contrast, the GUS assay requires processing of the test tissues that kills them. GFP produces bright green fluorescence when the tissues are illuminated with UV or blue light. Even the whole plants can be scanned for the expression GFP by exposing them to UV light produced by hand-held monitors during the evening hours.

The selectable marker gene *nptII* produces resistance to the antibiotic kanamycin. The gene *nptII* encodes the enzyme neomycin phosphotransferase, which detoxifies kanamycin. This gene has been used as a selectable marker with dicot plants, but it is of limited value in monocot plants. The transformed tissues are grown on medium containing 50-350 mg/l kanamycin sulphate. Another selectable marker gene, *bar*, encodes the enzyme phosphinothricin acetyltrans-ferase. This enzyme inactivates the herbicide phosphinothricin and, thereby, makes the cells resistant to this herbicide. The transformed tissues are grown on medium containing 1-10 mg/l phosphinothricin. A couple of other selectable markers are listed in Table 9.7.

The use of antibiotic resistance markers for the production of transgenic plants has aroused vocal opposition on the following grounds. The protein products of these marker genes could be harmful to humans. The resistance gene could be transferred to the microflora of the human gut when the crop produce is used as food. This will limit the usefulness of the concerned antibiotics for the management of human diseases. Therefore, several strategies were developed to produce transgenic plants without antibiotic resistance genes. One of these approaches uses non-antibiotic resistance selectable markers for the development of transgenic plants. For example, the gene xylose isomerase converts xylose to xylulose. The plant cells cannot use xylose, but they can utilize xylulose. The transformed tissues are grown on a medium containing xylose as the sole carbon source. The cells expressing xylose isomerase will be able to grow on this medium, while the rest of them will fail to multiply.

9.8. VECTORS FOR PLANT TRANSFORMATION

The vectors used for genetic transformation of plants are derived from either the genomes of viruses that infect plants or from plasmids of the two species of a plant pathogenic bacterium, viz., *Agrobacterium tumefaciens* and *Agrobacterium rhizogenes*. The virus genome vectors are used for expression of transgenes in plant cells for production of recombinant proteins, i.e., the proteins encoded by the transgenes. These vectors are not used for stable and heritable transformation of plants. In contrast, the vectors derived from the plasmids of *A. tumefaciens* and *A. rhizogenes* are used for stable and heritable transformation of plants. The plants produced by stable and heritable transformation can be used as transgenic or GM varieties of crops. In this section, we shall briefly discuss the virus genome vectors and the plasmid derived vectors for plants.

9.8.1. Vectors Derived from Virus Genomes

Most plant viruses have RNA genomes, but some viruses infecting plants have DNA genomes. Vectors have been developed from the genomes of both DNA and RNA viruses. In general, the genes that are not essential for infection of the host plants for the concerned virus and for virus multiplication are deleted and replaced by the desired transgenes. For example, the cauliflower mosaic virus (CaMV) has a double-stranded DNA genome of 8kb. Only two small genes of this virus, viz., genes *II* and *VII*, are not essential. Therefore, the size of DNA insert that can be accommodated in CaMV vectors is limited. In case the size of CaMV recombinant DNA is greater than the natural genome size of the virus by even few hundred base pairs, the recombinant DNA would not be packaged as virus particles. This would seriously reduce the infectivity of the recombinant DNA and its spread in the host plant.

In contrast, the Gemini viruses like maize streak virus (MSV) have single stranded DNA genomes. MSV is transmitted by leaf hopper, and it multiplies in high copy numbers in the nuclei of dividing cells. The single stranded genome is first made double stranded, and then is used to produce recombinant DNA. The recombinant DNA is introduced into the host cells with the help of *Agrobacterium*; this procedure is called **agroinfection**. In this procedure, the MSV recombinant DNA is integrated into the T-DNA (transferred DNA) of the Ti plasmid (tumor inducing plasmid) of *Agrobacterium*. The *Agrobacterium* is then used for coculture with the host cells, which results in the transfer of the T-DNA into these cells. Ultimately, the recombinant viral DNA is liberated into the plant cells. It multiplies within the host cells and may spread through the host plant tissues.

Table 9.8: The genes of the *vir* regulon of *Agrobacterium tumefaciens* that play important role in the transfer of T-DNA of Ti plasmid into the genomes of host plant cells

Gene	Operon	Function of the encoded Vir protein
virA	*virA* (single gene)	VirA protein is a protein kinase, and is produced constitutively at low levels; it interacts with the phenolic compound acetosyringone; it becomes phosphorylated, as a result; this activates VirA, which then phosphorylates protein VirG
virB	*virB* (11 genes)	These proteins form a tube between the *Agrobacterium* and plant cells; T-DNA is transported through this tube
virD2	*virD* (four genes)	Protein VirD2 nicks the T-DNA at is right and left borders; it binds at the 5' end of the single-strand copy of T-DNA and guides it to the plant cell nucleus
virE	*virE* (two genes)	The VirE proteins bind to the single-stranded copy of T-DNA and protect it from degradation
virG	*virG* (single gene)	It is a regulatory protein and is produced constitutively at low levels; it is phosphorylated by activated VirA; then it forms dimers, which bind the promoters of all *vir* operons and activates their transcription

As noted earlier, most of the plant viruses have RNA genomes. The use of genomes of these viruses comprises the following two steps. First, the virus genome is reverse transcribed to produce its cDNA that is double stranded. This cDNA version of the virus genome is used for construction of the recombinant DNA and its cloning in the bacterium *E. coli*. In the second step, the recombinant DNA is obtained from *E. coli*, and is transcribed in vitro. This step yields RNA copies of the recombinant DNA, which is used for infecting the host plant cells. The tobacco mosaic virus (TMV) has RNA genome, and the coat protein gene (*CP* gene) of this virus seems to be nonessential. Similarly, the *CP* gene of the brome mosaic virus (BMV) also seems to be nonessential. Both these viruses are promising candidates for use as vectors. TMV genome has been used to derive vectors for production of recombinant proteins (proteins encoded by transgenes) in tobacco.

9.8.2. Vectors Derived from Ti Plasmid

The plasmid vectors used for genetic transformation of plants are derived from the plasmids present in the plant pathogenic *Agrobacterium* spp. *Agrobacterium tumefaciens* produces the crown gall disease in dicot plant species, while *Agrobacterium rhizogenes* causes the hairy root disease. The disease causing properties of these bacteria reside in their respective plasmids, viz., Ti in *A. tumefaciens* and Ri in *A. rhizogenes*. Both these plasmids have a T-DNA region that is transferred into the plant nuclei and integrated into the plant chromosomes. These plasmids have a *vir* region (virulence region) that triggers the transfer of T-DNA. Both these plasmids have been used to develop vectors, but most of the vectors are based on the Ti plasmid.

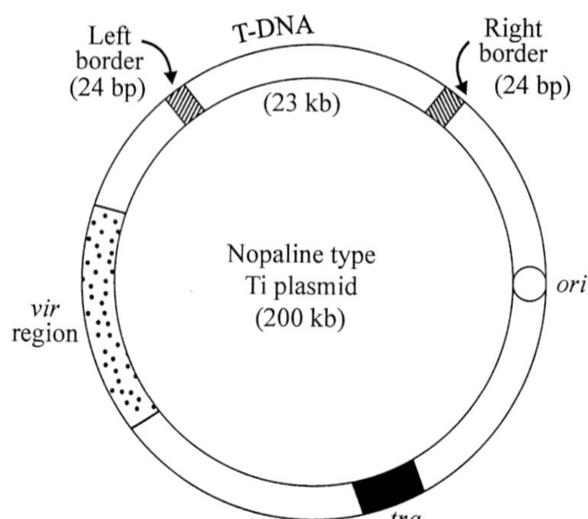

Fig. 9.17. A simplified representation of nopaline type Ti plasmid. T-DNA, transferred DNA; *ori,* origin of replication (functions in *Agrobacterium*); *tra*, conjugative transfer; *vir* region (virulence region).

The Ti plasmid is a large conjugative plasmid of about 200 kb. There are various types of Ti plasmids, each of which specifies the synthesis of a different opine. Opines are nitrogenous compounds produced by plant cells infected by *Agrobacterium* and utilized only by *Agrobacterium* cells. The various Ti plasmids are classified into two general groups: (i) nopaline type and (ii) octopine type plasmids. The general organization of the nopaline type Ti plasmid is shown if Figure 9.17. The detailed organizations of its T-DNA and *vir* region, the region most relevant for the genetic transformation of plant cells, are given in Fig. 9.18.

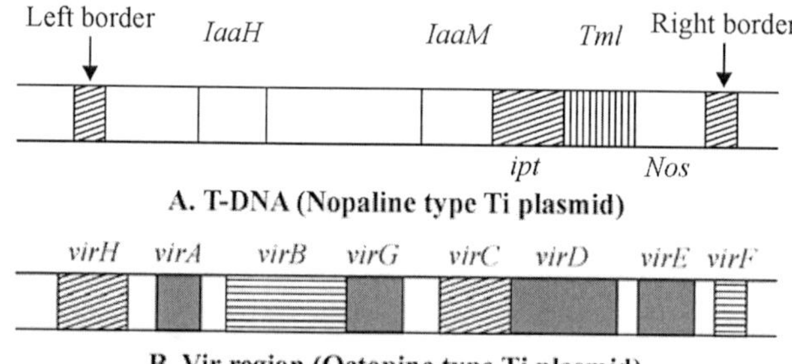

Fig. 9.18. A simplified representation of the T-DNA of a nopaline type Ti plasmid, and the *vir* region of an octopine type Ti plasmid. A. T-DNA of a nopaline type Ti plasmid. The loci *IaaM* and *IaaH* are involved in IAA synthesis, while the locus *ipt* is concerned with isopentenyl adenine synthesis. *Tml*, large tumour locus; *Nos,* encodes nopaline synthase. B. *vir* region of an octopine type Ti plasmid. *virA* to *virH* are the eight operons. Four of these operons (*virA, virG, virD* and *virB*) are essential for T-DNA transfer. The remaining four operons play an accessory role.

The T-DNA is about 23 kb in size. It has a 24 bp direct repeat as border sequence at its left and right ends; these border sequences are the only sequences necessary for the transfer of T-DNA into plant cells. The T-DNA has genes *IaaH* and *IaaM* for IAA synthesis and the gene *ipt* for synthesis of the cytokinin isopentenyl adenine. These three genes are involved in the production of crown galls. Therefore, they are deleted from the T-DNA of the vectors derived from the Ti plasmid; this is called **disarming of the Ti plasmid**. In fact, all DNA except the left and right border sequences can be removed from the T-DNAs of the Ti vectors. The genes to be transferred into the plant cells are integrated into the T-DNA between its left and the right borders. In case of the nopaline type Ti plasmids, the whole of the T-DNA is transferred as a single molecule.

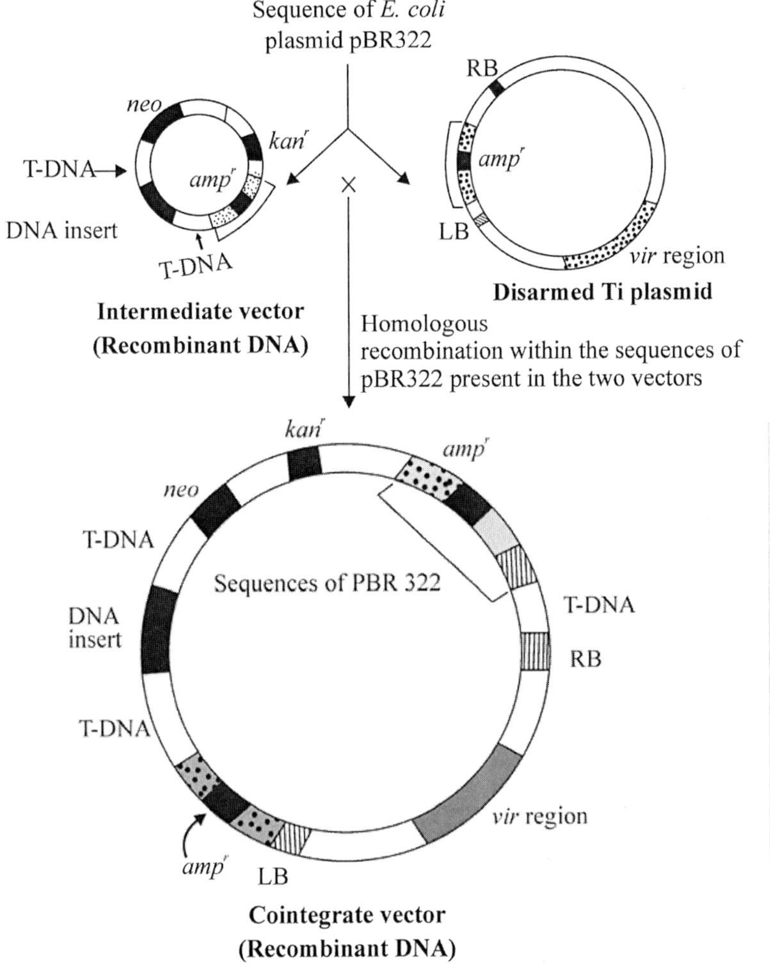

Fig. 9.19. A simplified schematic representation of the cointegrate vector system.

The vectors developed from the Ti plasmids are of two types: (i) cointegrate vectors and (ii) binary vectors. Both these vector systems comprise two separate vectors: one of these vectors is used for cloning of the target genes in *E. coli*, while the other is maintained in *Agrobacterium* and enables transfer of the T-DNA (containing the target genes) into the plant cells. In the case of **conintegrate vector system**, the vector used for cloning in *E. coli* is called **intermediate vector** (*IV*) (Fig. 9.19). The IV contains origin of replication for *E. coli*, T-DNA sequence without the borders, and a sequence of the *E. coli* plasmid pBR322. It also has selectable markers for selection of the transformed *E. coli* cells, *Agrobacterium* cells having the cointegrate vector, and the transformed plant cells. The other vector is known as **disarmed Ti**; it contains disarmed T-DNA with intact borders, the pBR322 sequence present in the IV, and an intact and functional **virulence region**. After integrating the desired gene in IV, the recombinant IV is multiplied in *E. coli*. The recombinant IV is then introduced into the *Agrobacterium* cells. This is usually achieved by conjugation between the *E. coli* and *Agrobacterium* cells induced by a plasmid like pRK2013 that is efficient in conjugation induction. The transfer of recombinant IV can also be achieved by electroporation. In the *Agrobacterium* cell, homologous recombination would take place within the pBR322 sequences of the IV and disarmed Ti. This will integrate the IV into the disarmed Ti, and yield the **cointegrate vector**. The bacterial cells containing the cointegrate vector are selected, and used for coculture with the plant cells to achieve gene transfer into them. In this system, the entire IV DNA is transferred into the plant genomes.

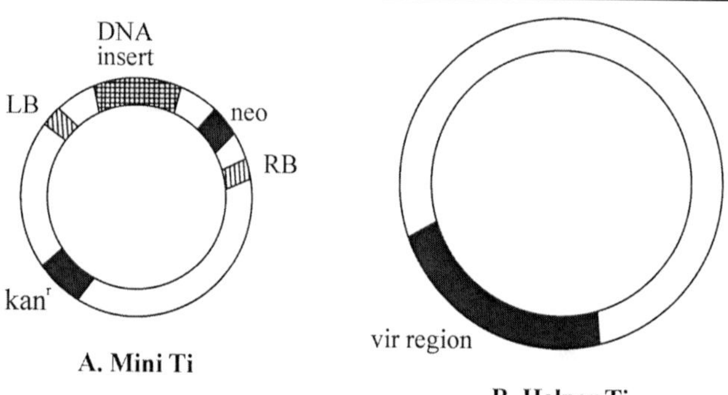

Fig. 9.20. A simplified schematic representation of a binary vector based on Ti-plasmid.

The **binary vector system**, comprises a mini-Ti or micro-Ti vector for gene cloning in *E. coli* and a helper Ti vector for assisting in gene transfer into the plant cells (Fig. 9.20). The **mini-Ti** has disarmed T-DNA along with intact borders and the genes to be transferred are placed within the T-DNA. It also contains replication origins for both *E. coli* and *Agrobacteruim*. The **helper Ti** plasmid lacks T-DNA (including the borders) but has a functional *vir* region. After cloning of the recombinant mini-Ti, it is transferred into *Agrobacterium* cells having the helper Ti. The *Agrobacterium* cells having both helper Ti and recombinant mini-Ti are selected and used for coculture with the plant cells to achieve gene transfer into them. In the binary vector system, only the T-DNA region of the mini-Ti is transferred into the plant genomes; this region has the target gene as well as the selectable marker for selecting the transformed plant cells.

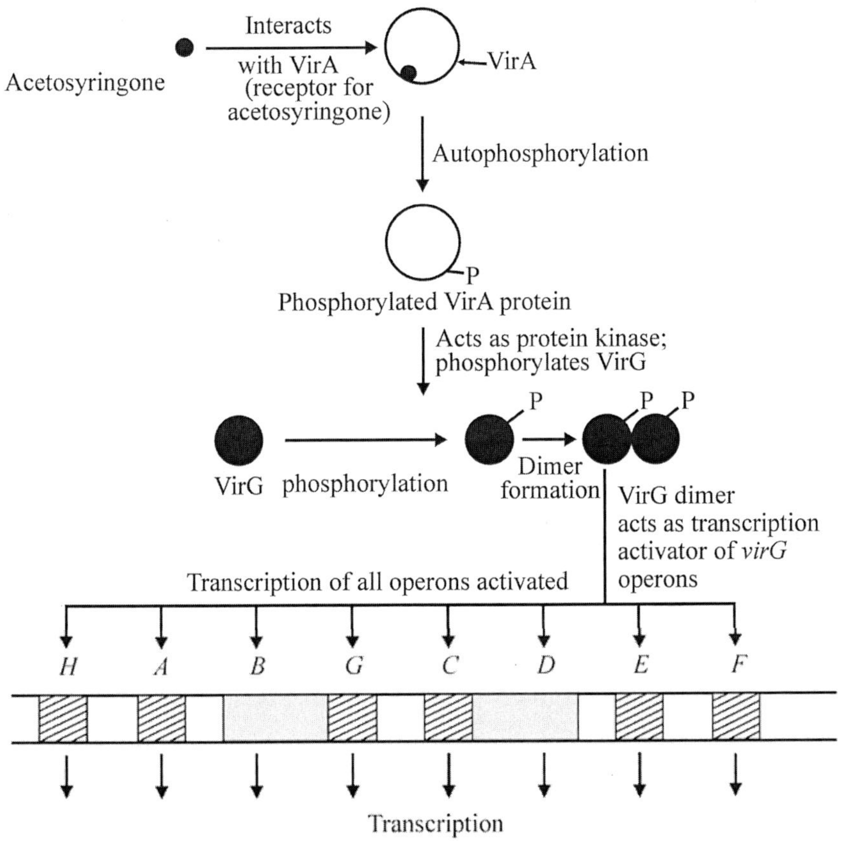

Fig. 9.21. Activation of transcription of *vir* operons in response to the phenolic compound acctosyringone liberated by plant cells at the wound sites. VirA and VirG are constitutively produced at low levels. VirA protein functions as receptor for acctosyringone. It phosphorylates itself following interaction with acctosy-ringone. It then phosphorylates VirG, which now foms a dimer. This VirG dimer activates the transcription of all the *vir* operons.

9.8.3. Transfer of T-DNA into Plant Cells

The *vir* region of the nopaline type Ti plasmids has eight operons named as *virA* to *virH* (Fig. 9.18B). Each of these operons has either one (e.g., *virA*, *virF*, *virG*) or more (e.g., 4 in *virD*, 11 in *virB*) transcription units. The proteins encoded by these operons perform specific functions in the T-DNA transfer. Two operons of the *vir* region, viz., *virA* and *virG*, are constitutively expressed at relatively low levels. Protein *VirA* acts as receptor for the phenolic compound acetosyringone, which is produced by the cells at wound sites in dicot plants. This interaction induces the VirA to phosphorylate itself, which activates this protein. The activated VirA protein phosphorylates protein *VirG,* which then undergoes dimerization. The dimerized *VirG* functions as the activator of all the *vir* operons. As a result, all the Vir proteins are produced at relatively high levels after the *Agrobacterium* cells are exposed to acetosyringone (Fig. 9.21).

Proteins *VirC* (a helicase) and *VirD1* (a topoisomerase) bind around the right border of T-DNA and make the DNA suitable for the action of *VirD2*. Protein *VirD2* is an endonuclease. It nicks the T-DNA at the right border and attaches to the 5' end of the T-DNA that is produced as a result of the nick (Fig. 9.22). The 3' end produced at the site of nick is used for DNA synthesis. This generates a single stranded copy of the T-DNA. The protein *VirD2* attached to the 5' end of this T-DNA single strand induces a nick at the left border as well. This liberates the single strand copy of the T-DNA. Protein *VirE2* binds the T-DNA single strand and keeps it suitable for transfer into the plant cells. A conjugal tube is formed between the bacterial cell and the plant cell. *VirB* proteins are involved in the formation of this tube and in the movement of the T-DNA into the plant cells. The protein *VirD2* guides the T-DNA into the plant cells and, ultimately, the nuclei of plant cells. Once inside the plant cells, the T-DNA becomes double stranded and becomes integrated into the plant chromosomes at random sites.

9.9. GENETIC TRANSFORMATION OF PLANTS

Introduction of DNA into cells is called **genetic transformation**. The introduced DNA may be lost from the cells after some time (**transient transformation**) or it may be stably maintained in them (**stable transformation**). In case of stable transformation, the introduced DNA may be integrated into the cell genome, e.g., in case of T-DNA; this is termed as **integrative stable transformation**. Alternatively, the introduced DNA may remain outside the genome as independent entity, e.g., in the case of viral vectors; this is known as **non-integrative stable transformation**. In case of plants, several different methods of transformation have been used. These methods are classified into two broad groups as (i) *Agrobacterium*-mediated and (ii) direct gene transfers. The most commonly used methods of gene transfers in plants are briefly described in the following sections.

9.9.1. *Agrobacterium*-Mediated Gene Transfer

In this approach, the gene is introduced into the plant cells via *Agrobacterium* cells. The gene is placed in the T-DNA region of a Ti vector system, and the

Agrobacterium cells containing the recombinant DNA are used for coculture with the plant cells. When the T-DNA from such *Agrobacterium* cells is transferred into the plant cells, the genes present in this T-DNA are also transferred. There are two methods of *Agrobacterium*-mediated gene transfers, viz., (i) *Agrobacterium* co-culture, and (ii) *in planta* methods.

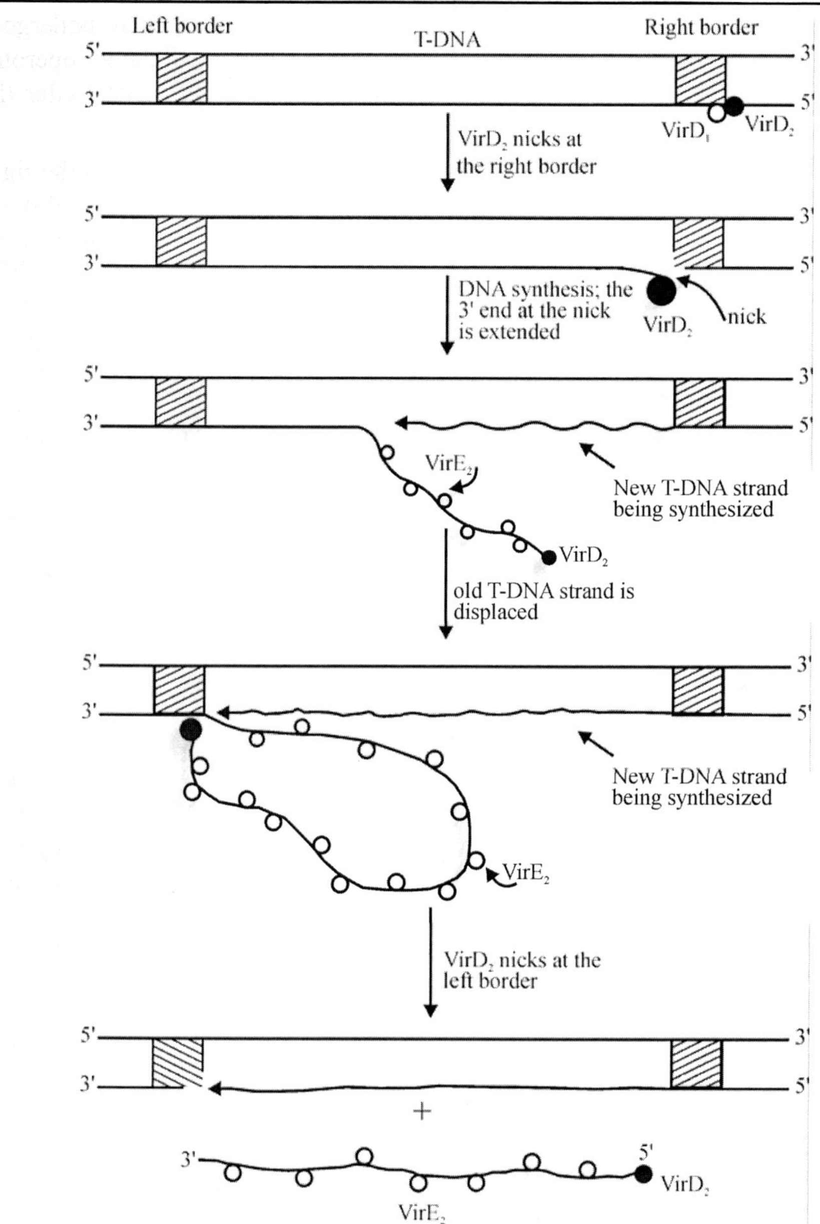

Fig. 9.22. A simplified representation of T-DNA transfer

The procedure of *Agrobacterium* co-culture is briefly described for transformation of leaf tissue of dicot plants (Fig. 9.23). The leaf is surface sterilized, and then small discs are cut out of it. The leaf discs are immersed into

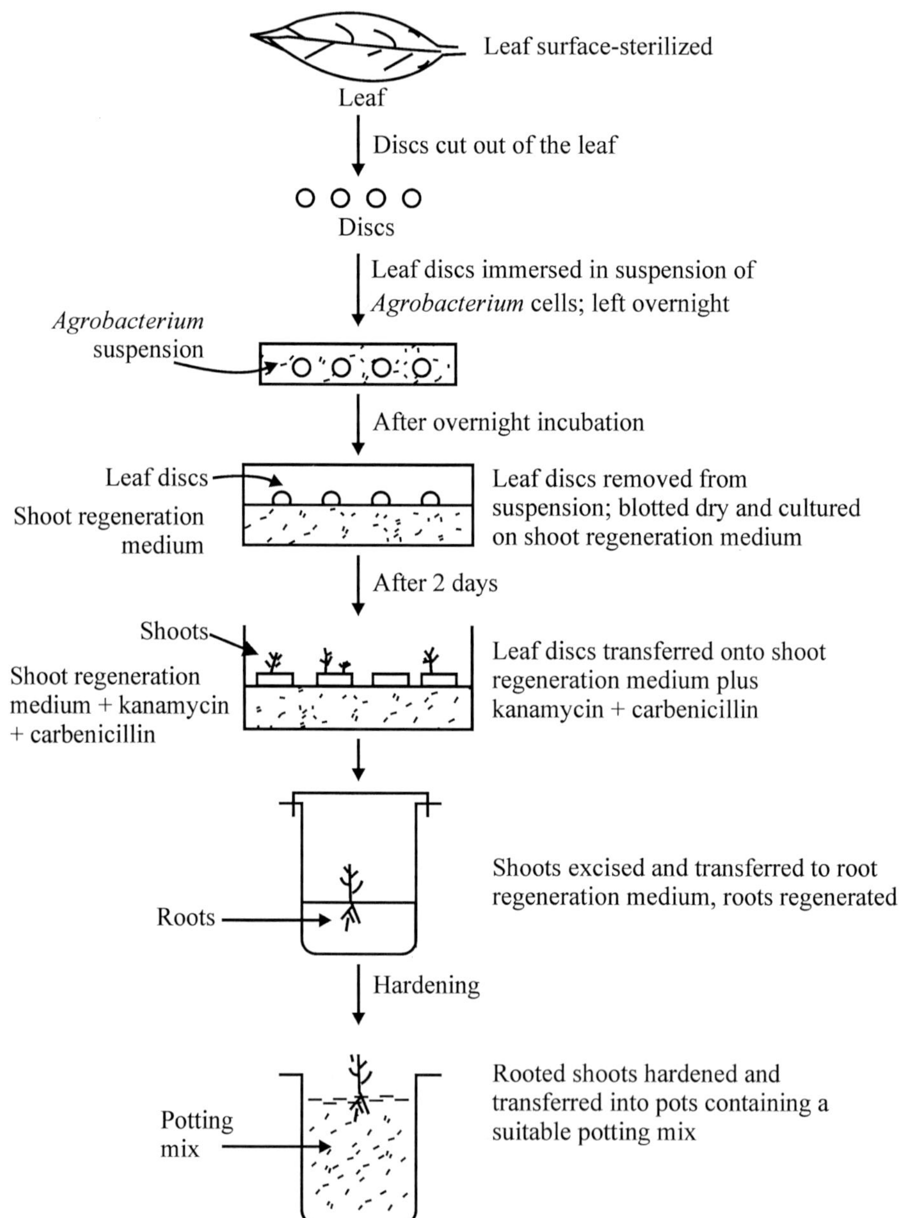

Fig. 9.23. A simplified schematic representation of genetic transformation of plants by *Agrobacterium* co-culture procedure

a culture of *A. tumefaciens* cells containing the recombinant Ti DNA, i.e., the Ti vector having the desired transgenes within its T-DNA. After a period of time (usually, the discs are immersed overnight), the discs are taken out, excess culture is removed from them by blotting with filter paper, and the discs are cultured on agar shoot regeneration medium for two days. During this period, the wounded leaf tissues will liberate acetosyringone, and the T-DNA containing the transgenes will be transferred into many plant cells and integrated into their chromosomes.

The transformed cells will express the transgene of interest and the selectable marker gene. Therefore, they will become resistant to the concerned selection agent, e.g., to kanamycin when the bacterial gene *neo* or *nptII* is used. The leaf discs are now transferred onto the shoot regeneration medium supplemented with the appropriate concentrations of kanamycin and carbenicillin. Kanamycin will select for the transformed cells, which will proliferate and regenerate shoots. Carbenicillin, on the other hand, will eliminate the *Agrobacterium* cells. If this is not done, the regenerated plantlets will be contaminated by the *Agrobacterium* cells. The shoots are now transferred to rooting medium, which also contains the two antibiotics. Rooting is more sensitive to kanamycin than shoot regeneration. Therefore, the use of kanamycin in the rooting medium prevents rooting of such shoots that are not transformed, but survived the exposure to kanamycin. The rooted plantlets are transferred into soil following a suitable hardening procedure. *Agrobacterium*-mediated transformation is almost routine for dicot plants; it can be used to transform leaf and other plant tissues, and cell cultures. This method has been successfully used to transform monocot crop species like rice, maize, etc. as well by supplying suitable concentrations of acetosyringone.

In the ***in-planta* method**, seeds or young inflorescence are immersed in fresh suspension cultures of appropriate strains of *Agrobacterium* (Fig. 9.24). For example, the seeds of *Arabidopsis* were imbibed in *Agrobacterium* culture, germinated, and the plants were raised to maturity. A small proportion of the produced by these plants gave rise to transformed plants that had T-DNA insertions at unique sites of their genomes. It has been suggested that the *Agrobacterium* cells enter the seeds and seedling during the imbibition and germination. These cells remain in the plants and transform either the gametes or the zygotes that give rise to the transformed progeny. In the floral dip method young inflorescence is dipped into the *Agrobacterium* culture either under partial vacuum or in the presence of a compound like Silvet L-77. After a period of time, the inflorescence is removed from the culture, and the plants are grown to maturity. A proportion of the selfed seeds from the treated inflorescences yield transformed progeny. The floral dip method has been successful in *Arabidopsis, Medicago*, rice, peanut, *Raphanus* and some *Brassica* spp. This approach is simpler, much easier, and independent of regeneration ability of the plant species. It also seems to be of general application.

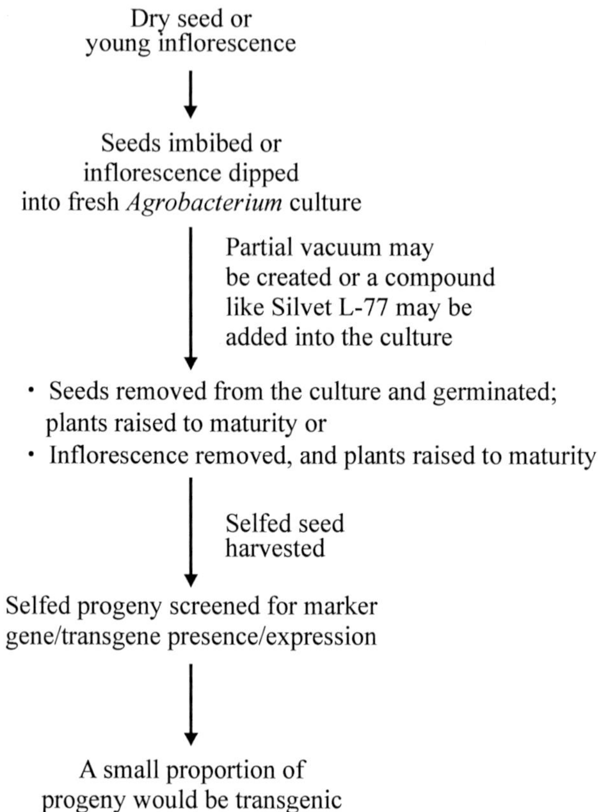

Fig. 9.24. A simplified representation of *in planta* transformation by seed inbition or floral dip method of *Agrobacterium*-mediated gene transfer in plants.

9.9.2. Direct Gene Transfer

The gene transfer in case of the direct methods is not mediated by a biological agent, and the DNA is directly introduced into plant cells. These methods use either a physical factor e.g., electric current (electroporation), particle gun, etc., or a chemical agent e.g., polyethylene glycol (PEG), for introducing the DNA into plant cells. Some of the various direct methods of genetic transformation are electroporation, particle gun, microinjection, lipofection, macroinjection, etc. Of these methods, the particle gun method is the most successful, and is widely used; this method is briefly described here. This method is also called the **biolistic** or the **ballistic method of DNA delivery**. In the **particle gun method**, a specifically designed device is used to accelerate gold or tungsten microparticles that become embedded into the tissue to be transformed. The microparticles are ~1μm (range 0.5 – 2.0μm) in diameter. The gold particles have more regular shape, are much less toxic but much more costlier than tungsten particles. The DNA is coated onto the surface of microparticles, and these particles are spread and adhered to a plastic membrane disc and dried. This plastic disc is called **macroprojectile** and is used only once. The coating of microparticles with DNA is a critical step. Generally, appropriate concentrations of calcium chloride and spermidine are added into the solution of plasmid DNA to achieve satisfactory coating of the DNA onto the microparticles. The microparticles are added to this solution and the suspension is vortexed (agitated vigourously) to allow uniform coating.

The particle gun is a compact device that has appropriate control mechanisms to regulate the process of microparticle delivery. The components directly involved in the delivery of microparticles are schematically represented in (Fig. 9.25). It has a gas acceleration tube that is blocked by a thin plastic membrane that would rupture when the pressure inside the acceleration tube reaches 1000 p.s.i. (pounds per square inch). Further down in the acceleration tube, the macro-carrier disc carrying the microparticles is located. At the end of the tube a stopping screen is situated. The tissue to be transformed is placed about 10 mm below the stopping screen. These components are enclosed in a robust chamber that has all the mechanisms necessary for regulated delivery of the microparticles into the target tissue. In simple words, the tissue to be transformed is placed below the stopping screen. The macroprojectile coated with the DNA-coated microparticles is loaded into the acceleration tube. The chamber of the particle gun is now closed and partial vacuum is created. The partial vacuum reduces the damage to plant tissues, and it also facilitates acceleration of the macroprojectiles.

After the partial vacuum is created, helium gas is released into the acceleration tube; the pressure of the gas is generally 1000 p.s.i. At this pressure, the rupture disc is disrupted; this generated shock waves that drive the macrocarrier down the acceleration tube. The macrocarrier ultimately strikes against the stopping screen, the microparticles pass through the screen, and become embedded into the plant tissue kept below. Some of these microparticles will enter into the nuclei of cells and the DNA present on their surfaces may lead to genetic transformation of these cells. This method can be used for transformation of shoot tips, somatic embryos or zygotic embryos. This makes this transformation method independent of regeneration ability of the plant species. It can also be used for transformation of any tissue like mature pollen grains, callus and suspension cultures, leaf discs, root and shoot section, etc. The frequency of transformation seems to be remarkably higher in meristematic than in non-dividing cells. Particle gun method has been used for production of transgenic plants of several plant species, including cotton, maize, rice, sorghum, sugarcane, tobacco, wheat, etc.

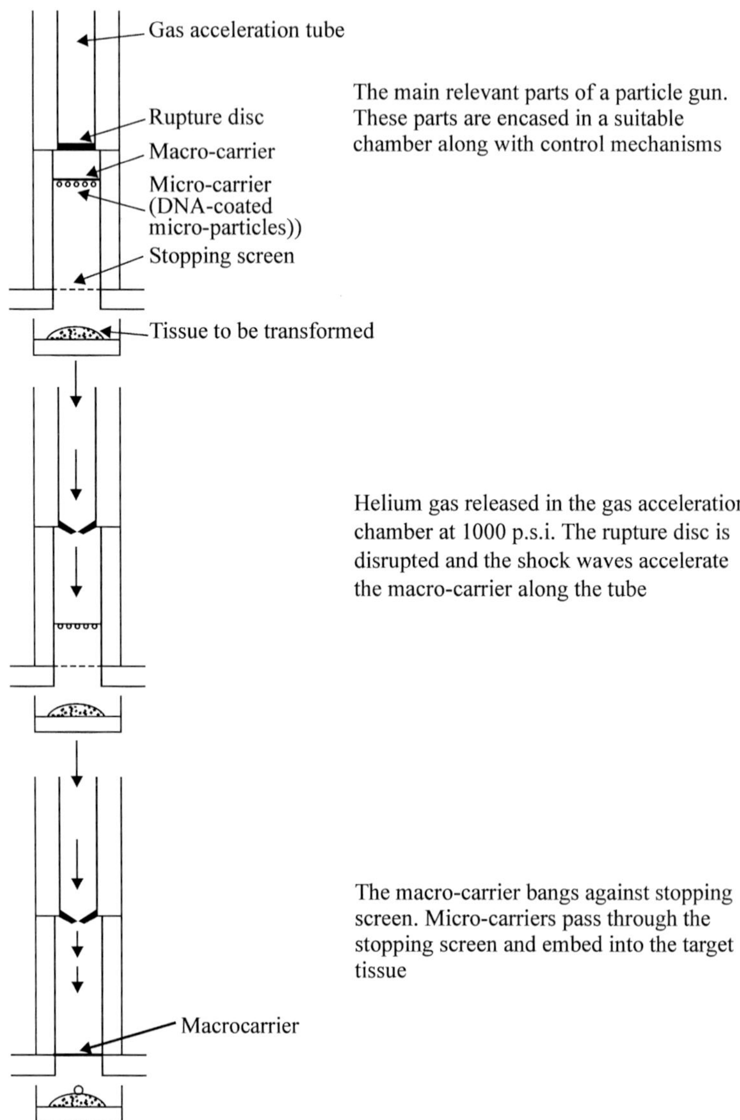

Fig. 9.25. The major parts of a particle gun that enable acceleration of micro-carriers coated with DNA. These particles embed into the target tissue and bring about genetic transformation.

9.10. GENETIC TRANSFORMATION OF CHLOROPLASTS

The methods of gene transfer into the nuclear chromosomes are not effective for transformation of chloroplasts and mitochondria. Efforts are being made to develop efficient methods for genetic transformation of mitochondria and, especially, chloroplasts. These efforts relate to vectors, selectable marker genes as well as methods of DNA delivery. Several vectors for DNA transfer into chloroplasts have been developed. For example, the pZS148 is a shuttle vector designed to replicate in *E. coli* as well as chloroplasts. This vector has been used to clone the target genes in *E. coli* and then to transfer them into chloroplasts by the particle gun methods. Similarly, some selectable antibiotic markers have been identified for the selection of transferred chloroplasts. Two of these markers are kanamycin resistance and streptomycin resistance. The methods of DNA delivery into chloroplasts include particle gun, PEG-mediated DNA delivery and galistan expansion femtosyringe methods. Out of these, the particle method is the most successful and the most widely used. However, the chloroplast transformation methods have been optimized for tobacco; in addition, some success has been obtained with tomato as well.

Transgene integration in chloroplast genomes offers several unique advantages. For example, the transgene would show maternal transmission. Therefore, there would be little or no risk of transgene dispersal through pollen grains. Since each chloroplast has several copies of the chloroplast genome, the transgene expression would occur at relatively high levels. Therefore, the intensity of the traits like stress resistance (herbicide, insect, disease, etc. resistance) would be much higher than that obtained by integrating the transgenes in nuclei. In addition, the recombinant protein would be produced at high levels, and this would be very useful for biopharming. The transgene expression is not affected by gene silencing. Finally, the expression of bacterial genes would be much better in chloroplasts than in nuclei. The chief limitations of chloroplast transformation are: (i) low frequency of transformation (~1% of the nuclear transformation rate), (ii) need for prolonged high intensity selection to recover transformants, (iii) relatively high cost of transgenic production, and (iv) limited success of the transformation methods with plant species other than tobacco.

9.11. TRANSGENE INTEGRATION AND INHERITANCE

The plants obtained from transformation experiments may or may not be truly transgenic plants. This is because the selection procedures used for plantlet regeneration are unable to eliminate the non-transgenic plantlets with 100% efficiency. As a result, a proportion of the putative or presumed transgenic plants are **escapes**, i.e., plants surviving the selection agent even when they do not have the concerned marker gene. In view of this, regenerated plants are subjected to a series of assays designed to provide the following information: (i) the presence of transgene in the plants, (ii) the integration of the transgene in the plant genome, (iii) the number sites of integration in the genome, (iv) the number of copies of the transgene integrated in the genome, (v) transcription of the transgene, (vi) accumulation of the protein encoded by the transgene, and (vii) production of the desired phenotype. There is usually more than one test that can be used to answer each one of most of these questions. Some of the assays that are commonly used for analysis and characterization of transgenic plants are listed below.

1. A small amount of leaf tissue is taken from each regenerated plant or putative transgenic plant, and DNA is isolated. PCR is carried out using this DNA as template and a pair of primers specific for the transgene transferred into the plants. All the plants showing amplification contain the transgene in question. This transgene may be integrated into the plant genome or it may be present in the *Agrobacterium* cells that may be present as contaminants within the plants in the intercellular spaces.
2. However, integration of this transgene has to be confirmed by Southern hybridization. This assay also allows the determination of the number of different sites where the transgene is integrated into the genome of each individual. Specific Southern hybridization procedure can be used to determine whether a given site contains a single copy or more than one of the transgene. Since Southern hybridization is a demanding procedure, this analysis is deferred to a later stage in the evaluation procedure, when the number of plants to be assayed is suitably reduced.
3. The transcription of the transgene can be assayed by the northern blot technique.
4. The production of the transgene encoded protein can be assayed by western blot assay, enzyme-linked immunosorbant assay (ELISA), or some other simpler approach.
5. The transgene specified phenotype can be scored in the suspected transgenic plants provide the evaluation for the concerned phenotype does not damage the plant.

If the assays listed at items 3, 4 and 5 are positive for a given plant, this plant is most likely to have the transgene integrated into its genome. This is because the transgene is linked to a plant promoter as a result of which it can be transcribed only when it is present in the nuclei of plant cells. The transgene integration will be confirmed by studying the inheritance of the transgenes in the progeny of the identified transgenic plants.

1. The selfed progeny of the identified transgenic plants are scored for the transgene determined trait to determine the inheritance of the transgene.

   In general stably integrated transgenes show Mendelian inheritance. The transgenes behave as dominant genes since the chromosome contributed by a non-transgenic individual will not have any allele for the transgene. Therefore, the F2 generation from a cross between a homozygous transgenic and a non-transgenic line will show a 3 transgenic : 1 non-transgenic ratio. But in some cases, the ratio might deviate from 3 : 1; generally, the frequency of transgenic individuals in such cases is lower than expected. This could be due to a variety of reasons, including reduced transmission of the transgene through the gametes, lower survival rate of the homozygous transgenic plants, transgene silencing, etc.

2. The transgenic progeny from each plant are then evaluated for performance in terms of both the transgenic trait and the agronomic characteristics. This allows the identification of the transgenic lines that hold promise for use as new transgenic varieties.
3. The promising transgenic lines are subjected to Southern analysis, and lines having a single copy of the transgene are further characterized. The genomic DNA surrounding the transgene is isolated and sequenced to determine the specific site of genome where the transgene is integrated.

CONCLUSION

The transgenic technology was developed to enable the use of genes that could not be accessed by plant breeders for development of improved crop varieties. This technology uses recombinant DNA technology to identify, isolate and clone the desired genes and to transfer and express them in plant cells. The technique of plant tissue culture facilitates the recovery of complete plantlets from the cells transformed with the desired transgenes. The plantlets are obtained from explants grown on tissue culture media that are induced to regenerate either shoots or somatic embryos. The shoots regenerated in vitro are rooted by a suitable procedure, while somatic embryos are germinated to yield complete plantlets. Shoot regeneration is promoted by cytokinins, while auxins are useful for somatic embryogenesis (high auxin concentrations) and rooting of shoots (low auxin concentrations). The in vitro raised plantlets are taken through a hardening procedure designed to enable their survival and establishment in the field.

Recombinant DNA technology uses restriction enzymes, several other specific enzymes, DNA molecules called vectors and host organisms (generally, *E. coli*), and a variety of specialized techniques for isolation and cloning of the desired genes so as to obtain multiple copies of the genes. Multiple copies of a gene can also be obtained by polymerase chain reaction or chemical DNA synthesis provided the base sequence information of the target gene is available. The coding sequence of the target gene is linked to appropriate 5' and 3' regulatory sequences and the resulting gene construct is placed in the T-DNA of a suitable Ti plasmid derived vector (cointegrate or binary vector). This recombinant DNA is introduced into *Agrobacterium tumefaciens*, which is co-cultured with the plant tissue to be transformed. The T-DNA along with the target gene is transferred into the plant cells and the same becomes integrated at random sites of the plant genome. The target gene construct can also be introduced into plant cells directly by using, say, a particle gun. The transferred cells are cultured onto a selection medium to allow these cells to regenerate complete plantlets and to select for such plantlets that are likely to be transgenic. Techniques are available for integration of target genes into the chloroplast genome.

The putative transgenic plants are subjected to molecular analysis to confirm their transgenic status, determine the site of transgene integration, and to assay the expression of the transgene. In addition, the confirmed transgenic plants are evaluated in the greenhouse and then in the field for their performance in terms of the modified trait as well as yield and other agronomic traits. The transgenic plants have to be subjected to biosafety evaluation. based on these data, The competent authority of the country would assess the biosafety data, and, if satisfied, would allow their commercial cultivation.

PRACTICE QUESTIONS

1. Define genetic transformation. List the various methods for gene transfers in plants and describe any one of these methods in detail.
2. Briefly describe the technique of tissue culture and discuss the regeneration of complete plantlets from callus/suspension cultures.
3. Briefly describe the recombinant DNA technology and its relevance to the production of transgenic plants.
4. Briefly describe the construction of genomic and cDNA libraries, and their relevance for identification and isolation of the desired gene from them.
5. Briefly describe the general organization of the gene construct for plant transformation. Discuss the various types of promoters used for production of transgenic plants.
6. Define reporter or marker genes. Briefly describe the different types of reporters used in plants and discuss the risks from antibiotic selectable marker genes and their remedies.
7. Briefly describe the organization of T-DNA and the *vir* region. Explain the mechanism of T-DNA transfer into the plant cells.
8. Write short notes on the following: (i) characterization of transgenic plants, (ii) chloroplast transformation, (iii) *Agrobacterium* coculture, (iv) *In planta* transformation, (v) particle gun, (vi) binary vector, (vii) cointegrate vector, (viii) T-DNA, (ix) *vir* region, (x) *Agrobacterium tumefaciens*, (xi) Ti plasmid, (xii) plant tissue culture, (xiii) callus and suspension cultures, (xiv) surface sterilization, (xv) shoot regeneration (xvi) somatic embryogenesis, (xvii) hardening of plantlets, (xviii) restriction enzymes, (xix) PCR, (xx) chemical synthesis of genes, (xxi) transgenic plants, (xxii) plant promoters, (xxiii) reporter genes.

RELEVANT LITERATURE

1. Gatehouse AMR et al. (eds) 1992. Plant Genetic Manipulation for Crop Protection. CAB International, Wallingford, UK.
2. Grierson D 1991. Plant Genetic Engineering. Plant Biotechnology, Vol. I. Blackie, Glasgow.
3. Lycett GW and Girerson D (eds) 1990. Genetic Engineering of Crop Plants. Butterworths, London.
4. Singh BD and Singh AK 2015. Marker-Assisted Plant Breeding. Springer, New Delhi, Heidelberg.
5. Singh BD 2015. Plant Biotechnology, 3rd edition. Kalyani Publishers, New Delhi.
6. Singh BD 2012. Biotechnology, Expanding Horizons, 4th edition. Kalyani Publishers, New Delhi.

■■■■

10 Transgenic Plants for Resistance to Biotic and Abiotic Stresses

10.1. INTRODUCTION

The various plant breeding schemes are based on genetic variation created mainly by segregation and recombination following hybridization between two or more individuals/lines differing for one or more traits. Therefore, useful genes present in only in the sexually compatible species can be utilized. For any crop species, the number of such species is limited. The techniques of embryo culture, *in vitro* pollination, *in vitro* fertilization, and somatic hybridization were developed in an effort to overcome the limitations imposed by the problems associated with sexual incompatibilities. These techniques have been helpful in enabling the use of genes present in only some of such related species that are otherwise sexually incompatible with the concerned crop species. The development of recombinant DNA technology and genetic transformation techniques, on the other hand, has allowed the transfer of desirable genes from any species to the crop species. The term **genetic engineering** is in common use and it includes both recombinant DNA technology and the techniques of genetic transformation. The plants produced by genetic engineering are called **transgenic plants**. In addition, even completely synthetic genes designed by humans can be used to produce transgenic plants. Another term, **transgenic technology** is now used to describe the techniques of genetic engineering used for the production transgenic plants as well as the procedures followed for enabling their commercial cultivation utilization.

In general, genetic engineering is considered to have progressed through the following three distinct phases. In the **first phase**, human genes were transferred and expressed in prokaryotes. The objectives of these gene transfers were production of pharmaceutically valuable proteins like insulin. This was followed by the **second phase**, in which prokaryotic genes were transferred and expressed in plants in order to make them resistant to various abiotic and biotic stresses like herbicides, insects, viruses, etc. In the **third phase**, genes from various sources are being introduced and expressed in plants with a view to enhance their produce quality, generate novel biochemicals, etc. We are currently witnessing the activities of second and third phases in the case of transgenic plants. In this chapter, we shall consider the production of transgenic plants with resistance to abiotic and biotic stresses. The production of transgenics with improved quality, for production of novel and valuable biochemicals etc. are discussed in the next chapter.

10.2. OBJECTIVES OF TRANSGENE EXPRESSION

A transgene may be expressed in a plant either to generate a useful phenotype or to obtain a valuable product. The useful phenotype can be generated in two diagonally opposite ways, viz., by expression of a transgene that provides new function or by suppression of an endogenous gene using the transgenic technology. The various ways in which transgenic technology is utilized in summarized as follows. (i) The transgene encodes a protein that has valuable applications as a therapeutic or diagnostic agent, research reagent, etc. (ii) The transgene encoded protein may itself produce the desired phenotype without any effect on the plant metabolism, e.g., insect resistance by the *cry* genes. (iii) The transgene encoded protein may participate in the plant cellular activities in one of the following ways: (a) activation of an endogenous pathway, (b) extension of an endogenous pathway, (c) improvement in the efficiency of an endogenous pathway, and (d) introduction of new branch in an endogenous pathway. (e) The transgene(s) may create a new pathway to generate a novel product. Finally, (f) the transgene may lead to suppression of an endogenous gene and, thereby, generate a useful phenotype.

10.3. ABIOTIC AND BIOTIC STRESSES

A **stress** may be defined as an environmental factor that imposes a restriction on the expression of the full genotypic potential of a crop plant. The various stresses are classified into two main groups, viz., abiotic and biotic stresses. An **abiotic stress** is generated by a non-living factor of the environment like moisture, salinity, alkalinity, acidity, heat, cold, etc. A biotic stress, on the other hand, is caused by living component of the environment, such as fungi, bacteria, insects, viruses, viroids nematodes, etc. Among the abiotic stresses, drought and salinity are two of the most important stresses limiting crop production. Herbicides are used to control weeds, and are not a component of the normal environment. However, they are an important component of the cropping environment since their application is necessary for the control of weeds. In his chapter we will discuss the strategies for the production of transgenic plants tolerant to herbicides; resistant to insect pests, viruses, and fungal and bacterial pathogens; and tolerant to salinity and drought stresses.

10.4. HERBICIDE RESISTANCE

Weed control in crop fields was greatly facilitated by the application of selective herbicides that act against specific groups of plants. For example, the herbicide 2,4-D acts against broad-leaved plants. But the selective herbicides are not readily degraded by microorganisms, and they tend to persist in the environment. In contrast, the herbicides like glyphosate, glufosinate, sulfonylureas, etc. are readily biodegradable, but they are all nonselective in that they kill all types of plants. Therefore, these herbicides can be used for weed control only when the crop varieties cultivated in the fields are resistant to the concerned herbicides. The development of herbicide tolerant crop varieties was the first important objective that was achieved by the transgenic technology. In fact, the information generated during the development of these varieties served as the basis for the methods and procedures for the development categories of transgenic crop varieties.

Herbicide tolerant crop varieties can be developed in one of the following two ways: (i) isolation of mutants resistant/tolerant to the herbicide, and (ii) introduction and expression of a transgene that confers resistance/tolerance to the herbicide. The mechanism of herbicide tolerance may involve one of the following three molecular events: (i) overproduction of the molecule that serves as the target for the action of the given herbicide, (ii) production a mutant form of the target molecule that is insensitive to the herbicide, and (iii) expression of an enzyme that inactivates the herbicide (Table 10.1). In addition, transgenic plants can be created that combine the molecular mechanisms listed in the items (ii) and (iii) above. An attempt will be made here to explain these strategies for herbicide tolerance using the herbicide Glyphosate or Roundup as an example.

Glyphosate is a broad spectrum herbicide that kills plants by inhibiting the synthesis of aromatic amino acids. This action of this herbicide is mediated by the enzyme 5-enolpyruvylshikimate-3-phosphate synthase (EPSPS), which is a critical enzyme that catalyzes the initial step in the biosynthesis of aromatic amino acids. Cell lines of carrot and tobacco were selected for resistance to the herbicide glyphosate. These cell lines overproduced the enzyme EPSPS due to gene duplication. As a result, they showed increased tolerance to the herbicide. On the other hand, the natural strain CP4 of *Agrobacterium* sp. and mutant strains of *Salmonella typhimurium* and *E. coli* are resistant to this herbicide. Similarly, site-directed mutagenesis has been used to produce glyphosate tolerant *Petunia* and *A. thaliana*. The basis of glyphosate tolerance in these cases is the herbicide insensitive enzyme EPSPS produced by these strains. Finally, *Achromobacter* sp. strain LBAA is tolerant to the herbicide glyphosate because it produces the enzyme oxidoreductase that oxidizes glyphosate and thereby inactivates the herbicide. Similar cases for resistance against some other herbicides are listed in Table 10.1.

Table 10.1: The mechanisms for herbicide tolerance

Mechanism	Molecular basis	Example	Genetic basis	Tolerance to herbicide
Over-production of the target molecule	Overproduction of the enzyme EPSPS*	Cell lines of tobacco and carrot	Gene duplication	Glyphosate
Target molecule tolerant to the herbicide	Enzyme acetolactate synthase (ALS) insensitive to the herbicides	*Arabidopsis thaliana,* tobacco, yeast, *E. coli*	Natural/ mutant genes	Sulfonylureas and imidazolinones
	EPSPS enzyme tolerant to glyphosate	Gene *aroA* from *Agrobacterium* sp. strain CP4	Natural gene	Glyphosate
		Mutant EPSPS; in *Salmonella typhimurium* and *E. coli*	Gene mutation	Glyphosate
		Petunia and *A. thaliana*	Site-directed mutagenesis	Glyphosate
Inactivation of the herbicide	Enzyme glyphosate oxidoreductase oxidizes glyphosate	Gene *gox* from *Achromobacter* sp. strain LBAA	Natural gene	Glyphosate
	Enzyme phosphinothricin acetyltransferase acetylates phosphinothricin	Gene *bar* from *Streptomyces hygorscopicus*	Natural gene	Glufosinate
	Conversion of the herbicide into an inactive compound	Gene *bxn* from *Klebsiella ozaene*	Natural gene	Bromoxynil

*EPSPS, 5-enolpyruvylshikimate-3-phosphate synthase

Attempts were made to develop glyphosate tolerant transgenic lines by expressing the mutant forms of the enzyme encoding EPSPS from *S. typhimurium, E. coli*, petunia and *A. thaliana*. The transgenic lines did show increased tolerance to the herbicide, but the level of tolerance was not enough for commercialization. In contrast, the expression of the gene encoding herbicide insensitive EPSPS enzyme from the natural strain CP4 of *Agrobacterium* sp. in soybean, cotton and several other crops was much more effective in conferring glyphosate tolerance. In these cases, the enzyme ESPS was transported into the chloroplast by linking it to the chloroplast-specific N-terminal transit peptide. Several transgenic lines with useful resistance to the herbicide glyphosate have been produced by expression of this gene (*aroA*). Finally, the introduction and expression of the gene *gox* from *Achromobacter* sp. strain LBAA into the crop plants produces glyphosate tolerant lines. The enzyme GOX was transported into the chloroplast where the EPSPS enzyme activity is located. However, the combined transfer and expression of the genes *gox* and *aroA* genes in crops like maize, soybean, etc. has resulted in several commercially cultivated glyphosate tolerant GM varieties.

10.5. INSECT RESISTANCE

Insect pests attack almost all the crops and cause some to considerable damage and yield losses. The importance of insect pests varies considerably from one crop to the other. In some crops like cotton, several insects cause extensive damage, and these crops cannot be cultivated without insecticide protection. Therefore, the development of insect resistant varieties has been an important objective in the breeding of several crops. But in some cases, useful sources of insect resistance are not available. Therefore, the development of insect resistant transgenic plants was one of the first transgenic projects taken up. Many proteins are known to produce deleterious effects in insects. Among them, the crystal (Cry) protein produced in the spores of soil bacterium *Bacillus thuringiensis* is lethal to specific groups of insects. This protein is encoded by the *cry* gene of the bacterium. The *cry* gene has been successfully deployed to generate insect resistant transgenic crop varieties.

10.5.1. Classification of Cry Proteins

The bacterium *B. thuringiensis* has been known to kill insects. *B. thuringiensis* preparations have been used since the World War I for insect control in several countries, especially in Europe. The insecticidal activity is due to the Cry protein, which is of 130 kDa and forms bipyramidal crystals. But some Cry proteins are only about 70 kDa; these are believed to be truncated versions of the Cry proteins and they do not form the typical Cry protein crystals. The Cry proteins were divided into four groups or first ranks on the basis of the insects affected by them. These four groups are denoted by Roman numerals as follows: CryI, CryII, CryIII, and CryIV (Table 10.2). Later, groups CryV and CryVI were created by including a diverse set of Cry proteins; these groups contain mainly nematicidal proteins. Each of the groups I to IV is further divided into subgroups denoted by capital letters, e.g., CryIA, CryIB etc. Finally, the subgroups are further subdivided by adding small letters written within parenthesis, e.g., CryIA(b), CryIA(c), etc. A much more elaborate classification for the Cry proteins was proposed by Crickmore et al. (1998). This classification is based on amino acid sequence homology among the Cry proteins, and classifies these proteins into 22 groups or first ranks, which are denoted by Arabic numerals, e.g., Cry1, Cry2, etc. The second and third ranks are denoted by capital and small letters, respectively, e.g., Cry1Aa, Cry1Ab, etc. Finally, the fourth ranks are again represented by Arabic numerals, e.g., Cry1A3.

Table 10.2: A simplified summary of the classification of Cry proteins produced by the soil bacterium *B. thuringiensis.*

The classical classification			Modern classification*
First rank and second ranks	**Molecular mass (kDa)**	**Active against**	
CryIA	132	Lepidoptera	Cry1 (secondary ranks A to K)
CryIB	138	Lepidoptera	
CryIC	134	Lepidoptera	
CryID	132	Lepidoptera	
CryIE	132	Lepidoptera	
CryIIA	70	Lepidoptera and Diptera	Cry2 (secondary ranks A to C)
CryIIB	70	Diptera	
CryIIIA	73	Coleoptera	Cry3 (secondary ranks A to C)
CryIIIB	74	Coleoptera	
CryIIIC	129	Coleoptera	
CryIVA	134	Diptera	Cry4 (secondary ranks A and B)
CryIVB	127	Diptera	
CryIVC	77	Diptera	
CryIVD	72	Diptera	
CryV	-	Nematodes	Cry5
CryVI	-	Nematodes	Cry6
CytA	27	No homology with Cry proteins**	

* The remaining 16 first ranks (from Cry7 to Cry22) of the modern system comprise few members from one or more of the first ranks of the old system or some new proteins.
** Cytotoxic to a variety of vertebrate and invertebrate cells.

10.5.2. Mechanism of Action of Cry Proteins

The Cry protein is a protoxin, the toxin fragment being located in the N-terminal half of the protein. The C-terminal part is highly conserved and is involved in formation of the typical crystal structure of these proteins. When the insect larvae feed on the Cry proteins, the crystal structure opens up in the alkaline environment of the midgut lumen yielding the fully extended polypeptide. The proteases present in the midgut cleave the Cry protein protoxin; cleavage of the protoxin at the two vertical arrows shown in (Fig. 10.1) liberates the toxin fragment. The toxin fragment has high affinity and binds highly specific receptors present in the plasma lemma of the brush border cells of the midgut epithelium. This binding creates pores in the plasma lemma that are most likely nonspecific in nature (Fig. 10.2). Ions and water enter into the affected cells causing them to swell and ultimately lyse. The Cry protein toxin has three domains as follows: domain I (forms ion channels and pores), domain II (recognizes the receptors present insect midgut epithelial cells), and domain III (involved in ion channel formation, receptor binding and specificity for the target insects). Thus hybrid *cry* genes and Cry proteins can be created by combining the desired domains from different *cry* genes.

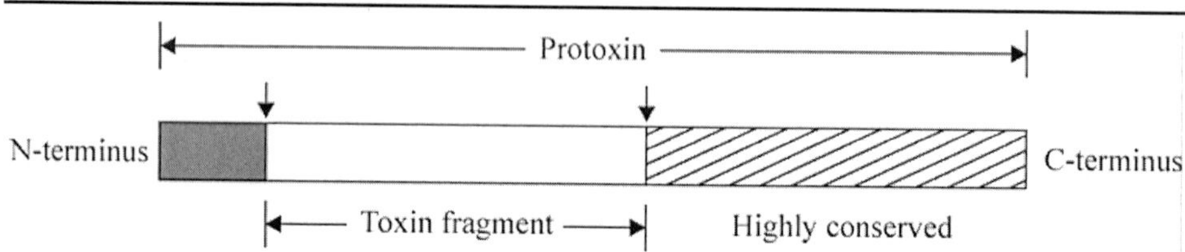

Fig. 10.1. Generalized organization of the full-length Cry proteins (~130 KDa) of *B. thuringiensis*. These Cry proteins are protoxins. The toxin fragment is liberated by proteolytic digestion of the protoxin at the sites indicated by vertical arrows. The C-terminus half of the Cry proteins is highly conserved and is involved in the bipyramidal crystal formation, which is typical of the full-length Cry proteins.

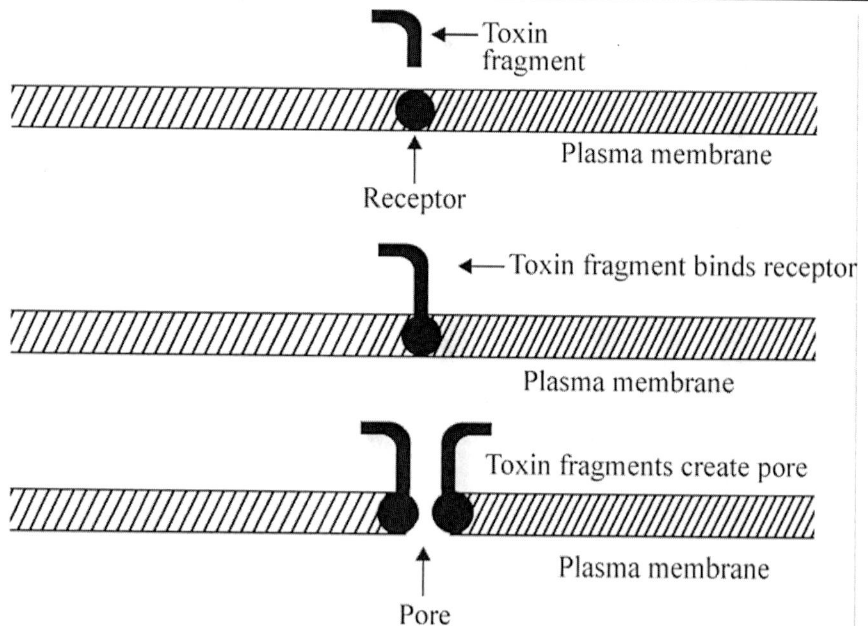

Fig. 10.2. The Cry protein toxin fragment binds to specific receptors present in the cell membrane of the brush border cells of the insect midgut epithelium. This generates pores in the membranes leading to influx of ions and water into the cells, which ultimately lyse.

10.5.3. Production of Insect Resistant Transgenic Plants

The first report on the production of insect resistant transgenic plants published in 1987 pertained to tobacco plants expressing the CryIA(c) protein. These plants produced enough Cry protein to kill all the *Maduca sexta* larvae in 6 days. Subsequently, many transgenic plants expressing the *cry* genes were produced. But in all these cases, the level of expression of the Cry proteins was low (around 0.001% or less of the total leaf protein). The low level of *cry* gene expression is the consequence of several features of this gene, including presence of sequences resembling plant introns, sequences similar to polyadenylation signal sequences, sequences that promote degradation of mRNA, and codons that are infrequently or even seldom used by plants. In addition, this gene is quite long.

Since only a part of the Cry protein is needed for insecticidal activity, the *cry* genes were truncated by deleting the unnecessary sequences at their C- and N-terminal ends. The truncated genes showed improved expression over the native genes. Further improvement in the level of expression was achieved by modifying the *cry* gene sequences to make them suitable for expression in plants. These modifications include replacement of all the disturbing sequences and inclusion of the plant-specific context for the translation initiation codon in the mRNA. In addition, the codons of the mRNA were modified to include only those codons that are frequently used by plants. *These modifications are done in such a way that the amino acid sequence of the concerned truncated Cry proteins is not altered, and it remains the same as that of the protein encoded by the native cry gene.* When the gene sequence was partially modified, the level of Cry protein in the transgenic plants increased to .01% of the of the total leaf protein. It further increased to 0.1% of the total leaf protein when the gene sequence was fully modified. Further increased expression levels were obtained by transporting the Cry proteins into chloroplasts (~1% of the total leaf protein) and by introducing and expressing the *cry* genes into chloroplasts (up to 2% of the total leaf protein).

Insect resistant transgenic varieties expressing *cry* genes have been produced in several crops; some of these varieties are listed in Table 10.3. These varieties have been used for commercial cultivation, and have become very popular among the farmers. The insect resistant varieties were the second most widely cultivated transgenic varieties, after herbicide tolerant crop varieties, for over a decade since the beginning of the commercial cultivation of the transgenic varieties in 1996. However, at present insect resistance varieties rank number three in popularity, while herbicide tolerant, and the combined or stacked trait varieties with herbicide tolerance and insect resistance rank number one and two, respectively. Thus, insect resistance trait remains the second most popular trait after herbicide tolerance.

Table 10.3: Selected examples of transgenic varieties of crops resistant to herbicide, insect pest, virus or drought that have been approved for commercial cultivation.

Transgenic variety	Modified trait	Developed by	Year of approval
Yield Gard (maize)	Resistance to European corn borer	Monsanto	1995
Nature Gard[+] (maize)	Resistance to European corn borer	Dow/Mycogen	1995
Bt brinjal	Lepidopteran insect resistance	MAHYCO, India	2013
Roundup Ready (maize)	Resistance to glyphosate	Monsanto	1998
Roundup Ready (cotton)	Resistance to Glyphosate	Monsanto	1996
Roundup Ready (soybean)	Resistance to glyphosate	Monsanto	1995
BXN cotton	Resistance to bromoxunil herbicide	Monsanto/Rhone-Poulenc	1995
Star Link * (maize)	Resistance to glufosinate and European corn borer	Aventis	1998
SYHTOH2 (soybean)	Herbicide (glufosinate and mesotrione) tolerance	Bayer CropScience and Syngenta	2014
T304-40 (cotton)	Glufosinate tolerance, insect resistance	Bayer CropScience	2011
Enlist™ Soybean	2,4-D and glufosinate tolerance	Dow AgroSciences LLC	2012
Sunup, Rainbow (papaya)	Resistance to ringspot Virus	Cornell Univ./Univ. of Hawaii	1997
Freedom (soybean)	Virus resistance	Asgrow	1995
Freedom II (squash)	Virus resistance (watermelon mosaic 2 and Zucchini viruses)	Asgrow	1995
Genuity® Drought Gard™ (maize)	Drought tolerance	Monsanto	2010
NXI-1T and two others (sugarcane)	Drought tolerance	P.T. Perkebunan Nusantara XI	2013

* Not being marketed at present.

10.5.4. Emergence of Resistance to Cry Proteins in Insect Pests

It may be pointed out that the specificity of a Cry protein for insecticidal action on a given host insect is due to the high affinity of the domain II of this protein for the receptors present in the midgut epithelial cells of the insect. Therefore, a mutation in the insect gene encoding this receptor would make the insect resistant to this Cry protein provided this mutation affected the binding of the Cry toxin fragment to the receptor. Emergence of resistance in the insect pests to Cry proteins has been reported by several workers. These reports were based on studies conducted either in the laboratory or in the field where *B. thuringiensis* preparations were used for insect control. The mutant receptor may, however, show increased affinity for another Cry protein toxin for which the wild type receptor would show low affinity. In such a case, the gene encoding the second Cry protein (for which the mutant receptor has high affinity) can be used to overcome the problem of insect resistance to the first Cry protein. In addition, several other strategies have been developed to manage the problem of insect resistance to the Cry proteins; some of these strategies are briefly described in the following paragraph.

One strategy for the management of emergence of resistance in the insect pests is to develop varieties that express two or more different *cry* genes that differ in their mode of action but are active against the same target insect pest. It is much less likely for the pest to develop resistance against both the Cry proteins. Alternatively, hybrid Cry proteins can be created by combining the desired domains from different Cry proteins to overcome the problems due to the development of the resistance to the Cry proteins in the insect pests. Finally, other genes that encode insecticidal proteins may be used either alone or in combination with *cry* genes. At the level of crop cultivation, the practice of planting refuge concept is being implemented (Section 12.14). There is evidence that the refuge crop strategy tends to delay the emergence of resistance to the Cry proteins in the insect pests.

10.5.5. Other Genes for Insect Resistance

The scientists have been making extensive and intensive efforts to identify and utilize genes other than the *B. thuringiensis cry* gene for the development of commercially successful insect resistant varieties. Some examples of the genes evaluated for this purpose are the insecticidal protein gene from *Streptomyces* sp., trypsin inhibitor gene from sweet potato, α-amylase inhibitor gene from common bean (rajma), lectin gene from snowdrop (*Galanthus nivalis*), chitinase gene from rajma, vegetative insecticidal protein (VIP) from *B. thuringiensis*, etc. Even viral genomes, e.g., stunt virus genome, have been transferred into plants to generate insect resistance. However, the only useful gene identified from these extensive searches is the *VIP* gene from *B. thuringiensis*. The *VIP* gene has been used in combination with *cry* genes to develop some transgenic varieties that have been released for commercial cultivation.

10.6. VIRUS RESISTANCE

Viruses constitute the fourth major group of biotic stresses ranking after fungi, bacteria and insects. These pathogens cause little economic loss to some crops like wheat, but are of considerable significance in Solanaceous crops like potato, tomato, etc. Viruses assume considerable significance in the case of vegetatively propagated crops since the viruses are transmitted through the propagules themselves if they are derived from infected plants. In these crops, development of virus resistant varieties is an important breeding objective. Sources of satisfactory resistance are available in the germplasm for many viruses, but for many other viral pathogens useful resistance may not be available. In such cases, genetic engineering approaches may be used to develop transgenic plants with resistance to the viral pathogens (Table 10.4).

Table 10.4: A summary of the various approaches for creation of virus resistance by genetic engineering

Method for creation of virus resistance	Efficacy	Applicable to all viruses	Effectiveness against unrelated viruses	Mechanism of action
Coat protein gene	High in many cases; transgenics commercialized	Yes	Some effect on their systemic spread	Interferes with viral genome replication and systemic spread in the plant
cDNA of satellite RNA	High	No	No	Hampers assembly of the virus particles
Defective viral genome	Medium	No	No	Hinders replication of the viral genomes
Antisense RNA	Low	Yes	Possible, if they share a conserved region in a critical gene and this sequence is used for creation of the transgene construct	Pairing with the RNA transcript/genome prevents translation/replication
Ribozyme/ RNase P	Possibly low	Yes		Pairing with the mRNA/RNA genome leads to their cleavage

10.6.1. Various Strategies for Transgenic Virus Resistance

Several different strategies for creation of transgenic plants resistant to viral pathogens have been conceived and evaluated. The most important of these approaches are listed in Table 10.4 and are briefly summarized here. (i) The gene encoding the coat protein of the concerned virus has been cloned and transferred into the host plant genome. (ii) Some RNA viruses have small RNA molecules, called **satellite RNA**, in addition to their genomes. Some of these satellite RNAs reduce the severity of the disease produced by the concerned viruses. The cDNA copy of such satellite RNAs is cloned and expressed in the concerned host plant. The satellite RNA molecules produced in the host cells would reduce the severity of the disease when the concerned RNA virus infects them. (iii) In case of some DNA and RNA viruses, defective or deleted copies of their genomes tend to interfere with the replication of their complete genomes. In the cases of such viruses, a defective copy of the viral genome is integrated into the host plant genome. This defective genome would interfere with the viral genome replication when the concerned virus infects these plants.

In addition to the above, (iv) viral genes other than the coat protein gene have been evaluated for generation of virus resistant plant. Some these genes are RNA-dependent RNA polymerase gene, genes encoding Potyvirus genome linked proteins (*VPg*), VPg-proteinase (*Nla*), cylindrical inclusion protein encoding genes, etc. (v) The antisense RNA technology is useful technology for suppression of native or endogenous genes. This technology has been used to prevent the expression of a critical viral gene like coat protein gene in an effort to generate resistance to the concerned viral pathogen. Finally, (vi) the ribozyme approach, another strategy for suppression of native or endogenous genes, has been evaluated for the development of virus resistant transgenic plants. As in the case of antisense RNA technology, this technique is also used to suppress the expression of some critical viral gene. These techniques are described in some detail in the following two sections.

10.6.2. Antisense RNA Technology

The mRNA is often termed as **sense RNA**. It is produced by transcription of that DNA strand of the given gene that is present in the 3’ → 5’ orientation in relation to the direction of the activity of the promoter of the concerned gene (Fig. 10.3A). Thus, the base sequence of the non-transcribed strand of the gene will be the same as that of the mRNA produced by this gene; therefore, this strand is called **sense strand** of the gene. The transcribed strand will, therefore, be termed as **antisense strand** since its base sequence is complementary to that of the sense strand of the gene and the mRNA encoded by the gene. As a rule, the antisense strand of a functional native gene is located in the 3’ → 5’ orientation in relation to the direction of promoter activity so that it is transcribed to yield functional mRNA. In general, the gene constructs used for **genetic engineering** contain the transgene in the proper orientation in relation to its promoter to ensure that it encodes functional mRNA. These gene constructs are called **sense constructs** (Fig. 10.3A), and their transcription generates functional mRNA or sense RNA (Fig. 10.3B).

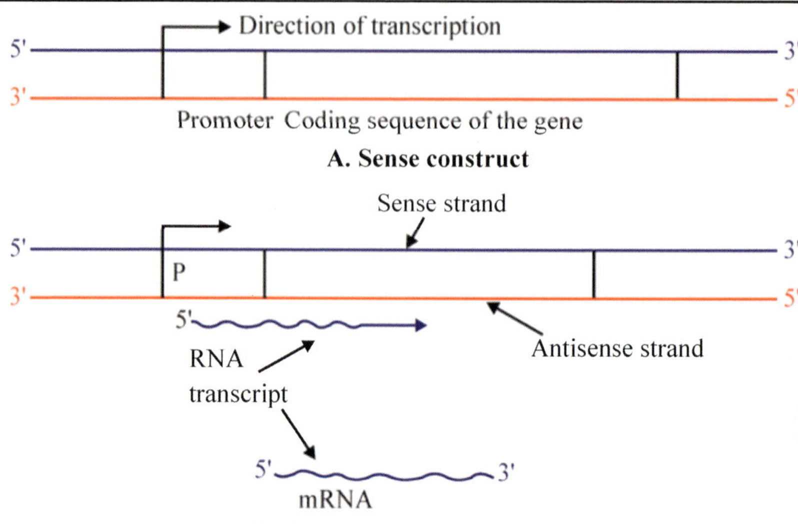

Fig. 10.3. A Sense gene construct. B. Transcription of a sense construct yields functional mRNA. The transcribed strand is termed as antisense strand, and is located in 3' → 5' orientation in relation to the direction of promoter activity of the concerened gene. P, promoter. The arrow above the promoter indicates the direction of transcription or promoter activity.

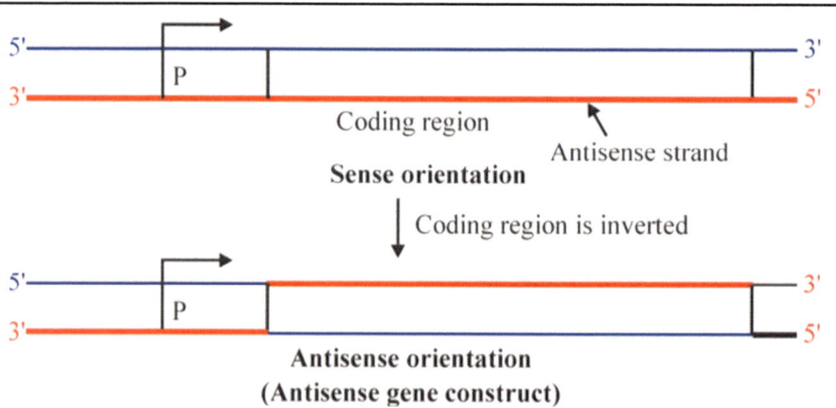

Fig. 10.4. An antisense gene construct is produced by inverting the direction of the coding region of the native gene in relation to the promoter of this gene. As a result, the sense strand of the gene is now located in 3'→5' orientation in relation to the promoter activity. Therefore, in an antisense construct, the sense strand of the gene will be transcribed.

The antisense construct is prepared by inverting, i.e., rotating by 180°, the coding region of a gene in relation to its promoter (Fig. 10.4). As a result, in the antisense construct of a gene, the sense strand of the native gene is now oriented in the 3' → 5' direction in relation to promoter. Therefore, the sense strand of the native gene will be transcribed in the case of the antisense gene construct. This transcription will yield **antisense RNA**, which will be complementary to the mRNA of the gene. The antisense RNA technology was developed with the objective of suppressing the expression of endogenous genes. In this technology, the antisense construct of the gene that is to be suppressed is introduced in the cells that already have the native gene in sense orientation. Both the native genes and the antisense gene constructs will be transcribed to yield mRNA and antisense RNA, respectively (Fig. 10.5).

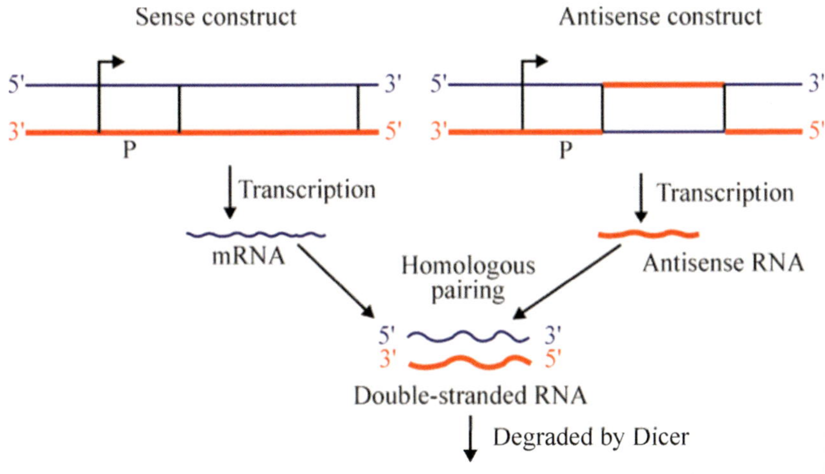

Fig. 10.5. Antisense RNA technology. The same cell has both native sense and antisense constructs of the gene. Therefore, both mRNA and antisense RNA are produced. The mRNA and antisense RNA are complementary. Therefore, they pair and form double-stranded RNA (dsRNA). The dsRNA is degraded by Dicer to generate small interfering RNA (siRNA) molecules. This prevents translation of the mRNA, and the siRNA may ultimately lead to gene silencing.

As a rule, the mRNA and the antisense RNA are complementary to each other. Therefore, the mRNA molecules will pair with antisense RNA molecules to yield double-stranded RNA (dsRNA) molecules. The dsRNA molecules are cleaved by the endonuclease dicer (Fig. 10.5). This digestion generates small dsRNA molecules called small interfering RNA (siRNA). The siRNA molecules would ultimately lead to the silencing of the concerned gene (Section 2.9; Fig. 2.16). This technology has been successfully used for suppression of some endogenous genes leading to the generation of transgenic varieties that have been approved for commercial cultivation. In case of virus resistance, the antisense technology would lead to the suppression of the targeted critical gene of the virus. As a result, the virus function will be hampered and the ability of the virus to produce the disease will be minimized.

10.6.3. The Ribozyme Technology

The ribozyme technology is based on RNA molecules that have enzymatic activities; these RNAs are called **ribozymes**. For example, the satellite RNA of the Tobacco Ringspot Virus (TobRV) has endoribonuclease activity. This activity is located in a part of the satellite RNA, and it cleaves other RNA molecules that come in contact with it. This catalytic sequence of the satellite RNA (referred to here simply as ribozyme) is joined with the 3' region of the mRNA that is desired to be degraded. This joining is achieved by producing cDNA of the ribozyme sequence and linking it with that region of the target gene sequence, which encodes the 3' half of the concerned mRNA. This region of the target gene is kept in the antisense orientation. This hybrid or chimeric DNA molecule is linked to a suitable promoter and other regulatory sequences (Fig. 10.6). Thus, this gene construct has a hybrid gene comprising the ribozyme sequence in sense orientation and the target gene sequence in the antisense orientation. This construct is introduced into the cell where the target gene is to be suppressed.

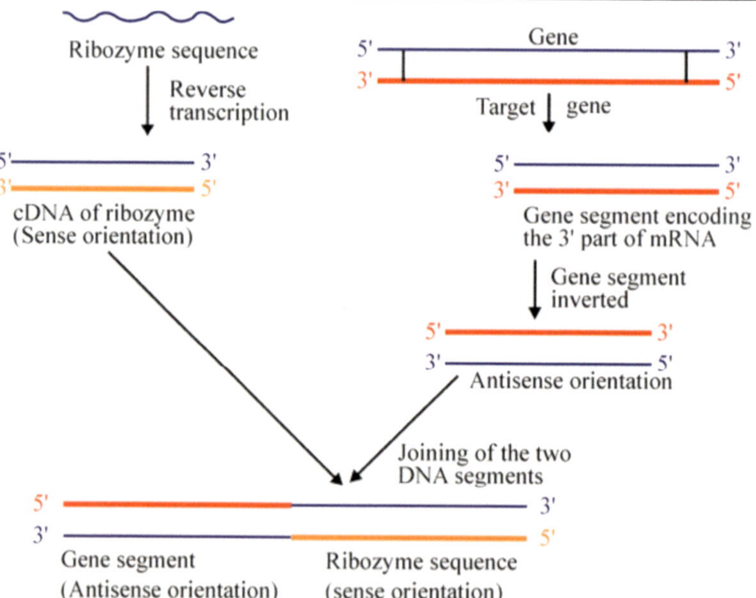

Fig. 10.6. The ribozyme approach for endogenous gene suppression. The target gene-ribozyme chimeric sequence is produced by joining the cDNA of ribozyme sequence (in sense orientation) with the target gene segment (in antisense orientation) encoding the 3' half of the target mRNA. This gene construct is expressed in the transgenic plants.

The native gene would produce mRNA, while the ribozyme-antisense gene construct will be transcribed to yield the 5' ribozyme-antisense RNA 3' hybrid molecule. This hybrid RNA will pair with the mRNA and thereby the ribozyme sequence will be placed close the 5' region of the mRNA. The ribozyme action will cleave the 5' region of the mRNA and effectively prevent its translation (Fig. 10.7). This will suppress the expression of the native gene. In case of virus resistance, the ribozyme may be targeted against the genomes of RNA viruses like tobacco mosaic virus (TMV). The ribozyme in this case, will degrade the viral genome and prevent multiplication of the virus in the host plant cells. This strategy was used to produce transgenic tobacco plants with resistance to TMV; these plants did show some level of resistance to the pathogen TMV.

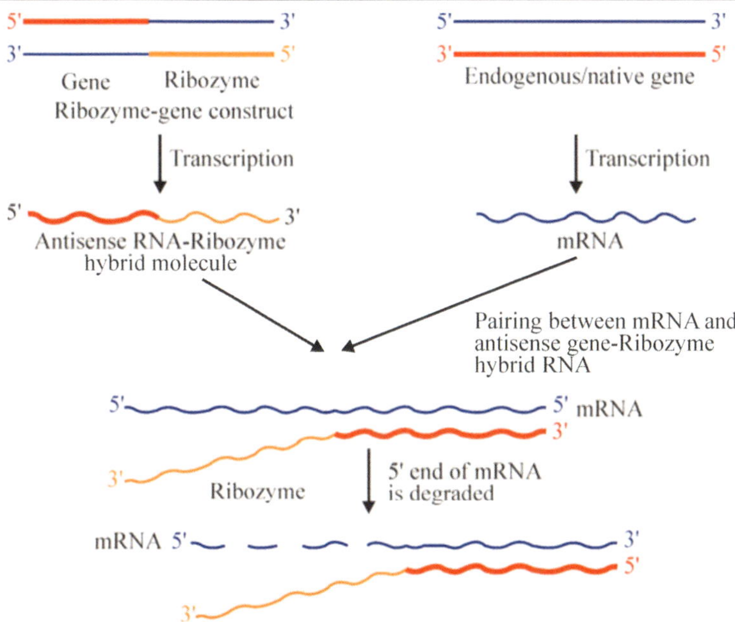

Fig. 10.7. Suppression of an endogenous gene by the ribozyme technology. The antisense gene-ribozyme sense construct encodes a hybrid RNA (antisense RNA-ribozyme RNA). The antisense RNA segment of this hybrid RNA pairs with the mRNA transcribed from the native gene. This paring places the ribozyme sequence close the 5' region of the mRNA. The ribozyme cleaves the 5' region of the mRNA and, thereby, prevents the translation of this mRNA.

10.6.4. Coat Protein Mediated Virus Resistance

The first transgenic plant created for virus resistance was a tobacco plant expressing the coat protein gene of TMV. These transgenic plants were produced by the procedure outlined in Figure 10.8. In simple terms, cDNA copy of the TMV genome was produced, and the gene encoding the coat protein of TMV was isolated and cloned. This gene was transferred and expressed in the tobacco plants. The transgenic tobacco plants obtained in this way either did not show the disease symptoms when inoculated with TMV or if the disease symptoms developed, their development was markedly delayed. These plants also showed delayed symptom development when inoculated with the related tomato mosaic virus and tobacco mild green mosaic virus. Thus, expression of the coat protein gene from a pathogenic virus generates resistance not only to this virus, but also to the viruses related to this virus.

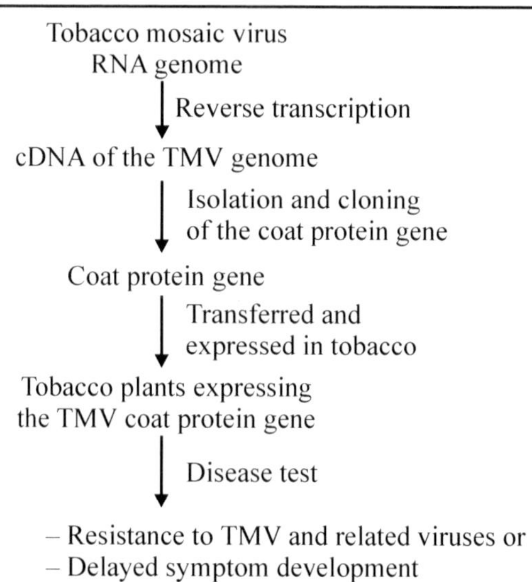

Fig. 10.8. A simplified representation of the procedure for development of TMV resistant tobacco plants by transferring and expressing the coat protein gene of TMV.

The expression of virus coat protein in plants seems to interfere with virus multiplication and the systemic spread of the virus in the infected plants. However, some other mechanisms of resistance, including RNA-mediated resistance, mediated by the virus coat protein gene have been proposed. In case of some viruses like Potyviruses, the transgenic plants do develop disease symptoms. But as the plants grow, the symptoms decrease in intensity and many plants even become symptom-free.

10.7. DISEASE RESISTANCE

The fungal and bacterial pathogens are the two most important groups of disease causing organisms. Ordinarily, these pathogens and their hosts show gene-for-gene relationship. This relationship means that for every resistance gene in the host, the pathogen has a virulence gene that specifically overcomes the host resistance due to this gene. In very simple terms, the mechanism of gene-for-gene system of host-pathogen interaction is as follows. The pathogen *avr* (avirulence) gene encodes a protein that itself or a molecule generated by it functions as the elicitor of host response. The resistance (*R*) gene encodes a protein, the R protein, that interacts with the elicitor molecules in a highly specific manner. In the case of obligate or biotrophic pathogens, the recognition of the elicitor by the R protein initiates a resistance response called **hypersensitive response (HR)**. The HR is initiated by the R protein that activates a number of genes that encode pathogenesis related proteins, which ultimately lead to the death of those cells that express these proteins. Thus, HR leads to the death of those cells that are infected by the pathogens and few cells surrounding these cells. The host cell death leads to the death of the invading pathogen since biotrophic pathogens require living host cells for their survival and proliferation.

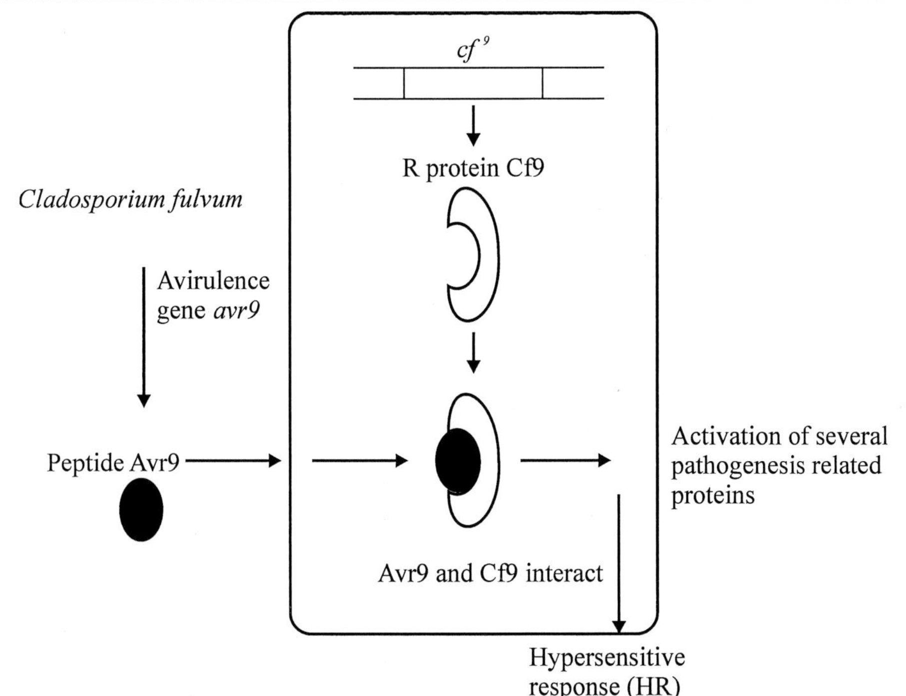

Fig. 10.9. A simplified schematic representation of hypersensitive response (HR) involving Avr9 peptide produced by the pathogen *Cladosporium fulvum* and the Cf9 R protein produced by tomato. In HR, the host cells surrounding the point of infection die, and the biotrophic (= obligate parasite) pathogen also dies along with these cells.

A well-known example of HR is provided by tomato and its fungal pathogen *Cladosporium fulvum*. The *avr9* gene of this pathogen encodes a peptide that functions as the elicitor of the host response. The tomato has the *R* gene *cf9* that encodes the *R* protein *Cf9*, which recognizes and interacts with the Avr9 peptide in a highly specific manner. After this interaction, the *Cf9 R* proteins activate several genes that encode pathogenesis related proteins; these proteins together give rise to the HR (Fig. 10.9). The *avr9-cf9* system has been used to create transgenic plants with resistance to fungal pathogens. The gene *cf9* is linked to a constitutive promoter, while the gene *avr9* is driven by a promoter that is activated by a nonspecific elicitor produced by various pathogens during host invasion. These genes are introduced into the host plant genome and expressed in the host cells. When a pathogen attacks such a transgenic plant, the gene *evr9* is expressed the elicitor peptide Avr9 is produced. The Avr9 elicitor interacts with the R protein Cf9 that is constitutively produced in the cells. This interaction activates the pathogenesis related proteins and these proteins together lead to the HR (Fig. 10.10).

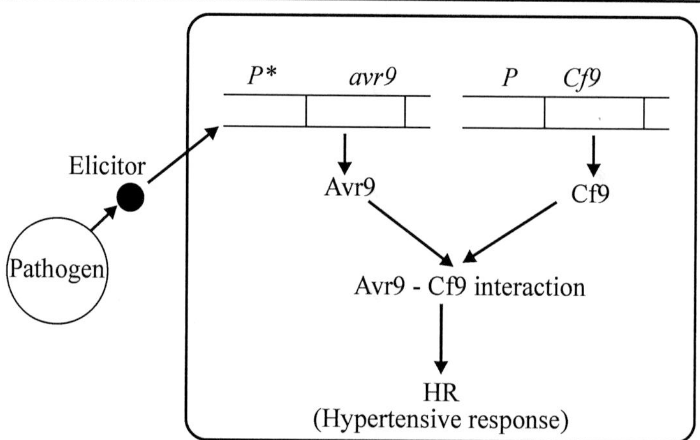

Fig. 10.10. A combination of *avr9* and *cf9* genes is used to produce disease resistant transgenic plants. The gene *avr9* is linked to the elicitor responsive promoter *P**, Gene *cf9* is driven by a constitutive promoter *P*. Gene *avr9* will be expressed in response to attack by a pathogen. *Avr9* protein will interact with the R protein *Cf9,* and this will eventually lead to HR (hypersensitive response).

In addition to the above, several other strategies for generation of resistance to fungal and bacterial pathogens have been evaluated with variable degrees of success. Some of these strategies are briefly outlined below. (i) Some pathogens produce disease by inhibiting a specific enzyme of the host plant. In such cases, a mutant allele of this gene that encodes a form of this enzyme that is insensitive to inhibition by the pathogen is introduced into the host genome. (ii) Some other pathogens cause disease by producing toxins. In such cases, genes encoding such proteins that inactivate the toxins are introduced and expressed into the host plant cells. (iii) In case of bacterial pathogens, genes encoding peptides with antibacterial activity have been expressed in the hosts. (iv) The expression of bacterial lysozyme in the host plants has been evaluated for resistance against bacterial pathogens. (v) The suppression of critical genes of the pathogens using

RNAi would inhibit pathogen multiplication and disease production. (vi) Introduction and expression of chitinase encoding genes has been used to create resistance against fungal pathogens. (vii) Some host proteins are required for the infection by some pathogens. In such cases, the gene encoding a modified form of this protein may be expressed in the host to interfere with pathogen infection process. (viii) Genes encoding thionins/producing phytoalexins that are different from those produced by the host plant have been expressed in the host. In all these cases, the transgenic plants showed increased levels of resistance to the concerned pathogens.

10.8. SALINITY AND DROUGHT TOLERANCE

Drought and salinity stresses are two important abiotic stresses affecting crop performance. There are two distinct pathways that lead to drought tolerance in plants. In one of the pathways, ABA is produced in response to the drought and salinity stress. The ABA, in turn, initiates an ABA signaling cascade that leads to the activation of stress-responsive genes. The other pathway operates independent of the ABA pathway. It operates through *DREB* or some other yet unknown genes, and eventually activates the expression of stress-responsive genes. These two pathways are interconnected at several points. A number of different genes have been isolated, cloned and evaluated for generation of resistance to abiotic stresses. In general, the transgenic plants produced using these genes showed improved tolerance to the concerned abiotic stresses. The following three examples will help appreciate the efforts to generate abiotic stress resistance in crop plants.

In plants, proline accumulation is associated with increased tolerance to moisture stress. Under stress situations, the proline is synthesized directly from glutamine in the place of ornithine that is used under the normal situations. The first two reactions of the proline synthesis under stress conditions are catalyzed by the enzyme pyrolline-5-carboxylate synthetase (P5CS), which is the rate-limiting enzyme of this pathway. The gene encoding P5CS is one of the stress-responsive genes that is activated by the ABA-independent pathway. This gene has been isolated from mothbean and expressed in tobacco. The transgenic lines overexpressing P5CS accumulated 10- to 18-times more proline and showed improved moisture stress tolerance than the control plants. Proline is proposed to induce the formation of strong H-bonded water layer around protein molecules. As a result, these protein molecules are protected from denaturation and tend to retain their native configuration under stress.

The wheat gene *TaCHP* encodes a zinc finger protein. This gene is normally expressed in the root-tip cells. The gene has been transferred and expressed in *Arabidopsis thaliana*. These transgenic plants of *A. thaliana* showed 20-30% higher peroxidase activity, and improved tolerance to salt stress. Similarly, transgenic wheat plants expressing the gene *TaCHP* performed better than the non-transgenic control plants under salinity stress. Salinity tolerance is also improved by introductions and expression of the tonoplast Na^+/H^+ antiporter genes like *OsNHX1* and *AtNHX1*. These antiporter proteins transport Na^+ from the cytoplasm into the vacuole. This prevents toxicity due to Na^+ accumulation in the cytoplasm, which is the chief cause of salinity-induced toxicity to plants. As a result, the K^+/Na^+ ratio remains high in the cytoplasm, and the plants are protected from salinity induced damage. The genes *OsNHX1* and *AtNHX1* have been transferred and expressed in several crop plants, including maize, wheat, tomato, etc. These transgenic lines invariably showed improved tolerance to elevated salt levels as compared to the non-transgenic controls.

10.9. CONCLUSIONS

Transgenic plants are created by transferring into plants the desired genes using recombinant DNA technology. Transgenic plants have been created for resistance to both abiotic and biotic stresses. Abiotic stresses are generated by nonliving factors like herbicides, drought, salt, etc. The biotic stresses are produced by living factors like insect pests, viruses, bacteria and fungi. Transgenic herbicide tolerance can be generated by transferring a gene that (i) overproduces the molecule targeted by the herbicide, (ii) produces herbicide insensitive target molecule, or (iii) encodes an enzyme that detoxifies the herbicide. Commercial transgenic varieties resistant to the herbicide glyphosate have been produced by combining the strategies listed as items (ii) and (iii) above.

Insect resistant transgenic plants are produced by transferring a gene that encodes an insecticidal protein. Several genes, e.g., trypsin inhibitor gene, α-amylase inhibitor gene, lectin gene, chitinase gene, the *cry* and *VIP* genes from *B. thuringiensis*, etc., have been evaluated for this purpose. The *cry* genes have been used to develop insect resistant commercially grown varieties of several crops. Generally, truncated and modified *cry* genes are used for this purpose. The Cry proteins are highly specific in their action against the larvae of various insects. This is because they bind in a highly specific manner to receptors present in the plasma lemma of brush border cells of midgut epithelium. This ultimately causes rupture of these cells, which eventually leads to the death of the larvae. Insects are known to develop resistance to the Cry proteins. This problem is sought to be resolve by developing varieties with two or more different *cry* genes, creation of hybrid *cry* genes, use of other genes for insect resistance, and cultivation of refuge crops.

Virus resistance can be generated by deploying the following: viral gene encoding its coat protein, cDNAs of suitable satellite RNAs of RNA viruses, defective genomes of certain viruses, antisense RNA technology and ribozyme technology. The last two technologies can be used to suppress endogenous genes of the plants as well. The expression of viral coat protein gene has been used to develop commercially cultivated varieties of some crops. In fact, the papaya crop of Hawaii was saved by the development of ringspot virus resistant varieties. A number of different strategies for resistance to bacterial and fungal diseases have been evaluated; these include expression of genes like those for bacterial lysozyme, antibacterial peptides, etc.

Resistance to abiotic stresses has been sough to be created by the transfer of various genes from different organisms. For example, wheat gene *TaCHP* and the rice anitporter gene *OsNHX1* has been found to increase salt tolerance of the transgenic plants. Similary, the mothbean gene encoding the enzyme P5CS of the proline biosynthesis pathway increased the moisture stress tolerance of the transgenic tobacco plants. Drought tolerant transgenic varieties of maize and sugarcane have already been approved for commercial cultivation.

PRACTICE QUESTIONS

1. Discuss the usefulness of transgenic plants in agriculture with the help of suitable examples.
2. Briefly describe the various strategies for production of transgenic plants resistant to herbicides. Explain the significance of herbicide tolerant plants in crop production.
3. Briefly explain the mechanism of action of Cry proteins. It is known that insect pests develop resistance to Cry proteins. How can this problem be managed?
4. Briefly describe the various strategies for the creation of transgenic virus resistant varieties of crops.
5. Outline the various approaches that have been evaluated for production of transgenic plants resistant to bacterial and/or fungal diseases. Explain one of these schemes is detail.
6. Describe the efforts for the development of drought and salinity resistant transgenic varieties of crops.
7. Briefly explain the antisense RNA and ribozyme technologies for the suppression of endogenous genes (generation of virus resistance).
8. Write short notes on the following: (i) suppression of endogenous gene expression, (i) antisense RNA technology, (iii) ribozyme approach for suppression of endogenous gene expression, (iv) coat protein-mediated virus resistance, (v) Cry proteins, (vi) insect resistant transgenic plants, (vii) herbicide tolerant transgenic plants, (viii) hypersensitive response, (ix) drought tolerant transgenic plants, (x) transgenic salt resistance, (xi) insect pest resistance to Cry proteins, (xii) mechanism of toxic action of Cry proteins.

RELEVANT LITERATURE

1. Chen M et al. 2007. Expression of *OsNHX1* gene confers salt tolerance and promotes plant growth in the field. *Plant Soil Environ.* 3: 490-498.
2. Crickmore N et al. 1998. Revision of the nomenclature for the *Bacillus thuringiensis* pesticidal proteins. *Microbiol. Mol. Biol. Rev.* 62: 807-813.
3. Gatehouse AMR et al. (eds) 1992. Plant Genetic Manipulation for Crop Protection. CAB International, Wallingford, UK.
4. Grierson D 1991. Plant Genetic Engineering. Plant Biotechnology, Vol. I. Blackie, Glasgow.
5. Hrerrera-Estrella L and Simpson J 1995. Genetically engineered resistance to bacterial and fungal pathogens. *World J. Microbiol. Biotechnol.* 11: 383-392.
6. Lycett GW and Grierson D (eds) 1990. Genetic Engineering of Crop Plants. Butterworths, London.
7. Moss JP 1992. Biotechnology for Crop Improvement in Asia. ICRISAT, Patancheru, India.
8. Sharma HC and Ortiz R 2000. Transgenics, pest management, and the environment. *Curr. Sci.* 79: 421-437.

■■■■

11 Transgenic Plants with Modified Quality and other Novel Traits

11.1. INTRODUCTION

The transgenic technology offers unique opportunities for desired modifications in the characteristics of the crop species. In the previous chapter, the application of this technology for the creation of transgenic plants with resistance to herbicides, insect pests, viral and fungal/bacterial/nematode-diseases, and to abiotic stresses like salinity and drought were discussed in detail. These activities were the focus of research efforts during the initial phases of the transgenic plant development. Subsequently, the focus moved on to the creation of plants with novel and useful features like enhanced shelf life, improved nutritional quality, traits like male sterility that are helpful in crop improvement activities, etc. In addition, transgenic plants are now being developed that will serve as replacements of the conventional fermenters used for the production of useful biochemicals. These transgenic plants are designed to produce and accumulate valuable biochemicals, which are otherwise produced through microbial fermentation processes or from animal cell culture systems. In this chapter, these aspects of transgenic technology and some of the techniques used for achieving these objectives are discussed.

11.2. ENDOGENOUS GENE SUPPRESSION

The quality of crop produce is affected by several genes. The expression of many of these genes has positive impact on produce quality, but that of some other genes has detrimental effect(s). Thus, it would be desirable to suppress the expression of such genes in an effort to enhance the quality of the crop produce. In addition, the produce from some crops contain certain anti-nutritional factors. In such cases, the nutritional value of the crop produce will be greatly enhanced by minimizing the production of the anti-nutritional factor by suppressing the expression of some critical genes involved in their biosynthesis. The expression of an endogenous or native gene can be suppressed in the following five ways: (i) overexpression of the concerned gene leading to its cosuppression, (ii) antisense

RNA technology, (iii) ribozyme approach, (iv) RNA interference (RNAi), and (v) gene disruption. The mechanism of RNAi has been explained in Chapter 2 (Section 2.9, Fig. 2.15); the strategy of RNAi for native gene suppression will be briefly outlined in Section 11.2.3. The antisense RNA and ribozyme technologies were described in some detail in Chapter 10 (Sections 10.6.2 and 10.6.3). The gene disruption and the overexpression strategies of endogenous gene suppression are described in the Sections 11.2.1 and 11.2.2 respectively.

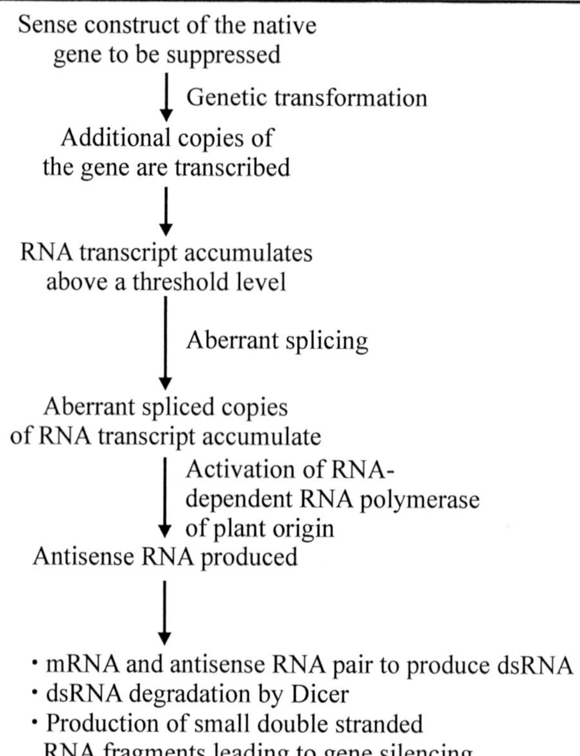

Fig. 11.1. The phenomenon of cosupprssion. According to one hypothesis, overproduction of the RNA transcript of the gene ultimately leads to production of antisense RNA, formation of double-stranded RNA, and gene silencing. dsRNA, double-stranded RNA.

11.2.1. Overexpression of the Gene (Cosuppression)

In this approach, a sense construct of the concerned gene is introduced into the plant genome. As a result, each cell of such a transgenic plant would contain four copies of the gene. The expression of these copies would produce about twice as much RNA transcript of the gene as the amount produced by the nontransgenic control plants. It would be expected that the intensity of the phenotype produced by this gene would be greater in the transgenic plants than that in the non-transgenic plants of the parental variety. But in the case of some genes, this leads to a drastic reduction in the intensity of the concerned phenotype and the level of expression of the concerned gene; this is called **cosuppression**. This phenomenon was first reported in the case of the *chalcone synthase* gene of petunia. It has been suggested that when RNA transcripts of a gene accumulate above a threshold level, some cellular mechanisms selectively degrade this RNA. One possibility is that this accumulation eventually activates RNA-dependent RNA polymerase activity of plant origin. This enzyme produces antisense RNA copies of the accumulated RNA transcripts of the gene (Fig. 11.1). Once the antisense RNA is produced, it would act in the same way as the antisense RNA technology described earlier in Chapter 10 (Section 10.6.2, Fig. 10.5).

The phenomenon of cosuppression is observed for only some of the plant genes. Even in case of the plant genes that show cosuppression, only a proportion of the transgenic events might show cosuppression. For example, only about 50% of the transgenic events show cosuppression in the case of chalcone synthase gene of petunia. Cosuppression approach has been used to inhibit ethylene production in carnation (*Dianthus caryophyllus*). The growth regulator ethylene is involved in fruit ripening and flower senescence. Ethylene is produced from S-adenosyl methionine (SAM) by the action of the enzymes ACC synthase and ACC oxidase, acting in the given order. The enzyme ACC synthase was cosuppressed in a carnation variety; this resulted in increased vase life of the transgenic variety. This transgenic variety has been commercialized.

11.2.2. Gene Disruption

Gene disruption consists of integration of a nucleotide sequence within the coding region of the target gene so that the target gene is unable to generate the protein product encoded by it. A highly simplified schematic representation of the gene disruption approach is given in Figure 11.2. The gene to be disrupted (gene *A* in this case) is isolated and cloned. A selectable marker gene is now integrated within the coding region of the cloned gene *A*. Thus, the cloned gene *A* is disrupted as it has a selectable marker gene within its coding region. The disrupted gene *A* is introduced into the plant cells, where it would become integrated into the plant genome. Most of the integration will be at random locations in the genome. But a small proportion of the integrations would occur by homologous recombination with the native gene *A*. This would place the selectable marker gene within the native gene *A*, and this would disrupt the native gene *A*. The disrupted gene *A* would be transcribed normally, but translation of the resulting mRNA would not produce the typical protein encoded by this gene. The technology of gene disruption has been used to create various knockout mice strains designed for various investigations. The frequency of homologous recombination can be increased considerably by combining this approach with the site directed nuclease technology (Section 2.11).

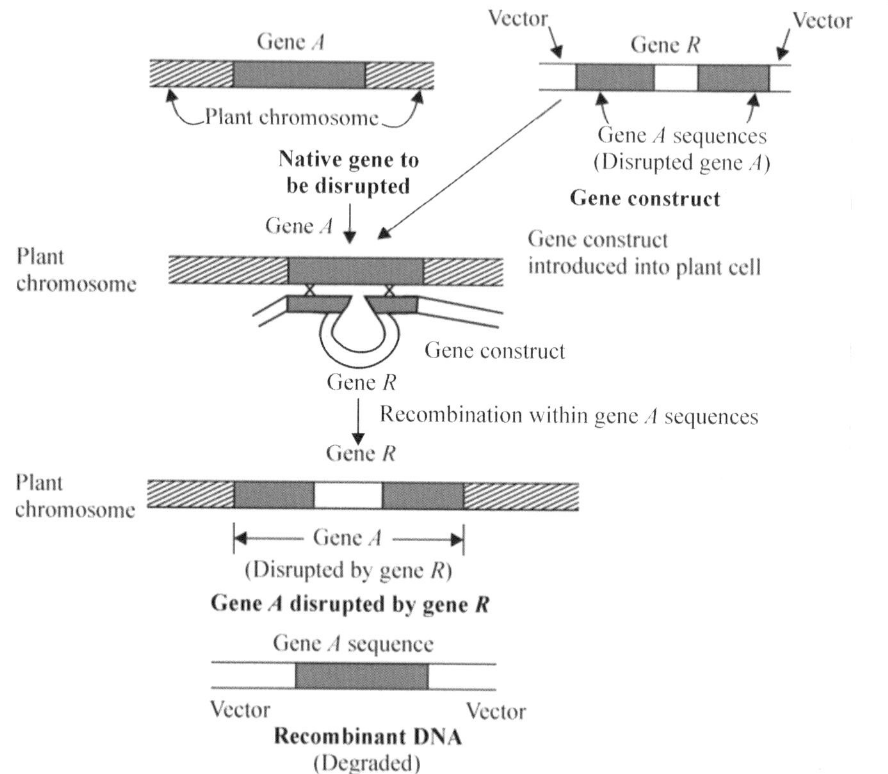

Fig. 11.2. A highly simplified schematic representation of gene disruption. Gene *R*, a selectable marker gene; gene *A*, the endogenous gene targeted for disruption (gene *A* should be isolated, cloned, and disrupted *in vitro*).

11.2.3. RNA interference (RNAi)

The RNAi technology can be used to suppress the expression of a specific endogenous gene present in the genomes of plants. The scheme for this procedure is outlined in (Fig. 11.3). A gene construct is created as follows: a 21 bp sequence from promoter or the coding region of the targeted gene is linked to a 6 bp spacer sequence, the other end of which is linked to the same sequence of the target gene placed in the antisense or reverse orientation. This gene is driven by a RNA polymerase III promoter like the U6 promoter. Transcription of this gene construct yields a RNA transcript that forms a double-stranded RNA (dsRNA) molecule with a hairpin loop formed by the spacer sequence. The dsRNA molecule is formed because of the presence of the same gene sequence in the reverse orientations separated by the spacer sequence. The hairpin is degraded by cellular ribonucleases, and the remaining 21 bp dsRNA molecule functions as siRNA. This ultimately leads to the silencing of the targeted gene.

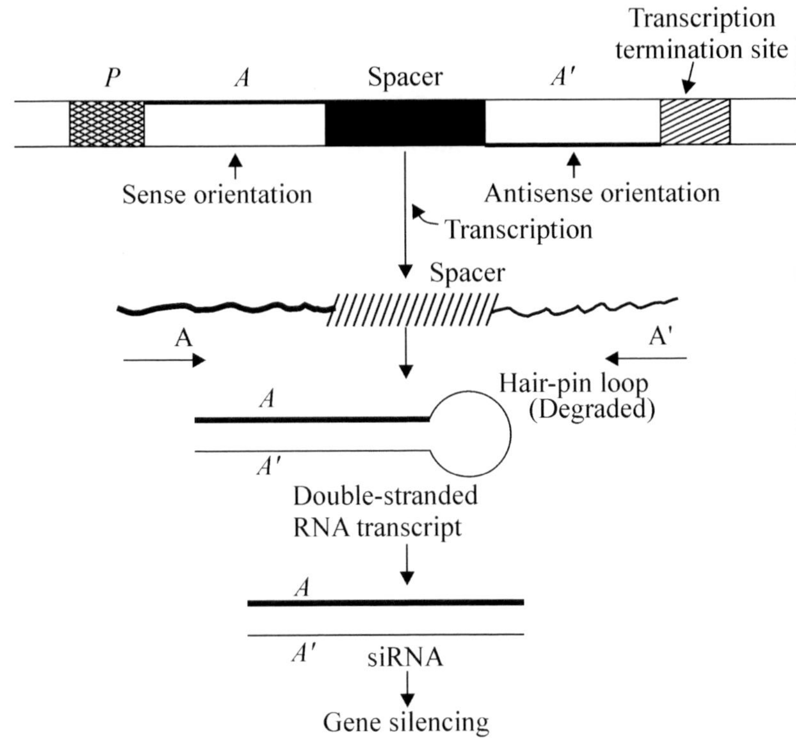

Fig. 11.3. A simplified strategy for suppression of a native gene by RNAi. *P*, a RNA polymerase III promoter; *A*, a segment of 21 bp from the promoter/coding region of the target gene; spacer, a random sequence used to facilitate hair-pin loop formation; *A'*, the *A* sequence in antisense orientation.

The RNAi strategy has been used to produce low caffeine coffee and *Meloidogyne incognita* resistant tobacco. RNAi was used to suppress the expression of δ-cadinene synthase gene(s) of cotton (*Gossypium hirsutum*) during seed development. These transgenic cotton lines had ultralow gossypol content in their cottonseed. However, the level of gossypol in the leaves and the floral parts of the transgenic lines was not affected. In addition, the oil content in the cottonseed of the transgenic lines was 4-8% higher than that of the nontransgenic line. The composition of the transgenic cottonseed was comparable to that of the nontrangenic line with respect to the remaining constituents. The gossypol is a toxic terpenoid that affects heart and liver functions. The presence of gossopol limits the usefulness of cottonseed, which is abundant in supply and otherwise in nutrition. The above transgenic lines were stable in performance over years and under field conditions, and are promising for commercialization.

11.2.4. Gene Silencing

In many transgenic lines, the level of transgene expression declines as the number of generations for which these lines have been grown increases this is called **gene silencing**. Gene silencing is often associated with DNA methylation involving cytosine (C) residues in the CG and CNG sequences, N being any of the four DNA nucleotides. The gene silencing has been classified into the following two groups: (i) transcriptional gene silencing and (ii) post-transcriptional gene silencing. The **transcriptional gene silencing** involves promoter methylation and prevents transcription of the affected transgenes. The **post-transcriptional gene silencing**, involves degradation of the RNA transcript after the transcription has taken place. The frequency of silencing of a transgene is affected by several factors, including the content and distribution of AT residues in the transgene, copies of the transgene integrated in a genome, etc. The problem of gene silencing can be minimized by the following approaches: (i) selection of transgenic events with single copy of the transgene, (ii) use of distinct promoters for different transgenes, (iii) keeping all the promoters in the same orientation, (iv) making the AT composition of the transgenes comparable to that of the concerned plant genome, and (v) inclusion of the AT-rich matrix attachment regions on both the sides of the transgene.

11.3. SLOW FRUIT SOFTENING TOMATO

The softening of tomato fruits is mediated by the enzyme polygalacturonase (PG). This enzyme digests pectin, the component of cell walls that cements together the walls of neighbouring cells. The degradation of pectin, therefore, leads to softening of the fruits. In view of the above, it is logical to expect that the softening of the tomato fruits will be slowed down if the level of expression of the enzyme PG were reduced. The gene encoding PG was isolated, cloned and an antisense construct of this gene was prepared. This antisense construct was introduced and expressed in tomato (Fig. 11.4). The transgenic tomato line showed much lower activity of PG and slower softening of the fruits. The best transgenic event was called FlavrSavr tomato; it was approved for commercialization in 1994. The fruits of FlavrSavr had about two weeks longer shelf life than the nontransgenic tomatoes. In addition, these fruits have superior flavor and higher total soluble solids than the nontransgenic control tomatoes.

The antisense RNA technology was used to produce a transgenic tomato line with slow and delayed fruit ripening. In this case, the gene encoding the enzyme ACC synthase of the ethylene biosynthesis pathway was suppressed using antisense RNA technology. As a result, the level of ethylene production was much lower in the transgenic line than in the nontransgenic control lines. The fruits of this line, therefore, ripen slowly over a period of time. This transgenic line was called endless summer tomato. However, it could not be commercialized due to intellectual property rights issues. Subsequently, a delayed ripening transgenic melon variety was released for commercial cultivation in USA.

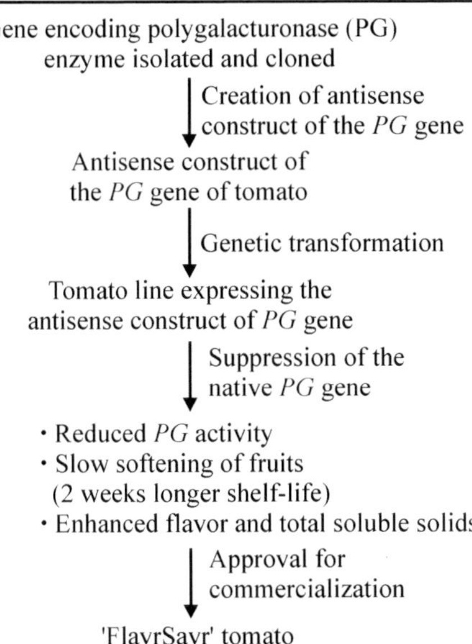

Fig. 11.4. A schematic representation of the development of FlavrSavr tomato with slow fruit softening.

11.4. MODIFICATION OF OIL QUALITY

The vegetable oils are triglycerides. In a **triglyceride** molecule, three fatty acid residues are linked to one glycerol molecule (Fig. 11.5). Vegetable oils are an essential component of human diet; a healthy person requires 30 g/day of oils and fats. They are rich sources of energy and the essential fatty acids, precursors of certain vitamins and hormones, and components of cellular structures. An **essential fatty acid** is that fatty acid, which cannot be synthesized by humans. Linoleic and linolenic acids are the essential fatty acids. In addition, oils and fats are used in many industrial products, including cosmetics, soaps, detergents, lubricants, etc. The suitability of a given oil for various purposes depends on the properties of the oil. The properties of oils and fats are determined by the type of the fatty acids and their relative proportions present in the oil. Therefore, the types of fatty acids present in the given oil and the proportions of these fatty acids in the oil are known as **oil quality**.

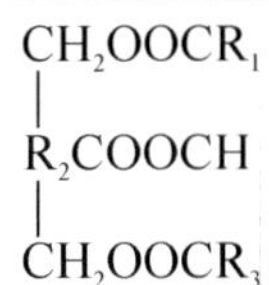

Fig. 11.5. A generalized representation of triglyceride (oil/lipid) molecule. R, R_2 and R_3, side chains of the fatty acids (representated by-OOCR).

The properties of fatty acids, on the other hand, depend on the length of their carbon chains and the number of double bonds in this chain. A fatty acid without a double bond is called **saturated fatty acid**. But the fatty acids having one, two or three double bonds are called **mono-, di-** or **tri-unsaturated fatty acids**. The fatty acids having 16 and 18 carbons in their structure are the predominant fatty acids in plants animals. In plants, fatty acids are synthesized in plastids, and acetyl-CoA is used as the building block for this purpose. The saturated fatty acids are synthesized by sequentially adding two carbons to the chain at a time, and the different saturated fatty acids are produced in this way. The unsaturated fatty acids are then produced by desaturation of the saturated fatty acids.

The oil quality is modified by genetic modifications in one or more of the enzymes of the fatty acid biosynthesis pathway. These changes can be brought about by either breeding or genetic engineering. Table 11.1 lists some of the enzymes of fatty acid biosynthesis that are possible targets for genetic modification by the transgenic technology. The various approaches for these modifications are as follows: (i) introduction of a new enzyme, (ii) suppression of an endogenous enzyme, (iii) site-directed mutagenesis to modify an enzyme, and (iv) synthesis of a hybrid gene for generation of a new enzyme activity.

Table 11.1: A list of some of the enzymes of fatty acid biosynthesis that are potential targets for genetic modification aimed at improvement in the oil quality

Enzyme	Role in fatty acid biosynthesis	Remarks
Acyl carrier protein (ACP)	Formation of thioester with acyl groups essential for fatty acid biosynthesis	The carbon chain length of fatty acids is regulated by this protein
β-Ketoacyl-CoA:ACP transacylase	Conversion of acetyl-CoA into acetyl-ACP	The chain length of fatty acids decreases as the activity if this enzyme increases
β-Ketoacyl -CoA:ACP synthase I	Condensation of acyl-ACP and malonyl-CoA	Participates in synthesis of fatty acids of up to 16 carbon chain length
Acetyl-CoA:ACP synthase II	Synthesis of stearic acid from palmitic acid by condensation reaction	The level of palmitic acid increases with an increase in the activity of this enzyme
Stearoyl-ACP desaturase	Desaturation of stearic acid to produce oleic acid	The level of oleic acid increases with an increase in the activity of this enzyme

Lauric acid (12:0; fatty acid chain of 12 carbon atoms long with zero double bond) is an industrially important fatty acid. It is abundant in coconut oil, but is not found in *Brassica napus* oil. But the seed oil of wild plant species California bay (*Umbellularia californica*) is rich in lauric acid. The gene encoding the enzyme lauroyl-ACP thioesterase from California bay was isolated and transferred into *B. napus*. Seed oils of some of the transgenic lines expressing the

California bay lauroyl-ACP thioesterase encoding gene contained up to 60% lauric acid. One of these transgenic lines has been approved for commercialization; this line has been named as Laurical (Table 11.2). Similarly, a transgenic variety of soybean with modified oil quality has been approved for commercialization and for use in human food and animal feed. This soybean variety has enhanced oleic acid content as compared to the nontransgenic control.

Table 11.2: Selected exampled of transgenic varieties of crops with modified quality or other novel traits that have been approved for commercial cultivation

Transgenic variety	Modified trait	Developed by	Year of approval
Laurical (canola)	Fatty acid composition of oil (high lauric acid)	Monsanto	1995
SeedLink (chicory)	Male sterility	Bejo Zaden	1997
Soybean; variety name not known	Fatty acid composition of oil (high oleic acid)	Du Pont	1997
Endless summer tomato	Altered ripening	DNA Plant Technology	1995
FlavrSavr tomato	Altered ripening	Monsanto/Calgene	1994
Innate™ potato	Reduced acrylamide production, modified starch quality, black spot bruise tolerance	J.R. Simplot Company	2014
HarvXtra™ alfalfa	Reduced lignin production	-	2014
Amflora™ potato	Reduced amylose and increased amylopectin	Developed by BASF	2010
Arctic™ Golden delicious apple	Non-browning phenotype	Okanagan Specialty Fruits, Inc.	2015
Vistive Gold™ soybean	Glyphosate tolerance, modified oil quality	-	2011
Argentine Canola	Herbicide tolerance, pollination control	Bayer	2012
Agrisure® Viptera™ 3111, Agrisure® Viptera™ 4 (maize)	Glufosinate tolerance, glyphosate tolerance, coleopteran insect resistance, lepidopteran insect resistance, mannose metabolism	Syngenta	2010

11.5. MODIFICATION OF PROTEIN QUALITY

The seed storage proteins of plants are primarily used as sources of nitrogen and sulphur during seed germination. They are also the chief sources of dietary proteins for vegetarian humans. In developing countries, these proteins are major sources of proteins of all humans. The storage proteins are encoded by a family of genes that may have merely four to around 100 members. The seed storage proteins are generally classified into the following groups: (i) water soluble albumins, (ii) globulins (soluble in dilute salt solution), (iii) alcohol-water mixture soluble prolamins, and (iv) dilute acid or alkali soluble glutelins (Table 11.3). **Globulins** are the major fraction of seed storage proteins in most legumes. These proteins are deficient in sulphur containing amino acids, but have adequate amounts of lysine. **Prolamins**, on the other hand, are the chief seed storage proteins in cereals (except in oats and rice), and occur only in cereal endosperms. These proteins are deficient in lysine. **Glutelin** is the major storage protein in rice, and is similar to prolamins in nutritional quality. **Albumins** are found in most dicot species, have good nutritional quality, and some of them, e.g., 2S albumins like SFA8 from sunflower, are rich in methionine.

Table 11.3: The important features of different groups of seed storage proteins

Seed storage protein	Soluble in	Occurrence	Deficient in	Remarks
Albunims*	Water	Found in most dicot species	Good nutritional quality	2S albumins, e.g., SFA8 from sunflower, are rich in methionine
Globulins**	Dilute salt solution	Major fraction in most legumes	Sulphur containing amino acids	
Prolamins	Alcohol-water mixture	Major fraction in cereals, except oats and rice	Lysine	Confined in cereal endosperm
Glutelins	Dilute acid or alkali	Cereals; major fraction in rice	Lysine	Similar to prolamins

* 2S albumins are rich in methionine, and have been used for improvement in protein quality
** 7S and 11S globulins are the major globulin groups; they are deficient in sulphur containing amino acids

The proteins are composed of 20 different amino acids; the proportions of these amino acids differ for different proteins. Humans can synthesize 12 of these amino acids from other amino acids. But they are unable to synthesize eight amino acids, viz., leucine (Leu), isoleucine (Ile), methionine (Met), phenyl alanine (Phe), threonine (Thr), tryptophan (Trp) and valine (Val), from other amino acids. Therefore, these eight amino acids are known as **essential amino acids**. These amino acids must be present in human diet in the appropriate amounts. The term **protein quality**, therefore, refers to the proportions of the eight essential amino acids in the given protein as compared to the proportions recommended by the World Health Organization (Table 11.4). The proteins having amino acid profile comparable to that of the WHO recommendation are said to have good nutritional quality or balanced amino acid composition. It may be reiterated that cereal seed proteins are deficient in Lys, Trp and Thr, while legume seed protein are deficient in sulphur containing amino acids and Trp. Therefore, neither the cereal nor the legume seed proteins have good nutritional quality.

Table 11.4: The amounts (as mol percent) of the eight essential amino acids in dietary proteins as recommended by the WHO

Amino acid	Amount (as mol percent)	Remarks
Ile	4.0	-
Leu	7.0	-
Lys	5.5	Deficient in cereals
Met	3.5	Deficient in pulses
Phe	6.0	-
Thr	4.0	Deficient in cereals
Trp	1.0	Deficient in cereals and pulses
Val	5.0	-

The nutritional quality of seed storage proteins can be improved by the following different strategies.

1. A transgene encoding a protein with the desired amino acid composition can be transferred into the concerned crop and expressed in the seed tissues. For example, the *SFA8* gene of sunflower encodes a protein that has 23% sulphur containing amino acids. This gene has been expressed in alfalfa and tall fescue to improve the forage protein quality. Since the protein SFA8 is suspected to be allergenic, it is not a good choice for improving the food protein quality. The gene *amA1*, on the other hand, is rich in sulphur containing amino acids and safe for humans. This gene has been expressed in seven commercial cultivars of potato leading to improved amino acid balance of the tuber proteins. In addition, tubers of the transgenic potato lines showed up to 60% increase in protein content and ~15-25% higher yields.
2. A synthetic gene may be designed to code for a protein that has the specified amino acid composition. This gene can be transferred into the concerned crop and expressed in the seeds. For example, the gene *CP3-5* synthesized by DuPont, USA encodes the protein CP3-5 that has 35% Lys and 22% Met. A sweet potato transgenic line has been developed using a synthetic transgene. The protein encoded by this transgene has high contents of the essential amino acids like Lys, Trp, sulphur containing amino acids, etc.
3. Another approach involves the modification of an endogenous gene. The base sequence of this endogenous gene is suitably modified and the modified gene is reintroduced into the crop where it would express in the seeds (Fig. 11.6). The rice gene *gt1* encodes glutelin, which is the major fraction of seed storage proteins in rice. Glutelins are deficient in Lys. The sequence of gene *gt1* was modified so that it now encodes a protein

rich in Lys. This modified gene has been transferred and expressed in rice seed endosperms.

4. An endogenous or native gene encoding a protein with good nutritional quality can be overexpressed to increase the proportion of this protein in the seed. One way to achieve this is to introduce additional copies of this gene driven by a strong seed-specific promoter. Another approach is to increase the stability of the mRNA encoded by this gene. For example, maize seed storage 10 kDa Zein is rich in Met. The expression level of this protein was increased by enhancing the stability of its mRNA. This modification resulted in increased levels of Met contents of the transgenic maize lines. Similarly, attempts to increase the wheat dough quality include over-expression of the gene encoding the high molecular weight glutelin subunit.

   The four strategies listed above will express such a gene in the seeds that will encode a seed protein having the specified amino acid composition. This will improve the average nutritional quality of the proteins obtained from these seeds.

5. Alternatively, the expression of the gene encoding protein with less desirable amino acid composition may be suppressed using RNAi or antisense RNA approach. This will allow increased expression of the genes encoding proteins with better amino acid balance. As a result, the nutritional quality of the seed proteins will be improved. The improved seed protein quality in the *opaque 2* mutants of maize results precisely from this phenomenon. This mutation reduces the accumulation of zein fraction of seed proteins, which is deficient in Lys and Trp. In case of *Brassica* spp., Cruciferin is the major seed protein; this protein is deficient in sulphur containing amino acids. The production of Cruciferrin was suppressed using antisense RNA technology. This led to increased accumulation of Napin, which has excellent amino acid balance. The seed proteins of the transgenic lines had 10% higher Lys, 8% more Met and 32% greater Cys than the nontransgenic controls.

11.6. MODIFICATION OF STARCH QUALITY

Starch molecules are polysaccharides produced by polymerization of glucose residues. Starch is stored in roots, seeds and modified shoots, and is the major storage polysaccharide. It serves as the chief source of energy for various biological processes. The starch grains are composed of two types of polysaccharides, viz., amylose and amylopectin. **Amylose,** in general, is a linear molecule with slight branching, and its chains form left-handed helices. It constitutes the core of starch grains and accounts for 30-35% of their mass. **Amylopectin**, on the other hand, is a highly branched polymer of glucose. It is the main constituent of starch grains and determines their basic. The properties of starch molecules depend primarily on the following factors: (i) the chain length of the molecule, (ii) the types of bonds between the glucose units constituting the chain, and (iii) the degree of branching of the chains and (iv) the lengths of the branches.

Starch is the main component in the foods of bulk of the human population. In addition, it is used in food, beverage, paper, textile, etc. industries. Different applications require starch with distinct and specific qualities. For example, high amylose starch is preferred for industrial food production; it is used as thickener and for production of snacks. The high amylopectin starch, on the other hand, is used in the industry for a variety of purposes, e.g., in food industry to improve uniformity, texture and stability, to enhance freeze-thaw stability of frozen foods, in paper industry, for production of gum, etc. In addition, foamed starch is an excellent packing material. The starch quality can be modified by the following methods: (i) genetic changes in the crops by plant breeding, (ii) genetic modification by genetic engineering, (iii) use of enzymes and chemicals during food processing, and (iv) use of microorganisms to bring about the desired changes in starch.

In this section, the genetic modification of starch yield and quality is briefly discussed. All the genes involved in starch biosynthesis have been identified, isolated and cloned. These genes have been evaluated for their usefulness in modification of starch yield and quality (Table 11.5). The enzyme AGPase (adenosine diphosphate-glucose pyrophosphorylase) is the first enzyme in starch biosynthesis. It converts glucose-1-phosphate to ADP-glucose that is used as substrate for starch biosynthesis. This enzyme is regarded as the rate-limiting enzyme in the starch biosynthesis pathway. Overexpression of this enzyme resulted in an increased starch yield in one study but not in another. GBSS I (granule bound starch synthase I) is the main enzyme responsible for the biosynthesis of amylose. The introduction and expression of the antisense construct of the gene *GBSS I* resulted in starch grains that were amylose-free and waxy. The enzymes soluble starch synthase I and II are the main enzymes of amylopectin biosynthesis. When these two enzymes were suppressed by antisense RNA technology, the chain length of the amylopectin was significantly reduced.

The branching of starch molecules in potato is catalyzed by the enzymes starch branching enzymes A and B. When antisense constructs of the genes encoding these enzymes were expressed in potato, the tubers lacked high molecular weight amylopectin. At the same time, the accumulation of amylose in the starch grains was greatly increased. The starch grains of this transgenic potato line were comparable to those of maize. In a transgenic line of potato, the enzymes GBSS, SSS I and SSS II were simultaneously suppressed by antisense RNA technology. The starch of this transgenic line was free from amylose and comprised short chain amylopectin. This starch was extremely freeze-thaw stable. This novel starch type has many applications, and is eco-friendly since its utilization does not require chemical modification.

Table 11.5: A list of the genes involved in starch biosynthesis that have been cloned and used for genetic modification of starch yield and quality in potato

Gene	Effect of transgene expression
AGPase (adenosine diphosphate-glucose pyrophosphorylase)	Increased starch content in one study but not in another
Anti-AGPase	Negligible starch accumulation in the tubers
Anti-GBSS I (anti-granule bound starch synthase I)	Amylose fraction reduced to almost zero; the starch grains are typically waxy
Anti-SSS I and II (anti-soluble starch synthase I and II)	The chain length of the amylopectin is significantly reduced
Anti-SBE A and B (starch branching enzyme A and B)	High molecular weight amylopectin reduced to zero; starch has high amylose content
Anti-GBSS I, anti-SSS I and anti-SSS II	Starch with increased freeze-thaw stability

11.7. GOLDEN RICE

The deficiency of vitamin A is estimated to be the primary cause of blindness in about 50 lakh children every year. Additionally, it leads to the death of about 10 lakh children annually. The vitamin A deficiency is known to be frequent in areas of the developing countries where rice is the staple or chief food. This is because the rice grain is deficient in β-carotene, which is often referred to as provitamin A. It was found that the rice endosperm contains the compound geranyl-geranyl pyrophosphate, which is an intermediate in the biosynthesis of β-carotene. This compound is converted into β-carotene by the following four enzyme activities: (i) phytoene synthase, (ii) phytoene desaturase, (iii) ζ-carotene desaturase, and (iv) lycopene cyclase. Initially, these four enzyme activities were provided by three transgenes transferred into rice genome and expressed in the rice endosperm.

The gene *psy* from daffodil (*Narcissus* sp.) encoded the enzyme phytoene synthase, which catalyzes the condensation of geranyl-geranyl pyrophosphate already present in the rice endosperm. A bacterial gene, viz., the gene *crtI* from *Erwinia uredovora*, encodes the enzyme phytoene desaturase. This enzyme provides both phytoene desaturase and ζ-carotene desaturase activities. The third gene, viz., the gene *lyc* from *E. uredovora*, encoded the enzyme that provided the lycopene cyclase activity (Fig. 11.6). The resulting transgenic lines accumulated β-carotene in the endosperm so that these rice grains had light golden color, giving these lines the popular name **golden Rice**. The grains from these lines accumulated up to 1.6 μg β-carotene per gram of rice. This level of β-carotene in the rice grain is not adequate for combating vitamin A deficiency since it would require the consumption of impractical amounts of rice per day. Therefore, efforts were made to create golden rice lines with higher levels of β-carotene in their grains.

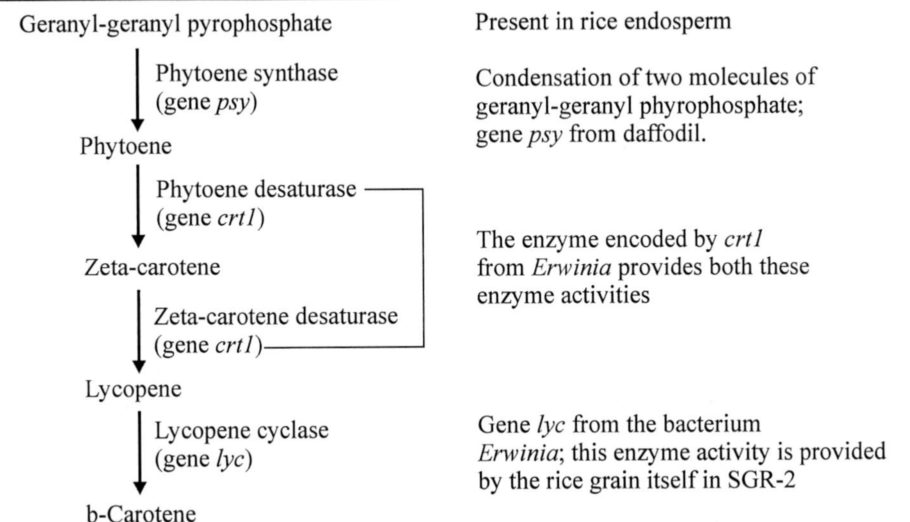

Fig. 11.6. The enzyme activities provided by the three transgenes used to create the initial lines of 'Golden Rice'. In the 'Golden Rice' SGR-2, the *psy* gene is derived from maize. The genes *psy* and *crt1* are driven by the *Glu1* promoter, which ensures endosperm specific expression. The lycopene cyclase activity is provided by the rice genome itself. This enzyme activity is present in the grains of the transgenic lines expressing the genes *psy* and *crt1*.

Efforts were, therefore, made to increase the level of β-carotene in the grains of golden rice. This resulted in the creation of two transgenic events, SGR1 and SGR2, by Syngenta. Inc. These lines had 8 μg (in the case of SGR1) and 37 μg (in the case of SGR2) of β-carotene per gram of rice. In the line SGR2, the gene *psy* is derived from maize in the place of daffodil. In addition, the genes *psy* and *crt1* (from *Erwinia* sp.) are driven by the promoter *Glu1*, which ensures its high expression in the endosperm tissues. Finally, this line does not contain the transgene *lyc*; this is because this enzyme activity is provided by the rice gene itself, which is activated in the lines expressing the genes *psy* and *crt1*. The lines SGR1 and SGR2 were donated to the Golden Rice Humanitarian Board for use in the development of suitable golden rice lines that would be commercialized. It may be pointed out that the golden rice is still to be approved for commercial cultivation and use as human food.

11.8. MALE STERILITY

Hybrid varieties are the F_1 generations from crosses between two or more lines. Therefore, the seeds of hybrid varieties have to be replaced every year by the fresh F_1 seed produced by crossing their parental lines. The production of hybrid seed is greatly facilitated by the use of male sterile lines as the female parents (Chapter 5, Section 5.4.5). The cytoplasmic-genetic system of male sterility has been the most successful strategy for hybrid seed production. But in many crops, suitable systems of male sterility are not available. Efforts have been made to develop transgenic male sterility systems in such crops where a satisfactory pollination control system is not available. Several different approaches have been evaluated for this purpose.

The first usable system of male sterility was produced by using two genes, viz., *barnase* and *barstar*, from the bacterium *Bacillus amyloliquefaciens*. The gene *barnase* encodes an RNase that degrades all the RNA present in the cell in which this gene is expressed. This blocks the protein synthesis in the cell as a result of which it dies. The gene *barnase* was linked with the tapetum-specific promoter *TA29* and transferred and expressed in tobacco. The anthers of this transgenic line did not produce pollen since the tapetum cells died due to the action of the protein Barnase. Thus, this transgenic line is male sterile and has to be maintained by crossing with a nontransgenic male fertile line.

The *barnase* male sterile plants will always be heterozygous. Therefore, 50% of the progeny of these male sterile plants produced after crossing them with the male fertile plants will be male sterile, and the remaining 50% will be male fertile. These male fertile plants will have to be removed from the field before they shed pollen. This is facilitated by linking the gene *bar* for phosphinothricin resistance with the *barnase* gene. In such a transgenic line, the male fertile plants can be easily eliminated by spraying phosphinothricin (Fig. 11.6). The fertility of this line is restored by crossing it with another transgenic line expressing the *barstar* gene. The Barstar protein is a specific inhibitor of the Barnase RNase. As a result, the activity of the Barnase RNase is inhibited in the tapetum cells of the hybrid plants. These tapetum cells function normally, and the hybrid plants are male fertile (Fig. 11.7). This system of male sterility has been used in *Brassica* spp. and some other crops.

A more convenient and attractive system of male sterility is based on the gene *bcp1*. The expression of this gene in the tapetum as well as the developing microspores is essential for normal development of the pollen grains. The expression of this gene was suppressed by the antisense RNA technology. The antisense construct of *bcp1* (*anti-bcp1*) was linked with a hormone inducible pollen-specific promoter. This *anti-bcp1* gene construct was transferred into cabbage (*Brassica oleracea*). The resulting transgenic cabbage line was male sterile when the plants were sprayed with the concerned hormone (Fig. 11.8A). However, the plants were all male fertile if they were not treated with the concerned hormone (Fig. 11.8A). Therefore, the multiplication of this male sterile line is really simple: the plants are grown without hormone spray and the seed obtained by selfing is harvested (Fig. 11.8B).

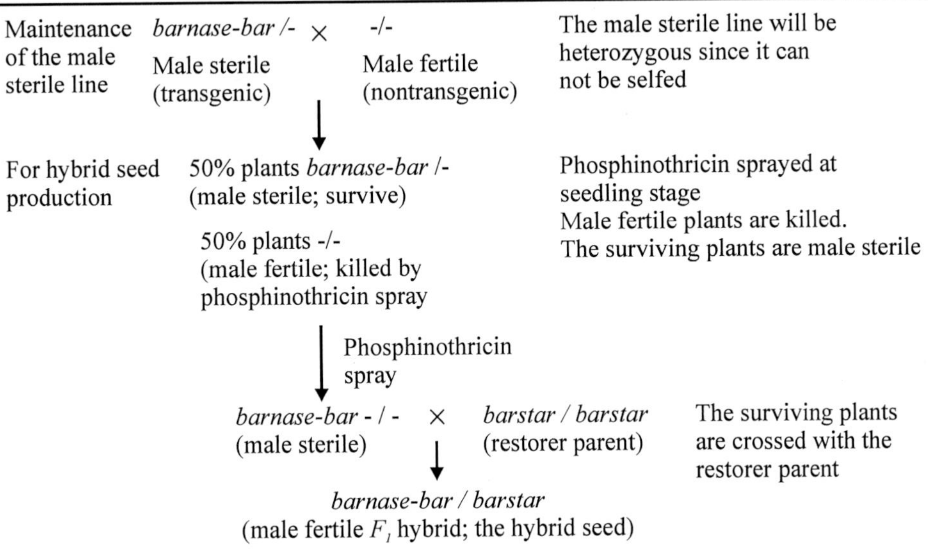

Fig. 11.7. The *barnase - barstar* system of transgenic male sterility.

When hybrid seed is to be produced, the plants of the male sterile line are sprayed with the concerned hormone (Fig. 11.8B). The *anti-bcp1* construct will now be expressed and all the plants in the line will become male sterile. The seed set on these plants will be produced due to pollination by the male fertile plants of the male parent of the hybrid variety. Thus, the seed produced on the plants of the male sterile line will be hybrid seed. These hybrid plants will be all male fertile as long as they are not treated with the concerned hormone. The chief advantages of this male sterility system are as follows. (i) The male sterile line is maintained simply by growing it without spraying it with the concerned hormone and allowing it to self-pollinate. (ii) All the plants of the male sterile line become male sterile when they are sprayed with the concerned hormone. Finally, (iii) any male fertile normal (not having the *anti-bcp1* gene construct) line can be used as the male parent of the hybrid. This system is being used in cabbage and some other crop species.

11.9. EDIBLE VACCINES

A **vaccine** is a preparation from a viral or a bacterial pathogen that is introduced into humans and animals to make them to immune to infection by the same pathogen at a later point of time. In general, the **conventional** or **traditional vaccines** are produced by either inactivation (killing) or attenuation (greatly reduced ability to produce disease) of the pathogens. But occasionally, the vaccine may contain few active/infective pathogens, which may cause the disease in the place of protecting from the disease. It is well recognized that the vaccine function is due to the presence of some critical antigenic proteins of pathogen origin. The genes encoding these antigens have been isolated from numerous pathogens and expressed in bacteria/ yeast/animal cells. The recombinant protein produced by the transgenic bacteria/ yeast/animal cells are purified and used to prepare vaccines; these vaccines are called **recombinant vaccines**. The recombinant vaccines are produced by growing the transgenic bacteria/yeast in fermenters or the transgenic animal cells in suitable large-scale culture systems. These production systems make the vaccines very costly. In addition, the vaccine preparations have to be kept at low temperatures to retain their effectiveness. This makes their storage as well as transport challenging, especially in the developing countries, and further increases the cost of vaccination.

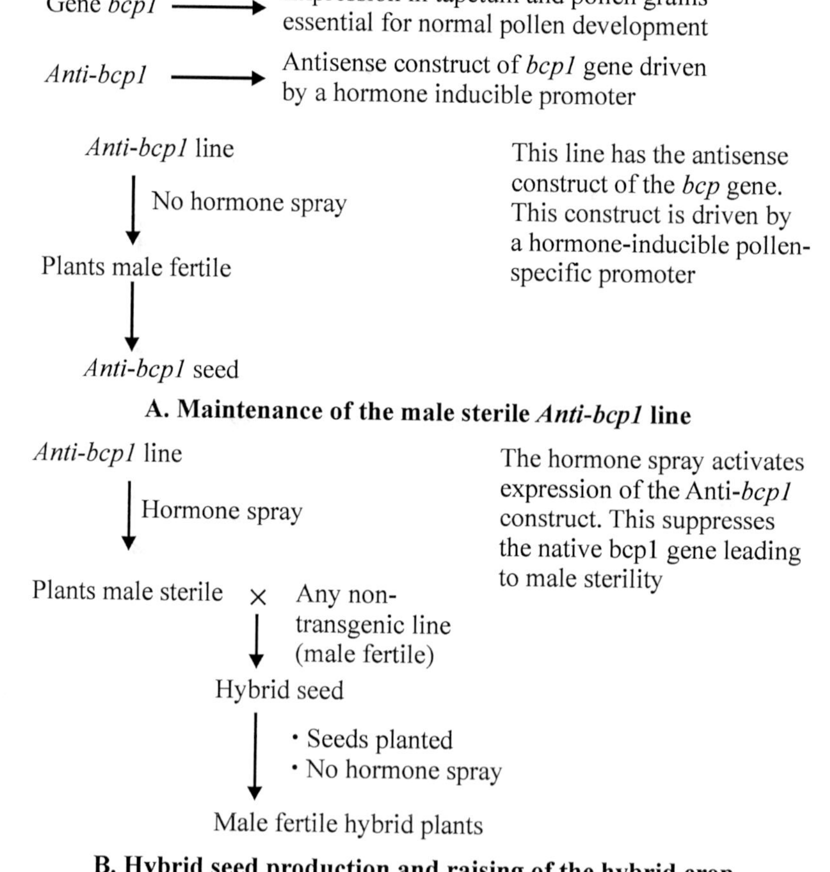

Fig. 11.8. The hormone-inducible system of male sterility

Therefore, transgenic plants are being developed as alternative systems for the production of antigenic proteins to be used for vaccine preparation. The recombinant proteins produced by such plants can be isolated, purified and used for the preparation of vaccines; these vaccines will be delivered to the subjects by the conventional approaches for vaccination. Alternatively, the produce from these plants may be directly consumed by the subjects leading to their vaccination. The plant parts consumed for vaccination are popularly termed as edible vaccines. Thus, an **edible vaccine** is the edible part of a transgenic plant that expresses the critical antigen from a pathogen. Therefore, this edible part contains adequate quantities of this antigen to cause vaccination of the individuals who consume this part as per the specified procedure. The edible vaccines are relatively inexpensive to produce, easy to store and painless when delivered to the subjects. It is proposed to develop edible vaccines for the animals in the first stage. The development of vaccines for humans will be taken up once the principles and the methods of vaccination using edible vaccines become well established.

Antigen encoding genes from several pathogens have been transferred and expressed in plants. For example, the *E. coli* gene encoding heat labile enterotoxin B (LT-B) has been expressed in potato. The LT-B is very similar to cholera toxin (CT) in terms of both function and immunogenicity. The LT-B produced in potato tubers was isolated, purified and fed to mice. The purified preparation generated immune response in the mice that was comparable to that produced by the LT-B of bacterial origin. In addition, feeding of 5g potato tubers per day per mouse for four days also generated immune response. This confirmed that the tubers from this transgenic potato could be used as edible vaccines.

Hepatitis B recombinant vaccine is prepared by expressing the gene encoding hepatitis B surface antigen (HBsAg) in yeast. This gene has been transferred and expressed in tobacco and potato. The tobacco transgenic line is reported to accumulate HBsAg up to 0.01% of the total leaf protein. The HBsAg is immunogenic only when it is present as virus like particles of about 22 nm size. It is important that the tobacco derived HBsAg was organized as 22 nm particles. In addition, this protein was comparable to the yeast-derived and human-derived HBsAg both in physical properties as well as immunogenic action.

The transgenic potato lines expressing the LT-B, hepatitis B surface antigen (HBsAg) and Norwalk virus capsid protein (NVCP) in their tubers are being evaluated as edible vaccines in human clinical trials. The results from these trials have been encouraging, and we might expect some edible vaccines to become commercially available in the near future.

11.10. BIOCHEMICAL PRODUCTION

Transgenic plants have been created with the specific objective of producing valuable pharmaceutical proteins and research reagents. The production of valuable biochemicals from transgenic plants is often called **molecular farming** or **biopharming**. The first commercial example of a pharmaceutical biochemical obtained through biopharming is the antithrombin protein Hirudin. The amino acid sequence of Hirudin protein was used to synthesize the gene encoding this protein. This gene was fused with the gene encoding the oil-body protein Olesin. The fusion gene was expressed in *Brassica napus* seeds, where it accumulated with the oil-bodies. This greatly facilitated the separation of Hirudin protein from *B. napus* seeds, and its purification. When the seeds are ground and extracted with water, the Hirudin-Olesin fusion protein floats with the oil bodies on the water surface. The fusion protein is easily separated and purified. Finally, the Olesin is removed by protease catalyzed cleavage, and Hirudin is purified.

Transgenic plants can be used to produce proteins and enzymes that serve as valuable research reagents. There are at least two commercial examples of research reagents produced from transgenic plants. The protein Avidin has been normally obtained from the chicken egg. The gene encoding Avidin was isolated and expressed in maize seeds. The transgenic maize line accumulated Avidin up to 3% of the seed dry mass. The recombinant Avidin is purified and marketed by Sigma, USA. A protein or biochemical produced from a transgenic line created for this purpose is called **recombinant protein** or **recombinant biochemical.** Sigma also markets another recombinant research reagent, namely, β-galacturonidase. This enzyme is traditionally obtained from bacteria. But in this case, the gene encoding the enzyme is expressed in maize seeds. The recombinant enzyme is isolated from seeds, purified and marketed for research applications.

Another approach for producing novel biochemicals from transgenic plants consists of modification of the biosynthetic pathways endogenous to the plants. In this approach, one or more transgenes are introduced into the plants that modify an endogenous pathway in such a way that it now produces a new end product. Alternatively, the introduced transgenes may generate a new branch in a pathway or give rise to a new pathway to produce a novel biochemical. The following two examples will help understand the basic strategy of this approach. Cyclodextrins are cyclic oligosaccharides, i.e., polysaccharides with a small number of subunits. They are used for a variety of purposes, including drug delivery, flavor enhancement, removal of undesirable molecules by trapping, etc. The cyclodextrins are not produced by plants. They are produced from starch by a process based on physical and enzymatic procedures, which is quite costly. A transgenic line of potato was produced by transferring into it a bacterial gene that encodes the enzyme cyclodextrin glycosyltransferase (CGT). This gene was expressed in the potato tubers. The tubers of this potato line accumulated cyclodextrin up to the level of 0.01% of the starch present in them.

The second example pertains to the production of polyhydroxybutyrate (PHB) in plants. PHB is obtained from bacteria and is used for the production of biodegradable plastic. Three genes, viz., *phbA*, *phbB* and *phbC*, concerned with the biosynthesis of PHB from the precursor acetyl-CoA were transferred and expressed in *Arabidopsis thaliana*. The transgenic plants did produce PHB at low levels (0.14% of dry weight). When the enzymes encoded by these genes were targeted into plastids, there was 100-fold increase in the level of PHB accumulation. Further refinements in the organization of the gene construct and the production of transgenic lines has increased the PHB yield to 7% of dry weight. These two biochemical production systems are yet to be commercialized.

11.11. CROPS FOR BIOFUEL PRODUCTION

Biofuels are high-energy materials produced from biomass, and include biodiesel and bioethanol. At present biodiesel is produced from vegetable oils, such as palm oil, soybean oil and Brassica oil. The bioethanol, on the other hand, is produced from fermentation of sugar (almost exclusively from sugarcane) and starch (primarily from maize). However, all these crops are traditional food crops. As a result, the diversion of their produce for biofuel production will adversely affect food supply and cause increase in food prices. Therefore, efforts are being made to develop nonfood crops suitable for biofuel production. This is proposed to be achieved by domestication and genetic improvement of wild species with promising attributes. The species that hold promise as biofuel crops include grasses like *Panicum* sp. and *Miscanthus* sp., and plant species producing oil, e.g., jatropha (*Jatropha curcas*), castor, etc. The plant species being domesticated for biofuel production are called **second generation energy crops**. In contrast, the food crops being used for biofuel production are known as **first generation energy crops**.

The second generation energy crops have several features that make them harmful for human health and/or the environment. Therefore, these crops must be improved by breeding and/or genetic engineering to make them safer to the environment and, more particularly, human health. Some examples of the traits of these crops that must be improved for their efficient use for biofuel production are given below.

1. Jatropha is projected as the most promising oil producing plant species suitable for biofuel production. But the presence of toxins and carcinogenic substances in its seeds are serious drawbacks associated with this crop. There have been newspaper reports of deaths due to inadvertent consumption of jatropha fruits by children. In addition, the jatropha fruits shatter on maturity. Genetic engineering can be used to suppress fruit shattering by transfer and expression of the Arabidopsis *SHATTERPROOF* gene. Similarly, antisense RNA technology or RNAi can be used to suppress the production to toxic and carcinogenic compounds making the plant safe for humans and animals.
2. In case of castor, the seeds contain several toxic proteins, including ricin. Ricin is highly toxic and is lethal to humans at a dose as low as 1 mg per

kg body weight. Plant breeding has been very successful in developing castor varieties having low contents of the toxic proteins. But the seed cake of these varieties still contains enough toxins to be unsuitable as cattle feed. In addition, the castor pollen has a variety of allergens that pose risks to human health. The production of toxic proteins as well as allergenic proteins can be suppressed by a suitable transgenic technology, such as antisense RNA technology and RNAi. This would make the castor seeds and pollen grains safe for humans and animals, and encourage the use of this plant as a biofuel crop.

3. *Brassica* spp. is known to liberate large amounts of the greenhouse gas methyl bromide. As a result, large-scale cultivation of these crops for biodiesel production is not eco-friendly. It is, therefore, important that the production of methyl bromide by the *Brassica* spp. should be reduced to minimize environmental concerns. The *Arabidopsis* gene *HOL* is involved in the biosynthesis of methyl bromide. Suppression of this gene in *Arabidopsis* reduced the methyl bromide production to less than one percent of that in the wild type. Therefore, the suppression of the orthologue of the *Arabidopsis* gene *HOL* may be expected to minimize methyl bromide production in *Brassica* spp. and make them eco-friendly biofuel crops.
4. The chain length of fatty acids in most vegetable oils is 16 or more carbons, which is not suitable for biodiesel production. The chain length of fatty acids should be around 14 carbons. This objective can be achieved by suppressing the elongation of C_{14} fatty acids to C_{16} fatty acids. The antisense RNA and the RNAi technologies can be used to suppress the elongases responsible for the addition of acetyl-CoA to C_{14} fatty acids. In addition, the desaturation of the fatty acids can be increased by expression of the appropriate fatty acid desaturases. These genetic modifications will greatly facilitate the use of vegetable oils for the production of biodiesel.
5. In case of energy crops like sugarcane, the lignocellulosic waste is generally burnt to produce energy. For example, the sugarcane bagasse is burnt in Brazil to produce energy needed for sugar processing. The ash resulting from bagasse is rich in silica, which is highly polluting, causes respiratory problems, and is difficult to manage. Silica is not essential for plant growth. Therefore, it would be desirable to develop transgenic lines with low silica content. The use of such transgenic lines for biofuel production would be safer to both human health and the environment.

11.12. CONCLUSIONS

The phenotype of a plant can be modified by either suppressing the function of an endogenous or native gene or by expressing one or more new transgenes. An endogenous gene can be suppressed by (i) overexpression of the given gene (this leads to cosuppression), (ii) the use of antisense RNA technology, (iii) the deployment of ribozyme technique, (iv) implementing the RNA interference (RNAi) technology, and (v) disruption of the concerned gene. The endogenous gene suppression approaches have been used to develop slow softening tomato; modification of oil, protein and starch qualities, production of male sterile lines, etc. Oil quality refers to the fatty acid composition of the oils. Modification of oil quality may use one of the following strategies concerned with fatty acid biosynthesis: (i) introduction of a new enzyme, (ii) suppression of an endogenous enzyme, (iii) site-directed mutagenesis to modify an enzyme, and (iv) synthesis of a hybrid gene for generation of new enzyme activity.

Humans are unable to synthesize eight amino acids, viz., Leu, Ile, Met, Phe, Thr, Trp and Val; these are termed as essential amino acids. Protein quality refers to the contents of the essential amino acids in relation to their recommended levels in the dietary proteins. The various approaches for modification of protein quality are as follows: (i) expression of a transgene encoding a protein with the desired amino acid composition, (ii) designing of a synthetic gene encoding a protein with the desired amino acid composition and expressing it in the seeds of crops, (iii) modification of the base sequence of an endogenous gene to enhance the amino acid balance of the protein encoded by it and expression of this gene in the concerned crop, (iv) overexpression of a native gene encoding the protein with the desired amino acid composition, and (v) suppression of the endogenous gene that produces a protein with unbalanced amino acid composition.

The starch quality is primarily dependent on the proportions of amylose and amylopectin present in the starch grains, and the lengths the chains and the extent on branching of the chains. Endogenous gene suppression has been used to produce potato lines producing typically waxy starch with negligible amylose, starch with high amylose, and starch with very high freeze-thaw stability. Expression *psy* and *crt1* transgenes has created the Golden Rice that accumulates high levels of pro-vitamin A in its endosperm. On the other hand, expression of transgenes selected for destroying the tapetum cells has given rise to transgenic male sterility. Similarly, suppression of a gene essential for pollen maturation has produced male sterile lines. This was achieved by the use of antisense RNA technology. The antisense gene construct is driven by a hormone inducible promoter, which makes the use of this male sterility system very convenient.

Transgenic plants are being created for the production of useful biochemicals; this is called molecular farming. The biochemical is generally the protein product of the transgene. But this could also be the product of a plant biosynthetic pathway that has been modified by the enzymes encoded by the transgenes. Plants expressing the antigenic protein of pathogen origin have been developed in an effort to produce vaccines in the transgenic plants. The recombinant antigen proteins may be isolated, purified and used for preparation of vaccines (plant derived vaccines). Alternatively, the plant parts containing the antigen may be consumed directly to effect vaccination (edible vaccines). It is proposed to use genetic engineering to develop safer and more efficient energy crops for the production of biofuels.

PRACTICE QUESTIONS

1. Briefly describe the various techniques for endogenous gene suppression. Briefly explain the usefulness of endogenous gene suppression.
2. Explain the meaning and relevance of oil quality. Discuss the application of transgenic technology for the modification or oil quality.
3. Explain the meaning and relevance of protein quality. Discuss the application of transgenic technology for the modification or protein quality.
4. Explain the meaning and relevance of starch quality. Discuss the application of transgenic technology for the modification or starch quality.
5. Transgenic plants are excellent systems for biochemical production. Discuss this statement in the light of available information.
6. Define edible vaccine. Discuss the use of transgenic plants as edible vaccines.
7. Briefly describe some approaches for the production of transgenic male sterile lines. Discuss the significance of these male sterile lines in crop improvement.
8. Write short notes on the following: (i) endogenous gene suppression, (ii) gene disruption, (iii) co-suppression, (iv) RNAi, (v) gene silencing, (vi) FlavrSavr tomato, (vii) Golden Rice, (vii) molecular farming, (viii) edible vaccines, (ix) genetic modification of new energy crops.

RELEVANT LITERATURE

1. Banga SS and Banga SK (eds) 1998. Hybrid Cultivar Development. Narosa Publ. House, New Delhi.
2. Chakraborty S, Chakraborty N, Agrawala L, Ghosh S, Narula K, et al. 2010. Next-generation protein-rich potato expressing the seed protein gene *AmA1* is a result of proteome rebalancing in transgenic tuber. *Proc Nat Acad Sci.* USA 107: 17533–17538.
3. Gress IJ 2008. Transgenics are imperative for biofuel crops. *Plant Sci.* 174:246-263.
4. Leemans J 1993. Ti to tomato, tomato to market. *Bio Technology.* 11:522-526.

5. Palle SR, Campbell LM, Pandeya D, Puckhaber L, Tollack LK et al. 2013. RNAi-mediated ultra-low gossypol cottonseed trait: performance of transgenic lines under field conditions. *Plant Biotechnol. J.* 11: 296–304.

6. Rathore KS, Sundaram S, Sunilkumar G, Campbell LM, Puckhaber L, Marcel S et al. 2012. Ultra-low gossypol cottonseed: generational stability of the seed-specific, RNAi-mediated phenotype and resumption of terpenoid profile following seed germination. *Plant Biotechnol. J.* 10: 174–183

7. Rishi AS, Nelson ND and Goyal A 2001. Molecular farming in crop plants. *J. Plant Biochem. Biotechnol.* 10:1-12.

8. Veisser RGF and Jacobsen E 1993. Towards modifying plants for altered starch content and composition. *TIBTECH.* 11: 63-68.

9. Vickers TA et al. 2003. Efficient reduction of target RNAs by small interfering RNA and RNase H-dependent antisense agents. *J. Biol. Chem.* 278:7108-7118.

■■■■

12 Biosafety Issues Related to and Adoption of Transgenic Crops

12.1. INTRODUCTION

Over the past two decades, we have been witnessing the success of plant biotechnology. This technology is helping farmers throughout the world produce higher yields, while reducing pesticide use and soil erosion. The benefits and safety of biotechnology has been proven over the past decade in countries with more than half of the world's population. What we need is courage in the leaders of those countries where farmers still have no choice but to use older and less effective methods. The green revolution and now plant biotechnology are helping meet the growing demand for food production, while preserving our environment for future generations (Norman E. Borlaug, quoted in ISAAA, 2009).

The transgenic technology has been used to develop a number of GM varieties that exhibit either novel traits like herbicide resistance or remarkable improvement in some very useful trait, such as drought tolerance. In fact, the transgenic technology is believed to be capable of generating such crop varieties that can lead to a number of benefits (Section 12.2), many of which have already been realized. The first transgenic or GM (genetically modified) variety was approved for commercial cultivation in 1994. In 1996, GM crop varieties were cultivated in six countries on 1.7 million hectares (mha); this area increased rapidly and surpassed 180 mha in 2014 distributed in 28 countries. This rapid adoption of GM varieties has taken place in spite of some to considerable resistance from various groups to the cultivation of GM crops and the consumption of GM food and feed products. A variety of concerns have been expressed about the safety of these genetically modified crop plants, and the debate on this aspect continues. These concerns have forced the stakeholders, including the government, to implement a regime of biosafety assessment and the regulation of the cultivation of GM crops and the utilization of their produce. In this chapter, we shall consider the adoption of GM crops by farmers and the biosafety issues relevant to these crops.

12.2. POTENTIAL BENEFITS FROM TRANSGENIC CROP VARIETIES

The transgenic crop varieties are called **first generation GM crops** when they are improved for resistance to a biotic or an abiotic stress. They generally have single transgenes and lead to yield increases by reducing the losses due to the concerned stresses. The **second-generation GM crops** are those trangenic varieties that have improved quality traits, including increased nutrient content, and improved digestibility. These varieties also include varieties that have two or more transgenes for resisting more than one abiotic/biotic stress, and those created to achieve higher yields. But the **third generation transgenic crops** are grown for production of biochemicals for various applications. The potential benefits from transgenic varieties of crops are many, and some of the important and most relevant advantages are given below.

1. Transgenic varieties resistant/tolerant to different biotic and abiotic stresses would minimize the year-to-year fluctuation in crop yields.
2. These varieties would also reduce the use of agricultural chemicals including pesticides. In addition, the farmer will be able to use biodegradable pesticides like herbicides, e.g., glyphosate and glufosinate; this would also reduce the environmental pollution and the pesticide residues in the crop produce.
3. They would also reduce the use of other agricultural inputs, including energy and irrigation water.
4. As a result of the above, not only the cost of cultivation would be reduced, the extent of environmental damage from agriculture would be minimized.
5. GM varieties with improved nutritional quality, e.g., Golden Rice, would promote human health.
6. The insect resistant Bt-varieties of maize are known to suffer much less insect damage. As a result, the level of fugal attack on them is much lower than that in non-Bt maize. Consequently, the risk of their produce containing mycotoxins is markedly reduced.
7. The production of valuable biochemicals from transgenic crop varieties is cost-effective, environment friendly and is based mainly on solar energy.
8. Transgenic varieties of crops can be developed for improved shelf life, reduced blemishes, reduced damage from storage pests, etc. Such varieties will reduce the losses of crop produce in transportation, storage, etc.
9. GM crop varieties may contribute to sustainable agriculture.

10. Transgenic varieties of crops are reported to increase economic level of the farmers in developing countries who, in general, have only limited access to various resources.

12.3. GM CROPS APPROVED FOR COMMERCIAL CULTIVATION

The debate initiated by the detrimental effects of the widespread use of pesticides like DDT brought in sharp focus the need for evolving appropriate regulatory procedures for new technologies. This debate has led to the development of regulatory procedures for commercial cultivation of GM crop varieties and the use of their produce as food or feed. The regulatory procedure pertaining to the GM crops is as follows. (i) The developer of a given GM variety prepares and submits a biosafety dossier (file or report) to the regulatory authority of the concerned country. (ii) The authority determines the risk of adverse effects of the GM variety on environment, human health, etc. on the basis of findings from the biosafety assays included in the dossier. (iii) The GM variety is approved for cultivation by farmers if the authority us satisfied that no adverse effects are associated with its cultivation. A GM variety has to be separately approved for commercial cultivation, as well as for use as food or feed. A GM variety approved for commercialization is said to be **deregulated**.

The tomato variety FlavrSavr was the first GM variety to be deregulated. Since then a large number of transgenic varieties have been approved for commercialization. The database on the approvals for commercialization of GM crop varieties is available at the following web site: *http://www.isaaa.org/gmapprovaldatabase/default.asp*. The GM crops are approved as events and not as varieties. An **event** may be defined as the integration of a transgene in a single transformed cell from which a complete plant is regenerated. Each event is characterized for the site of transgene integration, transgene function and other features of the concerned transgenic line. In June 2015, 166 singular and 211 stacked events (total of 377 events) covering 29 plant species were approved. A **singular event** is the integration of a single transgene into the crop genome, e.g., *B. juncea* Laurical™ Event 18. But in case of a **stacked event**, two or more singular events are brought together in a single variety, e.g., the VIPCOT™ Roundup Ready Flex™ Cotton event COT102 x COT67B x MON88913 combines three different events. The GM crops approved so far represent modification in 34 different traits. The herbicide tolerance is the most widely modified trait, followed by insect resistance, and modified produce quality. So far, only one GM crop (soybean) with modification of growth/yield has been approved.

GM varieties covering 29 crops have been approved in 40 countries for commercial cultivation and/or use as food or feed. The number of approvals is the highest in maize, cotton stands second highest, followed by potato, canola, and soybean in that order. Among the countries, Japan has granted the largest number of approvals; countries like USA, Canada, Mexico and South Korea rank second, third, fourth and fifth in terms of the number of GM crop approvals. India has granted approvals for merely 11 GM crop events. However, the number of GM crop event approvals granted in a given country does not necessarily reflect the extent of cultivation of these crops in that country. Most of the GM crop approvals are held by few private companies: the Monsanto holds the highest number of approvals (total 120 approvals covering 10 crops). In comparison, Syngenta, which ranks second in the number of approvals, has merely 56 approvals covering two crops. Finally, most of the recent GM event approvals relate to the second and the third generation GM crops. In additions, bulk of the approvals comprises stacked events; some of the approved GM crops have up to seven or even eight different transgenes.

12.4. ADOPTION OF GM CROPS BY FARMERS

Farmers readily adopt a new technology if they are convinced that the technology will lead to an increase in their net profits. The technology should be available to the farmers at an easily affordable price, should not be complicated, and the technology should be safe to the farm workers. Above all, the farmers should be able to sell the crop produce in domestic and/or international markets. In 1994, FLAVR SAVR™ tomato became the first transgenic variety to be approved for commercial cultivation. In 1996, transgenic varieties were grown on 1.7 million hectares in six countries. This area increased to 181.5 million hectares in 2014 (Table 12.1), which represents more than 100-fold increase as compared to the area in 1996. In the initial years, there was a rapid increase (20-fold increase in three years) in the area under GM crops. But the rate of increase declined greatly in the subsequent years, viz., only 2.5-fold increase in the next seven years. The number of countries growing transgenic crops also increased to 28 in 2014 as compared to merely six countries in 1996. However, the bulk (~90%) of the area under the GM crops is located in merely five countries, viz., USA, Argentina, Brazil, Canada, and India. Since 2012, the developing countries have been planting GM crops in larger areas than the developed countries.

The largest proportion of the GM crop area is occupied by soybean, followed by maize, cotton and Canola. In 2014, GM varieties of nine different crops were grown on a substantial scale, while approvals had been obtained for 29 crop species. The number of different traits modified in these crops was 34 in 2014, and included increased photosynthesis, modified starch, oil, protein, etc. quality, reduced nicotine content, etc. The most widely cultivated crops have been herbicide tolerant, followed by insect resistant varieties. But the stacked trait GM crops with herbicide tolerance and insect resistance traits have now become more popular than the insect resistant varieties and displaced the latter to the third place. But varieties with modified quality or virus resistance are grown on relatively small proportion (about 1% under each category) of the area under GM crops. The vast majority of GM approvals have been obtained by private multinational companies, and the number of approvals held by public institutions is barely more than two dozen. The Monsanto is the most dominant player with over 40% of the approvals, followed by Syngenta, Bayer CropScience, Dupont and Dow AgroSciences LLC.

Table 12.1: The area (in million hectares) under the GM varieties of crops in the five major countries (*Based on http://isaaa.org/resources/publications).*

Country	1997	2002	2006	2010	2012	2014	2016
USA	8.1	39.0	54.6	66.8	69.5	73.1	72.9
Argentina	1.4	13.5	18.0	22.9	23.9	24.3	23.8
Brazil	-	-	11.5	25.4	36.6	42.2	49.1
Canada	1.3	3.5	6.1	8.8	11.6	11.6	11.6
India	-	<0.1	3.8	9.4	10.8	11.6	10.8
Other countries	0.2	2.6	8.0	14.7	17.9	18.7	16.9
Total	**11.0**	**58.7**	**102.0**	**148.0**	**170.3**	**181.5**	**185.1**

12.5. AGRONOMIC IMPACT OF TRANSGENIC CROPS

Most of the GM varieties of crops have been designed to reduce yield losses due to various stresses. Herbicide tolerant varieties are the most widely grown GM varieties. These varieties allow more effective weed control and, thereby, minimize yield losses due to weeds. Similarly, the insect resistant and disease resistant varieties reduce losses due to insect infestations and disease attacks, respectively. These varieties also reduce the use of pesticides, which adds to the economic benefit to the farmers. For example, according to some estimates, the cultivation of the Bt-cotton hybrids in India resulted in 31% higher yields. At the same time, the use of insecticides was reduced by 39 per cent in comparison to that in non-Bt hybrid varieties of cotton. In addition, India was importing cotton before the commercial cultivation of Bt-cotton hybrids began in 2002. In contrast, India now has become a cotton exporting country.

A large number of studies have shown that the productivity of GM varieties of crops is generally higher than that of the non-transgenic varieties. The magnitude of the increased productivity, in general, is much higher in the developing countries than in the developed countries. However, there are some reports of lower yields of GM crops as compared to those of non-GM crops. But most of these reports have been published in newspapers, and not in refereed journals. Herbicide tolerant, insect and disease resistant GM crop varieties allow the adoption of zero tillage cultivation. The **low tillage** or **zero tillage** practice reduces the cost of cultivation by cutting down the use of energy, fossil fuels, etc.

At the same time, this practice eliminates the operation of land preparation and, thereby, simplifies the crop cultivation.

The use of GM crop varieties can also generate spillover effects on neighboring fields of the same crops, or on crops grown in the same field in the subsequent years. For example, the cultivation of Bt-crops generally results in lower intensity of insect infestation in the neighboring fields of the same crop. Similarly, when herbicide tolerant varieties are grown in a field, the weed control in the subsequent crop grown in the same field is much easier than before. Finally, transgenic maize varieties tolerant to drought are already in cultivation, and varieties with higher yield potential have already been developed. These varieties will lead to increased crop productivity from crops grown under drought stress and under normal conditions, respectively. It has been argued that GM crop varieties will enhance the sustainability of agricultural production systems in the following ways: (i) by supporting low tillage cultivation, (ii) by using lower amounts of nonrenewable resources like fossil fuels, pesticides, etc., (iii) by producing more from the same area of land leading to reduced pressure on forest lands, (iv) by performing better under conditions of abiotic stresses, etc.

12.6. ECONOMIC BENEFITS FROM TRANSGENIC VARIETIES

The transgenic varieties of crops have been rapidly adopted by the farmers world-wide, except for those in the European Union. In fact, the GM technology is one of the fastest adopted technologies despite the stiff and vociferous opposition from many groups. It has been suggested that this extent of adoption is an ample evidence for the economic benefits to the cultivators from the transgenic technology. The economic benefits to the farmers arise mainly from one or more of the following: (i) increased productivity, (ii) reduced use of inputs like fossil fuels and pesticides, farm labour, etc., (iii) simplified farm operations, and (iv) saving of the time required for farm operations. In addition, economic benefits also result from (v) reduction in the amounts of pesticide residues present in the crop produce, and (vi) the decreased risk of presence of mycotoxins in the grains of crops like Bt-maize due to the lower incidence of fungal attacks.

Several studies have shown that farmers benefit economically from the cultivation of GM crop varieties. It has been estimated that in the year 2006, the net benefit to the farmers world-wide from GM crop cultivation was as high as US$6.9 billion. Further, about 53% of this economic gain was to the farmers in the developed countries, while those in the developing countries had 47% of this profit as their share. In comparison, the global profit from GM crop varieties had risen to US$20.4 billion in the year 2013. The total profit during the period between 1996 and 2913 was estimated to be US$133.3 billion; this profit was almost equally shared by the farmers in the developed and the developing countries.

An analysis of 147 independent studies showed that during the 20 years between 1995 and 2014, the cultivation of GM crop varieties has resulted in, on an average, a 37% decline in pesticides use, a 22% increase in yield, and a 68% rise in average profit to the farmers. The benefits are greater in the case of insect resistant than herbicide tolerant crops and for the developing countries than those for the developed countries. Thus, the cultivation of GM crop varieties leads to substantial increases in crop yields and, as a result, economic returns to the farmers. Finally, the modified trait and the geographical region are the main factors affecting the quantum of economic benefits.

12.7. CONSUMER ACCEPTANCE OF TRANSGENIC CROP PRODUCE

There have been a number of studies on consumer preference to GM crop produce. In general, the consumers in developed countries, particularly European Union, have a greater preference for the non-GM food than that for the GM food. But in the developing countries, usually the reverse of this trend is seen. In some countries like China, the consumers may even be willing to pay a premium for some of the GM food. A recent study suggests that the consumers tend to pick and purchase GM food if they were displayed on the shelves of the stores. In European Union, it is mandatory to label a food product as GM food if it contains more than 0.9% GM ingredients. But the **organic food** is not allowed to contain any GM ingredients.

The national policies of different countries tend to differ on the emphasis placed on segregation of GM crops from the non-GM crops during cultivation as well as processing of the crop produce. During crop cultivation, the GM crop varieties can be separated from the non-GM varieties by keeping their fields at a minimum distance equal to the isolation distance for the concerned crop species. **Isolation distance** is the minimum distance that should be kept between two fields of the same crop to prevent chance pollination of the plants growing in one field by those in the other field. Alternatively, the fields may be permitted to be located closer than the isolation distance provided the crops in the two fields are planted at different dates so that they do not flower during the same period of time. The segregation of GM crops is considered to be responsibility of the farmer growing the GM crop. It has been suggested that separation of the GM crops during cultivation and food processing tends to increase food and animal feed costs.

12.8. SOCIAL IMPACT OF TRANSGENIC CROPS

The cultivation of GM crops is, in theory, expected to increase income of farmers, promote their health, and generate biochemicals of medicinal value. The health benefits would come from reduced use of pesticides, which appears to lead to lower incidence of poisoning to the farm workers. In addition, GM crops with enhanced nutritional value would increase the nutritional status of those who consume the produce from these crops. Some GM crop varieties have been designed to produce reduced amounts of harmful compounds. For example, the Innate transgenic potatoes generate reduced amounts of acrylamide than normal potatoes when they are cooked at high temperature. Acrylamide is a potential carcinogen when consumed by humans. Transgenic varieties are being designed to produce useful pharmaceutical compounds, including vaccines. These products will be useful in management of various health disorders and in promoting human and animal health.

The increased income of farmers would tend to increase the quality of life of farm communities. It would tend to generate more employment for the rural population and would encourage diverse business activities, including transport. There is some evidence that the adoption of Bt-cotton in India has increased the income of women workers. There have been suggestions that the introduction of GM crops has increased the likelihood of farmers incurring debts, which may lead to farmer suicides. However, many feel that the debt of farmers is not necessarily linked to the introduction of GM crops.

12.9. RISKS FROM GM CROPS

It may be pointed out that the varieties developed by conventional plant breeding methods are generally regarded as safe; we will refer to them as **conventional varieties**. The transgenic varieties differ from the conventional ones in several respects, viz., the genes introduced and expressed in them, the promoters used for controlling the expressions of these genes, the selectable marker genes used for their production, and the site in the genome where they are integrated, etc. In view of the above, the GM crop varieties are considered as **regulated articles**, and their field-testing has to follow specified procedure, and the cultivation and utilization of these varieties is subject to approval after the prescribed biosafety assessment. The biosafety assessment of the GM varieties is based on the assessment of risk from their cultivation and utilization.

Risk, in simple terms, is the magnitude of the negative effect (or harm) generated by an event multiplied by the probability of the occurrence of this event. In the case of GM crops, **risk assessment** is a stepwise procedure; it begins in the laboratory, is followed by confined field trials conducted on a small scale, and culminates with field tests on large scale. The risk assessment aims at ensuring the safe cultivation GM crops and utilization of the produce from these crops with the minimum negative effects on humans as well as the environment. It may be pointed out the level of risk estimated by one expert might differ from that by another. Similarly, the risk perception by the experts might often differ from that by the common people. In any case, the level of acceptable risk is likely to be influence by many social, economic and political considerations.

A variety of biosafety concerns have been expressed about the GM crops. These concerns are primarily related to their effects on human/animal health, on the environment including biodiversity, and on the agricultural production and food security. Some of the most important and the relevant concerns are as follows. (i) The transgene encoded protein might produce toxic effects or cause allergies. (ii) The antibiotic resistance markers used for the development of transgenic plants might produce toxic/allergic effects. (iii) Consumption of the GM food might make the gut microflora resistant to the concerned antibiotics, which would adversely affect the management of some of the human/animal health problems. (iv) The transgene integration in the genome might lead to activation of undesirable genes or it might inactivate some desirable genes. This would lead to unintended changes in the crop and might adversely affect the nutritional status (including the antinutritional compounds) of the crop produce. (v) The transgene expression might change with time due to gene silencing; this would alter the characteristic of the concerned GM crop varieties. (vi) The transgenic varieties may adversely affect the biodiversity, including microbial biodiversity. (vii) It is feared that the transgenic varieties of some crop plants might themselves become weeds. Finally, (viii) the transgenes might be transferred into related weed species and this may make the weeds difficult to control.

12.10. REGULATION OF GM CROP VARIETIES

Each transgenic event is considered as a **regulated article**. An **event** represents the integration of a given transgene construct into the genome of a single cell of the particular plant species. Thus, several distinct events would be generated by a single transgene construct. Each event is treated as a separate article since the same transgene is likely to have integrated at distinct genomic sites in the different events. Therefore, the GM crop varieties can be released for commercial cultivation and their produce used as food and feed only after the appropriate regulatory authorities grant their approval for the same. Once an event/a transgenic variety is approved for commercialization, it is generally referred to as **deregulated**. The decision for approval for commercialization is based on the information provided in the regulatory dossier submitted by the creator of the transgenic variety.

The regulatory dossier or file contains information on the assessment of risk posed by the GM crop to the environment during cultivation and to human/ animal health on consumption as food/feed. It is mandatory to include the required data on these two aspects. In general, there is a broad consensus among nations on the types of data required. But the amount of a given data needed and the importance attached to this data vary among different countries. The dossier may also contain data on the assessment of socioeconomic impact of the concerned GM crop. The inclusion of these data is optional, but in some countries it is a mandatory requirement. It has been suggested that the risk assessment data form the basis of the technical decision regarding approval of a GM event. In contrast, the socioeconomic assessment is used for political decisions concerning the approval.

The regulatory dossier contains information of the following features of the GM event/crop: life cycle of the crop, features related to reproduction, the transgene construct and the selectable markers used, the genetic transformation method employed, the toxic and allergenic effects of the transgenes, and the expected and unintended changes in the various features of the GM plant. In addition, data from molecular and biochemical analyses, animal feeding experiment, and various safety tests are also considered in this process. The dossier also contains data from field trials, which are conducted in a step-wise manner. Initially, the GM crops are tested in small scale confined field trials. Subsequently, they are permitted for evaluation in large scale field trials, if no safety concerns were detected in the confined fled trials.

In USA, three agencies are involved in the regulation of transgenic varieties and the products made from them. These agencies are the Food and Drug Administration (FDA), Environmental Protection Agency (EPA), and Animal and Plant Health Inspection Service (APHIS) of U.S. Department of Agriculture (USDA). APHIS oversees the field-trials of GM crops conducted under controlled conditions. It monitors and reports the occurrence of any unusual incidents during the trials. The EPA evaluates the effects of GM crops having GM pesticide traits, e.g., Bt- crops, varieties with DNA sequences from known pests or pathogens, etc., on the environment and human health. Any of these three agencies can order immediate withdrawal of a GM product from the market on the grounds of scientifically valid safety concerns (see Lemaux 2008).

The approaches to the regulation of GM crops in the United States of America and the European Union are quite divergent and in contrast to each other. The European Union follows a **precautionary approach**: it regards every GM crop and GM food to be unsafe, and adheres to the consumers' 'right to know'. As a result, the standards for approval of GM crops and of GM foods are quite stringent. In contrast, the US follows the **comparative approach** for regulation of GM crops. This approach compares a GM crop and/or food with the concerned conventional crop and/or food and attempts to determine if it is as safe as the latter. It also requires voluntary labeling of food products as GM food. The regulations followed by most other developed nations, such as Australia, Canada, Japan, etc. have some features of the above two approaches. It is desirable that the different countries follow essentially comparable regulations for the GM crops. This would eliminate the need for the preparation of fresh regulatory dossier for every country and, thereby, facilitate the regulatory approvals of the GM crops is different nations. This would, in turn, encourage investments in the research and development for GM crops resulting in the refinement of the GM technology. In essence, the harmonization of regulatory laws and procedures for the GM crops would permit the realization of the maximum benefits from these crops and minimize the potential risks from them.

12.11. SAFETY OF GM FOOD

The food prepared from the produce of transgenic crops is generally referred to as **genetically modified food** or **GM food.** A variety of safety concerns about the GM food have been expressed; the most relevant of these concerns will be briefly discussed later in this section. It may be pointed out that the GM crop produce are extensively evaluated for safety before they are allowed to be used as food and/or feed. These safety evaluations are done on the basis of the concept of substantial equivalence. Most of the foods used by humans have been consumed by them for a very long time so that they are **generally regarded as safe** (GRAS). Safety evaluation on the basis of the **concept of substantial equivalence** attempts to demonstrate that the GM food is as safe as the conventional foods. This simplifies the safety assessment since it is virtually impossible to demonstrate absolute safety of any food, including GM food.

The determination of substantial equivalence of GM crops with their conventional counterparts is based on the comparison of the two groups for their various characteristics, including chemical composition related to nutritional (e.g., protein, fat, starch, amino acid, vitamin, mineral, etc.) and antinutritional constituents. (i) In case the data from the GM and the conventional varieties are comparable, the substantial equivalence is taken to be established. (ii) But if there are some specific differences between the two groups of varieties, further safety tests focusing on the identified differences need to be carried out. This seems to be the most common finding. Finally, (iii) in some cases, the two groups might differ for several features. In such cases, safety tests would have to be designed for each individual case. It may be added that a large number of safety tests, including chemical composition estimation and animal feeding trials show that the GM foods are essentially comparable with the conventional foods. Now we briefly consider the important safety issues relevant to the use of GM crop produce as food.

1. It has been suggested that the antibiotic resistance markers used for the selection of transgenic cells and plantlets during the production of transgenic plants might be transferred to the microflora present in the digestive tracts of the animals and humans. This would make the gut microflora resistant to the concerned antibiotics and thereby pose problems in the management of several diseases. The available evidence indicates that the chances of a complete gene being transferred into gut microflora are remote. In any case, the techniques for removal of the selectable markers after the production of transgenic plants have been developed. In addition, some non-antibiotic selectable makers have also been developed that can be used for the production of transgenic plants.

2. It is feared that the proteins encoded by the transgenes may cause allergies. Many proteins are known to cause allergies. In general, these proteins have small size, their digestion is slow, and they are quite stable to heat. A large proportion of the transgenes are likely to be taken from such organisms that have been never used as food. Therefore, their allergenic potential is not likely to be known. In view of this, the allergenicity of transgene-encoded proteins should be carefully evaluated. This should be done even for such transgenes that are derived from a food crop. There is some evidence that the protein encoded by even such a transgene might show allergenicity when it is expressed in a different plant species.
3. Another fear related to the safety of GM food is the toxicity caused by the proteins encoded by the transgenes. For example, many biochemicals involved in resistance of plants to diseases and insects, e.g., glycoalkaloids and phytoalexins, can produce toxic effects. Therefore, even GM foods that contain products of transgenes from only food crops cannot be assumed to be safe. In view of this, the toxic effects of GM foods are evaluated in animal feeding trials by feeding the GM food as well as the pure transgene products. The use of GM crops as food is approved only after they are shown to have no toxic effects. But even then sometimes the safety of certain GM foods is not accepted by the consumers. For example, the safety of Bt- proteins is questioned even in the face of the experimental evidence indicating it to be nontoxic and non-allergenic to humans. The Bt- protein Cry9C, however, is slow to be digested and is stable to 90°C. Therefore, its presence in human food is not permitted, although it was found to be safe to animals.
4. There is always the risk of unintended contamination of non-GM food by GM food crop produce. Different countries have different maximum permissible limits for the GM ingredients. Therefore, the GM crops should be isolated from non-GM varieties of the same crop by the minimum isolation distance recommended for seed certification of the concerned crop species. Care should also be taken during packaging, transport, food processing, etc. to avoid mixing with the GM ingredients. However, contamination by produce of such GM varieties that are designed for plant molecular farming (PMF), for the production of biochemicals used for pharmaceutical and research purposes, is not allowed; this is called **zero tolerance**. Therefore, stringent precautions should be taken for the cultivation of such GM crops and during processing of their produce.

It may be pointed that complete safety of GM food cannot be established. But this is true for any food, including organic foods. The available evidence does not show the GM foods to pose a greater safety concern as compared to the concerned conventional or even organic foods. **Organic foods** are prepared from the produce of conventional crop varieties grown without the use of chemical fertilizers, chemical pesticides and growth enhancing chemicals. In any case, there are specific laws for labeling of the GM food and food containing GM ingredients. The laws for labeling are the most liberal in USA, where only such GM food is required to be labeled that differs from the conventional counterpart in terms of nutritional composition and/or food quality. But in EU, only such foods are exempt from labeling that have less than 0.9% GM ingredients.

12.12. ENVIRONMENTAL RISKS FROM GM CROPS

There has been a prolonged and active debate about the environmental effects of GM crops. It has been argued that these crops are likely to have a variety of adverse effects on the environment. The important risk perceptions include persistence of GM crops as weeds, transgene transfers to the related wild species including weeds, adverse effects of the transgene products on nontarget species, impact on microflora in root zone and on the leaf surfaces, adverse impact on biodiversity, etc. In this section, we shall briefly consider these objections in the light of the available experimental evidence.

1. It is feared that the plants of some transgenic varieties, such as herbicide tolerant varieties, might persist as volunteer plants in the field. Crop plants growing in a field and derived from seeds that were not deliberately planted in the field are called **volunteer plants**. Volunteer plants would pollinate the plants of conventional varieties grown in the same field and thus would contaminate the produce from the conventional varieties. Therefore, considerable care has to be taken during crop cultivation to minimize the risk of contamination of crop produce by the transgenic volunteer plants. However, it is unlikely that this risk can be completely eliminated.
2. Another perceived risk from transgenic varieties is that some of them could survive and persist under natural conditions and become difficult to control weeds. There is some evidence that herbicide tolerant Canola plants can survive and persist under natural conditions. However, there is little evidence that these plants can become weeds. It may be pointed out that most of our crops are highly domesticated, and they are not likely to proliferate under natural conditions. But fodder and forage crops and the crops being domesticated for biofuel production are not fully domesticated. The risk of GM varieties becoming noxious weeds is far greater in the case of such crops than in the cases of fully domesticated crops.
3. There has been considerable controversy regarding the adverse effect of the Bt toxins on nontarget species, including microorganisms. The findings from different studies have reported conflicting results. A survey of the published literature shows that the Bt crops have small

negative impact on some of the nontarget insects as compared to the fields without insecticide application. But in comparison to the fields with insecticide application, the Bt crops had beneficial impact on the nontarget insects.

4. There are controversial reports on the impact of transgenic crops on the microorganisms present in soil and on leaf surfaces. It appears that some transgenes have a negative impact on the microbial populations and diversity, while other transgenes have little or no effect. In addition, different studies have often reported contrasting findings for the same transgene.
5. It has been feared that the cultivation of herbicide tolerant crops would lead to the development herbicide resistant weeds. This would, in turn, result in greater application of herbicides for weed control. There is evidence for the emergence of herbicide resistance in weeds. The scientists are, therefore, developing effective strategies for the management of this problem.
6. It has been suggested that the GM crops would lead to a decrease in biodiversity, i.e., the total variability present among all the organisms. The available evidence shows that the adoption of GM crops tends to increase the diversity among the cultivated varieties of the concerned crops. The increased productivity of the GM crops will reduce the risk of conversion of forest areas to agricultural fields. This would reduce loss of biodiversity and the negative environmental impacts of agricultural activities.
7. It is feared that a GM crop might hybridize with a rare wild relative and thereby hasten the extinction of the latter. Therefore, care should be taken to avoid GM crop varieties in such areas that have rare relatives of the concerned crop.
8. The transgene specifying herbicide tolerance, insect/disease resistance or abiotic stress tolerance may be transferred from GM crops to wild and weedy relatives of the concerned crops. This could make the weeds difficult to manage, and these modified weeds could disturb the ecological balance. Therefore, the following should be carefully evaluated while considering the option of the cultivation of GM crops in areas having the wild relatives of the concerned crops: the modified trait, the chance that the transgene could be transferred to a wild relative, and the effects of this gene transfer on the ecosystem.

The available evidence shows that the cultivation of insect resistant and herbicide tolerant GM varieties of crops has resulted in a substantially lower use of insecticides and herbicides, respectively. In this context, the impact of herbicide tolerant GM varieties has been much greater that that of the insect resistant varieties. In addition, the herbicide tolerant varieties enable the use of biodegradable herbicides, which are environment friendly since they are readily removed from the environment. They also permit the adoption of low or zero tillage crop cultivation, which reduces the use of fossil fuels and CO_2 emission. Finally, the GM crops increase the sequestration of carbon from CO_2 in the form of the increased biomass generated by the GM varieties. But the emergence of resistance in insects and weeds will adversely affect the cultivation of insect and herbicide resistant varieties. Therefore, efforts should be made to develop effective strategies to minimize the risk of appearance of resistance in the insect pests and the weeds. There is some evidence that the higher yields of the transgenic varieties tend to reduce the pressure on forestland to be diverted agricultural uses.

12.13. TRANSGENE DISPERSAL VIA POLLEN AND DNA

The transgene transfers will occur from GM crop varieties to the conventional varieties if their fields are close enough and their flowering durations overlap in time. This would apply to all cross-pollinated and often cross-pollinated crops whose floral biology favors cross-pollination. The transgene transfers would contaminate the produce from conventional varieties with the GM produce. This might complicate the acceptance of the contaminated non-GM produce in certain markets, and would disqualify the produce from being sold as organic food. The various strategies devised for minimizing transgene transfers to conventional varieties include spatial separation or isolation, temporal separation or planting of the GM and non-GM crops at different dates, planting of non-GM border rows around the GM field, and integration of the transgenes into the chloroplast genomes. In addition, the transgenes may be driven by inducible promoters so that the transgene product will not be accumulated in the seeds produced by chance pollination by the GM pollen.

It is well known that some of our crops and their wild relatives exchange genes in those areas where the wild relatives are present in or near the fields of the concerned crops. In several cases, the weedy relatives have given rise to more aggressive weeds as a result of the genes transferred from the related crop species. In view of this, it is not surprising that gene transfers have been shown to take place when the GM crop was grown in an area, in which its sexually compatible wild relatives were present. Therefore, cultivation of GM varieties in such areas should be carefully considered to minimize the risk of the wild relatives becoming more aggressive and/or persistent as a result of transgene transfers. The GM varieties grown in such areas should be modified for a trait that would not increase the weediness of the related wild species in case the transgenes were transferred into them. In addition, the GM variety could deploy a suitable technology, e.g., one of the genetic use restriction technologies (GURT), to minimize the risk of gene transfers via pollen grains.

In nature, genes from plants do get transferred to soil microflora over a long time scale. The DNA from GM crop biomass would become liberated in the soil form where the transgenes may enter the microbe cells and become integrated into their genomes by homologous recombination; this would constitute **horizontal transgene transfer**. The experimental evidence suggests that such horizontal transgene transfers would take place at very low frequencies. The effects of such gene transfers on the microbial populations will largely depend on the following two factors: (i) the trait specified by the transgene, and (ii) the effect of this trait on the fitness of the cells receiving the transgene. The risk of transgene integration into the microbial genomes can be minimized by avoiding the use of such sequences in the transgene constructs that are homologous to the bacterial genome sequences.

12.14. ISSUES RELEVANT TO AGRICULTURE

A number of apprehensions concerning the agricultural performance of the GM crop varieties have been raised. Some of the important agricultural issues concerning GM crops are examined in this Section.

1. It has been suggested that the transgene expression might reduce the yields of the GM varieties as compared to the parental varieties, i.e., transgenes might cause **yield drag**. The evidence form a large number of field trials indicates the following. (i) Some GM traits might cause yield drag, while others would increase the yields of GM varieties. (ii) The effects of transgenes on yield are markedly affected by the genetic backgrounds in which they are placed, and the environmental conditions in which the GM varieties are grown. (iii) In general, newer events are likely to show much less negative impact on the yield than the older events involving the same transgene. Finally, (iv) stacked events are likely to show better performance including yield than single event varieties. **A stacked event GM variety** has two or more different transgenes affecting different or even the same trait.
2. The transgenes might be modified or deleted from genome subsequent to their integration into the crop plant genome. In addition, the expression of the transgenes may decline with time due to gene silencing, and it may be markedly influenced by the prevailing environment. These phenomena are known to occur, and they would pose problems to the cultivators. In view of this, the transgenic events are extensively evaluated in field trials. Only such transgenic lines are then advanced to the commercial stage that show stable transgene expression and production of the transgenic phenotype in a range of environments in which the GM varieties are to be cultivated.
3. It is well known that with time the target insect pests develop resistance to Bt toxins. Experimental evidence shows that resistant mutants of

insects tend to develop in the fields of Bt crops. There are a number of strategies to manage this situation. The first strategy that is widely used is the planting of refuge crops. A **refuge crop** comprises an insect susceptible non-Bt variety of the concerned crop grown next to the Bt crop without application of insecticides. The refuge crops tend to delay the mergence of the resistant mutants of the insect pests. The second approach consists of pyramiding two or more different Bt genes with different modes of action. Thirdly, the expression of the Bt gene could be made tissue-, time- or signal-dependent to reduce the exposure of the insect pests to the toxins. This is expected to minimize the risk for the development of resistance by reducing the selection for the resistant mutants. Fourthly, new hybrid Bt genes may be produced by combining of different Bt genes. Finally, genes other than Bt that confer insect resistance can be used to develop inset resistant GM varieties.

4. The repeated application of herbicides in known to promote the evolution of herbicide resistance in the weeds. This problem can be resolved by developing two or more varieties of a crop, each of which is resistant to a different herbicide. These varieties can be grown in rotation so that resistant weeds for one herbicide are eliminated by the other.

12.15. CONCLUSIONS

The transgenic varieties of crops offer several potential benefits, including increased and stable yields, protection from insects and diseases, reduced use of pesticides, conservation tillage, etc. So far GM varieties have been approved in 29 different crops for commercial cultivation in 40 countries. The commercial cultivation of GM varieties began in 1996 on 1.7 million hectares, while in 2014 the area under GM crops was 181.5 million hectares. GM crops were grown in six countries in 1996, but 28 countries planted them in 2014. USA, Argentina, Brazil, Canada and, India are the major contributors to the GM crop area (~90% of the GM crop area). Since 2012, the total GM crop area in the developing countries has been higher than that in the developed countries. The area under the GM soybean is the highest, followed by that under the GM maize, cotton and Canola. At present, GM varieties with herbicide tolerance are the most widely cultivated, followed by those with herbicide tolerance and insect resistance traits, while the insect resistant GM varieties rank the third.

The available evidence indicates that GM crops offer economic benefits to the farmers, particularly in the developing countries. The various benefits from the GM crops are as follows: (i) increased yield, (ii) reduced use of agricultural inputs, including fossil fuels, pesticides, farm labour, etc., (iii) simplified and reduced farm operations that take less time, (iv) reduction in the pesticide residues present in the crop produce, and (v) reduced chances of fugal infection and contamination the grains by mycotoxins. According to an estimate, the total economic benefit from the cultivation of GM crop varieties during the eighteen-year period between 1996 and 2013 was US$133.3 billion. It is noteworthy that the share of the farmers from developing countries was about 50% of the total global profit. In addition, it has been estimated that during the period between 1995 and 2014, there was about 37% reduction in the use of pesticides; this was coupled with around 22% and 68% increase in crop yields and farmer's profits, respectively, as a consequence of the cultivation of GM varieties of crops.

The safety evaluation procedures of GM crop varieties are designed to show that the GM varieties are essentially comparable to the conventional varieties, i.e., the two are substantially equivalent. The safety dossier submitted to the regulatory authorities contains information on a variety of aspects relevant to the GM crops. Information on food safety of the GM crop produce and the possible effects on the environment are mandatory requirements. The dossier may contain information on the socioeconomic impact of the GM crop; this is generally optional. In addition, the strategy for management of the emergence of resistance in the insects/weeds is provided, and a mechanism for post-release monitoring may be required in some countries. Each individual transgenic event involving the same transgene is treated as a separate GM product, and the safety evaluation is done for the different events on case-by-case basis. The available evidence indicates that the regulatory procedures have been effective in ensuring the safety of GM crops released for commercial cultivation.

The prolonged and costly evaluation and regulatory procedures, and the low consumer acceptance of GM crops and their produce have prompted the development of novel plant breeding tools including site-directed nuclease (SDN) technology described in Chapter 2 (Section 2.11). The varieties developed by these technologies are considered to be exempt from regulation by the regulatory authorities of USA as long as they do not contain any foreign DNA. With this condition, the SDN technologies enable inactivation of the targeted genes by inducing loss of function mutations in them. Several plant varieties produce by the zinc-finger nuclease (ZFN) technology and the transcription activator-like effector nuclease (TALEN) SDN system, and a mushroom strain developed by the CRISPR-Cas9 system have been exempted from the regulatory oversight in USA (Waltz 2016).

It is satisfying to note that several transgenic lines developed indigenously by Indian institutions are at various stages of development including biosafety research trials. The examples of such lines include insect resistant chickpea, insect resistant pigeonpea, insect resistant rice, and abiotic stores tolerant rice.

PRACTICE QUESTIONS

1. Transgenic crop varieties have been rapidly adopted by the farmers. Discuss this statement in the light of the available evidence.

2. Discuss the agronomic, environmental and socioeconomic impacts of the GM crops.
3. Briefly describe the procedure for regulation of the approval of GM varieties of crops for commercial cultivation.
4. Discuss the food safety issues related to the GM crops in the light of the available experimental evidence.
5. Examine the various apprehensions related to the environmental impact of large-scale cultivation of GM crops in the light of the available experimental evidence.
6. The large-scale cultivation of GM crops raises several agronomic issues that may adversely affect food production. Discuss this statement in the light of the currently available experimental evidence.
7. Write short notes on the following: (i) potential benefits from transgenic crop varieties, (ii) risk assessment, (iii) regulation of GM crop varieties, (iv) safety of GM food, (v) environmental impact of GM crops, (vi) agricultural issues relevant to GM varieties, (vii) economic benefits from the GM crops, (viii) impact of GM crops on resource poor farmers.

RELEVANT LITERATURE

1. Ellstrand NC 2003. Dangerous Liaisons? When Cultivated Plants Mate with their Wild Relatives. Baltimore, MD: Johns Hopkins Univ. Press.
2. Lemaux G 2009. Genetically Engineered Plants and Foods: A Scientist's Analysis of the Issues (Part II). *Annu. Rev. Plant Biol.* 60:511–559.
3. Talas-Ogras T 2011. Risk assessment strategies for transgenic plants. *Acta Physiol Plant* 33:647–657 DOI 10.1007/s11738-010-0624-5.
4. EFSA 2004. Guidance document of the scientific panel on genetically modified organisms for the risk assessment of genetically modified plants and derived food and feed. EFSA J 99:1–94. http://www.efsa.eu.int.
5. International Life Sciences Institute (ILSI) 2004. Nutritional and safety assessments of foods and feeds nutritionally improved through biotechnology. *Compr Rev Food Sci Safety* 3:35–104.
6. Carpenter JE 2010. Peer-reviewed surveys indicate positive impact of commercialized GM crops. *Nat Biotech* 28:319-21.
7. EFSA 2006. Opinion of the scientific panel on genetically modified organisms on the post market environmental monitoring (PMEM) of genetically modified plants. (Question No. EFSA-Q-2004-061). European Food Safety Authority (EFSA). EFSA J 319:1–27. http://www.efsa.

europa.eu/EFSA/Scientific_Opinion/gmo_op_ej319_pmem_en1,0.pdf. Cited 14 Dec 2007.

8. Lusser M, Raney T, Tillie P, Dillen K and Rodriguez-Cerezo E 2012. International workshop on socio-economic impacts of genetically modified crops co-organised by JRC-IPTS and FAO: Workshop proceedings. Publications Office of the European Union, Luxembourg.

9 Gruère G and Sengupta D 2011. Bt Cotton and farmer suicides in India: An evidence-based Assessment. *Journal of Development Studies* 47: 316-337.

10. Gusta M, Smyth SJ, Belcher K, Phillips WB and Castle D (2011) Economic benefits of genetically-modified herbicide-tolerant Canola for producers. *AgBioForum* 14: 1-13.

11. James C 2014. *ISAAA Brief No. 49–2010: Global Status of Commercialized Biotech/GM Crops: 2014.* http://www.isaaa.org/RESOURCES/publications/ briefs/default.asp

12. James C 2012. *ISAAA Brief No. 44–2012: Global Status of Commercialized Biotech/GM Crops: 2012.* http://www.isaaa.org/RESOURCES/ publications/ briefs/default.asp.

13. James C 2010. A global overview of biotech (GM) crops: Adoption, impact and future prospects. *GM Crops* 1: 8-12.

14. Krishna V, Qaim M and Zilberman' D 2015. Transgenic crops, production risk and agrobiodiversity. *European Rev. Agril. Econ.* pp. 1–28 doi:10.1093/erae/jbv012.

15. Waltz E 2016. Gene-edited CRISPR mushroom escapes US regulation: A fungus engineered using CRISPR–Cas9 can be cultivated and sold without oversight. *Nature* 532: 293.

■■■■

Part C.
Molecular Markers

13 Molecular Marker Systems and Trait Phenotyping

13.1. INTRODUCTION

In general, qualitative traits are governed by one to few major genes, are little affected by the prevailing environment, and are classifiable into a small number of clear-cut phenotypic categories. Thus, the qualitative trait phenotype is a dependable indicator of the genotype at the concerned loci. But most of the important traits show continuous variation and their phenotypes show some to considerable influence of the environment; such traits are called **quantitative traits**. Therefore, phenotypic selection for quantitative traits is generally challenging. The plant breeders have been trying to develop methods for indirect selection for these traits. The indirect selection for a trait would require a simply inherited and easily and reliably scorable trait (i.e., a marker trait) to be closely linked to the trait of interest.

A **marker** or, more specifically, **genetic marker** is a trait that is polymorphic, and easy to identify, score and follow in segregating populations. The following activities and developments would be required to make indirect selection a feasibility. (i) Development of a dependable genetic marker system that is distributed over the entire genome at sufficient density. (ii) Precise phenotypic evaluation of a suitable population for the traits of interest, and genotyping the population for the markers covering the entire genome. (iii) Analysis of the marker genotype and trait phenotype data to identify markers closely linked to the traits of interest. Once such markers are identified, they can be used for indirect selection for the concerned traits.

In this chapter, the salient features of various types of marker systems and the procedures for their genotyping are briefly described. In addition, the approaches for trait phenotyping are also considered. The next chapter (Chapter 14) is devoted to the detection of marker trait associations, and comprises the essential features of the important mapping populations, linkage mapping of genes and quantitative trait loci (QTLs), and the basics of association mapping. The various applications of molecular markers, including marker-assisted selection, genomic selection, diversity analysis, etc. are discussed in Chapter 15

13.2. GENETIC MARKERS

It was demonstrated long time ago that the genes for some qualitative traits show linkage with those controlling quantitative traits (Sax 1923, Thoday 1961), but the number of such qualitative traits was limited. Therefore, attempts were made to identify other easily scorable, stable and more abundant features, resulting in the development of markers based on proteins and, ultimately, DNA. The genetic markers can be grouped into the following four types: (i) morphological markers, (ii) protein markers, (iii) DNA markers and (iv) biomarkers.

The **morphological markers** are based on observable phenotypes of qualitative traits. These markers were the first to be used in genetic and other studies and are the easiest to score. However, these markers are limited in number and only few markers can be analyzed in a single cross. Further, they have to be scored during specific stages of development and environmental thresholds may be required for some of them.

A **protein-based marker** is generated by such variants of an enzyme or non-enzyme protein that differ from each other in electrophoretic mobility. These markers represent the coding regions of the genome. Isozymes constitute the bulk of protein-based markers. **Isozymes** are electrophoretic variants of an enzyme, have identical catalytic function, but they could be encoded by two or more different genes. When the isozymes are encoded by the same gene, they are often called **allozymes**. However, generally the term isozyme is used for both isozymes and allozymes. Detection of protein markers can often be done in the seedling stage, and several markers can be analyzed is a single cross. Isozymes serve as codominant markers and their analysis is relatively easy. But the number of good isozyme markers is limited, and their pattern may differ with the tissue used, the stage of development and the prevailing environment. Some isozyme markers have been used for marker-assisted selection for some useful traits.

The **DNA markers** are based on sequence variation in genomic DNAs of different individuals/lines. Their detection is based on differences in the following features: (i) amplification by PCR, (ii) fragment size reflected as differential migration in a gel, and (iii) ability to hybridize with the probes (short DNA sequences) fixed onto a microarray. The first use of DNA sequence variation dates back to 1974, when digestion with restriction enzyme was used for physical mapping of a gene in adenovirus. This approach was used to develop the marker called restriction fragment length polymorphism (RFLP) as the first DNA marker. Subsequently, several different DNA marker systems were developed, some of which became very popular with the researchers. In addition, a number of biomolecules are promising as **biomarkers** for the identification of desirable genotypes (Table 13.1). In the strict sense, the term **molecular markers** includes protein and DNA markers as well as the biomarkers, but this term is ordinarily used only for DNA markers.

Table 13.1: A summary of the salient features of the different types of genetic markers

Genetic marker	Marker system based on	Remarks
Morphological	Oligogenic trait phenotype	Easy detection, limited number
Protein/ Isozyme	Electrophoretic variants of isozymes and seed storage proteins	Limited number, developed the techniques used for DNA markers; replaced by DNA markers
DNA*	DNA sequence variation	Abundant, distributed throughout the genome, detection independent of developmental stage and the environment
Biomarker	Levels of cellular metabolites	Promising; in developing stages

*The term molecular marker is generally used for DNA markers.

An **ideal DNA marker system** should be abundant, codominant, multiallelic, and distributed over the entire genome. In addition, marker detection should require only small amount of DNA, and should be easy, rapid, cheap, highly reproducible and suitable for high throughput genotyping. However, no marker system has all of the above features; the SNP markers are close to being an ideal marker system (*Xu Y 2010, Jiang G-L 2013*).

DNA marker loci are very large in number, and several different marker systems can be scored in the same population. The markers can be scored at any stage of development, in any tissue and under any environment. Their assay is non-destructive, can be used for recessive alleles in the heterozygotes, and it does not require gene expression. Only small amount of tissue sample is needed and the DNA can be stored for future use. High density molecular maps have been developed in many crops, such as rice, wheat, maize, common bean, etc. DNA markers are used for a variety of purposes, including linkage mapping, marker-assisted selection, diversity analysis, phylogenetic studies, management of genetic resources, fingerprinting, positional cloning of genes, etc.

The DNA markers are classified in several different ways. One way of grouping them is based on the method of their detection; it classifies markers as hybridization-based (RFLP, DArT), PCR-based (RAPD, AFLP, CAPS, etc.) and sequencing-based (SNP) markers. Another approach is to classify markers based on their location in the genome. The **random markers** are located at random sites in the genome and their relevance to plant function is not known. The **gene-based markers** are located within genes, while **functional markers** are such gene-based markers whose alleles reliably reflect the functions of the alleles of concerned genes (Table 13.2).

Table 13.2: The salient features of random, gene-based and functional markers

Category of marker	Location	Cost of / effort in development	Marker-trait association	Remarks
Random	Anywhere in the genome	Low	Validation needed; may change due to recombination	Most of the DNA markers; must be tightly linked with the target trait
Gene-based	Within genes	Moderate	Validation needed; may change due to recombination	Relationship with the allelic effects on the phenotype not known
Functional	Within genes	Very high	Validation not needed; cannot change due to recombination	Marker alleles reliably related with the trait phenotype

There are over two dozen DNA marker systems. Some of them are widely used, and are listed here: (i) RFLP, (ii) random amplified polymorphic DNAs (RAPDs), (iii) amplified fragment length polymorphism (AFLP), (iv) sequence characterized amplified regions (SCAR), (v) simple sequence repeat (SSR) polymorphisms, (vi) cleaved amplified polymorphic sequences (CAPSs), (vii) diversity array technology (DArT), (viii) inter-SSR (ISSR) markers, (ix) start codon targeted (SCoT) polymorphism, and (x) single nucleotide polymorphism (SNP) (Table 13.3).

Table 13.3: The salient features of different DNA marker systems in common use

Marker	Developed by	Year	Detection based on	Remarks
RFLP	Botstein and coworkers	1980	Differential lengths of DNA fragments generated by restriction enzymes	Robust, not in common use, serves as a reference marker
RAPD	Williams and coworkers	1990	Amplification of genomic sequences using a single arbitrary sequence primer of 10 nt	Simplest DNA marker, low reproducibility
AFLP	Zabeau and Vos	1993		
SCAR	Paran and Michelmore	1993		
SSR	-	1990		
CAPS	Konieczny and Ausubel	1991		
DArT	Jaccoud and coworkers	2001	Hybridization of selectively amplified fragments with the genotyping array	Cheap, requires specialized software and equipment
ISSR	Zietkiewicz and coworkers	1994		
SCoT	Collard and Mackill	2009		

13.3. ISOLATION AND PURIFICATION OF DNA FROM PLANTS

Generally, DNA is isolated from leaf, but tissues like endosperm and other plant parts are also used. There are three types of procedures for plant DNA isolation. The CTAB (cetyltrimethylammonium bromide) method is the standard method, and it yields DNA preparation of high quality. It requires considerable time and effort, and costs about 1,000 times more than the rapid DNA isolation procedures. The rapid DNA extraction methods are cheap and fast, but DNA quality is low and storability is poor. The commercially available DNA isolation kits (Mini/Midi/Maxi kits) cost 2-3 times more than the CTAB method, but yield good quality DNA; the DNA yield is low and sample processing speed in moderate.

13.4. POLYMERASE CHAIN REACTION

Polymerase chain reaction (PCR) can produce millions of copies of a DNA segment in a matter of few hours. This procedure was developed and first used by Mullis and coworkers (Saiki et al. 1985). PCR uses a pair of primers complementary to the end sequences of the DNA segment to be amplified (called **amplicon**) and a thermostable DNA polymerase, usually Taq (source, *Thermus aquaticus*) polymerase, for DNA replication. The reaction mixture also contains the template or sample DNA from which the segment is to be amplified, and other reagents including the four deoxyribonucleotides (dNTPs) needed for DNA synthesis. The PCR procedure comprises the following three steps: (i) denaturation, (ii) annealing and (iii) primer extension (Table 13.4). In the **denaturation step**, the reaction mixture is heated to 94°C so that the two strands of the template DNA become separated. The duration of this step is 2 min in the first PCR cycle and only 1 min in the subsequent cycles. The mixture is then cooled to a suitable temperature (usually, between 40 and 60°C) to allow pairing of the primers with its complementary sequences present in the template DNA (Fig. 9.14). The duration of the **annealing step** is 1 min. The mixture is now heated to 72°C for 2 min; during this period the Taq polymerase extends the primers by adding nucleotides to their 3' ends. The completion of this step completes one PCR cycle. In this way, a new copy of the given target segment is produced at the end of the first PCR cycle. In the second and the subsequent cycles, both the old and the new copies of the target segment will function as templates. As a result, the number of copies of the target segment will be doubled after each PCR cycle. Thus, after n cycles, 2^n copies of the target segment are expected, but the actual number of copies generated is somewhat smaller. The Taq polymerase generally amplifies amplicons of up to 2 kb, but it can amplify longer fragments if special reaction conditions are provided.

Table 13.4: The three steps of PCR procedure

Step	Temperature (°C)	Duration	Objective achieved
Denaturation	94	First cycle, 2 min; later cycles, 1 min	Separation of the two strands of the template DNA
Annealing	40-60*	1 min	Pairing between the primes and the target DNA
Primer extension	72	2 min; 7 min in the final cycle	Replication of the target segment of template DNA

PCR procedure is carried out in fully automated thermal cyclers, and the researcher has only to specify the temperatures and the durations of the three steps of PCR. The PCR is usually run for 40 cycles. The PCR primers are generally 15-17 nucleotides (nt) long, and their sequences are complementary to the 3' ends of the target segment. Thus, a pair of primers is used to amplify one amplicon; as a rule, the melting temperatures of the two primers should be the same. The melting temperature of a DNA sequence is given by the formula 4(G + C) + 2(A + T), where G + C and A + T are the numbers of G.C and A.T base pairs in the DNA duplex. The annealing temperature during PCR is generally kept 1-2°C lower than the estimated melting temperature for the pair of primers used for PCR. But in the case of RAPD, the annealing temperature is set 5°C lower than the primer melting temperature since the primers are much shorter (generally, 10 nt) in this case. The primers are chemically synthesized by oligonucleotide synthesizers based on the phosphite triester method of DNA synthesis. The Taq DNA polymerase is the cheapest and the most widely used. But some other thermostable DNA polymerases like Pfu (source, *Pyrococcus furiosus*) and Vent (source, *Thermococcus litoralis*) are more efficient, and are used to achieve some specific objectives. In general, a single target sequence is amplified in one PCR tube, but two or even more different target sequences may be amplified in a single tube; the latter procedure is called **multiplex PCR**. For multiplex PCR, it is essential that all the primer pairs should have the same or nearly the same annealing temperature and their amplification products should preferably be of different lengths for easy separation and identification. Multiplexing increases the assay throughput, and decreases the cost, labor and time needed for the assay. **Throughput** is the actual number of different tests completed by an assay procedure in a unit time.

13.5. RESTRICTION FRAGMENT LENGTH POLYMORPHISM

Restriction fragment length polymorphism (RFLP) is detected by digesting the genomic DNAs from different individuals with the selected restriction enzyme and separating the DNA fragments by polyacrylamide (smaller fragments) or agarose (larger fragments) gel electrophoresis. The fragments in

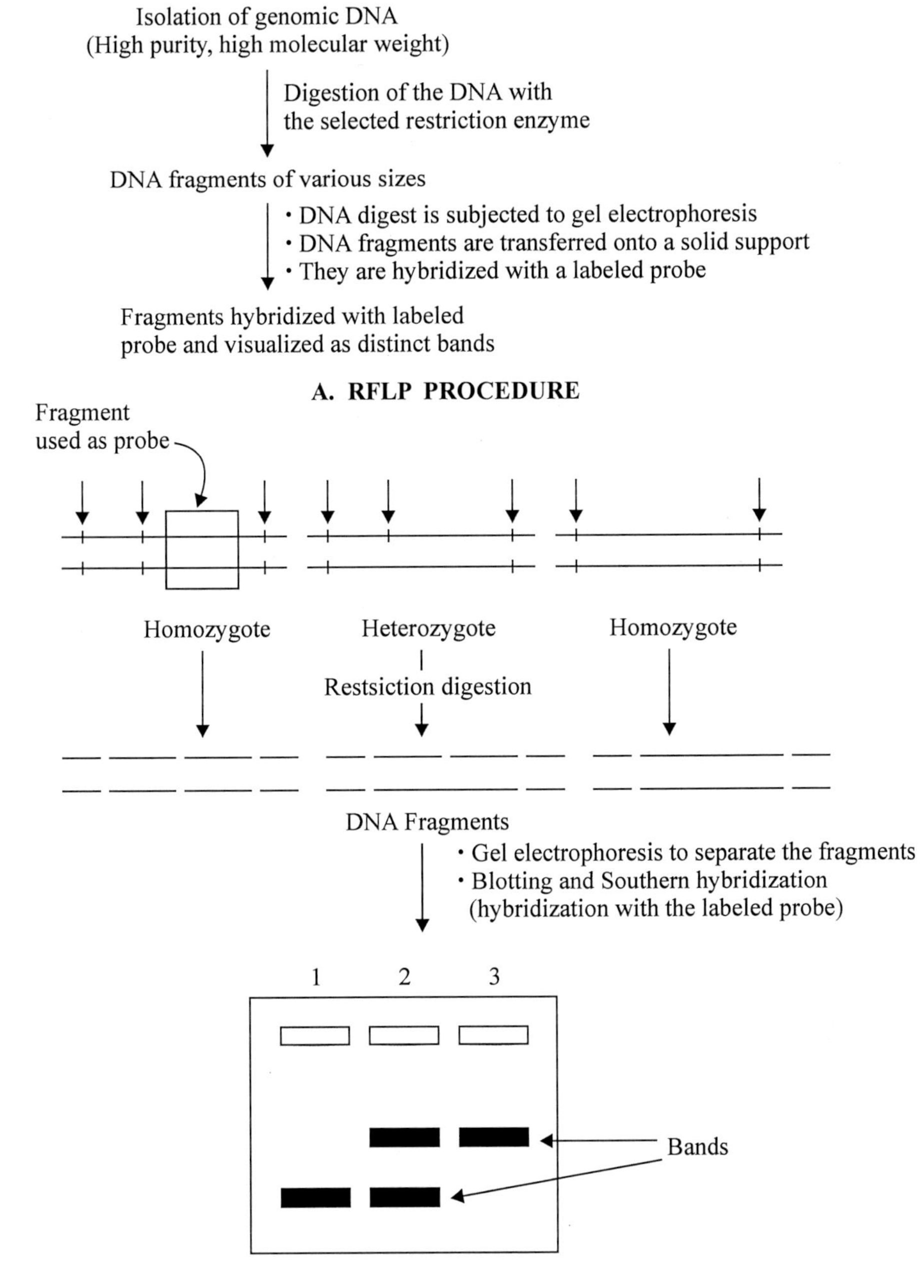

Isolation of genomic DNA
(High purity, high molecular weight)

↓ Digestion of the DNA with the selected restriction enzyme

DNA fragments of various sizes

↓
- DNA digest is subjected to gel electrophoresis
- DNA fragments are transferred onto a solid support
- They are hybridized with a labeled probe

Fragments hybridized with labeled probe and visualized as distinct bands

A. RFLP PROCEDURE

Fragment used as probe

Homozygote Heterozygote Homozygote

Restsiction digestion

DNA Fragments

↓
- Gel electrophoresis to separate the fragments
- Blotting and Southern hybridization (hybridization with the labeled probe)

1 2 3

Bands

B. MOLECULAR BASIS OF RFLP MARKERS

Fig. 13.1. A schematic and simplified representation of the RFLP procedure. A. simplified procedure for RFLP detection. B. The molecular basis of RFLP markers and their codominant nature. Arrows (→) indicate the recognition sites for the restriction enzyme used for digestion of the DNA. Lanes 1 and 3, the two homozyotes; Lane 2, the heterozygote.

the gel are denatured, transferred and fixed onto a solid support (**Southern blotting**), and hybridized with a suitably labeled DNA probe (**Southern hybridization**). The probes not hybridized to the DNA fragments are removed by washing and the fragments to which the probe is hybridized are detected as distinct bands by autoradiography or development of color following specific procedures (Fig. 13.1). RFLPs were the first DNA marker, and they were once widely used for genetic mapping. They were later replaced by more user-friendly SSR markers, but at present SNPs are the markers of choice.

Enzymes that produce internal cuts or cleavages in DNA molecules are known as **endonucleases**. A class of endonucleases cleaves DNA only within or near such sites that have specific base sequences. These enzymes are called **restriction enzymes** or **restriction endonucleases;** they DNA molecules within or near highly specific sequences called **recognition sequences** or **restriction sites**. These enzymes can be of type I, II or III; the type II enzymes have palindromic sequences with rotational symmetry as recognition sites (Fig. 9.8) with an even number (usually, 4, 6 or 8) of base pairs, and cut either within or just outside these sites. The term restriction enzyme ordinarily refers to the type II enzymes since only they are used in genetic investigations, including generation of markers like RFLP, AFLP, DArT, etc., and genetic engineering. Most restriction enzymes cannot cut at methylated recognition sites; therefore, these enzymes cut repetitive DNA into much larger fragments than unique DNA since the former is generally much more methylated than the latter.

During electrophoresis, the fragments are separated on the basis of their size, the smaller fragments moving faster than the longer ones. The size of fragments in the gel lanes is estimated by comparing them with a mixture of fragments of known size (generally called **marker DNA**), usually run in the first (the extreme left) lane of the gel. During blotting, the fragments are transferred from the gel onto a solid support in such a way that the relative positions of the fragments remain the same as in the gel. The solid support initially comprised nitrocellulose filter membranes. Subsequently, nylon and other membranes with superior features were developed and are generally preferred. The blotting may be based on capillary action (original procedure) or on electrophoresis (the preferred procedure).

Southern hybridization is DNA: DNA hybridization that uses DNA fragments of typically 500-3,000 bp; these fragments are called **probes**. Probes enable the detection of those fragments having sequences complementary to the probes from among a large number of diverse fragments fixed onto the membrane. Probes are usually selected from genomic or cDNA (copy DNA or complementary DNA) libraries, but they may also be obtained by asymmetric PCR or some other suitable approaches. Single-copy sequences are the source of the best probes, but low copy number or even multiple-copy DNA sequences have been used as probes. Usually, the probes are obtained from DNA sequences of the same species **(homologous probes)**, but probes from other related species

(heterologous probes) are also used. The probes are suitably labeled for easy detection. Initially, radioactively labeled probes **(hot probes)** were used, but later chemically labeled probes **(cold probes)** became available and are preferable. Chemical labeling involves attaching a suitable ligand like an enzyme, a fluorophore, etc. with a nucleotide, and to use this nucleotide for labeling of the probe.

The differences in the lengths of restriction fragments result from the following alterations in the concerned genomic region: (i) change in the recognition site for the restriction enzyme as a consequence of a SNP, and (ii) a detectable large deletion or insertion. In addition, inversion and translocation may also produce RFLP. The available evidence reveals that the bulk of RFLPs result from mutated restriction sites. In a diploid species, a unique probe ordinarily detects two bands, one fast and slow moving band. These two bands represent the two alleles at this RFLP locus. The RFLP alleles are always codominant since the two bands are readily separated and identified (de Vienne 2003). The RFLP markers can be used for construction of marker linkage maps, which can be integrated with conventional linkage maps. RFLPs can also be used for mapping of genes and quantitative trait loci (QTLs) of interest. RFLPs were extensively used during the 1980s and 1990s, and are still used as anchor markers. There are numerous RFLP loci, the marker system is highly reproducible, but 5-10 μg of DNA of high purity is needed from each line for RFLP analysis. RFLP analysis is technically demanding, costly, tedious and time consuming. But many of the RFLP markers can be converted into user-friendly PCR-based markers like CAPS (cleaved amplified polymorphic sequences).

13.6. RANDOMLY AMPLIFIED POLYMORPHIC DNA

Randomly amplified polymorphic DNA (RAPD) markers are detected by using high molecular weight genomic DNA for PCR amplification using a single 10 nucleotide long RAPD primer. Since a single primer is used, the primer must bind at two neighboring sites of the template DNA to enable amplification. In addition, the two binding sites should be oriented towards each other and should be, usually, within 2 kb. The PCR products are separated by agarose gel electrophoresis and visualized as distinct bands in the gel by staining the gel with ethidium bromide. It may be expected that a 10 nucleotide long primer would produce about 2 bands in rice, 9 fragments in maize, and 19 PCR products in barley. The RAPD markers are dominant since they are scored as presence/ absence of the amplified fragments. These markers are the simplest since their development involves only designing of the primers. As a result, they were once extensively used for linkage mapping, diversity analysis, etc. Some groups like Operon Technologies had designed standard set of RAPD primers for use by research workers. But the interest in RAPD markers has declined in view of its poor reproducibility. In case a useful RAPD marker has been discovered, it can be converted into a highly reproducible SCAR marker (Section 13.9).

13.7. AMPLIFIED FRAGMENT LENGTH POLYMORPHISM

Amplified fragment length polymorphism (AFLP) procedure combines some features of each of the RFLP and RAPD procedures. It involves restriction digestion of genomic DNA, reduction in the number of different fragments by selective PCR amplification and detection of the amplified fragments by gel electrophoresis (Fig. 13.2). About 100-500 ng genomic DNA from each individual is digested with one frequent cutter (4 bp recognition site), e.g., *Mse*I, and a rare cutter (6 bp recognition site), e.g., *Pst*I, restriction enzyme. The ends of the resulting restriction fragments are modified by ligating them to appropriate adapters. An **adapter** is a short nucleotide sequence (oligonucleotide), one end of which is single stranded and complementary to the single stranded region produced by the concerned restriction enzyme. It may be pointed out that most restriction enzymes cleave DNA within the recognition sites, and generate short single stranded regions of the fragments. The single stranded regions generated by a single restriction enzyme are complementary with each other. This property greatly facilitates construction of recombinant DNA.

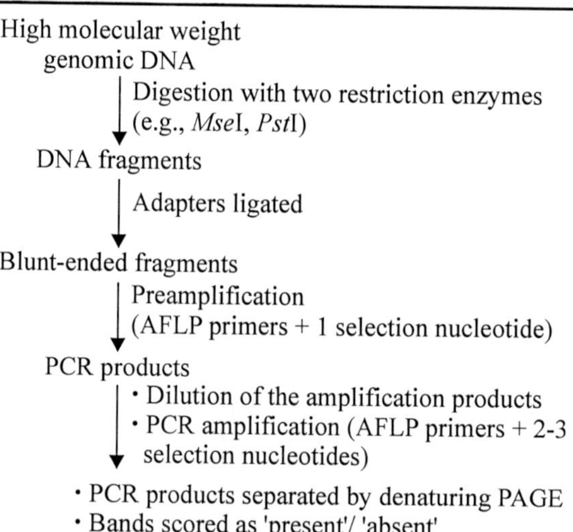

Fig. 13.2. A schematic representation of the AFLP procedure. The DNA fragments will be *Mse*I-*Mse*I (both ends generated by *Mse*I), *Mse*I-*Pst*I and *Pst*I-*Pst*I. In case of the *Mse*I-*Mse*I and *Pst*I-*Pst*I fragments, both the ends will be produced by digestion with the *Mse*I and *Pst*I, respectively. Similarly, the two ends of *Mse*I - *Pst*I fragments will be produced by different enzymes. The AFLP primer for *Pst*I end is labeled to permit scoring of the bands. AFLP, amplified fragment length polymorphism, PCR polymerase chain reaction; PAGE, polyacrylaride gel electrophoresis.

The fragments are now amplified by a two-step PCR procedure using AFLP primers. An **AFLP primer** is complementary to the concerned adapter and the restriction enzyme recognition sequences, and has up to three arbitrary nucleotides, called **selection nucleotides**, at its 3' end. In the first step, the AFLP primers have a single selection nucleotide, so that only $1/16^{th}$ (= 1/4 x 1/4) of the total fragments are amplified. The mixture of fragments is suitably diluted after the pre-amplification step, and used for the second step, the amplification step. The AFLP primers in this step have 2 or 3 selection nucleotides, so that only $(1/4)^4$ or $(1/4)^6$ of the fragments is amplified. The primer for the ends generated by the 6 bp cutter is labeled, preferably with fluorescence. The fragments are separated by denaturing PAGE, and about 50 to 100 distinct bands are visualized by fluorescence.

The AFLP generates dominant markers that are scored as present or absent. This marker system is highly reproducible, less cumbersome and more informative than the RFLP marker system. It does not require a marker development step needed for the RFLP, SSR and the SNP marker systems. But it uses restriction enzymes that are costly and require high quality genomic DNA. In some plant species, these markers are mostly concentrated in the centromeric regions. A PCR-based STS marker can be derived from an AFLP marker found to be linked to a target gene much in the same way as RAPD markers are converted into SCAR markers. The AFLP method can be modified in several different ways, e.g., use of three restriction enzymes, use of one AFLP and one SSR sequence primer, use of hybridization to microarray in the place of PAGE for detection of polymorphism (DArT marker system; Section 13.8) etc. AFLPs can be used for gene/QTL mapping, MAS, diversity analysis, germplasm characterization, etc.

13.8. DIVERSITY ARRAY TECHNOLOGY

Diversity array technology (DArT) is a microarray-based, low cost and high throughput marker system. The first step in DArT is to construct a genotyping array by pooling ~5 ng DNA from several diverse individuals of a species. The DNA pool is digested with two restriction enzymes; the fragments are ligated to appropriate adapters and amplified following the AFLP procedure (Section 13.7). The amplified fragments are cloned, and the DNA inserts from all the clones are amplified, and spotted onto a suitable solid support, e.g., a microscopic slide; this array is called **discovery array**. The DNA from any two of the lines/individuals used for construction of the discovery array is separately processed by the above procedure, and the PCR products are labeled with different fluorescent dyes. These PCR products are together hybridized with the discovery array, and the spots showing hybridization with the fragments from only of the lines/individuals are identified. These spots represent DNA fragments that detect polymorphism between the two lines/individuals; these fragments are spotted onto a solid support to produce the genotyping array. The DNA (50–100 ng) from each line/individual to be genotyped for DArT markers is processed by the same procedure that was used for the development of the arrays. The PCR products are labeled with fluorescence, and hybridized with the genotyping array. Hybridization is detected by fluorescence, and the data are processed using a suitable software like DArTsoft (Fig. 13.3). The marker genotype data are shown as 1 (fluorescence signal present) and 0 (signal absent) (Jaccoud et al. 2001).

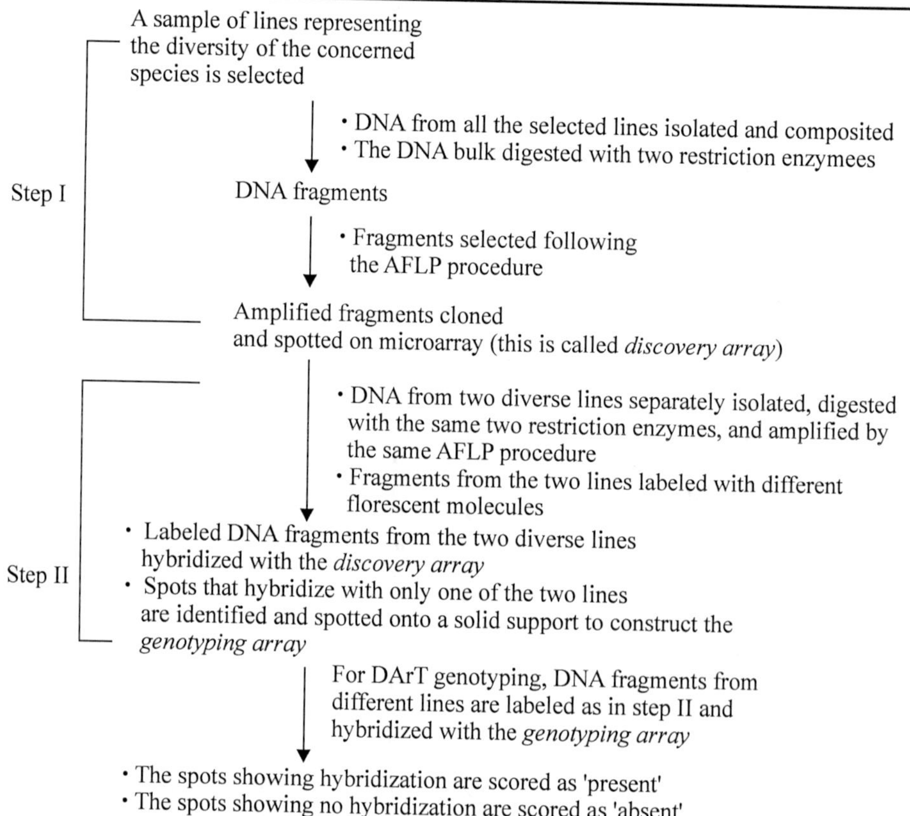

Fig. 13.3. A schematic reperesation of the DArT marker system. AFLP, amplified fragment length polymorphism; DArT, diversity array technology.

The polymorphism detected by DArT arises from the same events as those for RFLP markers. DArT markers are dominant, highly reproducible, and distributed throughout the genome. They are used for fingerprinting, diversity studies and linkage mapping of genes and QTLs. The DArT procedure has been fully automated; it genotypes thousands of loci in each assay and costs only few cents per data point. Genotyping arrays for several crop species like wheat, rice, chickpea, etc. have been developed. The main limitations of DArT are the use of restriction enzymes, and the need for specialized equipment and software, which may not be affordable by many researchers.

13.9. SEQUENCE CHARACTERIZED AMPLIFIED REGIONS

The sequence characterized amplified region (SCAR) markers were developed to increase the reproducibility of RAPD markers. Their development involves the isolation and purification of the fragment representing the desired RAPD marker from the gel. This fragment is cloned, its two ends are sequenced and this sequence information is used to design a pair of 20-24 nt long primers (Fig. 13.4). These primers are expected to amplify a single fragment corresponding to the concerned RAPD marker. A proportion of such primer pairs will amplify polymorphic fragments. Such primer pairs give rise to SCAR markers, which can be used for MAS for the genes linked with them. SCAR markers are generally dominant, but some of them are codominant. Although SCAR markers were originally developed with reference to RAPD markers, the STS markers derived from AFLP and other marker systems are also often referred to as SCAR markers. A **sequence tagged site (STS)** is a locus that is reliably and reproducibly amplified by a pair of primers specific for the two termini of this locus. A STS marker represents a STS locus that shows polymorphism on PCR amplification using the specific primer pair.

An RAPD marker linked to a gene of interest is identified

↓

The band representing this RAPD marker is excised from the gel, DNA fragment is isolated, purified and cloned

↓

The two ends of the fragment are sequenced; these end sequences are used to design a pair of 20-24 nt long primers

↓

Primers are used for PCR amplification; if polymorphism is detected with these primers, a SCAR marker is obtained

Fig. 13.4. The development of a SCAR marker from an RAPD marker. nt, nucleotide.

13.10. SIMPLE SEQUENCE REPEAT (SSR) POLYMORPHISMS

Simple sequence repeats (SSRs) or **microsatellite sequences** are made up of generally 5-50 tandem repeats of 1-6 bp long sequences that function as repeating units. The most common repeating units have two nucleotides, of which the CA, GA and AT repeats are the most common and highly polymorphic. In addition, three and four nucleotide long repeating units are also common in plant genomes. The site of the genome having a specific SSR sequence is considered as a locus for the concerned SSR sequence. The different loci for a given SSR sequence are distributed throughout the genome at every 30 to 100 kb. But in some plant species like tomato, the SSR loci seem to be concentrated around the centromeric regions of the chromosomes. The SSR loci are highly polymorphic and they show multiple alleles. The alleles at a given SSR locus differ for the number of the repeating units present in them (Fig. 13.5A). Further, there is no sequence variation among the alleles of a single SSR locus. The SSR loci can be discovered as follows: (i) identification of the clones with DNA inserts having SSR sequence, and sequencing of the DNA insert, (ii) analysis of genome sequences for discovery of SSR loci and designing of the primers, and (iii) screening of the expressed sequence tag (EST) databases for SSR locus discovery and designing of primers.

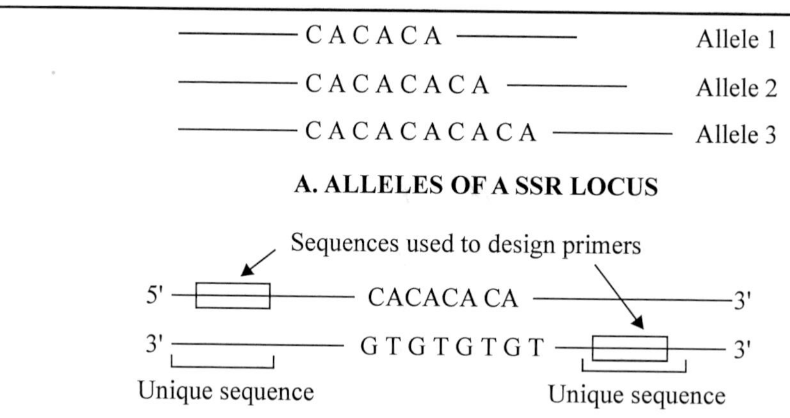

Fig. 13.5. SSR markers. A. Different SSR alleles show variation in the number of repeat units (CA in this case). B. A SSR locus is amplified by designing a pair of primers based on unique DNA sequence flanking the SSR locus. All the alleles of this locus will have the unique sequences used to design the primers. SSR, simple sequence repeat.

The SSR loci are used for the development of PCR-based markers generally called **simple sequence repeat markers (SSR markers)**. The SSR markers are developed as follows: the unique sequences on the two sides of a SSR locus are determined, and used for designing a pair of PCR primers usually of 17-20 nt (Fig. 13.5B). This pair of primers is used to amplify the concerned SSR locus from the genomic DNAs of different individuals/lines. The amplification products are subjected to PAGE and the different alleles of the SSR locus are seen as bands migrating to different positions in the gel, representing the fragments of different lengths. In general, a pair of SSR primers amplifies a single SSR locus, and a single band is seen in case of homozygotes, while two bands are observed in the heterozygotes at the concerned SSR locus (Fig. 13.6). Further, only a single SSR locus is amplified in each PCR tube and the PCR products of a single tube are run in a single gel lane. But more than one SSR locus can be amplified in a single PCR tube and the PCR products are separated in a single gel lane; this is termed as **multiplex PCR**. It is important that all the primer pairs used in a single PCR tube should have the same or almost the same annealing temperature, and that their amplification products should be of different lengths. Suitable strategies can be devised to achieve multiplexing even when the different primers generate PCR products of overlapping sizes. For example, the different primers can be labeled with different fluorophores and the PCR products can be analyzed with the help of automated DNA sequencers. The unique sequences flanking most of the SSR loci are generally highly conserved within a species and even between related species. Therefore, SSR primers designed for a species can often be used in some related species as well.

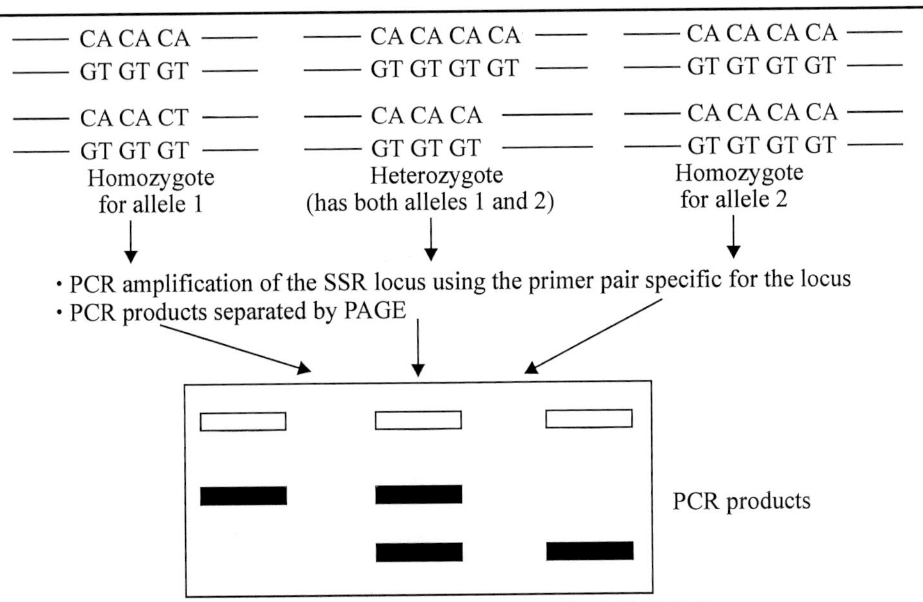

Fig. 13.6. Detection of SSR alleles by PCR using the primer pair specific for the locus and visualization of the PCR products following PAGE (polyacrylamide gel electrophoresis). SSR, short sequence repeat; PCR, polymerase chain reaction.

The SSR markers are user-friendly, can be automated, have medium throughput, and are distributed throughout the genome. They are highly polymorphic, generally multiallelic and show co-dominance. They require only ~100 ng DNA per sample, are relatively cheap and often transferable across species. This marker system used to be the marker of choice till few years ago, when SNPs came on the scene and rapidly occupied the pristine position among the DNA markers. The development of the SSR markers is technically demanding as considerable effort is needed to generate the necessary sequence information. Further, the cost of automation for the SSR markers is rather high, and they are not as abundant as the SNP markers. Finally, some SSR loci may have null allele. In case of **null allele**, the primers fail to amplify a detectable fragment, while they successfully generate fragments in case of the other alleles of the locus.

13.11. INTER-SSR (ISSR) MARKERS

Inter-simple sequence repeat markers (**ISSR markers**) detect length polymorphism in the genomic regions flanked by the same SSR sequence that is used as the primer for amplification of the genomic DNA. The amplification is based on a single 18-20 nt long primer consisting of a SSR sequence. The primer may have the SSR sequence only; this marker system is in no way superior to RAPD markers. Alternatively, the primer may have few arbitrary nucleotides usually at their 3' ends (Fig. 13.7); these primers are called **anchored primers**. The 3' anchored primers will amplify the genomic regions flanked by the SSR sequence constituting the primers; these SSR sequences should preferably be located within 2 kb of each other and they must be oriented opposite to each other (Fig. 13.9). Further, amplification will take place only when this genomic region has bases complementary to the arbitrary nucleotides added at the 3' ends of the primers. The ISSR primers generate several bands from a single sample, which are scored as present/absent, i.e., as dominant markers. This procedure can be modified in several different ways. For example, an ISSR primer may be used with an AFLP primer during the amplification step of the AFLP procedure. ISSR markers have better reproducibility than the RAPD marker system. They are easy to use, relatively inexpensive and have high throughput and a single primer generates multiple polymorphic markers. This marker system does not require sequence information of the template DNA, and can be modified in various ways to suit the need of the researchers. However, ISSR markers do not have high reproducibility, and some of the primers do not generate clear bands.

13.12. START CODON TARGETED POLYMORPHISM

Start codon targeted (**SCoT**) **polymorphism** refers to the sequence polymorphism in the short genomic region around the start codons of the genes. The SCoT markers are designed to detect the polymorphism in the start codon region (Collard and Mackill 2009). This marker system uses a single primer, called **SCoT primer**, of 18 nt for PCR amplification of the genomic DNAs of the test individuals/lines. The PCR is carried out using annealing temperature of 50°C and the primer extension time of 2 minutes or more. The PCR products are separated by agarose gel electrophoresis, and between 2 and 6 clear bands are generated by each primer. A SCoT primer has some fixed nucleotides that are common to all primers, and some variable nucleotides that differ from one primer to the other. Different SCoT primers should differ from each other for at least one nucleotide at their 3' ends. The amplification of a genomic region will take place when the start codons of two genes are located within 2 kb of each other, and the two genes are oriented towards each other. These markers are dominant and are scored as presence/absence of the concerned bands. The markers are based on polymorphism in the genic regions and are highly reproducible. There is no need for marker development, and the design of primers does not require any sequence information. They can be used for linkage mapping of genes/QTLs, diversity analysis, germplasm characterization, etc.

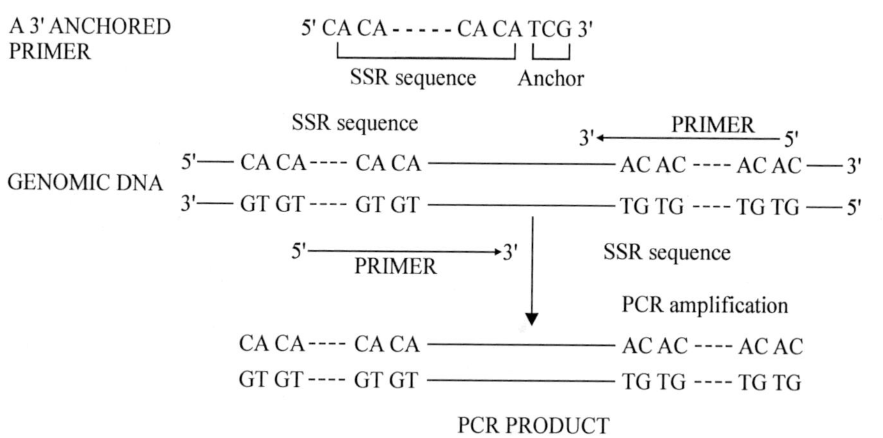

Fig. 13.7. A schematic representation of an ISSR marker. The two SSR loci should be within 2kb and oriented opposite to each other. SSR, simple sequence repeat; ISSR, inter-simple sequence repeat; PCR, polymerase chain reaction.

13.13. SINGLE NUCLEOTIDE POLYMORPHISM

Single nucleotide polymorphism (**SNP**) is the presence of different nucleotides at the same genomic site of different individuals. Thus, a SNP locus can have up to four alleles (Fig. 13.8), but most methods of SNP genotyping identify only two alleles at each locus. A SNP locus comprises a single polymorphic base pair, and is identifiable by the sequences surrounding this base pair. The SNPs are produced by base substitutions, and transitions are the more frequent causes of SNPs than are transversions. SNPs are distributed throughout the genome at very high density; in plant genomes, one SNP may be present at every 100-300 bp. SNPs are present in both coding and noncoding regions of the genome, but are more frequent in the latter. SNPs have been classified on the basis of several criteria, e.g., location in the genome (coding, noncoding, exonic, intronic, and promoter SNPs), the type of change produced in the amino acid sequence of the encoded protein (synonymous and nonsynonymous SNP), etc. Initially, the SNPs had to be first discovered, following which they were used for genotyping. But methods that combine SNP discovery and genotyping have been developed and are being refined.

Allele 1	5' —— A —— 3'	3' —— T —— 5'		Individual 1
Allele 2	5' —— T —— 3'	3' —— A —— 5'		Individual 2
Allele 3	5' —— G —— 3'	3' —— C —— 5'		Individual 3
Allele 4	5' —— C —— 3'	3' —— G —— 5'		Individual 4

Fig. 13.8. The four possible alleles at a single SNP locus. The same genomic segment from four different individuals differs for a single nucleotide at the given site (called SNP locus) in the segment. The 5'→3' strands in the four individuals contain the nucleotide A, T, G and C, respectively.

The SNP discovery may involve one of the following approaches. (i) Whole genomes of two or, preferably, more individuals may be sequenced and their sequences analyzed to discover SNPs. Obviously, this approach will identify SNP loci distributed over the entire genome. (ii) A pair of specific primers may be used to amplify desired genomic regions; the amplification products are sequenced and their sequences are compared with genome sequences to detect SNP loci. (iii) The genome sequences and the EST sequences available in the databases are analyzed to identify SNP loci. Finally, the RNA transcripts from several individuals may be sequenced and compared with each other or with a reference genome sequence for SNP discovery. Once the SNP loci are identified, they are validated using a suitable assay procedure, and the validated SNPs are then used for genotyping of individuals/lines. There are over 30 different methods that can be used for SNP genotyping at small to moderate scale, but some of them have been adapted for high throughput genotyping systems (Table 13.5). These methods are based on DNA : DNA hybridization, extension of specifically designed primers, ligation of specific oligonucleotides, or DNA replication. One of these methods, allele-specific PCR, is described in Section 9.15. The approaches for combining SNP discovery with SNP genotyping are briefly outlined in Section 9.16.

The SNP markers are abundant, distributed over the entire genome at high density, and their detection procedures are highly amenable to automation and, as a result, high throughput genotyping. They are highly reproducible and the cost per data point is the lowest among DNA markers. The SNPs are rapidly becoming the most preferred marker system, and have displaced the SSR markers from being the marker of choice. However, their discovery requires sequence information that demands financial and other resources. Further, the cost of genotyping equipment is high particularly in the case of high throughput genotyping platforms. Finally, the use of this marker system is technically demanding, and for many research workers it may be more practical to use commercial genotyping facilities.

Table 13.5: Some selected techniques for genotyping of known SNPs

Detection method	Basis of detection	Remarks
Cleaved amplified polymorphic sequences	Restriction digestion of PCR product	It was developed for PCR-based genotyping for RFLPs
Allele-specific PCR	Amplification of the genomic region having the SNP	Low throughput; modified as high throughput KASP™ assay
5'-Nuclease assay	Release of fluorophore by 5'nuclease action of Taq polymerase during DNA synthesis	Modified as high throughput TaqMan OpenArray Genotyping system
Molecular beacons	Fluorescence generated only when the molecular beacon base pairs with the SNP allele	Some degree of multiplexing is allowed by labeling of different molecular beacons with separate fluorophores
Primer extension	Size or mass differences among the primer extension products	Developed as high throughput homogeneous MassEXTEND (hME) assay based on MALDI-TOF MS
Invasive cleavage or invader technology	Fluorescence generated due to cleavage after formation of invasive structure	Developed as high throughput assay system
Oligonucleotide ligation	Presence/absence of the ligation product	Used in the Illumina's highly multiplexed GoldenGate™ assay

13.14. CLEAVED AMPLIFIED POLYMORPHIC SEQUENCES

The **cleaved amplified polymorphic sequence (CAPS) markers** were developed for PCR-based detection of RFLP markers (Williams et al., 1991). The procedure for detection of CAPS markers is as follows. The genomic DNAs from different individuals are used for PCR amplification using the pair of primers specific for the given CAPS marker. The PCR products from the different individuals would of identical length, but they would differ for the presence of a recognition site for a restriction enzyme. This restriction enzyme is used for digestion of the PCR products, as a result of which the PCR products from some individuals will be cleaved into two fragments, while those from others will remain intact (Fig. 13.9). The PCR products from the different individuals are subjected to agarose gel electrophoresis, and the bands are scored. The homozygotes will show either one (no cleavage of the PCR product) or two (cleavage of the PCR product), while the heterozygote will show three bands. CAPS markers are produced by elimination of restriction sites in some of the individuals/lines. This change in the recognition site is produced by a SNP or insertion/deletion (InDel) within the restriction site. Initially, CAPS markers were developed on the basis of known RFLP markers; hence they are sometimes called **PCR-RFLP**. They can be also developed by analysis of, especially, EST sequences using computer programs designed for the purpose.

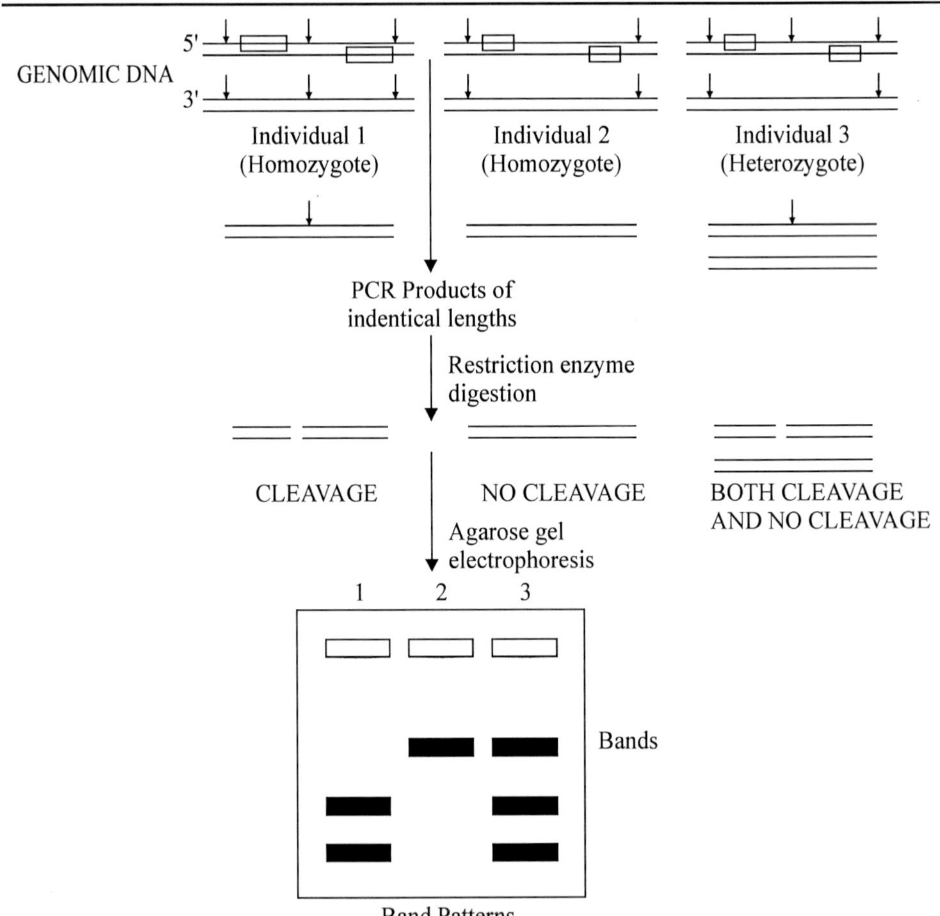

Fig. 13.9. A schematic representation of a CAPs marker. The arrows (in the `GENOMIC DNA' row) indicate the recog-nition sites for the restriction enzyme. The sequences enclosed within boxes are used as PCR prmers (18-24 nt long). The PCR products from all the individuals are of the same sizes, but they differ for the presence of a recognition site for the restriction enzyme. CAPS cleaved amplified polymorphic sequences products.

CAPS markers are codominant, easy to use and relatively economical. They are used for scoring known SNP loci of interest and for positional cloning of genes/QTLs. However, this marker system is based on restriction digestion of the PCR products, which requires additional work and adds to the cost; it also prevents automation of the marker detection.

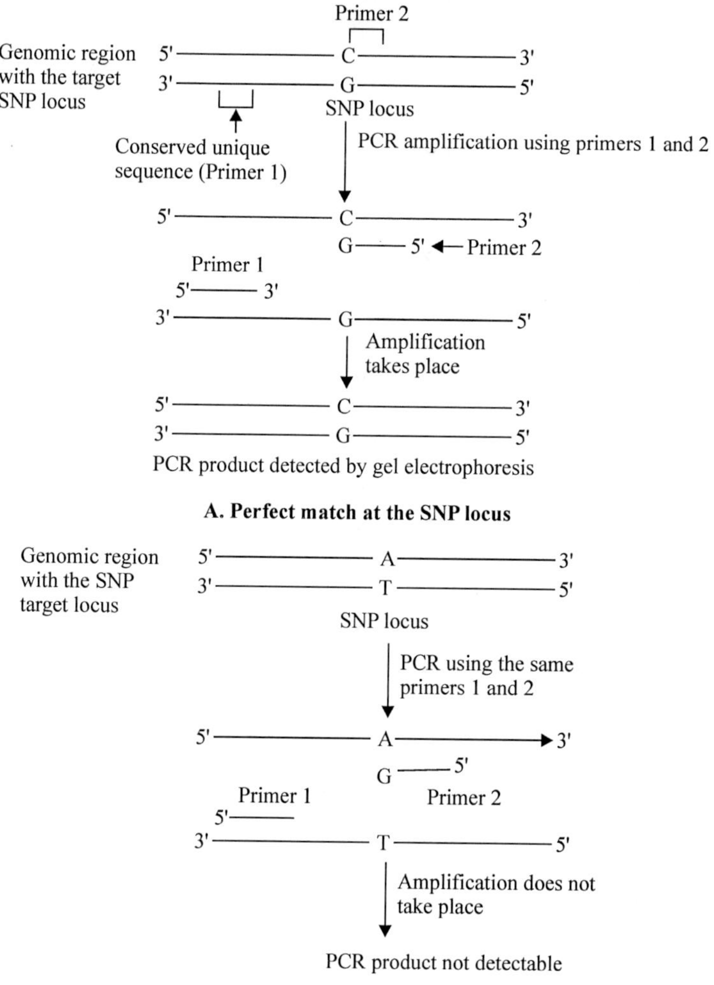

Fig. 13.10. A simplified representation of the allele-specific PCR procedure for the detection of SNP alleles. **A.** PCR amplification takes place when there is perfect match between the SNP locus and the 3' terminal base of the primer 2. As a result, a detectable PCR product will be produced. This clearly indicates the SNP allele to be C since the terminal base of the primer 2 is G. **B.** But when there is mismatch at the SNP locus, the 3' terminal base of primer 2 will not be properly aligned. As a result, amplification will not take place, and a PCR product will not be detectable.

13.15. ALLELE-SPECIFIC PCR

Allele-specific PCR uses a pair of PCR primers that are so designed that the PCR product will be generated only when the base at the 3'end of one of the primers is complementary to the SNP allele at the target locus. One of these primers binds to the genomic sequence at a suitable distance from the SNP locus; this sequence is conserved in all the alleles of the SNP locus and is located at the 5' side the locus. The other primer binds to the genomic region having the SNP locus, and its 3' terminal base is placed precisely at the SNP locus. The genomic DNAs from the different test individuals are isolated and used separately for PCR amplification using this pair of primers. In case the 3' terminal base of the second primer is complementary to the SNP allele present in the genomic DNA, this base will pair with the SNP allele and a PCR product will be produced (Fig. 13.10A). But when the base at the 3'end of the primer is not complementary to the SNP allele, there will be no pairing between the two and no PCR product will be generated (Fig. 13.10B). But sometimes amplification may occur even when the 3' terminal base of the second primer is not complementary to the SNP allele. This problem can be resolved by creating a mismatch at the base located next to the 3' terminal base. It may be pointed out that the mismatch at the second base will not prevent amplification when the 3' terminal base is complementary to the SNP locus. However, it will always eliminate amplification when the SNP allele is not complementary to the base at the 3'end of the primer.

This method is simple, user-friendly and can be used in any laboratory that has PCR and gel electrophoresis facilities. However, it has low throughput, and permits genotyping for a limited number of SNP loci at a time. The assay procedure has been refined to develop a high throughput commercial genotyping system called KASP™.

13.16. HIGH THROUGHPUT GENOTYPING

A **high throughput genotyping system** simultaneously analyzes a large number of samples for a very large number of markers. For example, a single Sentrix Array Matrix of the Illumina GoldenGate SNP genotyping platform can simultaneously analyze 1,536 SNP loci in 96 different samples. The high throughput genotyping systems have been designed for SNP loci that are already identified and validated as true SNPs. The various genotyping systems are basically of the following three types: (i) low multiplex, (ii) moderate multiplex and (iii) high multiplex systems. The examples of **low multiplex** systems are invader technology, pyrosequencing and KASP™ assays. Some examples of assays amenable to **moderate multiplexing** are as follows: TaqMan OpenArray system, SNP genotyping using MALDI-TOF MS, nanofluidic dynamic arrays, and the Array Tape procedure. **High level multiplexing** is permitted by the following procedures: the Illumina GoldenGate, and Affymetrix's whole genome genotyping platforms. In addition, there are several **whole genome based array platforms**; these include Illumina Infinium HD assay, Affymetrix's GeneChip and Axiom® genome-wide arrays, Agilent's SurePrint arrays and Beckman Coulter's SNPstream genotyping system.

The methods that combine SNP discovery with SNP genotyping are based on DNA sequencing. These methods became possible due to tremendous improvements in the speed and throughput of DNA sequencing and drastic reduction in sequencing costs. The various methods of DNA sequencing are grouped into the following three categories: (i) first generation, (ii) second or next generation and (iii) third generation DNA sequencing methods (Table 13.6). Of the two first generation methods, Sanger Coulson method was the most successful, was automated, widely used for genome sequencing projects, and remains the gold standard. The three important NGS methods are 454 DNA sequencing, the Solexa or Illumina sequencing method and the ABI SOLiD technology. The 454 method uses pyrosequencing procedure that uses DNA synthesis for sequencing of DNA molecules. The Illumina method also sequences DNA by synthesis, while the SOLiD technology employs oligonucleotide ligation for this purpose. The NGS procedures sequence millions of different DNA fragments and usually generate short reads. However, they generate huge amounts of sequence data in relatively short period of time (Table 13.6). Specialized computer programs are required to process and assemble the sequence data. The third generation DNA sequencing methods do not generate multiple copies of the DNA fragments; instead they sequence single molecules of DNA. Two of these methods, viz., Helicos genetic analysis system and single-molecule real-time (SMRT) technology have been commercialized, and some other technologies may soon reach the market. In addition, new more efficient and rapid techniques have been developed for RNA sequencing; the RNA sequence data can also be analyzed for detection of SNPs. The DNA/RNA sequence data from the test individuals are compared with a high quality reference genome sequence to detect the presence of SNPs.

Table 13.6: Salient features of some of the important DNA sequencing methods

Sequencing method	Strategy	Remarks
First generation sequencing methods		Multiple copies of DNA fragments used for sequencing are obtained by cloning in bacteria; relatively long sequence reads
Chemical method	Chemical modification of bases	
Enzymatic method	DNA synthesis; ddNTPs* used to block DNA synthesis	Procedure automated; extensively used for genome sequencing; yields high quality sequence data; read length** up to 1,000 bp; remains the standard reference sequencing method

Sequencing method	Strategy	Remarks
Second or next generation sequencing (NGS) methods		Millions of different DNA fragments sequenced simultaneously; multiple copies of each fragment produced by PCR; short read lengths; much faster and cheaper
454 DNA sequencing	DNA synthesis	Read length 500 bases or more; 400 Mb sequence data in 10 h; equipment, e.g., Genome Sequencer (GS) FLX System
Solexa technology or Illumina sequencing method	DNA synthesis	Read length 35-250 bases; over 400 Gb sequence data in 7-8 days; equipment, e.g., Genome Analyzer 1 Gb and HiSeq 600 Gb
ABI SOLiD technology	Oligonucleotide ligation	Read length ~ 50 bases; 20 Gb (even 300 Gb) of sequence data in 6-7 days; equipment, e.g., SOLiD 3.0
Third generation sequencing (TGS) methods		Single DNA molecules sequenced; often termed as single molecule sequencing (SMS) methods; two methods commercialized
Helicos Genetic Analysis System	DNA synthesis	Read length about 35 bases; 35 Gb sequence data per run; equipment , e.g., HeliScope™
Single-molecule real-time (SMRT) technology	DNA synthesis by single DNA polymerase molecules	Read length ~1,000 bp (even up to 20 kb); 20 Gb sequence data per 30 minutes; equipment , e.g., PACBIO RS
RNA Sequencing (RNA-Seq)	cDNA[@] used for sequencing	cDNA sequenced by a suitable NGS technology; used for analysis of gene expression, SNP discovery
Single molecule direct RNA sequencing	Direct sequencing of RNA molecules by synthesis	Read length up to 55 bases; up to 12 Gb sequence data per run; equipment , e.g., Helicos® Genetic Analysis System

*dideoxynecleotides.

**The average sequence length of a single fragment.

[@]cDNA, complementary DNA; produced by reverse transcription of RNA by reverse transcriptase.

It may be pointed out that SNP discovery by complete whole genome sequencing requires considerable amount of work, but sequencing of the genomes the individuals in a sample of reasonable size is still very expensive. Therefore, different strategies have been devised to reduce the amount of sequencing work required for the discovery and genotyping of the SNPs (Table 13.7). These strategies either detect SNP loci distributed over the entire genome or identify them in specific genomic regions. There are three genome-wide methods, viz., reduced representation sequencing, restriction site-associated DNA sequencing (RAD-Seq), and low coverage genotyping. In **reduced representation sequencing**, only a fraction of the genome is sequenced in each individual of the sample, but the complete genome is sequenced when all the individuals are considered together. The reduction in the genome fraction used for sequencing is achieved either by selecting fragments of a specified size or by using the AFLP approach for selecting only a fraction of the fragments for sequencing. In the **RAD-Seq approach**, only small segments of DNA surrounding the recognition sites of the selected restriction enzyme are sequenced, but these recognition sites are distributed over the whole genome. In **low coverage genotyping**, the whole genome is sequenced at low (2x – 4x) coverage and appropriate computer programmes are used to identify SNPs. There are two basic schemes for low coverage genotyping, viz., genotyping by sequencing and multiplexed shotgun genotyping. In addition, some strategies have been devised for SNP detection in targeted regions of the genome. The three main techniques for **targeted SNP discovery** are as follows: SureSelect from Agilent, SeqCap from Nimblegen, and Targeted Sequencing System from RainDance. In addition, the use of RNA sequence data for SNP discovery and genotyping limits the marker search to the genomic regions transcribed in the tissue from which the RNA was isolated during the specific developmental stage of the organism and under the prevailing environment.

Table 13.7: The chief approaches for high throughput combined SNP discovery and genotyping

Genotyping method	Salient feature	Remarks
Genome-wide methods		
Reduced representation sequencing	A fraction of fragments sequenced in each individual; sequences of all the individuals together represent the entire genome	Suitable for many types of studies; methods: reduced representation libraries (RRLs), complexity reduction of polymorphic sequences (CRoPS).
Restriction site-associated DNA sequencing (RAD-Seq)	Sequencing of short regions surrounding all recognition sites for the selected restriction enzyme(s)	Used for various studies including linkage mapping; modifications: genotyping by sequencing (GBS), 2-enzyme GBS, ezRAD, etc.
Low coverage genotyping	Sequencing of the whole genome at 2x – 20x compared to the normal 30x coverage*	Suitable for genotyping of recombinant populations; techniques: genotyping by sequencing and multiplexed shotgun genotyping
Targeted marker discovery methods	SNP discovery and genotyping in specific genomic regions	Three main techniques for capturing the target sequences: SureSelect from Agilent, SeqCap from Nimblegen, and Targeted Sequencing System from RainDance

*30x sequencing coverage means that each base of the genome has been sequenced, on an average, 30 times.

13.17. TRAIT PHENOTYPING

Phenotyping comprises scoring of the different individuals/lines of a population/sample for the expression of the traits of interest. It must be kept in mind that a precise phenotyping is essential for linkage mapping of the genes/QTLs governing the traits of interest. Since the oligogenic traits show very high (~100%) heritability and there is complete correspondence between phenotype and genotype, they can be evaluated on the basis of individual plants grown in nonreplicated trials. But the expression of quantitative traits is affected by the prevailing environment; therefore, their phenotyping should always be done in replicated trials conducted according to an appropriate experimental design. Whenever possible, the replicated trials should be conducted at more than one location and, if feasible, during two or more years; this can be done when inbred lines, homozygous lines or clones are used for phenotypic evaluation. The lines included in a trial should not be adapted to different growing conditions since their performance under uniform conditions will not reflect their genotypic potential. These lines should not differ markedly for flowering time because this trait influences the expression of several other traits as well. Finally, evaluation for some traits may interfere with the proper expression of other traits. For example, evaluation of a population for tolerance to biotic or abiotic stress will affect the expression of many other traits, including yield.

In case of many studies, large populations have to be evaluated for several traits in a short period of time. In view of this, techniques for high throughput precision phenotyping are being developed; these together constitute the discipline of **phenomics**. Phenomics is essentially a twenty-first century discipline that began in relation to human neurobiological studies. But today, it means different things to different people, and is extensively applied to plant studies. A simple definition of **plant phenomics** would be collection of huge amounts of data pertaining to plant growth, architecture, composition, performance, etc. using high throughput techniques, and analysis and interpretation of these data. The techniques used in phenomics are automated, high throughput procedures that are either improvements of the classical methods or have been newly developed for the purpose. One of the most promising approaches in plant phenomics uses light reflectance, transmission or fluorescence for either imaging or estimation of various indices that enable evaluation of a variety of plant phenotypes and responses (Araus and Cairns 2014, Fiorani and Schurr 2013, Munns et al. 2010, Chaerle et al. 2009). Aerial vehicles like drones or unmanned aerial vehicles would enable phenomics easy and reliable data collection from large tracts of fields.

The imaging technology analyzes various types of images for estimation of a variety of plant phenotypes, including responses to biotic and abiotic stresses. In most cases, images are acquired at one time, but for some traits like growth rates a series of images of each plant need to be taken at successive periods of time. In case of plants that tend to grow in the horizontal plane, e.g., rosette plants, only top view image is acquired. But in case of most crop plants that also grow vertically, one top view and two side view images of each plant are taken. The images are acquired using one or more of the following types of video cameras: (i) visual cameras or RGB (red, green and blue) cameras, (ii) near infrared (NIR) cameras, (iii) infrared (IR) cameras, (iv) fluorescence cameras, and (v) hyperspectral cameras. The images may be taken either manually (on small scale) or by an automated imaging system (in the large scale phenomics facilities). The automated imaging system may be based on either moving cameras (the plants remain static) or on moving plants (plants are grown in pots kept on a conveyor belt). The images are stored in a computer and processed using a suitable software package (Table 13.8) to derive the desired information.

Table 13.8: Some computer programs used for image processing

Software package	Application	Available at
ImageJ	Basic image analysis software; most image analysis software are plugins of ImageJ	*http://rsbweb.nih.gov/ij/index.html*
HTPheno	Estimation of plant height, plant width, projected shoot area etc.; plugin of ImageJ	*http://htpheno.ipk-gatersleben.de/*
Rosette Tracker	Plant growth etc. of Arabidopsis; plugin of ImageJ	*http://telin.ugent.be/~jdvylder/RosetteTracker/*
Martrack Leaf	Leaf growth rate	*As additional files in Mielewczik et al. (2013).*
High-throughput Plant Growth Analysis	Plant area, AGR and RGR of rosette plants	*http://www.msu.edu/~jinchen/HPGA*
SmartRoot	Analysis of root system	*http://www.csc.univie.ac.at/rootbox/rsa.html*
RootReader3D	3-D root system reconstruction	-

Image analysis can generate information on the phenotypes of various plant traits including responses to biotic and abiotic stresses (Table 13.9). For example, the analysis of visual images allows estimation of leaf area, height, volume and biomass of plants as well as their architecture, seed and fruit morphology, growth rate, stay-green period, flowering period, fruit ripening, etc. It also allows determination of senescence, necrosis, infections by diseases, certain kinds of nutrient deficiencies, etc. Similarly, IR imaging (thermography) allows determination of canopy temperature, which depends mainly on stomata closure; therefore, canopy temperature reflects the drought tolerance of the different genotypes. In addition, ratios between reflectance (or transmittance) at two specific wavelengths provide useful information about certain traits. For example, normalized difference vegetation index (NDVI) is perhaps the most widely used to measure responses to abiotic stresses like drought, salinity, etc.; it has also been used to predict yield (Table 13.10). NDVI is estimated as the ratio of reflectance in the red and NIR regions [NDVI = $(R_{NIR} - R_{red})/(R_{NIR} + R_{red})$], where R_{NIR} and R_{red} represent the reflectance at 800 nm and 680 nm, respectively. Similarly, the ratio of transmittance of red and NIR wavelengths is measured by SPAD chlorophyll meters and used for the estimation of chlorophyll contents of plants. In addition, IR thermometers are being used in many breeding programs to monitor canopy temperature that is used to estimate canopy temperature depression (CTD). **CTD** is the difference between the canopy temperature and the temperature of the surrounding air at the same point in time. A positive value of CTD indicates lower canopy temperature. CTD has been used to identify drought stress tolerant genotypes in crops like rice and wheat is the measurement of a trait that is significantly and predictably correlated with the trait of interest.

Table 13.9: The various imaging technologies and their applications

Imaging technology	Wavelength range	Applications
Visual imaging	400 – 950 nm (with filters: 400 – 700 nm)	Plant morphology, health and architecture, including growth rate, nutrient deficiency, disease incidence, senescence, etc., and soil water content; used as reference for the other imaging techniques
Near infrared (NIR) imaging	900 – 1700 nm	Root system, water status, etc
Infrared imaging	8000 – 14000 nm	Canopy temperature, stomatal conductance, drought stress, etc
Fluorescence imaging	Same as visible imaging	Photosynthesis, chlorophyll content, stress (biotic and abiotic) response
Hyperspectral imaging	Visible to NIR range	Biochemical features like pigment composition, nitrogen use efficiency
Magnetic resonance imaging		Visualization of bound and free water distribution in plant tissues
Laser scanning		Construction of 3-D images of plants

The phenomics is a rapidly developing field and is the area of intensive research. It may be expected that more efficient, faster and cheaper phenotyping techniques will be developed in the near future and plant phenotyping may become a much less challenging task in the near future.

Table 13.10: Some selected surrogate measurements in common use for phenotyping of plants.

Surrogate measurement	Trait of interest	Measurement technique	Remarks
NDVI*	Chlorophyll content; response to drought, salinity or nutrient deficiency	LED-based spectrometers	Being used in many plant breeding programs
Chlorophyll fluorescence	Chlorophyll content	Fluorescence imaging	-
Transmittance in the red and NIR regions	Chlorophyll content	SPAD chlorophyll meters	Being used by many large scale breeding programmes
Canopy temperature depression (CTD)	Heat and drought stress tolerance	IR thermometers; thermal imaging	Temperature measurements in the mornings and the afternoons
Normalized difference water index**	Drought stress tolerance	NIR reflectance measurement	Allows the onset of water stress to be detected
Stay-green phenotype	Drought stress tolerance	Visual imaging	Combination of visual, fluorescence and thermal imaging is more informative

* NDVI, normalized difference vegetation index. NDVI = $(R_{NIR} - R_{red})/(R_{NIR} + R_{red})$, where R_{NIR} and R_{red} represent the reflectance at 800 nm and at 680 nm, respectively.

** Normalized difference water index = $(R_{860} - R_{1200}) / (R_{860} + R_{1200})$, where R_{860}, and R_{1200} denote reflectance at 860 and 1200 nm respectively.

13.18. CONCLUSIONS

For a long time, plant breeders have been in search of a suitable marker system primarily for easy indirect selection for such traits that are difficult-to-evaluate and/or select. This search culminated in the development of DNA-based markers, which are generally termed as molecular markers. Scientists have developed over two dozen molecular marker systems, of which AFLP, DArT, SSR and SNP markers are widely used. SNPs are currently the most favoured markers in view of their dense distribution over the entire genome and the availability of extremely high throughput genotyping platforms. SNPs are rapidly replacing the SSR markers that were once the most widely used markers in biology.

A breeder would like to have a DNA marker linked to the trait that is to be subjected to indirect selection. It is the simplest if this information were already available to him. Otherwise, the breeder would need to select a suitable marker system, and then identify the specific marker linked to the trait of his interest. In addition, a suitable mapping population would have to be developed. This population has to be genotyped for a large number of markers of the selected marker system. Further, the population has to be reliably phenotyped for the trait(s) of interest. The data on marker genotype and target trait(s) phenotype are used for linkage mapping to identify the markers linked to this trait/these traits.

PRACTICE QUESTIONS

1. Define the term genetic marker. Briefly describe the various types of genetic markers, and the various activities necessary for their application for crop improvement.
2. What are DNA markers? List the important features of an ideal DNA marker. Give a list of the various types of DNA markers, and describe any one of them in some detail.
3. Briefly describe the RFLP markers. Explain the method by which at least some of the RFLP markers can be converted into PCR based markers.
4. Define the terms genotyping and phenotyping. Briefly describe the method for genotyping and phenotyping, and discuss their relevance.
5. Write short notes on the following: (i) genetic marker, (ii) isozyme markers, (iii) DNA markers, (iv) an ideal marker, (v) RAPD, (vi) RFLP, (vii) PCR, (viii) AFLP, (ix) DArT, (x) SCOT markers, (xi) SNP, (xii) genotyping, (xiii) high throughput genotyping, (xiv) phenotyping, (xv) phenomics.

RELEVANT LITERATURE

Araus JL, Cairns JE 2014. Field high-throughput phenotyping: the new crop breeding frontier. *Trends Plant Sci* 19:52-61.

Chaerle L, Lenk S, Leinonen L et al 2009. Multi-sensor plant imaging: towards the development of a stress-catalogue. *Biotechnol J* 4:1152-1167.

Collard BCY, Mackill DJ 2009. Start codon targeted (SCoT) polymor-phism: a simple, novel DNA marker technique for generating gene-targeted markers in plants. *Plant Mol Biol Rep* 27:86–93.

de Vienne D (ed) 2003. Molecular markers in plant genetics and biotechnology. Science Publishers, Inc, Enfield (NH), USA.

Fiorani F, Schurr U 2013. Future scenarios for plant phenotyping. *Annu Rev Plant Biol* 64:267–291.

Jaccoud D, Peng K, Feinstein D et al 2001. Diversity arrays: a solid state technology for sequence independent genotyping. *Nucleic Acids Res* 29: e25.

Jiang GL 2013. Molecular markers and marker-assisted breeding in plants. *InTech* http://dx.doi. org.10.5772/52583.

Muller-Starck G 1998. Isozymes. In: Karp A, Isaac PG, Ingram DS (eds) Molecular tools for screening biodiversity. Chapman and Hall, London, pp 75-81.

Munns R, James RA, Sirault XRR et al 2010. New phenotyping methods for screening wheat and barley for beneficial responses to water deficit. *J Exp Bot* 61:3499-3507. doi:10.1093/jxb/erq199.

Saiki R, Scharf S, Faloona FA et al. 1985. Enzymatic amplification of b-globin genomic sequences and restriction site analysis for diagnosis of sickle cell anemia. *Science* 230: 1350-1354.

Sax K 1923. The association of size differences with seed coat pattern and pigmentation in *Phaseolus vulgaris* Genetics 8: 552-560.

Singh, BD and Singh AK 2015. Marker-Assisted Plant Breeding, Principles and Practices. Springer India, New Delhi.

Thoday JM 1961. Location of polygenes Nature 191: 368-370.

Williams MNV, Pande N, Nair M et al 1991. Restriction fragment length polymorphism analysis of polymerase chain reaction products amplified from mapped loci of rice (*Oryza sativa* L.) genomic DNA. *Theor Appl Genet* 82:489-498.

Xu Y 2010. Molecular plant breeding. CAB International, Wallingford, U.K.

■■■■

14 Marker-Trait Associations

14.1 INTRODUCTION

The phenomenon of linkage, first reported by Morgan in 1910, is the tendency of two or more genes being inherited together. Linkage is the consequence of two genes being located close to each other in the same chromosome. New combinations of linked gene are produced by crossing over between them (Fig. 14.1A). In general, the frequency of crossing over depends mainly on the distance between the linked genes, and it can be estimated from the frequencies of the new gene combinations produced in the progeny. This property has long been used to construct linkage maps depicting linked genes (Fig. 14.1C), and this procedure is called **linkage mapping**. The procedure of linkage mapping has been gainfully used to prepare linkage maps of molecular markers and to identify useful marker-trait associations that could be used for marker-assisted selection in breeding programs. **Marker-trait association** signifies linkage between a molecular marker and the gene or quantitative trait locus (QTL) involved in the expression of the trait of interest (or the target trait). In addition to linkage mapping, marker-trait associations can also be detected by the procedure of association mapping. Both linkage mapping and association mapping are carried out in carefully constructed plant populations as described in Sections 14.2 and 14.3. The linkage mapping of genes and QTLs, and the procedure for association mapping are explained in the following sections (Sections 14.5, 14.6 and 14.7).

14.2 POPULATIONS FOR LINKAGE MAPPING

A **mapping population** comprises of a suitably large number of either individuals or lines and can be used for either linkage or association mapping. The populations used for linkage mapping are produced by crossing generally two completely homozygous lines, but more lines are involved in crosses for creation of some of the mapping populations. The parents of a mapping population must differ for at least the traits of interest. But it would be highly desirable that they differ for as many traits as possible so that the same mapping population could be used for analysis of many traits. In addition, they should differ for as many marker loci as possible to allow the construction of adequately dense marker linkage maps. Generally, the parents will be different genotypes of the same species, but they may also belong to different species if it is required for some reason. The progeny from the cross are handled as per the procedure appropriate for the type of mapping population to be produced. Mapping populations allow construction of linkage maps of molecular and other markers, and the identification of markers linked to the trait of interest. There are several different types of mapping populations suitable for linkage mapping. Some of the important mapping populations as follows: (i) F_2 population, (ii) backcross population, (iii) doubled haploids, (iv) recombinant inbred lines (RILs), (v) near-isogenic lines, (vi) chromosome segment substitution lines (CSSLs), and (vii) interconnected populations. The salient features of these mapping populations are briefly described in the following sections.

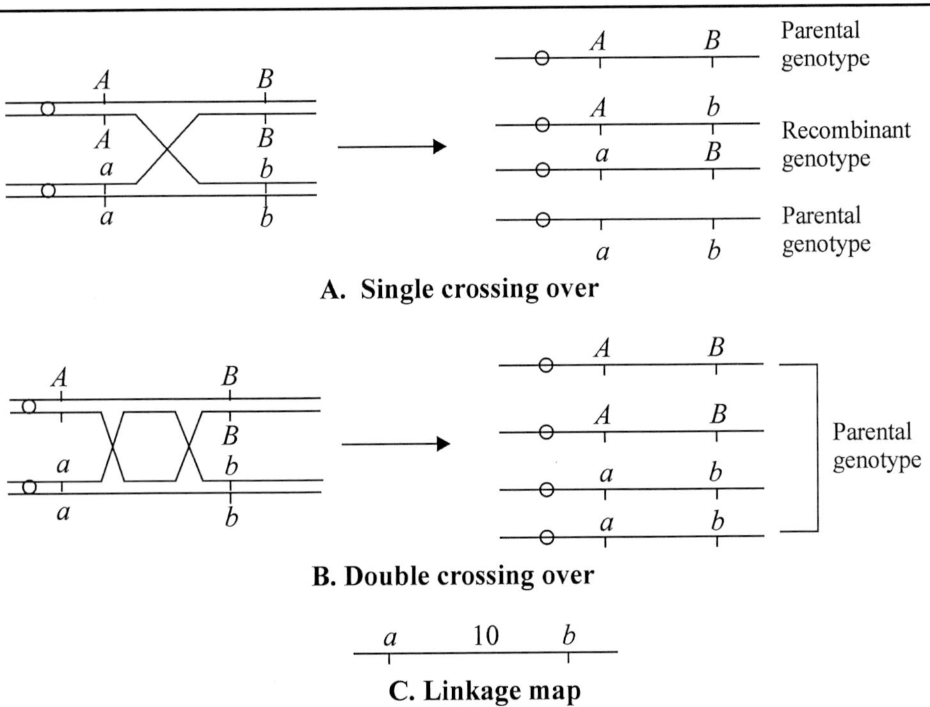

Fig. 14.1. A schematic representation of the basis of linkage map. A. A crossing over produces recombination between linked genes. B. Double crossing over, however, does not lead to recombination. C. A linear map of genes linked together can be prepared on the basis of genetic distance estimated from recombination between the linked genes. In this figure, the genes *a* and *b* are located at 10 map units (denoted by cM, centimorgans) from each other. Crossing over is the exchange of strictly homologous chromatid segments between homologous chromosomes. Recombination, on the other hand, is the production of new gene combinations of linked genes as a result of recombination.

14.2.1 F_2 Population

A **F_2 mapping population** is produced by crossing two homozygous lines and either selfing or sib mating the F_1 plants produced from this cross (Fig. 14.2). Thus, F_2 populations are produced after two generations, which can be completed in one year with the use of greenhouse or off-season nursery facilities. In addition, creation of these populations requires the minimum effort. The segregation ratio in the F_2 generation will be 3:1 for a dominant trait or marker,

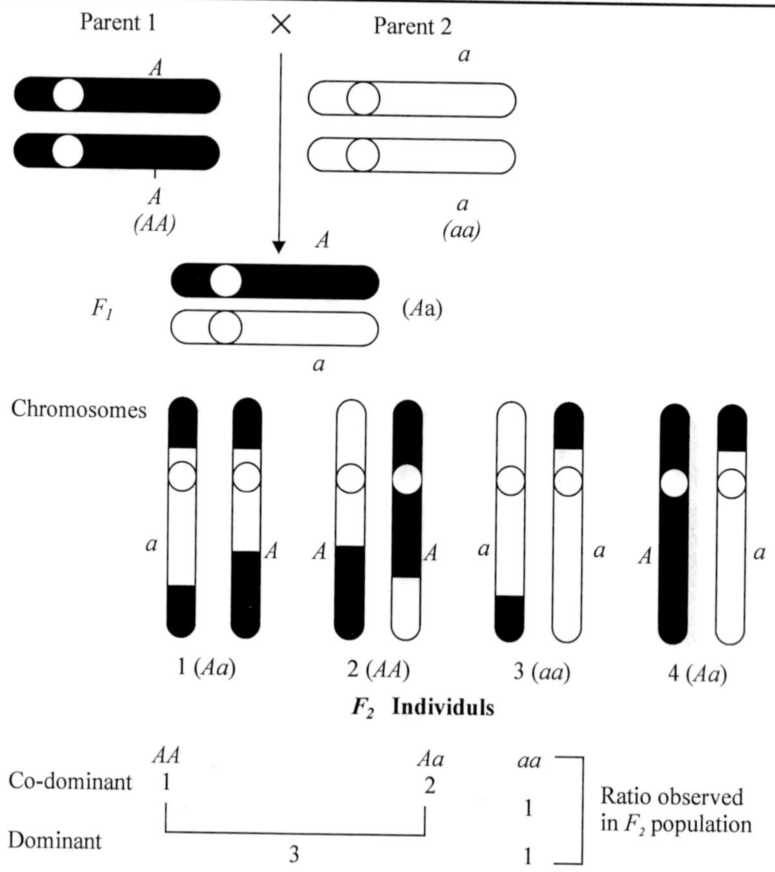

Fig. 14.2. A schematic representation of an F_2 population. In F_2 generation, the ratio for codominant markers/traits will be 1:2:1, while it will be 3:1 for dominant markers/traits. The chromosome composition of any F_2 individual will be a random combination of the chromatids produced during meiosis in males and females. Only one chromosome pair of each individual is shown in this figure.

while it will be 1:2:1 for a codominant trait/marker. Since the F_1 plants will be heterozygous for all the loci for which their parents differ, they will produce an extremely large number of different types of gametes. As a result, each F_2 plant may be expected to possess a genotype that is different from those of the others.

These populations are generally grown for evaluation of qualitative traits with oligogenic control. This is because the phenotypes of such traits can be easily and reliably assessed on the basis of individual plants. In general, these populations are not used for analysis of quantitative traits and mapping of quantitative trait loci. Therefore, F_2 populations are used for preliminary mapping of oligogenes and preparation of marker linkage maps. It may be pointed out that the F_2 generation is produced after only one round on meiotic division in the F_1 generation. As a result, there will be opportunity for only one round of recombination between the markers and the genes of interest. Therefore, the markers that would be identified to be linked to the genes of interest would, on an average, be at a greater distance than the markers identified in some other mapping populations like RILs.

14.2.2 Backcross Population

Backcross is the cross between F_1 hybrid and one of its parents. A backcross generation is denoted by *BC*, and the number backcrosses made to obtain a given generation is depicted by the appropriate number added after this symbol as suffix, e.g., BC_1, BC_2, BC_3 etc. When the F_1 (say, *Aa*) is crossed with the parent having the recessive trait or marker allele (say, *aa*), the backcross progeny exhibit a 1 (*Aa*) : 1 (*aa*) ratio (Fig. 14.3).

But when it is mated to the other parent that has the dominant trait phenotype or marker allele (*AA*), the ratio obtained in the backcross generation is 1 (*AA, Aa*) : 0. Therefore, the backcross population to be used for mapping should always be derived by crossing the F_1 to the parent with the recessive trait phenotype or marker allele. It may be pointed out that this is not necessary in the case of codominant markers and traits, since in such cases the heterozygotes can be readily distinguished from the homozygotes. Therefore, in the case of dominant markers, only those markers that are linked in coupling phase with the target trait can be identified. This population also requires only two generations to be created, which can be completed in one year by using off-season crops or greenhouses. The same population can be utilized for further backcrossing to enable the introgression on the gene being mapped into the other parent of the cross, e.g., the advanced backcross QTL method proposed by Tanksley and Nelson (1996). As in the case of F_2, there is only one round of recombination. Crosses between the F_1 and the parent have to be made to generate the backcross population; this may require considerable effort and limit the population size, particularly in crops like wheat, barley etc.

Both F_2 and backcross populations are ephemeral, i.e., they cannot be perpetuated beyond one generation. Therefore, they have to be created afresh every time it is desired to use them. In addition, it is not feasible to evaluate them in replicated trials and, for that reason use them for QTL mapping.

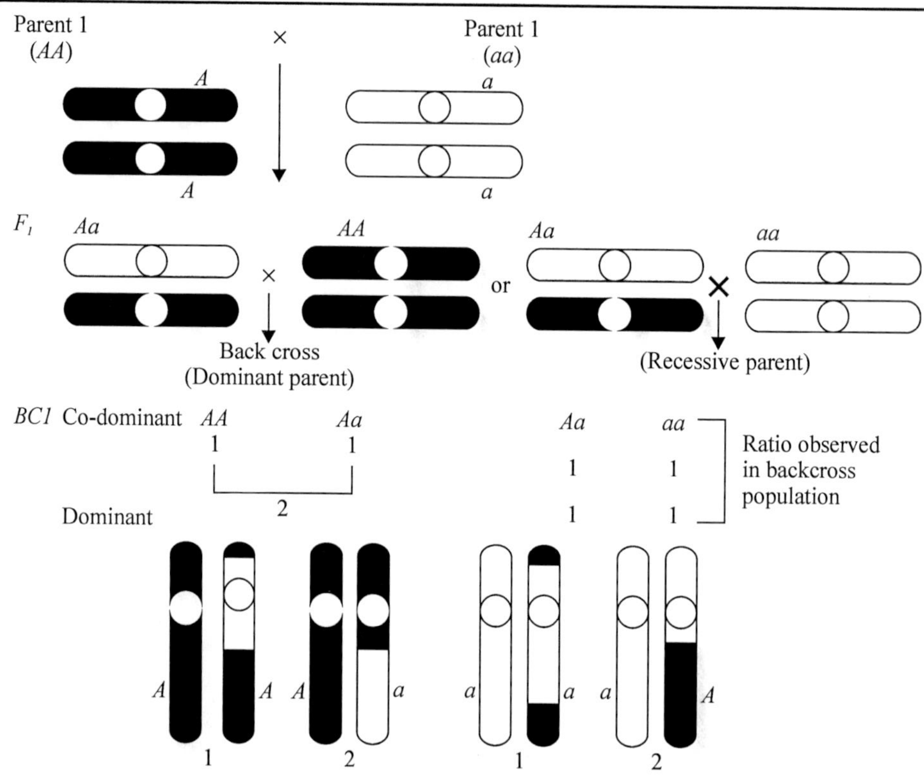

Random Individuals

Fig. 14.3. A schematic representation of the backcross (*BC*) population. Segregation will not be detectable for dominant markers/trails in the *BC* population produced by crossing with the parent with the dominant trait/marker. All the *BC* progeny will receive one chromosome from the parent used for the back-cross, while the other chromosome will come from the F_1 and will have segments from both the parents due to crossing over.

14.2.3 Doubled Haploids

A **doubled haploid (DH)** plant is obtained by doubling the chromosome number of a haploid (*n*) plant. As a result, such plants are completely homozygous and there no residual heterozygosity in them. A DH mapping population is constructed as follows (Fig. 14.4). The two parents selected for developing the mapping population are crossed and the F_1 is used to produce haploid plants. The chromosome numbers of the haploid plants is doubled by a suitable treatment, e.g., colchicines treatment, and the seeds from each plant are harvested separately. The progeny from every haploid plant will be a single homozygous line, and all such homozygous lines constitute the DH population. The haploid plants can be produced by any one of the following approaches: (i) anther culture, (ii) interspecific hybridization and (iii) pollination with a haploid inducer line. Anther culture successfully produces larger numbers of haploid plants in many crops like rice, barley etc. In some crops, some interspecific crosses, e.g., barley x *Hordeum bulbosum*, yield haploids of crop species. In case of maize, when some haploid inducer strains are used as pollinator, they generate 8-20% haploid progeny of the female parent. The segregation ratio for markers as well as qualitative traits is 1:1 irrespective of the dominance relationships the concerned loci. There is only one round of recombination before the isolation of DH lines. These populations are perpetual and they can be evaluated in replicated trials. However, the production of DH populations requires much more technical skill and various resources than that of F_2 and backcross populations.

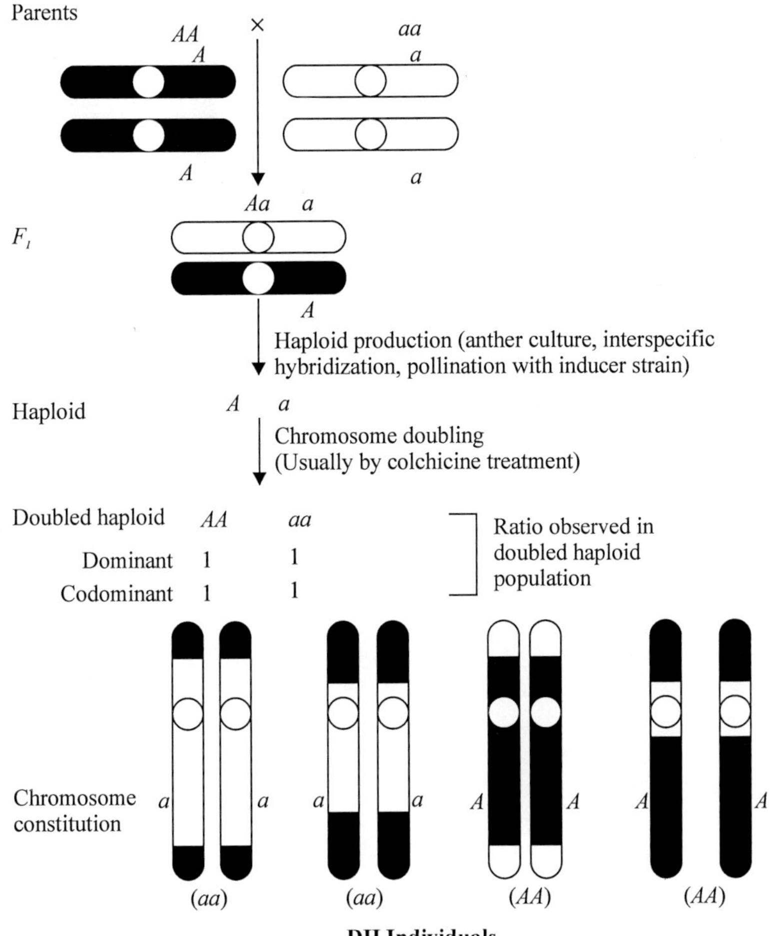

Fig. 14.4. A schematic representation of the DH population. Each DH individual will have identical homologues for each chromosome. Only one chromosome pair is shown here for each individual. DH, doubled haploid.

14.2.4 Recombinant Inbred Lines

A **recombinant inbred line** (**RIL**) is a homozygous line isolated from the F_1 from a cross by continued selfing of the individual F_2 plants obtained from the F_1. In such cases, where selfing is not possible, sib mating can be used for inbreeding process. A **recombinant inbred line** (**RIL**) **population** consists of a random sample of the RILs developed from a cross (Fig. 14.5). It is important that there should be no selection of any kind during the inbreeding process. In general, the RIL populations are developed following the single seed descent (SSD) procedure described in some detail in Section 4.6. In simple terms, the SSD procedure involves selfing of each *F2* plant harvesting a single seed from each plant. The seeds are mixed and grown in F_3, and the process of F_2 is repeated. This procedure is continued till, usually, eight or more generations, but RIL populations developed by SSD up to even four or five generations have been used for linkage mapping. At this stage, all the selfed seeds produced by each F_5 or F_6 plant are harvested and the next generation is grown as individual plant progeny. Each individual plant progeny is an RIL, and the RILs are maintained by harvesting the selfed seeds from all the plants of each RIL in bulk and keeping a random sample for raising the next generation of the RIL. As a result of selfing, the heterozygosity declines to ½ of that in the previous generation and there is proportionate increase in the level of homozygosity (Table 14.1). There will be some plant loss during the SSD procedure. Therefore, the F2 population should be suitably larger than the size of the RIL population desired at the end of the SSD procedure.

Table 14.1: The level homozygosity and heterozygosity in the RILs at different staged of the inbreeding process.

RIL	Heterozygosity (%)	Homozygosity (%)
$F_{3:4}$*	25.00	75.00
$F_{4:5}$	12.50	87.50
$F_{5:6}$	6.25	93.75
$F_{6:7}$	3.125	96.875
$F_{7:8}$	1.562	98.437
$F_{8:9}$	0.781	99.218
$F_{9:10}$	0.391	99.609

*The RIL population is handled up to F3 as per the SSD procedure, after which it will be maintained as an RIL, i.e., by harvesting the selfed seeds from all the plants of the RIL in the bulk and retaining a random sample of this for raising the next generation of the RIL.

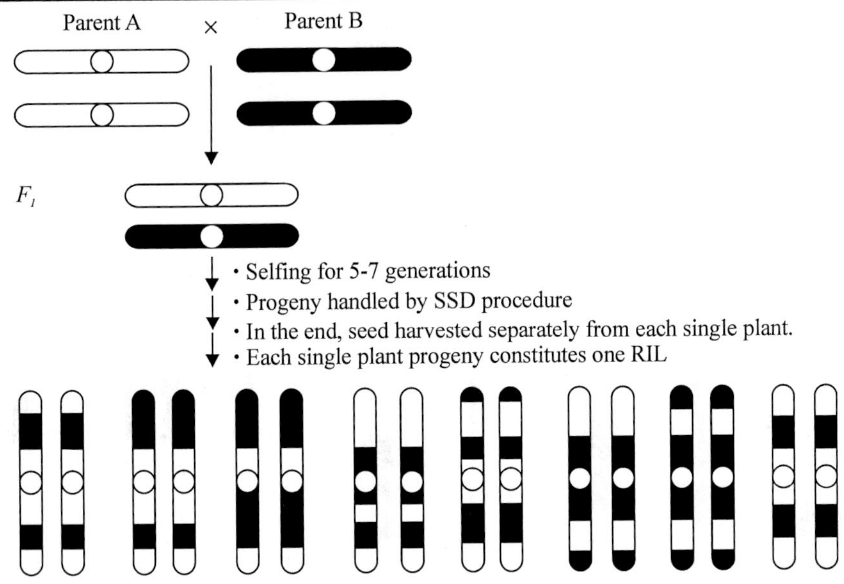

Fig. 14.5. A schematic representation of recombinant inbred lines (RILs). Chromosome composition of a single chromosome pair of each RIL is depicted. Each chromosome of the different RILs has variable segments from both the parents, and the two homologues in an individual are identical (as is the case for DH individuals). SSD, single seed descent; RILs, recombinant inbred lines.

An RIL population comprises almost homozygous lines derived from a cross. These lines will be true homozygous for most of their loci/genes, they might be heterozygous for few to several loci/genes Thus, for a given locus/gene, say, the gene with alleles *A* and *a*, the vast majority of RILs will be homozygous, i.e., an RIL will be either *aa* or *AA*, but some of the RILs will be *Aa*. In an RIL population, the ratio of the lines homozygotes for the two alleles of the gene, viz., line with the genotype *AA* and *aa*, will be nearly 1:1. This ratio will be obtained for both dominant as well as codominant markers and traits. In addition, there will be, on an average, about two rounds of recombination or crossing over between any two linked loc/genes; this will be the case when the RIL population is developed by selfing. It may be pointed out that most RIL populations are developed by selfing. The linked loci may be two genes, two markers or one gene/QTL and one marker. Therefore, the distance between two loci as measured by recombination frequency will be twice as much in an RIL population as that in the an F_2 population. As a result, it may be expected that markers located closer to the gene/QTL of interest will be identified by linkage mapping using RIL populations than BC, DH and F_2 populations. The reason for about two rounds of recombination if the case of RILs is as follows. Recombination will take place in the double heterozygotes *Aa Bb*. All the plants will be double heterozygotes in the F_1 generation. The F_1 generation will lead to one round of recombination. The double heterozygotes will be present in the F_2 as well as the later generations, but their frequency will go on rapidly declining with the advancing generation. If the population size remained constant generation after generation, the total number of the double heterozygote plants in F_2 and the subsequent generations will add up to about one generation provided the two loci under consideration were in close linkage. Thus, the total number of rounds of recombination in an RIL population will be about two.

Since the RIL populations are homozygous, they can be evaluated in replicated trials, and can be maintained indefinitely. However, RIL populations should be maintained in such a manner that there is no risk of out-crossing and mechanical mixture. The construction of RIL populations take up to 8 to 10 generations, and the development of RILs presents problems in species showing sever inbreeding depression. Further, in self-incompatible species, considerable effort may be required to achieve the desired level of inbreeding. In several crop species, large RIL populations have been developed and some RIL populations (e.g., an RIL population comprising 300 RILs in the case of *Arabidopsis thaliana*) have been used by many research workers for various studied including linkage mapping.

14.2.5 Near-Isogenic Lines

Isogenic lines comprise pairs of lines that have the same homozygous genotype, except for a single gene/locus. But such pairs of lines ordinarily differ from each other for a variable amount the genomic regions surrounding the locus in question, and some other small regions of the genome. Therefore, such line pairs are not exactly isogenic, and for this reason, they are termed as **near isogenic lines (NILs)**. Near isogenic lines are generally produced by backcross program that has been discussed in some detail in Section 4.7. The backcross program is routinely used by plant breeders to transfer desirable traits from a donor variety (the donor parent, DP) into an otherwise popular variety deficient in this trait (the recurrent parent, RP). The F_1 from the cross between the DP and the RP is backcrossed to the RP (Fig. 14.6). The progeny from this backcross and the subsequent backcrosses are backcrossed to the RP. During the backcross and the subsequent selfing generations, a strict selection is done for the trait being transferred and the plant type and other features of the RP. Generally, after 5-6 backcrosses, the plants having the desired trait and most similar to the RP are selected and selfed. Plants homozygous for the target trait are selected from the progeny and the best progeny from these plants is used as the improved version of the RP (denoted here as RP'). The RP and RP' constitute a pair of NILs.

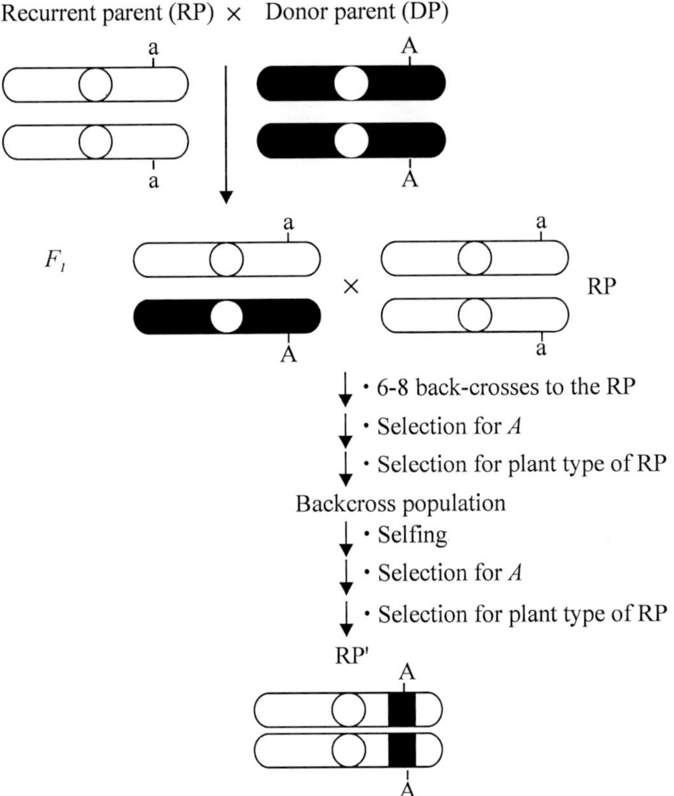

(Modified version of RP; the RP' has the allele *A* in the place of allele *a*)

Near - isogemic line (NIL)

Fig. 14.6. A schematic representation of the development of an NIL (denoted by RP'). The RP' may have small DP genome segments (depicted as black coloured segments) in one or more of its other chromosomes. RP, recurrent parent; DP, donor parent; NIL, near isogenic line.

As mentioned earlier, the two members of a pair of NILs differ for the alleles of the target gene, and for the alleles of the genes linked to the target gene. In addition, they would also differ for the alleles of those genes that are present in the genomic regions of the DP that could not be eliminated during the backcross programme. In case several pairs of NILs were developed using the same DP but different RPs, this group of NILs would show the following features. (i) Members of all the NIL pairs will differ for the alleles of the target gene and the markers linked to them. (ii) The ratio between the alleles of these markers and genes will be 1:1 irrespective of whether they are codominant or dominant. (iii) They will also differ for some other marker alleles that are not linked to the target gene. (iv) But the ratio between the alleles of these marker loci will not be 1:1 since the different NIL pairs are expected to from the RP differ for different genomic regions that are not linked to the target gene. These features would allow identification of the markers most likely to be linked to the target gene, which needs to be validated by linkage analysis. In addition, the genomic position of the target gene has to be determined by linkage mapping.

NILs are homozygous, and can be maintained indefinitely. Often the NILs produced in the breeding programs can be used as a mapping resource. In such cases there will be no need for developing the NILs; this will save considerable time and effort and other resources. The modified member of the NIL pair (RP') can be crossed to the RP and a very large F_2 population can be raised and used for fine mapping of the target gene. The fine mapping will facilitate positional cloning (Section 15.6), if desired the target gene. However, creation of a NIL pair requires at least 5-6 generations of backcrossing, which makes the total duration of 8-9 generations. Thus, it involves considerable amount of time as well as effort. Further, NILs can be used for mapping of only such genes that were the targets of introgression in the concerned NILs.

14.2.6. Chromosome Segment Substitution Lines

A **chromosome segment substitution line** (**CSSL**) is a line that carries a chromosome segment from another variety of the same species or of a related species in place of its own chromosome segment. A **CSSL population** comprises a set of introgression lines or CSSLs derived by a suitably designed backcross programme; these lines are selected in such a way that each line has a different chromosome segment from the DP (Fig. 14.7), and that the sum total of DP chromosome segments present in the set of CSSLs, ideally, represents to entire DP genome. The CSSLs are often called **introgression line library**. The DP may be a variety/germplasm line from the same species as RP or it may be a genotype/line from a related species. In the case of latter the term alien may be added before the term chromosome segment substation lines. CSSLs are produced by ~6 backcrosses of the F_1 between the selected DP and RP to the RP, followed by two generations of selfing in the same way as are the NILs (Section 14.2.6). The homozygous lines are selected using a set of markers evenly distributed over the entire genome and polymorphic between the DP and the RP. Each homozygous line has a different but slightly overlapping segment of the DP chromosome. It may be pointed out that a sufficiently large number of plants are used for each backcross so that nearly all the different segments of the DP genome are included in the backcross.

CSSLs are homozygous and once created can be maintained indefinitely. They can be evaluated in replicated yield trials, and are suitable for mapping of both oligogenes as well as QTLs. CSSLs can be used for linkage as well fine mapping in the same manner as are NILs, since each CSSL is a modified version of the RP just as the RP' member of a NIL pair. They allow detection of QTLs with small effect and also the use of a CSSL with superior features in breeding programmes. They are very useful for gene introgression especially when the DP is an unadapted germplasm, including a related species. The development of CSSL requires as much time as that of NILs, but it necessitates considerably more work in terms of the larger population size and the analysis with a relatively large number of markers. In addition, undesirable genes/QTLs may be linked to the favorable genes/QTLs present in the introgressed genomic segment; this phenomenon is called **linkage drag**. Whenever linkage drag is encountered, additional breeding effort would be required to eliminate the linked undesirable genes/QTLs.

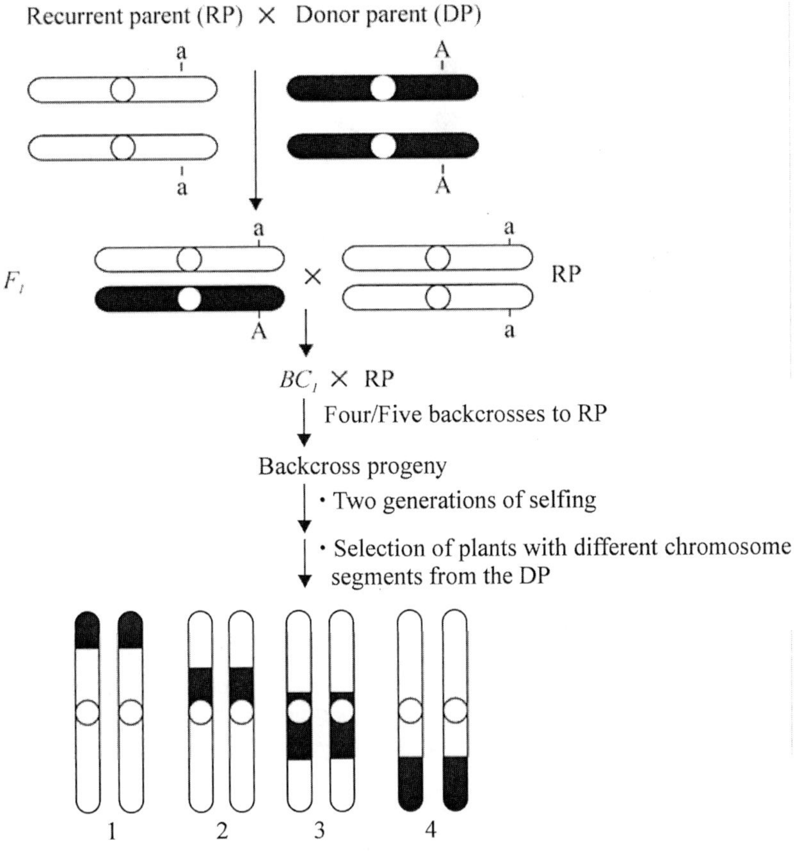

Fig. 14.7. A schematic representation of development of chromosome segment substitution lines (CSSLs). The chromosome composition of the introgression lines or SSLs is depicted for a single chromosome pair. Each of the other chromosomes of these (SSLs) will be from the RP and will have no segment from DP. The other SSLs will have a single distinct DP segment from some other chromosome of the DP genome. RP, recurrent parent; DP, donor parent; BC_1, first back-cross generation.

14.2.7. Interconnected Populations

The construction of **interconnected populations** involves mating among a set of homozygous lines in such a way that two or more crosses have one parent in common. The crosses may be made according to a suitable mating design, e.g., full diallel, half diallel, factorial, nested, round robin, etc. Generally, half diallel mating design excluding the selfs has been used for creating interconnected mapping populations. In the case of nested mating scheme, a set of homozygous lines are mated to one to few homozygous lines and, generally, many RILs are isolated from each of the crosses (Fig. 14.8). The RILs from the individual crosses can be analyzed separately as RIL populations, while those from all the crosses can be analyzed together following the nested scheme. In addition to RILs, the interconnected populations can be F_2, backcross or DH populations as well. Suitable computer programmes, e.g., MCQTL, have been developed for the analysis of data from multi-cross populations.

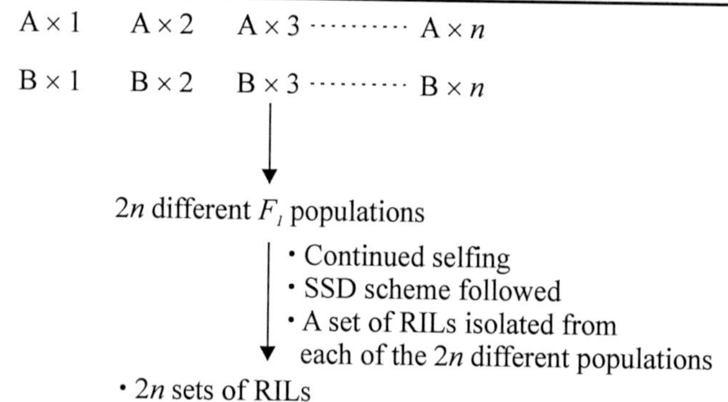

Fig. 14.8. A schematic representation of the development of a nested association mapping (NAM) population. The inbreds A and B are reference inbreeds, while inbreds 1, 2, 3, 4,...... *n* are diverse inbreds representing the desired diversity in the concerned crop species. SSD, single seed descent; RILs, recombinant inbred lines.

14.2.8 Multiparent Advanced Generation Inter-Cross Populations

The **multiparent advanced generation inter-cross (MAGIC) populations** are a collection of RILs generated from a complex cross produced by crossing a number (usually, 8) of parental lines. The parental lines used for producing the complex cross may be inbred lines, clones or individual plants selected from populations of interest. The complex cross is subjected to continuous selfing and the progeny plants in the selfing generations are handled according to SSD. In the end, a suitably large number of RILs are isolated, which together constitute the

MAGIC population (Fig. 14.9). Some workers have used more elaborate progeny handling schemes to develop the RILs comprising the MAGIC populations. MAGIC populations are homozygous and can be maintained indefinitely. They can be used for mapping of both genes and QTLs, and they are suitable for both linkage and association mapping. Since their development involves several parental lines, they are likely to show segregation for many genes/QTLs, and many of these loci may exhibit multiple alleles. In some cases the parental lines may be elite breeding lines; in such situations, the MAGIC lines might be used for variety development without any further breeding effort. The construction of MAGIC populations requires considerable effort and resources, but once created, they can be multiplied, distributed and maintained perpetually.

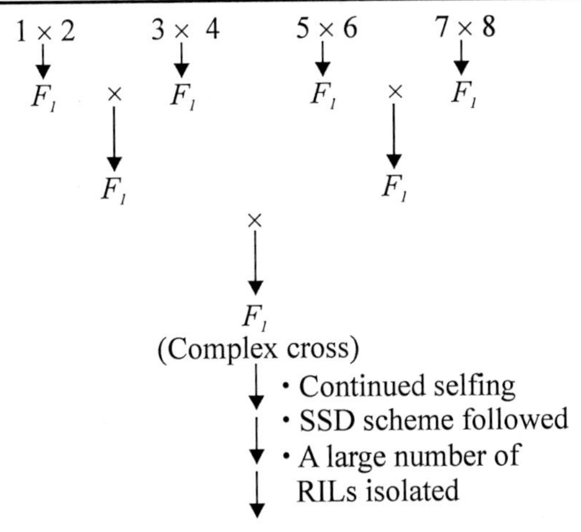

Fig. 14.9. A schematic representation of a simple approach for the development of a multiparent advanced generation inter-cross (MAGIC) population. The parental lines 1,2,3, ... 8 can be inbreds, clones or individual plants (selected from a natural population). SSD, single seed descent; RILs, recombinant inbred lines.

14.3. POPULATIONS FOR ASSOCIATION MAPPING

Association mapping (AM; Section 14.7) is based on linkage disequilibrium (LD) between the target gene and the marker estimated from a suitable population, called **association mapping population**, **association mapping panel** or **simple association panel**. The association panel may consist of a random collection of plants/lines drawn from natural populations, germplasm collections, breeding lines, including varieties, or developed from multiparent crosses. The association panels are broadly grouped into the following two categories: (i) population-based association panels and (ii) family-based association panels. The **population-based association panels** consist of individuals randomly drawn from natural populations, germplasm collections or inbreds, breeding lines and varieties developed in breeding programmes. AM panels derived from natural populations will be relevant for plant breeding programmes only when they concern a wild relative of a crop species. The AM panels based on unadapted germplasm, will detect marker-trait associations that can be used only after considerable breeding effort. But the information obtained from panels derived from elite germplasm, breeding lines and cultivated varieties would be directly useful for breeding programmes. The **family-based association panels**, on the other hand, consist of lines derived from either multiparent complex crosses or from multiple interconnected crosses. The nested association mapping (NAM) populations are produced by the nested mating scheme (Section 14.2.7), and the MAGIC populations (Section 14.2.8) are two good examples of family-based AM populations. These populations are widely used for AM since they have little population structure and familial relatedness.

14.4 SOME RELEVANT FACTS ABOUT MAPPING POPULATIONS

In case of such cross-pollinated species in which inbred development in not feasible, the F_1 from a cross between two selected clones or plants can be used as a mapping population. This is because the parents themselves will be heterozygous and the F_1 and will show segregation for many loci. Further, the F_1 may be backcrossed to one of the parents and the BC population may be used for mapping. This approach is followed in crops like potato, sugarcane, etc. The size of mapping population should be as large as possible so that the findings from mapping studies are reliable. It has been suggested that the size of mapping population should be at least 200, but some research workers feel it should be 500 more. In case the objective of a study is fine mapping of genes and QTLs, the size of mapping population of several thousand plants/lines would be desirable. In some populations, the segregation ratios for some of the markers may deviate significantly from the ratio expected in the population. In contrast, other markers may show the expected ratio in the same population. Therefore, the segregation ratio should be tested for each marker used for genotyping of the population. A simple option for tackling the problem of some markers showing significant deviation from the expected segregation ratio is to exclude the data for such markers from further analysis.

14.5 LINKAGE MAPPING OF MARKERS AND GENES

As mentioned earlier, **linkage** is the tendency of two or more genes to be inherited together due to their location close to each other in the same chromosome. However, linkage between two genes is disrupted by **crossing over**, which is exchange of homologous segments of chromatids between the homologous chromosomes. In general, the frequency of crossing over is positively associated with the distance between the linked genes. The frequency of crossing over between two genes can be estimated from the frequency of recombination between these genes. The **frequency of recombination** is the proportion of gametes with the new allelic combinations of the linked genes; these new allelic combinations are, as a rule, produced by crossing over (Fig. 14.1). The simplest approach for estimating recombination frequency is to analyze the progeny from a test cross, e.g., *Aa Bb* (F_1) x *aa bb*. It is assumed here that allele *A* is linked with allele *B*, and that allele *a* is linked with allele *b*. In the progeny of this test cross, the total of the frequencies of *Aa bb* and *aa Bb* genotypes will be the frequency of recombination (r) between the genes *a* and *b*. Therefore,

$$r = \frac{Number\ of\ the\ recomobinant\ testcross\ progeny}{Total\ number\ of\ testcross\ progeny}$$

The value of r can be expressed as percent. The value of r can be similarly estimated from DH populations and from the backcross populations produced by crossing the F_1 with the dominant parent. But in the case of the latter, this can be done only when the markers/traits are codominant. In the cases of F_2, RIL and other populations, r is estimated by the maximum likelihood procedure. It may be mentioned that in the case of RILs, the value of r is approximately $1/2R$ when the value of R is low. Here, R is the frequency of those recombinant inbred lines that have new allelic combinations for the genes a and b.

14.5.1. Genetic Distance

The recombination frequency is converted into genetic distance by using a suitable mapping function. Generally, either Haldane (1919) or Kosambi (1944) mapping function is used for this purpose. The **Haldane mapping function** is based on the assumption that the occurrence of crossing over at a genomic site does not interfere with the occurrence of another crossing over at a nearby site, i.e., there is no interference. The Haldane mapping function converts the value of r into genetic distance (m) in centiMorgans by following formula.

$$m = 50 \ln (1 - 2r)$$

The genetic distance of one centiMorgan is that distance in the chromosome that allows one percent of crossing over between the pair of genes under consideration. The Kosambi distance, on the other hand, assumes interference, and converts r into m by the following formula.

$$m = 25 \ln \left[\frac{(1+2r)}{(1-2r)} \right]$$

In general, the value of the m estimated by the Haldane function is higher than that obtained by the Kosambi function for any given value of r.

The estimates of generic distance between any two given genes are affected by a variety of factors like the sex, the genotype, the type of cross (interspecific or intraspecific cross), etc. In general, the genetic distance is proportional to the physical distance. **Physical distance** is the distance between two points in the genome/chromosome in terms of the number of base pairs (bp; often expressed as kb, kilobase pairs or Mb, megabase pairs) present between them. However, the exact relationship between the genetic and physical distances is quite variable and is affected by several factors, e.g., the species, the genomic region, etc.

14.5.2. Mapping of Molecular Markers

A linkage map is specific for the mapping population used for its construction. For mapping of molecular markers, the two parents are genotyped for a large number of markers. When a group of plants or lines are evaluated for a set of markers, this is called **genotyping**. The markers that show different alleles, i.e., are **polymorphic**, between the two parents are used for genotyping each individual of the mapping population. The marker genotype data are analyzed to detect linkage between marker pairs, and the frequencies of recombination between the linked markers are estimated. The decision whether two loci are linked is made on the basis of the estimated LOD (logarithm of odds) between them (Section 14.5.3). There are several computer software packages that can be used for this purpose and also to construct the linkage map (Table 14.2). MapMaker/Exp is perhaps the most widely used program for this purpose. It is command-driven programme that can analyze very large data pertaining to dominant as well as codominant markers and traits. It converts recombination frequency into genetic distance using Haldane mapping function, but the Kosambi function can also be used. Once the linked loci are identified, they are classified into probable linkage groups. Finally, the most likely order of the loci in each linkage group is determined and the loci are depicted in form of a linkage map.

Table 14.2: Some computer software used for mapping of molecular markers and oligogene

Software program	Application	Source
MapMaker/Exp	Primary map construction	genome.wi.mit.edu/distribution/mapmaker
MultiMap	Linkage mapping; also physical mapping	tara@chimera.hgen.pitt.edu, perlin@cs.cmu.edu
AntMap	Primary map construction	http://lbm.ab.a.u-tokyo.ac.jp/~iwata/antmap/
MergeMap	Integration of linkage maps into a consensus map	http://www.cs.ucr.edu/~yonghui/mgmap.html
JoinMap	Integration of linkage maps into a consensus map	A commercial package

Software program	Application	Source
MultiPool	Mapping using pooled DNA sequence data from NGS	http://cgs.csail.mit.edu/multipool/
MMAPPR	Mapping of recessive mutations using RNA-Seq data	http://yeast.genetics.utah.edu/software.php
MapPop	Selective mapping and bin mapping	athttp://ars-genome.cornell.edu/software.html
NGM	Mapping of recessive mutations using NGS sequence data	http://bar.utoronto.ca/NGM

14.5.3. LOD score (B)

The LOD score is used as the basis for deciding whether two loci segregating together are linked or not. The value of LOD score is calculated by using the following formula.

$$LOD\ score = log10\ (P1/P2)$$

where, P1 is the probability of getting the given data with the assumption that the two genes are linked and there is recombination between them at the specified frequency, while P2 is the probability of getting the same data when the two genes are segregating independently. Thus, LOD score not only provides a test for linkage, it also provides the most likely estimate of the frequency of recombination between the pairs of genes. Further, a LOD score of 1 indicates that the given data are 10-times more likely with linkage and the specified recombination frequency than with independent assortment. In general, a LOD score of 3 is considered as the threshold value. A **LOD score threshold** is the minimum value that should be equaled or surpassed by the calculated LOD score to be considered as significant. Therefore, if the calculated value of LOD score is either equal to or greater than 3, the two loci are considered to be linked. An extensively used method for estimation of the LOD score threshold value is based on **permutation test**. In this method, the marker genotypes of the individuals included in the sample are kept fixed or unchanged. The data on trait phenotype are, however, shuffled so that all the individuals in the sample now have a random trait phenotype from the original data. The genotype and phenotype data are now analyzed and the LOD score is computed. This process is repeated 1000 times; all the LOD scores so obtained will be due to false association between the marker and the target trait. The threshold value of LOD score for 0.05 probability will be the one that is lower than 5% of the LOD score values obtained by this process. Often the LOD score threshold value may be chosen for 0.01 or even 0.001 probability values. The LOD score values are estimated for a range of recombination frequencies, and these values may be plotted as a curve. The recombination frequency giving the peak or the highest value of LOD score is taken to be the most likely frequency of the recombination between the two loci.

14.5.4. Mapping of Genes

The term **gene** in this discussion is considered to govern a qualitative trait and means the same as an oligogene. Since there is complete correspondence between the qualitative trait phenotype and the genotype at the locus governing the trait, genes may be treated in the same way as molecular markers for the purpose of mapping. In fact, qualitative traits were the first marker types and were widely used for preparation of linkage maps. Therefore, the purpose of mapping oligogenes is to identify molecular markers linked to them so that these markers can be used for indirect selection for these genes. A simple approach to achieve this will be to phenotype the mapping population for the trait of interest or the target trait, and to genotype all the plants in the population for all the markers that are polymorphic in the population. Analysis of the phenotype and genotype data will reveal the marker linked to the gene governing the target trait. However, this will require considerable amount of genotyping work, which can be minimized by one of the following approaches.

1. In case the target trait has been transferred from a DP into an RP to produce a pair of NILs, the RP, RP' and DP are genotyped for a number of markers. The markers that have different alleles in the RP and the RP', but have the same alleles in the RP' and DP are identified. These markers will be located in the genomic regions transferred into the RP' from the DP. The RP' is crossed with RP to generate a mapping population like F_2. This population is phenotyped for the trait, and genotyped for the small number of markers identified earlier; these data are analyzed to detect the marker linked to the target gene.
2. When NILs for the target trait are not available, which is likely to be the case for most traits, the bulk segregant analysis (BSA) approach can be used. In case of BSA, the mapping population is phenotyped for the target trait, say, disease resistant. Ten most resistant plants/lines and 10 most disease susceptible plants/lines are selected. The DNAs from the resistant plants/lines are isolated and composited to create the **resistant bulk.** Similarly, DNAs from the susceptible plants are isolated and mixed to constitute the **susceptible bulk**. The DNAs from the two parents of the concerned population and from the two bulks are screened with a large number of markers (Fig. 14.10). The markers that distinguish the resistant parent and the resistant bulk from the susceptible parent and the susceptible bulk are likely to be linked with the target gene. These markers are used to genotype the entire mapping population, and the genotype and phenotype data are analyzed to identify the marker(s) linked to the gene of interest.

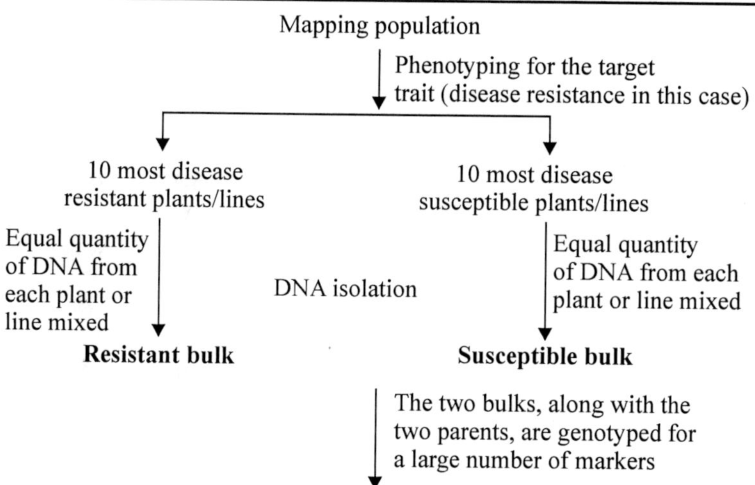

- Markers that differentiate the resistant parent and the resistant bulk from the susceptible parent and the susceptible bulk are identified.
- The whole mapping population is genotyped for these identified markers, and phenotyped for the target trait.
- Phenotype and genotype data are analyzed to identify markers linked to the gene for disease resistance.

Fig. 14.10. A schematic representation of the bulked segregant analysis (BSA). The trait considered here is disease resistance. The scheme is applicable to any qualitative trait, as well as quantative trait. In the case of quantitative traits, one bulk will comprise 10 individuals/lines with the extreme high and the other bulk will have 10 individuals/lines with the extreme low phenotype for the trait.

3. BSA approach can be used with RNA-Seq data. The two phenotypes bulks are created as above, and RNA bulks are created from them. The RNA bulks are sequenced and the sequence data are analyzed using a suitable computer programme, e.g., MMAPPR (Table 14.2), to identify SNP markers that differ between the two bulks and are linked to the target gene.
4. The BSA approach can be used for QTL mapping as well. A **high bulk** is created by pooling DNAs from 20–50 plants with the extreme high phenotype for the target trait. Similarly, a **low bulk** is created from 20–50 plants showing the extreme low expression for the target trait. The QTL linked markers can be identified using one of several approaches, including RNA-Seq data analysis.

The linkage detected between a marker and a gene must be confirmed as well as validated. **Confirmation** involves analysis of the same mapping population by some other researchers. Alternatively, the same researchers may create another mapping population from the same parents and use it for analysis. If the marker- and the gene (or trait) are found to be linked in these populations also, it shows that the detected linkage is real. In case of **validation**, the marker-trait linkage is evaluated in materials unrelated to the mapping population. If the marker shows linkage with the trait in these materials also, it can be used for marker-assisted selection.

14.5.5. Complete and Consensus Linkage Maps

In a **complete linkage map** every point in the genome of the species is linked to one or more markers. Thus, the number of linkage groups in such a map is the same as the haploid chromosome number of the species. Further, the terminal loci in each linkage group correspond to the subtelomeric regions of the concerned chromosomes. Finally, any new marker added to the map will be linked to one or more already mapped markers. A **consensus** or **integrated linkage map** is created by merging two or more different maps constructed for the same species. Two or more linkage maps can be merged together provided they share some minimum number of anchor or core markers. An **anchor** marker is highly polymorphic and is evenly distributed over the entire genome. Generally, RFLP and SSR markers are used as anchor markers. Merger of linkage maps is carried out using suitable computer software, e.g., JoinMap and MergeMap. In general, a consensus map allows identification of markers linked more closely to the target genes than do the individual linkage maps that were merged to derive the consensus map. Consensus maps have been developed in several crop species like maize, potato, melon, wheat, etc.

14.5.6. Comparative Mapping

Comparative mapping refers to the comparative analysis of linkage maps of different species. Maps of different species belonging to the same tribe or even family can be compared if they share a set of molecular markers. For this reason, the species whose maps are to be compared are subjected to mapping employing a set of common markers usually derived from single copy DNA sequences. Comparative mapping has been done for many groups of species. One general observation is the conservation of synteny across considerable distances. **Synteny** means that a set of genes are located in the same chromosome of the different species. Therefore, **conservation of synteny** means that almost the same set of genes is located in a single chromosome of the different species. For example, the general organization of the chromosomes in the grass family or Graminae is largely conserved. Further, very distantly related taxa like dicot and monocot species also show considerable degree of synteny. The available evidence suggests that genes present in the corresponding positions in the syntenic chromosome of related species may be expected to have similar function. Therefore, if the function of a gene is known in one species, this would indicate the function of the gene located in the corresponding chromosomal position of a related species.

14.5.7. High-Resolution Mapping

Linkage mapping generally identifies markers located at ~10 cM and occasionally at <5 cM from the target gene. But for some procedures, e.g., positional cloning, markers located at about 0.1 cM or less would be required. Identification of such markers is called **fine mapping** or **high-resolution mapping**. Fine mapping is based on very large populations of few thousand plants/lines, which have to be genotyped for an extremely large number of markers if the whole genome were to be covered. Therefore, mapping strategies have been devised to reduce the genotyping work. The **pooled-mapping technique** is one such strategy. In this technique, plants with the recessive phenotype for the target trait are selected from a very large mapping population segregating for the trait. These plants are randomly divided into several small pools. The plants in each pool are genotyped for a large number of markers. The pools in which at least one plant is recombinant for a marker and the target gene are identified. This data is subjected to suitable analysis that allows the marker to be mapped in relation to the target gene. The pooled mapping technique is able to identify markers located very close to target genes, and it can be used for fine mapping of QTLs as well. Some other strategies for fine mapping are selective mapping, use of genomic resources like high density linkage maps and genomic sequences, analysis of RNA-Seq data, etc.

14.6 LINKAGE MAPPING OF QTLS

A **quantitative trait locus** (**QTL**) is a genomic region that is involved in the control of a quantitative trait. A QTL may have one or, mostly, several genes, some which may be relevant to the expression of the concerned quantitative trait. It is now recognized that the various QTLs are of several different types, and they have been grouped into a number of categories on the basis of diverse criteria. For example, QTLs are classified as main effect QTLs and epistatic QTLs on the basis of the type of effect generated by them. The **main effect QTLs** directly affect the expression of the trait, while the **epistatic QTLs** influence the trait phenotype by interacting with the main effect QTLs. The main effect QTLs are further classified as major and minor QTLs. A **major QTL** explains 10% or more of the phenotypic variation of the concerned quantitative trait, while a **minor QTL** accounts for less than 10% of the phenotypic variation. The main effect QTLs are also classified as stable and unstable QTLs. The phenotypic effect produced by a **stable QTL** on the concerned trait is little affected by the environment. In contrast, the effect generated by an **unstable QTL** is sensitive to environmental influences and such a QTL may be expressed in only some of the environments tested. In general, major QTLs have stable expression, while the minor QTLs show unstable expression pattern.

14.6.1. QTL Detection and Mapping

QTL detection and mapping requires a reasonably dense molecular maker map of a mapping population that is suitable for the target trait phenotype evaluation in replicated trials. In view of the above, mapping populations comprising homozygous lines, e.g., RILs, DH etc., are the most widely used for QTL mapping. The mapping population is evaluated for the target trait in a replicated trial conducted, preferably, over locations and years. A suitably large number of markers distributed over the entire genome are used to screen the two parents of this population, and the markers having different alleles in them are identified. The population is now genotyped for a large number of markers that detect polymorphism between its two parents. The marker genotype data are used to construct a molecular marker map for the population; generally, MapMaker/Exp is used for this purpose. The marker genotype and trait phenotype data are now analyzed using a suitable QTL detection model to identify the genomic regions having QTLs governing the trait. QTL mapping is carried out using a suitable computer program, some of which are listed in Table 14.3. **QTL Cartographer** is one of widely used QTL mapping software packages. There are several methods for QTL mapping, such as single marker analysis (SMA), simple interval mapping (SIM), multiple interval mapping (MIM), composite interval mapping (CIM), inclusive composite interval mapping (ICIM), joint inclusive composite interval mapping (JICIM), etc. (Table 14.4). Out of the various methods, composite interval mapping is the most commonly used method for QTL mapping.

Table 14.3: Some computer software used for mapping of QTLs

Software program	Application	Source
MapMaker/ QTL	Interval mapping, etc.	http://hpcio.cit.nih.gov/lserver/ MAPMAKER_QTL.html
QTL Cartographer	SIM*, CIM, MIM, etc.	http://statgen.ncsu.edu/qtlcart/WQTLCart.htm
MapManager QT/QTX	SMA, SIM, CIM, etc.	http://mapmanager.org/mmQTX.html
R/QTL	SMA, SIM, CIM, MQM, etc.	http://www.r-project.org/
R/QTLBIM	Bayesian interval mapping	http://qtlbim.org/
FlexQTL	Bayesian QTL mapping	www.flexqtl.nl
MCQTL	QTL mapping in multi-cross and biparental populations	http://www.genoplante.com
QGene 4.0	SIM, CIM, MIM, multitrait methods, etc.	http://coding.plantpath.ksu.edu/qgene

*SIM, simple interval mapping; SMA, single marker analysis; CIM, composite interval mapping; MIM, multiple interval mapping; MQM, multiple QTL mapping.

Table 14.4: A summary of the most common approaches for QTL mapping

Method	Salient feature	Remarks
Single marker analysis (SMA)	Evaluates each marker separately for association with the target trait	Used before applying the other methods for QTL analysis mainly to detect missing data
Simple interval mapping (SIM)	Searches QTLs within the genomic regions flanked by two markers	Detects two linked QTLs as a single QTL; extended as CIM, ICIM, JICIM
Composite interval mapping (CIM)	It uses the markers linked with QTLs present in marker intervals other than the interval being searched as cofactors; markers linked to QTLs identified in a step-wise manner	The most widely used method of QTL detection and mapping
Inclusive composite interval mapping (ICIM)	A modification of CIM; uses all marker information in analysis; markers linked to QTLs identified only once	It is superior to CIM in detecting more true QTLs and fewer false QTLs that CIM
Joint inclusive composite interval mapping (JICIM)	The effects of the makers not showing significant effect on the trait in step-wise regression analysis are set at zero	Extension of ICIM for analysis of data from NAM populations
Multiple interval mapping (MIM)	Simultaneous search of QTLs in multiple marker intervals	Takes into account epistatic interactions; computationally intensive
Bayesian QTL mapping	Based on reversible jump Markov Chain Monte Carlo (MCMC) procedure	Designed for the detection on multiple QTLs; complex computation

Single marker analysis is the simplest QTL detection method. This method does not require a marker linkage map, and evaluates one marker at a time for its association with the trait of interest. Single marker analysis is almost always used before applying the other mapping methods. This is done mainly for detection of missing data from the data sets. The **simple interval mapping method** is based on a marker linkage map. The adjacent markers in the map are used to define marker intervals, and QTL detection is carried out at several locations within each of the marker intervals. In addition, SIM procedure analyzes each marker interval independent of the other intervals. SIM has become a standard QTL mapping procedure. It permits the QTL to be located onto the maker linkage map. SIM has been further developed as CIM, ICIM, etc. However, SIM usually detects only large effect QTLs and will detect a single QTL when two QTLs are located close to each other. The **composite interval mapping (CIM)** also considers the effects of QTLs in other marker intervals as well while it searches for QTLs in the given marker interval. In this method, first of all SMA is carried out. After this, a multiple QTL model is built up by selecting the markers showing successively high LOD scores and using them as cofactors in the analysis. The use of the other markers showing significant association with the trait as cofactors tends to increase the power of QTL identification and to separate QTLs that are linked but present in separate marker intervals. CIM procedure is relatively simple, is implemented by many computer software programs, including QTL Cartographer, and is the most widely used method of QTL analysis. However, it cannot detect epistatic QTLs and the method of selection of marker cofactors is essentially arbitrary.

14.6.2. QTL Confidence/Support Interval

The position of a gene (oligogene) is shown by a point on the linkage map. In contrast, the position of a QTL is not depicted by a point but it is represented by a bar drawn beside the line denoting the concerned linkage group (Fig. 14.11). This bar depicts the QTL confidence interval or QTL support interval. It represents the genomic region in which the QTL is the most likely to be located. The QTL confidence interval generally covers the region located between two points, on the either side of the peak point in the LOD score curve, at which the LOD score declines by a value of 1. For this region, this bar is often termed as **one LOD support interval**.

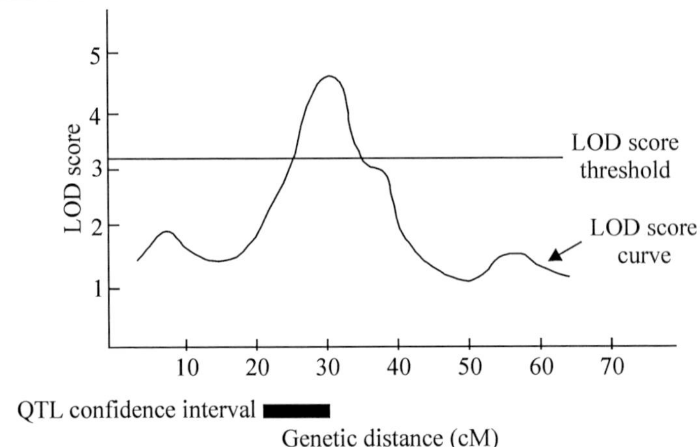

Fig. 14.11. A schematic representation of the LOD score curve, and the QTL confidence interval (shown here as solid horizontal bar). The QTL positions, and confidence intervals can also be depicted along side the linkage map of the concerned linkage group.

14.6.3. Concluding Remarks on QTL Mapping

The results obtained from a study on QTL mapping should be confirmed in the same way as the linkage between markers and genes (Section 14.5.4). This would reveal whether the detected QTL is real or not. Similarly, the results from QTL mapping should be validated by the same approach that is used for validation of marker-trait associations (Section 14.5.4). QTL fine mapping may be carried out to identify markers located at distances of <1 cM from the QTL. There are several approaches for QTL fine mapping, some of which are as follows: (i) QTL mapping in intercross recombinant inbred lines, (ii) recurrent selection backcross (RSB) QTL mapping, (iii) multi-parent advanced generation inter-cross (MAGIC) populations, (iv) combination of QTL mapping and transcriptome profiling, etc. It may be pointed out that different mapping studies in a given species may detect different QTLs for the same trait. Some of these different QTLs may be true QTLs and represent the differences in the QTLs present in the populations used for these studies. However, some other QTLs may be the result of factors like population size, experimental errors, etc. A QTL meta-analysis can be used to identify the true QTLs, also called **meta-QTLs**. The meta-analysis is based on the linkage maps, the QTL positions and the QTL confidence intervals generated in the different studies and determines a much smaller number of meta-QTLs. Computer software programmes like MetaQTL and BioMercator ver. 3 has been designed to carry out QTL meta-analysis.

It should be kept in mind that different QTLs may segregate in different populations. The expression of QTLs is generally affected by the genetic background as well as the prevailing environment. In addition, when smaller populations are used for mapping, the number of QTLs detected for a trait is smaller and the effects of the detected QTLs are larger than when larger populations are used for mapping. In addition, QTLs often occur in clusters, and a single QTL generally has several genes, but few QTLs are known to have a single gene each. QTL fine mapping has allowed positional cloning of several QTLs. Many QTLs were found to code for transcription factors, some QTLs encode enzymes, and some others code for a variety of proteins performing important functions.

14.7 ASSOCIATION MAPPING

Association mapping (**AM**) is based on linkage disequilibrium (LD) estimated from a variety of populations (Section 14.3). AM offers certain advantages over linkage mapping. For example, development of the time-taking construction of mapping populations is not necessary. Further, the markers detected to be associated with the traits of interest tend to be much closer to the concerned genes/QTLs than those identified by linkage mapping. The general procedure of association mapping is as follows (Fig. 14.12). A suitable AM population, also called AM panel, is created either by assembling a random sample of plants/lines from natural populations, germplasm collections or breeding materials, or by creating a multi-cross population (Section 14.3). This populations is phenotyped for the target trait(s) and genotyped for a large number of markers (generally, SSR and SNP markers) covering the entire genome or a specific segment of the genome. When AM uses markers distributed over the entire genome it is referred to as **genome-wide association study (GWAS)**. GWAS involves considerable amount of genotyping work, but tests for all the loci involved in the control of the target trait and showing polymorphism in the given AM panel. The marker genotyping work can be reduced by limiting the genotyping to a specific genomic region if there were reason to believe that the loci of interest are located in a particular region of the genome. This approach of AM is called **candidate gene approach for AM**. The genotype and phenotype data are analyzed to detect markers associated with the trait, and to prepare LD maps for the loci governing the trait.

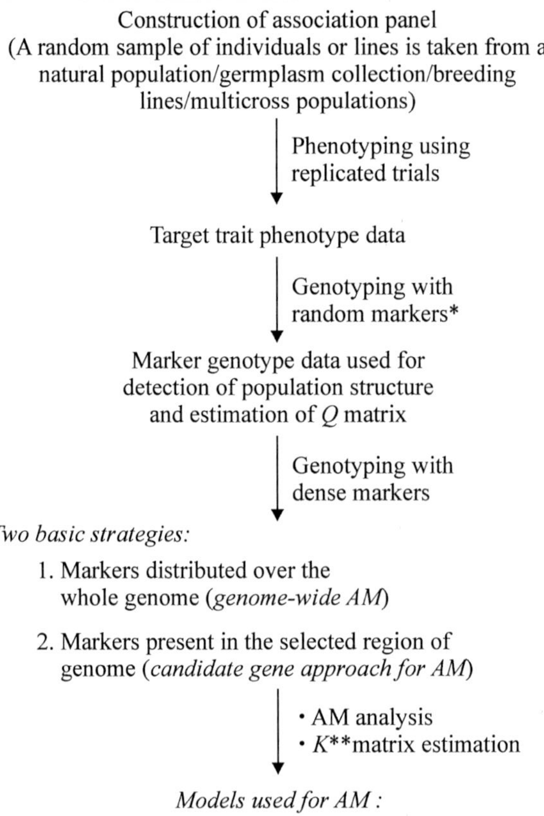

Fig. 14.12. A schematic representation of association mapping in plants. *Markers expected to segregate independent of each other, i.e., located over 40 cM apart, and distributed over the entire genome. **K_I Kinship (familial relatedness) matrix. Q matrix, matrix based on population structure. AM, association mapping; GLM, general linear model; MLM, mixed linear model.

14.7.1. Linkage Disequilibrium

In a random mating population, the gene and genotype frequencies do not change unless they are disturbed by selection, mutation, migration and/or random drift; these factors are called **evolutionary forces.** In the absence of evolutionary forces, the population will be at Hardy-Weinberg equilibrium; this means that the frequencies of the three genotypes for a gene, viz., *AA, Aa* and *aa*, in the given population will be p^2, $2pq$ and q^2, respectively. When we consider two genes (say, genes *a* and *b*) that are segregating independently, the following four types of gametes (gametic combinations of the two genes) will be produced: *AB*, *Ab*, *aB* and *ab*. The expected frequencies of these four gametic combinations will be, *pA.pB*, *pA.pb*, *pa.pB* and *pa.pb*, respectively, where *pA*, *pa*, *pB* and *pb* are frequencies of the alleles *A*, *a*, *B* and *b*, respectively, in the population. In case of equilibrium, *pAB.pab* would equal *pAb.paB*, where *pAB*, *pab*, *pAb*, and *paB* are the frequencies of the allelic combinations *AB*, *ab*, *Ab* and *ab*, respectively.

Table 14.5: Some common measures of linkage disequilibrium (LD)

LD estimate	Formula	Remarks
D	$= pAB^* - (pA.pB)$ $= pab - (pa.pb)$ $= (pAB.pab) - (pAb.paB)$	Considerable effect of allele frequencies
D'	$D/\min(pA{\cdot}pb, pa{\cdot}pB)$** when the value of D is > 0 $D/\min(pA{\cdot}pB, pa{\cdot}pb)$ when the value of D is < 0	Range, 0-1; much less affected by allele frequencies than D
r^2	$(D)^2/(pA{\cdot}pa{\cdot}pB{\cdot}pb)$	Range, 0-1; most suited for AM

**pA*, *pB*, *pa* and *pb* are the frequencies of alleles *A*, *B*, *a* and *b*, respectively, and *pAB, pab, pAb* and *paB* are the frequencies of allelic combinations *AB, ab, Ab* and *aB*, respectively.

** min $(pA{\cdot}pb, pa{\cdot}pB)$, and $(pA{\cdot}pB, pa{\cdot}pb)$ mean that the lower of the two estimates will serve as the denominator

In case of linkage, the value of *pAB.pab* would not equal that of *pAb.paB*. The difference between the values of *pAB.pab* and *pAb.paB* provides a simple estimate of linkage disequilibrium (LD). A simple definition of **linkage disequilibrium (LD)** is that the allelic combinations of two or more loci occur in frequencies different from those expected from their random association. Thus, LD is a nonrandom association of the alleles of two or more loci. The simplest measure of LD is *D*, which is estimated as follows (Table 14.5).

$$\begin{aligned} D &= pAB - (pA.pB) \\ &= pab - (pa.pb) \\ &= (pAB.pab) - (pAb.paB) \end{aligned}$$

The numerical value of *D* is of little value in determining the strength of LD as it is markedly affected by the allelic frequencies. The calculation of LD is explained in Table 14.6 using numerical examples.

Table 14.6: Some selected software used for association mapping. All packages, except *GenStat*, are available free of charge

Software	Application	Source
TASSEL	LD estimation, association mapping using GLM, MLM, etc. methods, analysis of population structure and kinship, etc.	http://sourceforge.net/projects/tassel
EMMAX	AM studies with large samples, takes into account population structure and kinship	http://genetics.cs.ucla.edu/emmax/
GenAMap	Structured association mapping	http://sailing.cs.cmu.edu/genamap/
FaST-LMM	AM in very large samples (max. 120,000 individuals)	http://fastlmm.codeplex.com/
STRUCTURE	Detection of population structure; *Q* matrix computation	http://pritch.bsd.uchicago.edu/structure.html
SPAGeDI	Kinship (*K*) matrix computation	http://www.ulb.ac.be/sciences/ecoevol/spagedi.html
GenStat	AM using GLM and MLM, accounts for population structure	http://www.vsni.co.uk/software/genstat

In view of the above, the estimates of *D* are standardized to obtain more useful estimates of; *D'* is one such standardized version of *D* and it is estimated as follows.

$$|D'| = D/\min(pA{\cdot}pb,\ pa{\cdot}pB) \text{ when the value of } D \text{ is} > 0$$

$$|D'| = D/\min(pA{\cdot}pB,\ pa{\cdot}pb) \text{ when the value of } D \text{ is} < 0$$

where, min (*pA·pb, pa·pB*) means that the lower of the two values, viz., *pA·pb* and *pa·pB,* will be used as denominator. Similarly, the term min (*pA·pB, pa·pb*) signifies that the lower of *pA·pB* and *pa·pb* will serve as the denominator for *D*. The values of *D*' range between 0 and 1, and show minimum effect of allele frequencies. But values of *D'* are not easy to interpret, and they show strong effect of small sample size. Another estimate of LD derived from the estimates of *D* is r^2, which is estimated as follows.

$$r^2 = (D)^2/(pA{\cdot}pa{\cdot}pB{\cdot}pb)$$

The values of r^2 range between 0 and 1, are easy to interpret, and are much less affected by allele frequencies than are D' estimates. The *D'* and r^2 estimates are the most widely used LD measures. Further, r^2 is perhaps the most suitable for AM studies.

The above considerations relate to LD estimation between two loci each having two alleles. Formulae for the estimation of LD between two loci with multiple alleles and between multiple loci have also been developed. The estimate of LD between two loci is influenced by a number of factors, e.g., the mating pattern of the population, selection population structure, familial relationships, admixture or migration, genomic region, etc. LD estimates are used to construct LD maps, in which different loci are depicted at distances estimated as LD units. LD mapping is carried out with the help of using suitable computer software like ALLASS (allele association) and *LDMAP* Version 0.1. The findings from one association mapping study should be confirmed by replication studies in the same way as are the results from linkage mapping studies (Section 14.5.4).

14.7.2. Methods for Association Mapping

Association mapping was initiated for identifying genes responsible for genetic diseases. The first method for AM was case and control scheme. In this approach, the marker genotypes of a small group of individuals having the disease (the cases) is compared with those of a large group of individuals without the disease (the control). This permits the identification of the marker(s) linked to the disease(s) in question. Subsequently, much improved family-based designs were developed, and some of them have been used for a limited number of AM studies in plants. The **general linear model (GLM)** was developed for AM in plants. GLM can take into account population structure and minimize false positive marker-trait associations due to population structure. When a random mating population is subdivided into few to several groups, and each of these groups has its own gene and genotype frequencies, the population is said to be structured. This situation tends to increase the frequency of false positive marker-trait associations. Population structure can be detected by the computer software SRUTCTURE, which also generates a *Q* matrix. This matrix can be used for correction in the GLM to minimize the effects of population structure. The computer software TASSEL implements GLM and uses the *Q* matrix to improve the reliability of marker-trait associations (Table 14.6).

Another factor that increases the frequency of false positive associations is kinship or familial relatedness. The mixed linear model takes into account both population structure and kinship (estimated as *K* matrix) to minimize the frequency of false positive associations. The *K* matrix can be estimated by *TASSEL* or *SPAGeDi* software programmes (Table 14.6). The chief limitation of MLM is that it is computation intensive. Several modifications of MLM have been designed to address the above and other limitations of MLM, but so far we do not have a completely satisfactory model for AM analyses.

14.7.3. Conclusions from Association Mapping in Plants

The following important conclusions can be drawn from the numerous AM studies that have been carried out with plants.

1. In case of not even a single quantitative trait, QTLs accounting for the total phenotypic variation have been discovered by all the genome-wide AM studies taken together. In some cases, a very large part of the variation remains unexplained; this has been called **missing heritability**.
2. Gene interaction is involved in control of all the traits that have been investigated.
3. The genomic regions that are most often identified to be associated with various traits are not the protein encoding regions.
4. Generally, the extent of LD in self-pollinated species is greater than that in cross-pollinated species.
5. In some genomic regions, the LD extends to much longer distances than in other genomic regions of the same species.

14.8. CONCLUSIONS

Molecular markers facilitate indirect selection for traits, and enable mapping of QTLs. Mapping of genes and QTLs is based on analysis of a mapping population constructed for the purpose. Linkage mapping uses populations created by crossing two or, sometimes, more parents differing for as many traits as possible, including the trait(s) of interest. Some of the common mapping populations are F_2, backcross, doubled haploid, recombinant inbred line (RIL), and near isogenic line populations. But association mapping is based on either natural populations or propulations comprising germplasm accessions, breeding lines, etc.

Several computer programmes have been developed for mapping of markers, genes, and QTLs. Mapmaker/Exp is perhaps the most widely used programme for linkage mapping of markers and genes. This programme estimates the likelihood of genes/genes and markers being linked with the specified distance (in cM) between them. This likelihood is expressed as LOD score; a LOD score of 3 or more is generally regarded as sufficient evidence for linkage. A number of software are available for QTL mapping, and QTL Cartographer is one of most popular QTL mapping tools. There are several models for QTL mapping; of these, composite interval mapping strategy in widely used. The position of a QTL in the genome is usually depicted by the QTL support interval. Association mapping involves estimation of linkage disequilibrium between the gene/QTL and marker. A number of software programmes are available for association mapping, e.g., TASSEL, which is quite popular.

The marker data and the information on marker-trait associations can be used for a variety of applications, including identification of varieties and hybrids, selection of parents for hybridization programmes, marker-assisted selection, and positional cloning of genes.

PRACTICE QUESTIONS

1. Define a mapping population. List the different types of mapping populations and describe the construction and relevant features of one of these populations.
2. Briefly describe the association mapping approach to identify marker trait associations.
3. Briefly explain the linkage mapping of markers and/or oligogenes.
4. Explain the meaning of high resolution mapping and briefly describe the procedures for developing comparative, consensus and high resolution maps.
5. Briefly discuss QTL mapping, and discuss its applications.
6. Comment on any two of the following statements:
   (a) Linkage mapping populations take two or more years for their construction.
   (b) RIL populations are a perpetual genetic resource.
   (c) The frequency of recombination between a pair of genes is indicative of the physical distance between them.
   (d) QTLs detected in a population are expected to be found in all other populations of the same species.
7. Write short notes on the following: (i) linkage map, (ii) genetic distance, (iii) linkage disequilibrium, (iv) association mapping, (v) comparative mapping, (vi) consensus maps, (vii) high resolution mapping, (viii) QTL mapping, (ix) LOD score, (x) doubled haploid population, (xi) RIL population, (xii) near isogenic lines.

RELEVANT LITERATURE

Singh BD and Singh AK 2015. Marker-Assisted Plant Breeding, Principles and Practices. Springer India, New Delhi.

Tanksley SD and Nelson JC 1996. Advanced backcross QTL analysis: a method for simultaneous discovery and transfer of valuable QTL from unadapted germplasm into elite breeding lines. Theoretical and Applied Genetics 92: 191-203.

■■■■

15 Applications of Molecular Markers in Plant Breeding

15.1. INTRODUCTION

The molecular markers have found a variety of applications in various fields of biology, including plant breeding. The chief feature of conventional plant breeding is that all the decisions about the genotypes of different individuals/ lines are based on their phenotypes. This works almost fine for the qualitative traits, but is relatively much less satisfactory for most of the quantitative traits. For this reason, linkage mapping of the genes governing quantitative traits could not be achieved using the conventional approach. The advent of DNA markers has enabled scoring for variation at the DNA level, and even at the gene level when gene-based and functional markers are used for genotyping. As a result, it has now become almost a routine procedure to map quantitative trait loci (QTLs), and to find markers closely linked to genes of interest. In addition, molecular markers are used to facilitate a variety of plant breeding activities, some of which are listed here: (i) germplasm characterization, (ii) diversity analysis, (iii) analysis of phylogenetic relationships, (iv) clear-cut identification of genotypes/ lines/varieties (DNA fingerprinting), (v) selection of parents for hybridization, (vi) evaluation of genetic purity of seed lots, plant produce, etc. (vii) indirect selection for traits of interest (marker-assisted selection), and (viii) positional cloning of genes (Table 15.1). These applications of molecular markers are briefly considered in the following sections.

15.2. GERMPLASM CHARACTERIZATION AND CONSERVATION

Molecular markers can be used to genotype all the accessions in a given germplasm collection (Section 15.5). The marker genotype data can be analyzed to assess the extent of genetic diversity (Section 15.3) present in a germplasm collection. This would permit the identification of gaps as well duplications in the germplasm collections, and appropriate steps may be taken to rectify them. Marker genotype data may be used to monitor genetic drift in the germplasm accessions during the conservation procedures. Finally, molecular markers would be quite useful during introgression of desirable genes/QTLs form unadapted germplasm into crop species. Suitable breeding schemes have been proposed for even such situations where we have no information about the genes/QTLs involved in the control of the target traits.

Table 15.1: The various applications of molecular markers in plant breeding

Application	Population	Objective	Remarks
Diversity analysis	Germplasm collection	Identification of gaps, duplications, genetic drift, characterization, etc.	Genome-wide markers used
	Breeding materials	Phylogenetic analysis, selection of parents for crosses, etc.	Genome-wide markers used
DNA fingerprinting	Varieties	Identification, genetic purity of seed lots	A selected set of genomic markers used
	Hybrids	Identification, genetic purity of seed lots	A selected set of genomic markers used
Positional cloning	Selected line with the target gene	Identification, isolation and cloning of the target gene	Markers closely linked to the target gene
Linkage mapping	Populations developed from biparental crosses, multiple biparental crosses or multiparent crosses	Preparation of linkage maps of markers, genes and QTLs, and identification of markers linked to genes/QTLs	Genome-wide markers used
Association mapping	Natural populations, breeding populations or multi-cross populations	Identification of markers linked to genes/QTLs	Genome-wide markers used; candidate gene approach may be used where feasible
Marker-assisted selection (MAS)	Breeding populations	Indirect selection for the target trait and other genomic regions	Markers linked to the target genes/QTLs and/or distributed over the genome
(i) Marker – assisted selection	Backcross generations of selected crosses	Introgression of target gene/QTL with the minimum of the donor parent genome	Markers closely linked to the target gene/QTL, and distributed over the genome
(ii) Marker-assisted recurrent selection	Segregating generations of self- and cross-pollinated crops	Increasing the frequencies of the QTLs controlling the target traits	All the markers significantly associated with the target trait
(iii) Single large scale MAS	F_2 or F_3 generations of self-pollinated crops	Ensuring that the plants selected in the advanced generations have the target genes/QTLs	Markers linked to the target genes/QTLs of interest
(iv) Genomic selection	Training and breeding populations of crops	Selection for all the genomic regions contributing favourably to the target trait(s)	Markers distributed over the entire genome used to compute genomic estimated breeding values

15.3. DIVERSITY AND PHYLOGENETIC ANALYSES

Genetic diversity may be defined as the sum total of genetic variation present within a species/population. It may be pointed out that only a part of the total genetic diversity may be expressed at the phenotypic, while the rest of it may not be expressed and may be present at the genotypic level only. The analysis of genetic diversity may be based on data on a variety of traits, including qualitative and quantitative traits and molecular markers. These data may be collected from a sample comprising individuals, inbreds, purelines or clones, each of which is called an **entity**. The first step in genetic diversity analysis involves estimation of genetic distances (GD) for all possible pairs of entities in the sample. The modified Rogers' distance (GD_{MR}) is perhaps the most widely used estimate of GD. The following formula is used for the estimation of GD_{MR} from the molecular marker data.

$$GD_{MR} = \left[N_{10} + N_{01} / (2N)^{1/2}\right]$$

where, N_{10} is the number of bands present only in the first entity of the pair of entities being used for estimation of GD, N_{01} is the number of bands present only in the second entity of the pair, and N is the total number of bands scored in the sample.

The GD estimates are then used to classify the entities of the sample into several smaller groups; this classification is often termed as **phylogenetic analysis**. A given group has entities that are more closely related to each other than they are with those in the other groups. There are several methods of grouping the entities, the dendrogram or phylogenetic tree being perhaps the most easily understood representation. The dendrogram may be either rooted or unrooted (Fig. 15.1). In addition to the grouping methods based on GD, there are other methods for clustering that are model based. In addition, principal component analysis, principal coordinate analysis and multidimensional scaling are also used for diversity analysis. Menu-driven computer software packages like NTSYS-pc, PHYLIP and DARwin are available for diversity analysis using a variety of data sets. PHYLIP has a large group of programs for construction of rooted and unrooted trees from a vast variety of data.

15.4. DNA FINGERPRINTING

The term **DNA fingerprinting** was first used by Sir Jeffreys in 1985 for a DNA-based assay designed to identify human individuals with extremely high degree of confidence. The present-day methods of DNA fingerprinting are based on simple sequence repeat (SSR) markers. The other marker systems available for finger-printing of plant materials are amplified fragment length polymorphism (AFLP), sequence tagged sites (STSs), sequence characterized amplified regions (SCARs), etc. DNA fingerprinting of plant varieties would be useful for their registration and for the protection of plant breeders' rights (PBR; Section 16.4.14). However, there are some difficulties in the use of marker genotype data for variety registration and other PBR applications. For example, some varieties that are 'uniform' at the phenotypic level may show some amount of plant-to-plant variation at the marker genotype level. Such varieties would fail to satisfy the uniformity criterion of the variety registration and seed certification. As a result, the molecular marker data are at present not used as the basis of plant variety registration, and they are used only as additional data. Efforts are being made to develop appropriate procedures and guidelines that would persuade to policy makers and the plant breeding community to accept marker genotype data as evidence in resolving the various PBR related issues.

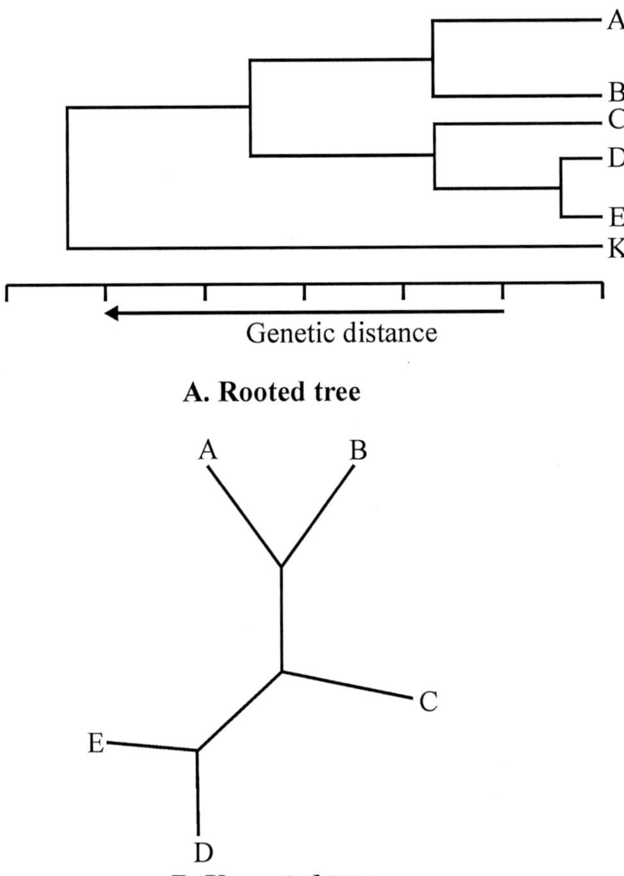

Fig. 15.1. Phylogenetic trees or dendrograms. A. A rooted tree; a distantly related entity (entity K) is included to enable the creation of rooted tree. B. Unrooted tree; the entities that are more closely related are joined by shorter branches

15.5. EVALUATION OF GENETIC PURITY OF PARENTAL LINES AND SEED LOTS

Genetic purity of a seed lot means that the given percentage of seeds in the seed lot belongs to the variety in question. The seed lots of a given variety may be contaminated by seeds of other varieties/hybrids. The genetic purity of seed lots can be assayed by some chemical tests, few biochemical assays, protein banding patterns following electrophoresis, and morphological evaluation in grow-out tests. Molecular marker profiles, however, enable unequivocal identification of plant varieties, including hybrid varieties, and of the parental lines of hybrid varieties (Table 15.2). Generally, PCR-based markers like SST, STS, SCAR, etc. are used. The purity of hybrid seeds depends primarily on the purity of their parental lines. A well-selected set of these markers will need to be assayed to differentiate the concerned variety from other popular varieties/hybrids. This set of markers can also be used to establish the identity of the variety/hybrid from which the given plant-produce was obtained and also to determine the purity of the produce.

Table 15.2: The use of molecular markers for detection of contamination in seed lots

Seed lot	Contamination	Detection by	Remarks
Varieties	Other varieties	A selected set of genomic markers	The set of markers should distinguish all the relevant varieties
Hybrids	Other hybrids/ varieties	A selected set of genomic markers	The set of markers should distinguish all the relevant hybrids and varieties
	Male parent seed	One codominant/two dominant genomic markers	The dominant allele of one dominant marker present in each parent
	Female parent seed	One codominant/two dominant genomic markers	The dominant allele of one dominant marker present in each parent
CMS* parent of the hybrid	Maintainer of the CMS line	One marker	Based on mitochondrial or chloroplast genome

*CMS, cytoplasmic male sterile.

One of the limitations in exploiting the full potential of hybrid varieties arises due to the contamination of the hybrid seed by seeds from other genotypes. For example, the hybrid seeds may contain variable proportions of the seeds produced by the male and/or the female parents of the concerned variety. The contamination by the male parent seed can be detected and quantified by a single SSR marker that differentiates the male parent from the female parents of the hybrid. In case of dominant marker systems, at least two markers will be needed, but only one of these markers should be present, i.e., produce band, in each parent. But one of the chief problems in hybrid seed production is the maintenance of the purity of the male sterile female parent of the hybrid. The seed of the male sterile line is often contaminated by the seed of its maintainer line, and the plants of these two lines can be distinguished from each other only after the maintainer line plants begin to shed pollen. This, of course, is too late as a part of the seed set on the male sterile line will be due to pollination by the maintainer line pollen. The plants of the male sterile and the maintainer lines will have identical phenotypes and nuclear genome marker genotypes. Therefore, these plants can be differentiated only on the basis of markers based on mitochondrial or chloroplast genomes. In case of rice, the *drrcms* marker is a SSR marker based on the mitochondrial genome. This marker is able to differentiate between the male sterile lines based on the wild abortive, *Oryza nivara* as well as *Oryza rufipogon* cytoplasms from their respective maintainer lines.

15.6. POSITIONAL CLONING OF GENES

Gene cloning involves the identification of the gene governing a trait of interest, its isolation, and cloning and propagation of the gene in a suit suitable host like *Escherichia coli.* Once a gene is cloned, it can be sequenced, introduced into plants to produce transgenic plants, and analyse these plants to determine the function of this gene. In case the transgenic plants have desirable performance, they can be used as new varieties with novel traits. Gene cloning is facilitated by knowledge of its RNA/DNA sequence or the protein encoded by this gene. But often this information may not be available for the gene of interest and the only information available may be the phenotypic effect generated by a mutation in this gene. In such cases, the strategy of positional cloning is used to isolate the gene involved in the mutation. **Positional cloning** or **map-based cloning** of genes uses linkage map information to identify, isolate and successfully clone a desired gene. Positional cloning can be undertaken for a gene that has mutated to generate a distinct phenotype and the mutant is available for analysis. In addition least one molecular marker linked to this gene as closely as possible should be known. It is preferable that one marker located at less than one centiMorgan (cM) on either side of the target gene is known. This is very important because 1 cM genetic distance may mean a considerable length of DNA, e.g., about 140 kb in case of *Arabidopsis* to approximately 4,600 kb in the case of wheat.

The positional cloning of genes has the following the well-defined steps (Fig. 15.2). The **first step** comprises the isolation of a mutant allele of the target gene, which produces a distinct and easily recognizable phenotype so that it can be easily tracked down during mapping. This mutant allele is mapped using a suitable mapping population to identify markers linked to the gene. A fine

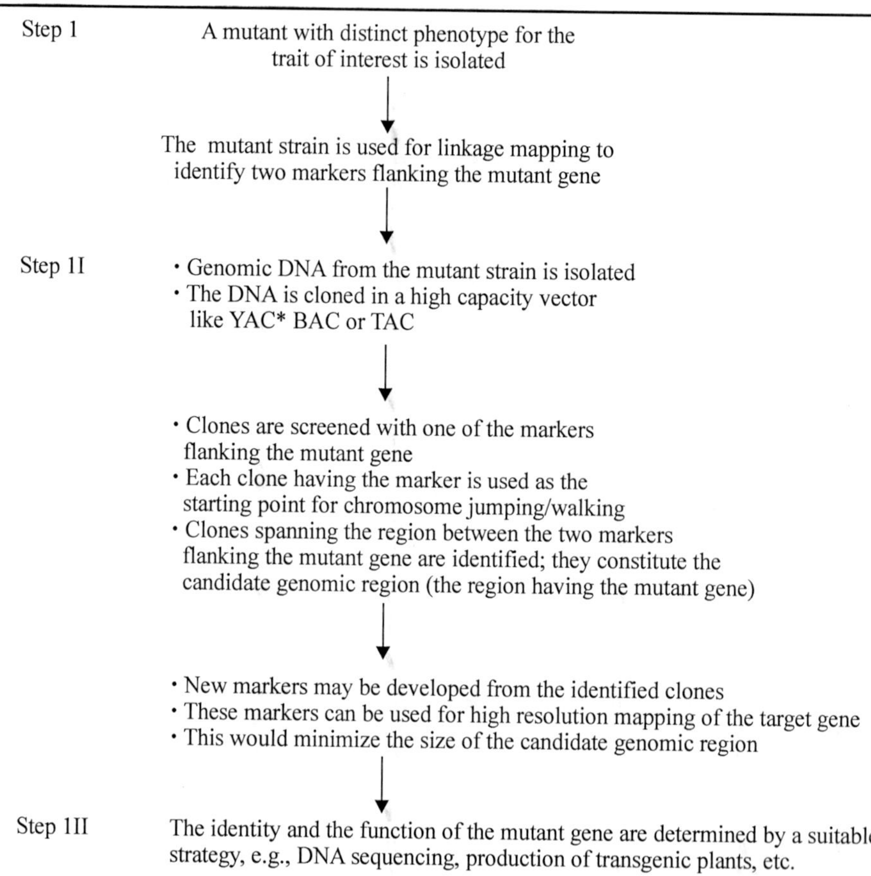

Fig. 15.2. A highly simplified schematic representation of positional cloning. *YAC, yeast artificial chromosome; BAC, bacterial artificial chromosome; TAC, transformation competent artificial chromosome.

mapping strategy may be used to identify markers located within, preferably, <1 cM of the target gene. The **second step** consists of isolation of the genomic DNA from the mutant strain, construction of genomic library using a high capacity vector like yeast artificial chromosome (YAC) or bacterial artificial chromosome (BAC) vector. The genomic library is screened with one of the markers located closest to the gene of interest. The clone having this marker provides the starting point for searching the clones that together would be most likely to have the genomic segments with the target gene. The genomic segment present in this clone is used for analysis following a suitable technique like chromosome jumping/walking to reach the marker located on the other side of the gene. The genomic segments identified in this manner make up the **candidate genomic region**. The size of the candidate genomic region can be minimized by undertaking a high resolution mapping effort based on a suitably large mapping population. In the **third step**, the candidate genomic region may be sequenced to identify the gene(s) present in it. In case this region has more than one gene, a suitable strategy, including their introduction into suitable strains of the concerned species to identify the gene that is the most likely to be responsible for the mutant phenotype. The positional cloning approach has been successfully used to clone, identify and characterize a number of plant genes as well as QTLs. Some examples of such genes/QTLs are gene *ABI1* of *A. thaliana* (it specifies abscisic acid insensitivity, the fertility restorer gene gene *Rf-1* of *Oryza sativa* var. *japonica*, the QTL *Hd1* governing heading date in rice, etc.

15.7. SELECTION OF PARENTS FOR HYBRIDIZATION

Selection of parents for hybridization is the first and the most critical step in a breeding programme. Generally, it is based on pedigree information on and/or phenotype data of the parental lines that may or may not be subjected to diversity analysis for this application. Molecular marker data provide additional information that is often useful in the identification of the desirable parents. Marker data are particularly useful in the case of quantitative traits since lines with similar phenotype may have entirely different QTLs for the concerned trait. For example, molecular marker data have been quite helpful in the identification of different sources of resistance to *Fusarium* head blight of wheat, which is an important disease of wheat throughout the world. Efforts are being made to develop suitable strategies for the use of marker genotype data for the selection of inbreds that would produce superior hybrids. In crops like maize, inbreds and populations can be classified into distinct heterotic pools. The members from the same heterotic pool exhibit little or no heterosis when crossed together, but considerable heterosis is observed when members from different pools are hybridized. Marker genotype data are able to assign inbreds and populations to the appropriate heterotic pools. However, maker genotype data are so far not able to predict the extent of heterosis that will be obtained by crossing any two selected parental lines. Studies on heterosis using molecular markers have revealed the existence of heterosis QTLS or loci distinct from the loci governing the concerned traits. It is hoped that as greater information would become available about the heterosis loci, it might become feasible to predict the level of heterosis expected in various cross combinations.

15.8. MARKER-ASSISTED SELECTION

The selection of desirable plants/lines in the conventional plant breeding methods is, as a rule, based on their phenotypic performance. This requirement places certain limitations that either enhance the time period of an activity or preclude the possibility of reliable evaluation. For example, selection for fruit/grain characteristics can be done only after the fruits/seeds are produced. As a result, the selected plants can be used for hybridization only in the next generation. Similarly, traits like yield cannot be evaluated in off-season nurseries or greenhouses. Therefore, selection has to be performed in populations grown in the main season only, which delays the progress of the breeding programme. Evaluation of quantitative traits is reliable only when it is based on replicated trials, and the traits measured on single plants cannot be depended upon. Many of these limitations can be resolved by using molecular markers linked to the desired genes/QTLs for their indirect selection. The method of selection in which a gene or QTL is selected for on the basis of the genotype of the molecular marker linked to this gene/QTL is known as **marker-assisted selection (MAS)**. The idea of MAS was first advocated by Tanksley (1983). MAS can be used in combination with the conventional breeding methods or novel breeding schemes (Sections 15.12, 15.13) could be devised to take full advantage of the marker data. MAS has been widely used for introgression of the desired genes and QTLs using backcross programmes.

15.9. MARKER-ASSISTED BACKCROSSING

The backcross programme is used to transfer genes and QTLs for the desired traits from varieties/germplasm lines/related species (donor parent DP) into a popular variety (recurrent parent, RP) that is deficient in the concerned trait. The transfer of a gene/QTL from one line into another is generally referred to as **gene/QTL introgression**. The objective of backcross programmes to introgress the genes/QTLs from the DP so that the end product of the programme is the same as the RP except for the introgressed trait(s). Therefore, in a backcross programme, the F_1 from the cross RP x DP and the subsequent generations are backcrossed with the RP. In the backcross generations, strict selection is made for the trait being transferred and for the plant type of the RP. In conventional backcross programs, these activities are based on phenotypic evaluation, but in marker-assisted backcross programmes marker genotypes are used to accomplish the various objectives of the programmes. The backcross programme based on molecular markers is generally called **marker-assisted backcrossing (MABC)**. MABC has the following three well-defined activities based on markers: (i) foreground selection, (ii) background selection and (iii) recombinant selection (Table 15.3).

Table 15.3: A summary of objectives achieved by the use of molecular markers during marker-assisted backcrossing

Activity	Objective	Markers used
Foreground selection	Indirect selection for the gene/QTL being introgressed	Markers linked to the target gene/QTL, preferably one marker on either side of the gene/QTL
Background selection	Maximum recovery of the recurrent parent genome	Markers distributed over the entire genome
Recombinant selection	Minimizing the donor parent genomic region surrounding the introgressed gene/QTL	One marker located on each side and as close to the target gene/QTL as possible

15.9.1. Foreground Selection

The use of molecular markers to select for the gene/QTL being introgressed is known as **foreground selection** (Fig. 15.3). The objective of foreground selection is to select for the target gene, i.e., the gene being introgressed, without phenotypic evaluation for the trait governed by the target gene. Foreground selection for a gene/QTL will be highly desirable in a variety of situations, some of which are listed below.

1. The phenotypic evaluation of the trait is time-taking, costly, burdensome or technically demanding.
2. The phenotypic evaluation of the trait has low reliability.
3. The desired trait phenotype is governed by the recessive allele of the concerned gene.
4. Introgression of both vertical and horizontal resistances in a single variety.
5. Introgression if multiple QTLs.
6. Introgression of genes/QTLs for multiple traits.
7. The trait can be assessed only after flowering.
8. The trait expression has a threshold requirement that is difficult to provide in a uniform and dependable manner.

The foreground selection will be one hundred percent effective when functional markers are used for MAS. The **effectiveness of selection** is the likelihood that a plant selected on the basis of marker genotype will have the same allele of the gene/QTL that is predicted by the marker genotype. The effectiveness will be nearly one hundred percent when MAS is based on gene-based markers. But in case the MAS is based on markers linked to the target gene, the effectiveness of selection will decline with an increase in the genetic distance between the marker and the concerned gene. This is because crossing over between the gene and the linked marker will change the linkage relationship between their alleles, and the frequency of this happening will increase with the genetic distance between them. Therefore, when a marker is situated at more than 5 cM distance from the target gene, it is advisable to use an additional marker located on the other side of this gene for foreground selection. When one marker is located on one side of a locus and another marker is located on the other side of the locus, this pair of markers is called **flanking markers**. The use of flanking markers minimizes the risk of selection of the wrong allele of the gene/QTL on the basis of marker genotype and, as a consequence, enhances the effectiveness of foreground selection.

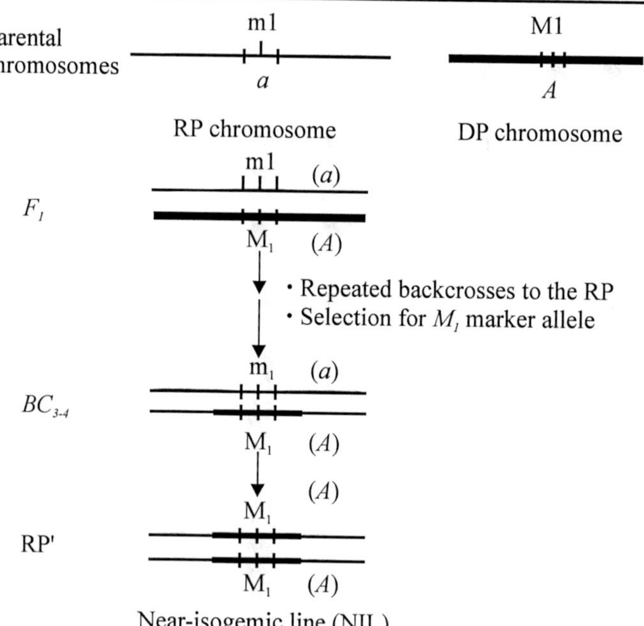

Fig. 15.3. A schematic representation of foreground selection. For simplicity, the marker is assumed to be located within the target gene, *a*. RP, recurrent parent; DP, donor parent; RP', the modified version of RP having the target allele *A* introgressed from the DP and the DP genomic regions surrounding the allele *A; BC*, backcross generation; M_1, M_2 the markers linked to the alleles *A* and *a*, respectively.

15.9.2. Background Selection

The objective of a backcross programme is to introgress only the target gene/QTL into the genome of the RP. Therefore, the RP', i.e., the new version of the RP obtained at the end of the backcross programme, should ideally have only the target gene/QTL and it should not contain any other genomic segment from the DP. But this ideal is rarely, if ever achieved, and the RP' contains variable lengths of the DP genome surrounding the target gene as well as in the other chromosomes of the RP. Since in most backcross programmes the DP is an inferior performing line, it may be expected that the additional amount the DP genome retained in the RP' genome will adversely affect the performance of RP'. Therefore, it is important that the genome of the RP is recovered in the RP' to the maximum extent possible. In a conventional backcross programme, a selection for the plant type of the RP is practiced during the backcross and the selfing generations to maximize the recovery of the RP genome. In this context, molecular markers may be used to increase the effectiveness of selection for the RP genome; this is referred to as **background selection** (Fig. 15.4). Background selection is based on markers distributed throughout the genome at sufficient density to ensure the selection for the entire RP genome, except for the genomic region containing the gene/QTL of interest. The use of molecular markers for background selection increases the rate of RP genome recovery in the backcross progeny. For example, the recovery of the RP genome following two to four backcrosses combined with molecular marker-based background selection may be expected to be comparable to that after up to six backcrosses in the conventional scheme where phenotypic selection for RP plant type is practiced.

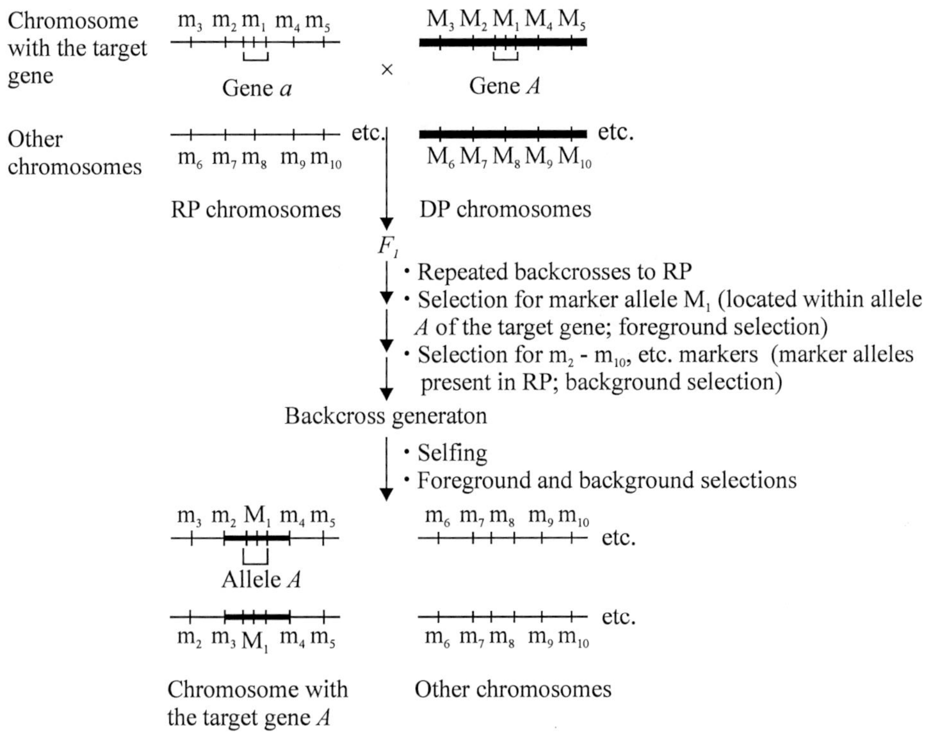

Near-isogenic line (RP')

Fig. 15.4. A simplified schematic representation of background selection. RP, recurrent, parent; DP, donor parent; RP', modified version of RP. For simplicity, marker M1 is assumed to be located within the target gene.

It may be pointed out that background selection increases the marker genotyping work to a considerable extent, and this has to be completed within a relatively short period of time. All the plants in a backcross generation must be screened for the markers used for the background selection before they flower so that this marker genotype information is used to select the plants for the next backcross. This may not be feasible for many breeding programmes. As a result, in most MABC programs the selection for RP genome is largely or exclusively based on the phenotypic evaluation for the RP plant type. The genome-wide markers for the background selection are generally used to analyze the RP genome recovery in the plants obtained at the end of the programme. In some studies where markers were indeed used for background selection, two backcrosses were able to recover genome RP to the extent that would be expected after five backcrossed without any background selection.

15.9.3. Recombinant Selection

The foreground selection for the target gene/QTL tends to retain in the RP' a considerable length of the DP genome surrounding the target gene/QTL. The length of this DP genome may be particularly large when the target gene is transferred from a related species. For example, a tomato variety was developed by transferring the gene Tm_2 from a related species through 11 backcrosses coupled with strict phenotypic selection for the plant type of this variety. When this variety was analysed using molecular markers, it was found to contain >51 cM of the genomic region containing the Tm_2 gene; this represented the whole chromosome arm in which the gene Tm_2 is present. The genomic region surrounding the target gene may have some genes/QTLs producing undesirable effects on the performance of the RP'; this phenomenon is known as **linkage drag**. It may be pointed out that many products of gene transfers have not been used for commercial cultivation mainly due to linkage drag. Linkage drag is particularly important when inferior lines and, especially, related species are used for gene transfers. Molecular markers can be used to minimize the risk of linkage drag by selecting against the DP genomic regions flanking the target gene/QTL. This application of molecular markers is termed as **recombinant selection** (Fig. 15.5). Thus, recombinant selection is a special type of background selection that focuses on the genomic regions linked to and flanking the target gene. Further, recombinant selection is based on a pair of markers that are located on either side of the target gene/QTL at <5 cM. It is, however, preferable that these flanking markers are located within 1 cM of the gene being introgressed, particularly when the DP is an inferior germplasm accession, including a related wild species.

Recombinant selection is more convenient when it is done according to the two-step scheme outlined in Figure 15.5. The marker M1 is a gene-based marker located within the gene being introgressed and is used for foreground selection. The markers M2 and M3 are located within 1 cM of the target gene *A*, and are used for recombinant selection. The **first step** comprises identification of at least one plant in the BC1 generation that shows recombination between the markers M1 and M2; such a plant will have the marker genotype M1m2 (m2 is the marker allele present in the RP). This plant is used to produce the BC2 generation. In this generation, all the plants will have the marker genotype M1m2. Therefore, search is made for a plant that shows recombination between the markers M1 and M3. Such a plant will have the genotype M1m3 for these two markers, and this plant will be the double recombinant, i.e., m2M1m3. This plant is termed as **double recombinant** because it is produced as a result of two recombination events, viz., recombination between the markers M1 and M2, and between M1 and M2. This double recombinant will carry up to only 2 cM of the DP genome flanking the introgressed gene. It may be pointed out that in case the recombinant M1m2 is not found in the BC1 generation, the recombinant between the markers M1 and M3 should be selected. In this situation, the recombinant between M1 and M2 will be selected in the next backcross generation. Another scheme for recombinant selection is to select for the double recombinant in a single backcross generation. But this approach will require a backcross population that is over 50-times larger than the total number of the backcross plants needed for both the steps of the two-step approach described above.

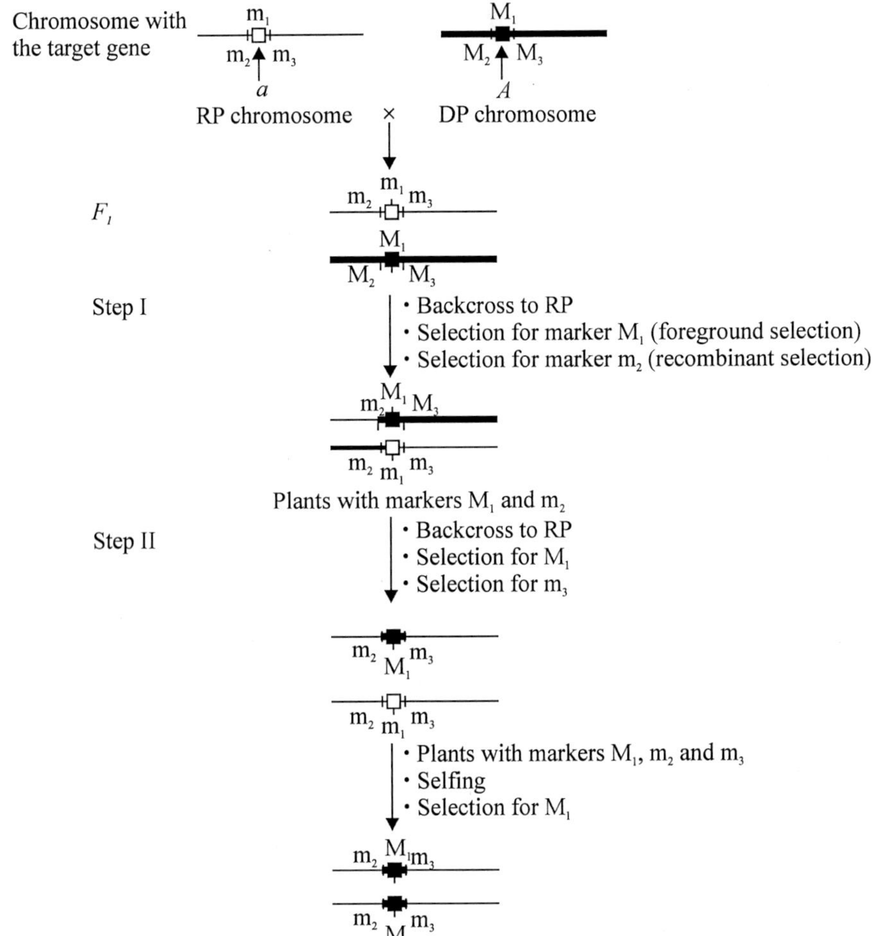

Fig. 15.5. A simplified schematic representation of recombinant selection. The objective of recombinant selection is to minimize the DP genomic region flanking allele *A* of the target gene. This is achieved by selection for the marker alleles m_2 and m_3 that are located very close to and on either side of the target gene allele *A*.

15.9.4. A Generalized Scheme for effective MABC

In view of the above discussion, a generalized scheme for an effective MABC programme can be outlined as under (Fig. 15.6).

1. The F_1 from the cross between the selected DP and RP is backcrossed to the RP to produce a sufficiently large RP population.
2. In the BC1 generation, the plants are screened with the markers M1, M2 and M3 (Fig. 15.5). All BC1 plants with either m2M1M3, or M2M1m3, or m2M1m3 maker genotype are selected.
3. In case, a plant has the marker genotype m2M1m3, it will be backcrossed to RP. Further, if more than one plant were obtained with the other two genotypes, all the plants will be screened with the set of markers selected for background selection. The plant with the largest proportion of the recurrent parent marker alleles will be selected and backcrossed to the RP to produce a sufficiently large backcross population.
4. In the BC2 generation, all the plants are screened with the marker M1 and either marker M2 or M3. In case the BC1 plant used for backcrossing was m2M1M3, the marker M3 will be used for screening. But the marker M2 will be used if the plant selected in BC1 were M2M1m3. In case the BC1 plant was m2M1m3, only M1 will be used to enable foreground selection for the desired allele of the target gene. All the plants that are recombinant for the concerned marker pairs used for screening are selected.
5. These selected BC2 plants will be double recombinants, i.e., will be m2M1m3. If there were more than one such plant, all of them will be screened with the set of markers selected for background selection. The plant with the highest proportion of the RP marker alleles will be selected and used for either one or, preferably, more backcross.
6. All the backcross progeny will be subjected to foreground (using marker M1) and background selections. The plant having M1 allele and the highest proportion of the RP marker parent shall be identified and selfed.
7. Foreground and background selections will be carried out during the selfing generations, and, in the end, a line homozygous for the introgressed allele (M1M1) and with the highest proportion of the RP genome is isolated.
8. The selected line is an NIL of the RP (RP') modified with the introgression of the desired gene. This line will be evaluated for the introgressed trait and for agronomic performance and might be released for commercial cultivation if found to be desirable.

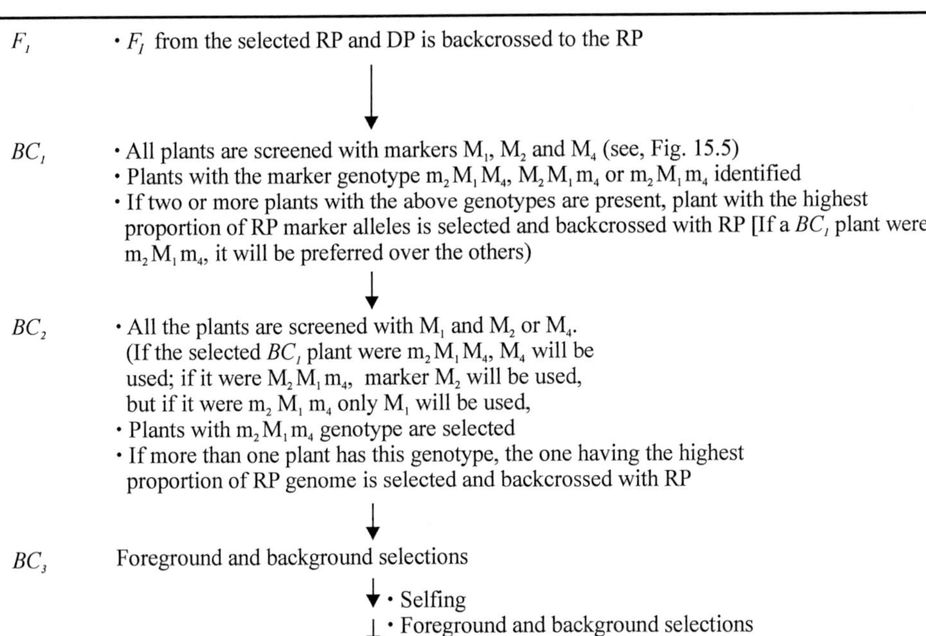

Fig. 15.6. A generalized scheme for effective marker-assisted backcross MABC. DP, donor parent; RP, recurrent parent; RP', the modified version of RP, *i.e,* RP having the introgressed gene *A; BC_1, BC_2* and *BC_3*, first, second and third backcross generations, respectively.

15.9.5. Gene Pyramiding

The idea of gene pyramiding was put forward in relation to disease resistance. The disease resistance is of two broad types, vertical or oligogenic and horizontal or polygenic. The **vertical disease resistance** is governed by oligogenes, and is race or pathotype specific. It produces complete resistance or immunity, but is generally overcome by the pathogen in a few (often) to several (in some cases) years of time. The **horizontal disease resistance**, on the other hand, is governed by polygenes, and is race nonspecific. It does not produce immunity, i.e., there is some disease symptom development, but is generally durable (lasts for a long time). It was suggested that durable vertical resistance could be generated by bringing together several different genes for resistance to the same disease in a single variety; this procedure is called **gene pyramiding**. Gene pyramiding was carried out by using ingenious breeding and disease testing schemes, but this was always a demanding and time taking task. The main difficulty was to reliably identify the resistance genes present in the different plants in the segregating generations. This problem is easily resolved by using molecular markers linked to the different resistance genes for selection of the plants having the target resistance genes.

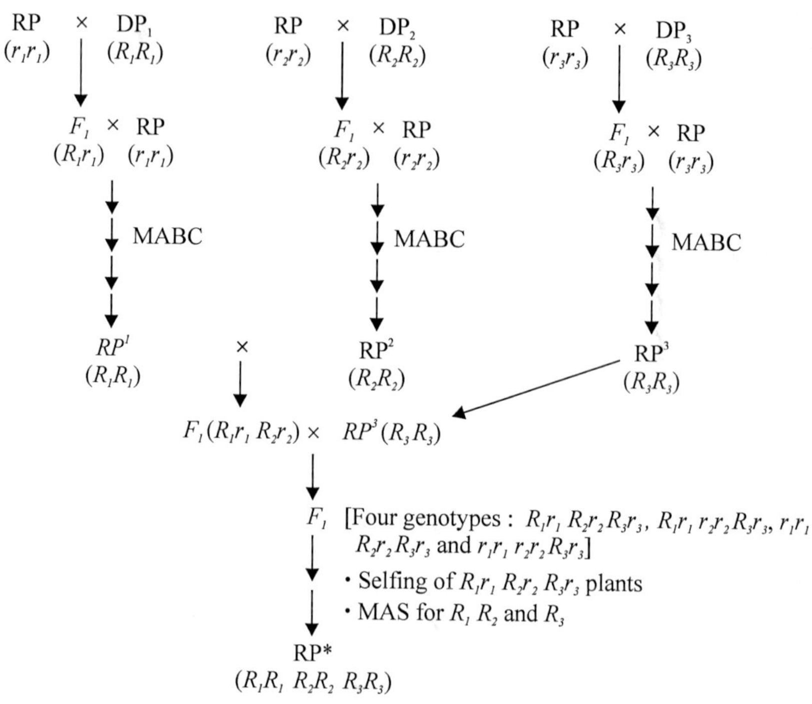

Fig. 15.7. A simple schematic representation of gene pyramiding using MABC. The RP has the recessive alleles of genes r_1, r_2 and r_3 (genotype, $r_1r_1\,r_2r_2\,r_3r_3$), but only the alleles relevant to the concerned DP are shown. RP^1, RP^2 and RP^3 are the modified versions of RP produced by the introgression of resistance genes R_1, R_2 and R_3, respectively. RP* is the modified version of RP having all the three genes (R_1, R_2 and R_3). RP, recurrent parent; DP, donor parent; MABC, marker-assisted backcrossing; MAS, marker-assisted selection.

Gene pyramiding is based on MABC. When all the genes to be pyramided in the selected RP are present in a single DP, the procedure for gene pyramiding is the same as MABC for a single gene. The only difference between the two is that a larger BC population would be needed to be able to select a plant having all the target genes and the desired level of RP genome recovery. But when the different genes are present in different DPs, they all have to be mated to the RP according to a suitable mating scheme. A simple and perhaps the most effective scheme for gene pyramiding is outlined in Figure 15.7. The RP is susceptible to a disease for which different resistance genes are available in three different DPs. The DP_1 has the resistance gene R_1; DP2 has the resistance gene R_2, while DP3 has the resistance gene R_3. The RP is crossed separately to the DPs and the F_1 from each of the three crosses is used for separate MABC programs. As a result, three different modified versions of the RP (RP^1, RP^2 and RP^3) are obtained at the end of MABC. Each of these modified versions is essentially the RP that has a distinct gene for resistance to the concerned disease. The three genes are brought together in the RP by first crossing RP^1 with RP^2. The F_1 from this cross is then mated with RP^3, and the F_1 from this complex cross is selfed to produce the F_2. MAS is done for the three resistance genes and the plants having all the three genes are selfed to raise the F_3 generation. The line homozygous for all the three resistance genes is selected as the pyramided version of the RP (RP*). MABC has been extensively used for gene pyramiding. Many of the gene pyramided versions of the various RPs have been used for commercial cultivation; some such examples are listed in Section 15.14.

15.10. QTL INTROGRESSION AND OTHER ISSUES

Molecular markers can be used to facilitate introgression of QTLs governing the traits of interest. In general, each QTL represents a genomic region that is equivalent of the genetic distance of 10 cM or more. Therefore, it is desirable that a set of three markers be used for foreground selection for each QTL. One of these markers should be located within the QTL (preferably in the middle of the QTL), and the remaining two markers should be located on either side of the QTL. In general, the QTLs selected for introgression are those that produce a large effect on the phenotypic expression of the target trait, e.g., the major QTL *Sub-1* for submergence tolerance in rice. But most QTLs may not produce sufficiently large phenotypic effects so that they are unable to generate useful improvement in the trait of interest. In such cases, two or more QTLs affecting the trait would need to be introgressed at the same time. This would require much larger backcross populations than that needed for introgression of a single QTL. Often the introgression of QTLs may fail to produce the expected improvement in the target trait. This is often the case with the introgressed QTLs that have relatively smaller effects on the trait. The possible reasons for such failures are QTL x genetic background interaction, and QTL x environment interaction. Efforts are being made to develop breeding and selection strategies for enhancing effectiveness of QTL introgressions. There are several successful examples of QTL introgression, and some of these have led to the development of improved versions of the RP used in the MABC programmes.

Selection for horizontal disease resistance is hampered mainly by the environmental influences on the expression of the disease. Molecular markers are used to map the QTLs involved in horizontal resistance and to identify markers linked to these QTLs. These markers are then used for indirect selection of the concerned QTLs, i.e., for MAS for horizontal resistance. It will be highly desirable to combine vertical disease resistance with horizontal resistance to increase the effectiveness and the durability of resistance. But it is impossible to select for horizontal resistance in the presence of vertical resistance. This difficulty is easily overcome by the use of molecular markers to select for the concerned genes and QTLs governing resistance to the disease in question. For example, MAS was used to combine vertical resistance to barley stripe rust (caused by *Puccinia striiformis* f. sp. *hordei*) with QTLs for horizontal resistance to the disease.

In some situations, it may be desirable to combine MAS with phenotypic selection/screening. This strategy is generally referred to as **combined MAS.** Combined MAS would be more desirable than MAS alone under the following situations: (i) the marker is not tightly linked to the target gene, (ii) phenotypic expression of the gene/QTL is likely to be affected by genetic background and/or the prevailing environment, (iii) it is desirable to select for such genetic background that would enhance the phenotypic expression of the gene/QTL, (iv) it is desirable to reduce the number of plants for marker genotyping, etc. In any case, the product of MAS has to be evaluated for the phenotype specified by the introgressed gene/QTL, and for agronomic performance before it can be used for commercial cultivation

15.11. MARKER-ASSISTED RECURRENT SELECTION

MAS can be used in combination with the recurrent selection scheme to increase the effectiveness of the scheme. In recurrent selection schemes, superior plants are selected from a genetically variable population, intermated with each other, and the seeds from all the crosses are composited to raise the next generation. The basis of selection may be phenotypic performance of the plants themselves or of their test-cross progeny (Section 5.3). This process is continued for few to several cycles with a view to accumulate the favorable QTL alleles for the trait of interest. In 1990, Lande and Thompson proposed the following scheme for using marker genotype data as the basis for selection in a recurrent selection scheme called **marker-assisted recurrent selection** (**MARS**; Fig. 15.8). In simple terms, the plants in F_2 or BC_1 generation of a desired cross are phenotyped for the trait. They are also genotyped for a large number of markers distributed genome-wide. The marker genotype and trait phenotype data are subjected to multiple regression analysis to identify the markers associated significantly with the trait. The trait phenotype data and the genotype data of the significant markers are used to construct a combined selection index. The F_2/BC_1 plants with superior combined index are selected and intermated. Equal amounts of seed from all the crosses are composited to grow the next generation. This completes one selection cycle.

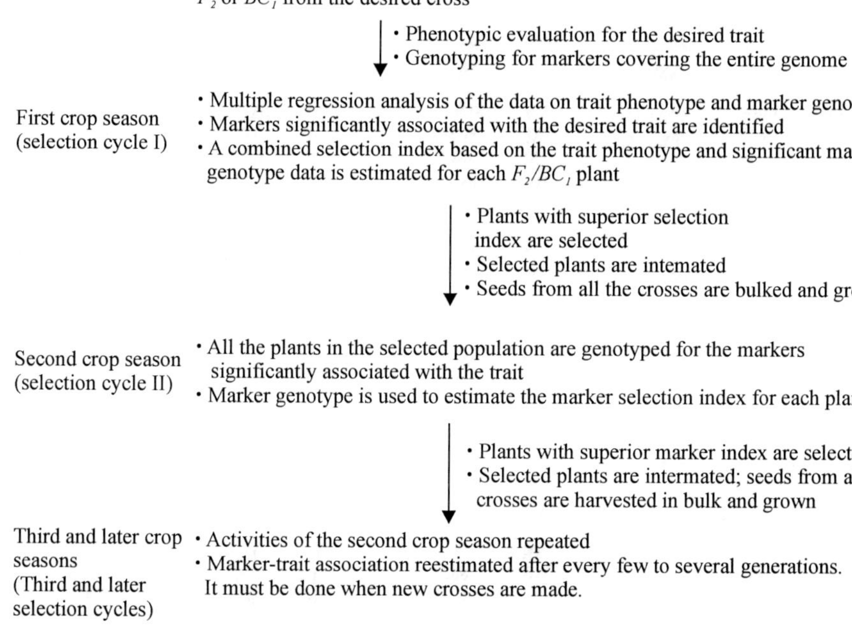

Fig. 15.8. A simple schematic representation of the marker-assisted recurrent selection. A crop season refers to the crop grown in the main crop season or in off-season nursery or in a greenhouse. The 'first crop season' has to be the main crop season. The marker genotype used for computation of selection index pertains to the markers significantly associated with the trait of interest.

The progeny plants from crosses among the selected plants are genotyped for the significant markers. The marker genotypes are used to compute the marker index for each plant, and plants with superior marker index are selected. The selected plants are intermated and the seeds from all the crosses are composited. This completes the second selection cycle. The selection based on marker index can be continued for few to several cycles. After this, the markers showing significant association with the trait are identified afresh. This step will be necessarily taken whenever new crosses are made. It may be pointed out that the use of marker data for selection of the superior plants allows the selected plants to be used for crossing in the same season. This is because marker genotypes can be estimated in the seedling stage itself. In addition, marker-based selection can be effectively done in crops grown in the off-season nurseries or greenhouses. Therefore, 2-3 selection cycles can be completed in a single year as compared to one selection cycle in about 2 years in the case of the conventional recurrent selection schemes. The MARS scheme is being widely used by private seed companies for the improvement of both cross- and self-pollinated crops.

15.12. NOVEL BREEDING SCHEMES

It has been realized that the conventional breeding schemes are not able to make full use of the marker genotype data. Therefore, several innovative breeding schemes have been proposed by various research workers. These novel breeding methods include (i) inbred enhancement and QTL mapping, (ii) advanced backcross QTL (AB-QTL) analysis, (iii) single large scale MAS, (iv) pedigree MAS, (v) single backcross doubled haploid scheme, (vi) breeding by design, (vii) mapping as you go, (viii) marker-evaluated selection, and (vii) genomic selection or genome-wide selection. The AB-QTL analysis scheme was proposed by Tanksley and Nelson in 1996, and is used for QTL introgression from inferior germplasm accessions without prior information about the QTL. In fact, QTL mapping is a part of this breeding scheme. The RP is crossed with the selected germplasm, and genetations the F_1 is backcrossed to the RP. In the first and the subsequent backcross generations, the plants with inferior phenotypes are eliminated, and only those having superior features are retained. The selected BC_2/BC_3 plants are subjected to appropriate analysis to identify the favorable QTL alleles. The plants with superior features can be used to develop improved lines. These lines can serve either as new varieties or as parents in breeding programmes.

In case of **marker-evaluated selection**, large amount of seed of the F_3 generation of the selected cross is produced. The F_4-F_6 generations are grown as bulks in the selected agricultural ecosystems on farmers' fields and/or experiment stations. These bulk populations may be subjected to mass selection. The F_7 generations is analyzed along with the parents of the cross to identify the marker alleles selected for in these populations. This information allows the breeders to identify the marker allelic combinations most suited for the different agricultural ecosystems. MAS is then used in the segregating generations of desired crosses develop lines expected to be most adapted to these environments. The breeding by design and genomic selection schemes are highly ambitious applications of marker technology. The genomic selection scheme (Section 15.13) is being used in breeding programmes, while breeding by design is an ambitious concept yet to be implemented by a breeding programme.

15.13. GENOMIC SELECTION

In 2001, Meuwissen and coworkers proposed a special type of MAS, called **genome-wide selection** or **genomic selection (GS)**. The GS scheme uses for selection information derived from genotype data of a large number of markers distributed over the whole genome irrespective of their associations with the trait being non-significant. In addition, the GS scheme employs two separate populations, namely, training and breeding populations. The **training population** should adequately represent the genetic diversity present in the breeding population. It should be sufficiently large because the accuracy of the GS model increases with the size of the training population. The training population is maintained separately from the breeding population, and is used only for training of the GS model (Fig. 15.9). The GS model is mathematical model that is used to estimate the effects associated with each of the markers on the target trait. Model training is the process of fine-tuning of the GS model using marker genotype and trait phenotype data from the training population. The fine-tuning of the model increases its accuracy of prediction of the trait phenotype on the basis of marker genotype. Therefore, the GS model should be trained for 2-3 generations on the training population. A variety of different mathematical approaches have been evaluated for use as model in GS, e.g., ridge regression, Bayesian approaches, semi-parametric regression methods and the machine learning methods. However, an entirely satisfactory GS model is yet to be developed.

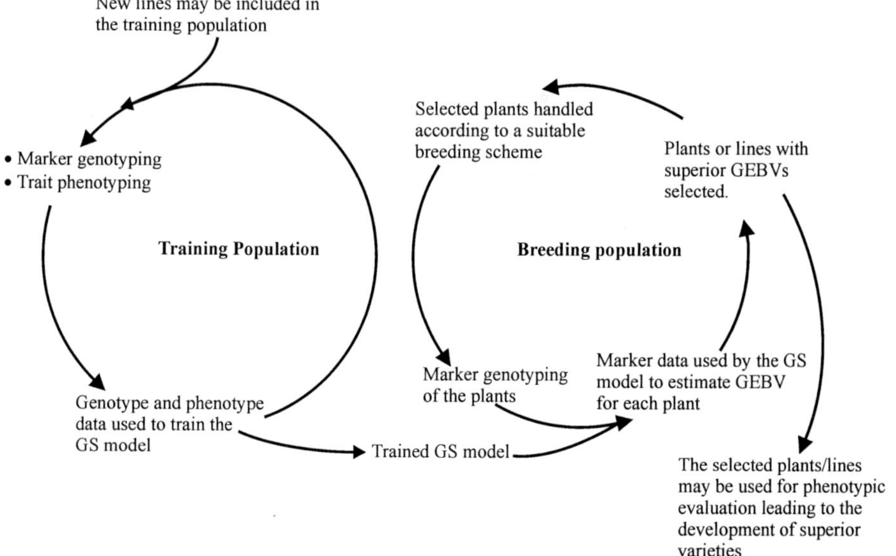

Fig. 15.9. A simple schematic representation of the genomic selection scheme. The training population is related to the breeding population. Genotyping is done for a large number genome-wide markers. Phenotyping is based on replicated trials. GS model training refers to fine-tuning of the GS model. GEBVs, genomic estimated breeding values; GS, genomic selection.

The general procedure for GS may be described as follows. First of all, a suitable training population is created. The plants/lines included in the training population are evaluated for the desired trait(s) using a replicated trial. The phenotypic evaluation should be as precise as possible. For this reason, the evaluation should, preferably, be conducted over locations and years. The plants/lines of the training population are also genotyped with a very large number of markers covering the entire genome. These markers should be distributed over the genome as evenly as possible. The trait phenotype and the marker genotype data are used for GS model training. This involves estimation of the various parameters of the GS model; these estimates are retained for application to the breeding population. The GS model in the end generates estimates of the effect on the target trait associated with each marker. The plants/lines in the breeding population are genotyped with the same set of markers that was used to genotype the training population. It may be emphasized that there is no phenotypic evaluation of the breeding population. As a matter of fact, only the lines taken out of the breeding population for the development of new varieties are subjected to phenotypic evaluation. The marker effect estimates obtained from the training population are used to estimate the genomic estimated breeding value (GEBV) for each plant/line in the breeding population on the basis of its marker genotype. ***GEBV*** of a plant/line is the sum of the effects associated with all the marker alleles present in this plant/line. Thus, GEBV of an individual is the estimate of its breeding value obtained from genome-wide marker genotype data. The **breeding value (BV)** of an individual/line is the expected phenotype of the progeny produced by this individual/line. The GEBV is used as the basis of selection in the breeding population. The plants/lines with superior GEBV are selected and handled according to the desired breeding scheme (Fig. 15.9).

GS has been highly successful in the breeding of dairy cattle. This prompted the adoption of GS in crop breeding. Recurrent selection schemed using GS have been devised for cross-pollinated as well as self-pollinated crop species. In addition, a futuristic breeding scheme based on GS has been proposed. This scheme would allow extraction of lines for use as new varieties at regular intervals. It would also allow continuous up-gradation and expansion of the breeding and training populations. Finally, these populations can be continued for long periods of time to support the variety development needs. It has been advocated that GS would encourage the use of exotic germplasm in crop breeding programmes. This would happen mainly because it will reduce the time required for their effective utilization from over 10 years to merely 3-4 years. Several programmes based on GS have been initiated to achieve a variety of objectives.

GS does not require QTL discovery and mapping. The GS scheme greatly reduces the time needed for one selection cycle. Further, it produces greater gains in the improvement of the traits, including yield, than phenotypic selection. The GS scheme is expected to reduce inbreeding as well the loss in genetic variability. However, there is still insufficient data to inspire confidence in the GS scheme. Further, the GS models are based on additive gene action and they do not take into account the epistatic gene effects. But in case of complex traits like yield, epistasis may be of considerable importance. The GS scheme would be useful only when dependable off-season and/or greenhouse facilities are available. This is because GS is not superior to phenotypic selection on the basis of per cycle of selection. It is superior only on per unit time basis since it can be used in combination with off-season or greenhouse facilities so that up to three selection cycles are completed in a single year.

15.14. ACHIEVEMENTS WITH MAS

MAS was first used for the selection for resistance to soybean cyst nematode. MAS is being extensively used by the private seed industries for various activities, including marker-assisted recurrent selection and genomic selection. Monsanto, USA was the first to release, in 2006, a maize hybrid developed by using MAS for commercial cultivation in USA. Subsequently, two rice varieties (Cadet and Jacinto) of rice having unique cooking qualities developed by MAS were released in USA. The rice varieties Angke and Conde were developed in Indonesia by transferring genes for bacterial blight resistance; these varieties give about 20% more yield than the popular rice variety IR64. In India, rice varieties have been developed by introgression of genes for bacterial blight resistance and the major QTL *Sub-1* for submergence tolerance. The examples of bacterial blight resistant rice varieties are Improved Sambha Mahsuri (produced by introgression of three bacterial blight resistance genes into Sambha Mahsuri), and Improved Pusa Basmati 1 (produced by introgression of two bacterial blight resistance genes into Pusa Basmati 1). Similarly, the rice varieties Swarna-Sub 1 and Sambha Mahsuri-Sub1 were produced by introgression of the QTL *Sub-1* into the mega rice varieties Swarna and Sambha Mahsuri, respectively. In case of maize, the single cross hybrid maize Vivek QPM 9 has been produced by MAS for the recessive gene *o2* (*opaque 2*); the *o2* gene was introgressed into the parental inbreds of the hybrid Vivek Hybrid Maize 9 to obtain this quality protein maize hybrid.

15.15. ADVANTAGES OF MAS

1. Foreground selection is simpler and more convenient than phenotypic selection for many traits.
2. In cases of introgression of recessive genes, the backcrosses can be made one after the other without any need for raising a F_2 generation.
3. The selected plants can be used for making crosses in the same season.
4. Recombinant selection minimizes the risk of linkage drag.
5. Selection for even traits like yield can be done in off-season nurseries or greenhouses. Therefore, 2-3 generations can be subjected to selection in a single generation.
6. It greatly facilitates gene pyramiding, and permits horizontal resistance to be combined with vertical resistance.
7. Background selection can be used in the backcross programmes to ensure a high level of recovery of the RP genome.

15.16. LIMITATIONS OF MAS

1. For many traits of interest, tightly linked markers are not available.
2. Introgression of QTLs often does not produce the desired results.
3. Marker genotyping adds to the overall cost of the breeding programmes.
4. The use of markers is technically demanding, and many plant breeders may not be fully familiar with the marker technology and terminology.
5. When large numbers of plants or markers are to be genotyped, high throughput facilities are needed, which may not be available to most of the breeding programmes in the public sector. One approach to resolve this limitation would be to develop centralized genotyping facilities to be used by a community of plant breeders.
6. The use of markers greatly increases the information generated during the crop season. This information has to be processed and used for decision making in the breeding programmes. As a result, there is considerable increase in the workload of the breeders.
7. The results from linkage and association mapping need to be confirmed and validated, which requires additional resources, including the time of the research workers.
8. The full utilization of molecular marker technology in plant breeding requires the use of off-season nurseries and/or greenhouse facilities. Dependable off-season nursery and/or greenhouse facilities may not be available to many breeding programmes.

15.17. CONCLUSIONS

The molecular markers have provided a novel and enabling approach for achieving a variety of plant breeding objectives. They the realization of even such objectives that could not be achieved by conventional breeding schemes, e.g., combining of horizontal and vertical resistance, carrying out 2-3 cycles of recurrent selection in one year, etc. Therefore, MAS is becoming increasingly popular among breeders, particularly in the developed countries. The private seed industry has integrated the marker technology as a regular component of their breeding activities. The industry has adopted high throughput genotyping technology, developed systems for tracking plant samples from the field into the laboratory, and then relate the marker genotype information back to the field plants with very high degree of accuracy. In addition, statistical and bioinformatics capabilities for acquisition, processing and storage of the huge amounts of data have been developed.

One of the main limitations of the marker technology remains the high cost of marker genotyping. This often puts this approach out of reach of many breeding programs in the developing countries. In the case of marker systems like SNPs, the cost per data point is quite low, e.g., around US $ 0.09. But when this cost is worked out for one plant it comes to US $ (for 384 SNPs), while for a breeding population of 500 plants it would be a staggering US $ 17,280.00. Further, many individual laboratories cannot afford high throughput genotyping platforms that are needed for fast genotyping work. In view of the above, centralized genotyping facilities may be developed. The United States Department of Agriculture has established four regional genotyping centers. These facilities have been of great help to small grains breeders and allowed them to routinely use the marker technology in their breeding activities. Another option is to use commercial services for marker genotyping. The use of marker technology will become considerably attractive if cost free and user-friendly software packages were available for analysis and interpretation, and storage and retrieval of the various types of marker data.

PRACTICE QUESTIONS

1. Define marker-assisted selection. Discuss the achievements, advantages and the limitations of marker-assisted selection.
2. List the various applications of molecular markers in plant breeding. Briefly describe the use of markers for germplasm conservation, genetic diversity analysis, and DNA fingerprinting.
3. Define recurrent selection. Briefly describe the procedure for marker-assisted recurrent selection and discuss its advantages and limitations.
4. Define maker-assisted backcrossing. Briefly describe the procedure for maker-assisted backcrossing with special reference to foreground, background and recombinant selections, and gene pyramiding.
5. Write short notes on the following: (i) marker-assisted selection, (ii) positional cloning, (iii) marker-assisted backcrossing, (iv) genomic selection, (v) markers in germplasm conservation, (vi) marker-based genetic diversity analysis, (vii) DNA fingerprinting, (viii) foreground selection, (ix) background selection, (x) recombinant selection, (xi) marker-assisted recurrent selection, (xii) novel breeding schemes based on molecular markers, (xiii) gene pyramiding.

RELEVANT LITERATURE

Singh, BD and Singh AK 2015. Marker-Assisted Plant Breeding, Principles and Practices. Springer India, New Delhi.

Tanksley SD 1981. Molecular markers in plant breeding Plant Molecular Biology Report 1: 3-8.

■■■■

Part D.
Miscellaneous

16 Variety Release, Seed Multiplication and Intellectual Property Rights

16.1. INTRODUCTION

The tools and techniques of plant breeding are designed to create new combinations of alleles of various genes present in different lines/accessions of the concerned crops and/or their related species. The transgenic technology, however, enables the utilization of genes from any source to create new genotypes of crop species. The lines with the new allelic combinations or the transgenic lines that are superior to the existing commercially cultivated popular varieties of the concerned crop to be identified and released as new varieties. This would require not only selection of superior lines from the breeding populations, but also their evaluation in replicated trials over locations and years to confirm the superior performance of the new strains. After the release of a new variety, good quality seed has to be produced in sufficiently large quantities so that the same can be made available to the farmers at a reasonable cost for commercial crop production. In plant breeding context, the term **seed** is used for any plant part that is used to propagate the concerned plant species, i.e., is used as **propagule**. Finally, the commercialization of a new variety needs to be regulated as per the prevailing intellectual property rights (IPR) laws of the country. These laws are designed to ensure that the individuals/organizations involved in the development or creation of a new variety get appropriate financial returns from their efforts. This chapter is devoted to a brief discussion of the procedure for variety release and seed multiplication followed in India, and the relevant Indian IPR laws.

16.2. RELEASE OF NEW VARIETIES

The procedure followed for the release of new varieties has been designed to ensure that the new variety consistently performs better than the popular checks over years in the region for which it is recommended for cultivation. This procedure comprises three different steps as follows: (i) evaluation, (ii) identification for release, and (iii) release and notification. A brief description of these steps is given in the following sections.

16.2.1. Evaluation

A **strain** of **line** is a group of plants that differs from other such groups in various characteristics and genotype. A single strain may have plants of a single genotype (this is the more common situation) or more than one genotype. A strain/line can be released as a variety only after its evaluation following a specified procedure. In India, the evaluation generally comprises (i) a station trial and (ii) a set of multi-location trials. The specific details of these trials vary depending upon the crops, but their salient features are briefly described here. The **station trial**, also called **preliminary yield trial** is a replicated trial conducted by the breeder for the new lines developed by him. It enables him to determine if any of the lines developed by him is superior in performance to the popular varieties of the concerned crop. Therefore, the popular varieties of the crop are used as checks in these trials. It is ordinarily conducted at one location for one year, and the lines that perform better than the checks are promoted to the next stage of evaluation, viz., the multi-location trials.

Table 16.1: The nine agroclimatic zones recognized for the evaluation of wheat entries

Zone	Areas included in the zone
Northern Hills Zone (NHZ)	Hills of Uttar Pradesh, Himachal Pradesh, and Jammu and Kashmir
Northern Plains Zone (NPZ)	Punjab, Eastern Haryana, Western U.P., Delhi, Ganganagar, Alwar, Bharatpur (Rajasthan), Gwalior, Bhind, Morena (M.P.), Una and Panota Valley (H.P.); Jammu (Jammu & Kashmir)
North-Western Plains Zone (NWPZ)	Rajasthan (except Kota and Udaipur), Western parts of Haryana, parts of Gujarat
North-Eastern Plains Zone (NEPZ)	Eastern U.P., Bihar
Far Eastern Zone (FEZ)	West Bengal, Chhota Nagpur, Assam, Sikkim, Arunachal Pradesh, Tripura, Manipur, Mizoram, Nagaland, Meghalaya
Central Zone (CZ)	Kota and Udaipur (Rajasthan), Jhansi (U.P.), parts of Gujarat, M.P. (except Eastern and South-Eastern parts)
South-Eastern Zone (SEZ)	Orissa, Southern and South-Eastern M.P., some parts of Maharashtra, A.P. (east coast region)
Peninsular Zone (PZ)	Karnataka, Western A.P., plains of Tamil Nadu, remaining parts of Maharashtra
Southern Hills Zone (SHZ)	Hills of Tamil Nadu

The **multi-location trials** are conducted at several locations generally for three years under the concerned All India Coordinated Crop Project. A line included in the multi-location trials is generally referred to as **entry**. An entry is evaluated at 10 or more locations within the agroclimatic zone for which it was developed. An **agroclimatic zone** comprises regions that have similar agricultural and climatic conditions with reference to the cultivation of a given crop. The country has been divided into one (in case of sorghum and pearl millet) to nine zones (for wheat and gram) depending on the following: distribution of the concerned crop; area under the crop, soil type, climate, etc. of the different regions. For example, the number of zones is 8 for rice and mungbean, 7 for pigeon pea and pea, 6 for maize and groundnut, 5 for barley and mustard, 4 for linseed and soybean, etc. The agroclimatic zones for evaluation of wheat entries are outlined in Table 16.1. The details of the evaluation procedure differ to some extent from one crop to the other. In case of wheat, all new entries are evaluated for one year under one of the conditions in the initial evaluation trials (IET). The IET for wheat is conducted under the following conditions: (i) irrigated, timely sown; (ii) irrigated, late sown; and (iii) rainfed, timely sown. Ordinarily, an entry is evaluated under any one of these conditions. An entry superior to the check is advanced to the uniform regional trials conducted under the same conditions as the IET. An entry that is superior to the checks in the first year of URT is evaluated for one more year in the URT, and is also included in the agronomic trials. The objective of agronomic trials is to determine the optimum agronomic practices for the entries.

16.2.2. Identification for Release and Prerelease Multiplication

On the basis of the performance of an entry for one year in IET and two years in URT, the entries that are superior to the checks are proposed for identification for release and for pre-release multiplication. The proposal is made by the breeder who developed the entry on a prescribed proforma. This proposal should also include data from disease and insect tests, and quality tests are also considered. The disease and insect tests are done for all the entries in IET as well as URT; they are conducted both under natural epidemic as well as artificial epiphytotic conditions. The quality tests are done on all entries included in the URT. The identification of an entry for release is done by a committee during the annual workshop of the concerned All India Coordinated Project. The identified entry is then evaluated in model agronomic trials, adoptive trials, and minikit trials. The data from adoptive trials are considered for the release of the entry as a variety. The minikit trials are done on farmers' fields and aim to popularize the varieties among the farmers by the time they are officially released.

16.2.3. Variety Release and Notification

After an entry has been identified, it is evaluated for one year in adoptive trials. In additions, disease, insect, and quality tests are also done. The breeder submits the proposal for release of the entry as a variety. This proposal contains data from

URT (two years), adoptive trials and disease, insect and quality tests. The breeder may give this entry a name like Padma, Jaya, etc. The proposal is considered by the Subcommittee on Crop Standards and Release of Varieties (SCSRV). A variety that is approved for release is then notified by the Director, High Yielding Varieties (DHYV). The notification of a variety is essential for the production of its certified seed. In addition, a state can release a variety through its State Variety Release Committee, for which separate trials are conducted by the state. But the variety released by state can be notified by the DHYV only if this variety was identified for release by the concerned All India Coordinated Project.

16.3. SEED MULTIPLICATION

In the context of crop production, **seed** signifies any part of the plant that is used for propagation of the concerned crop, i.e., is used as **propagule**. Thus, potato tubers and sugarcane stems used for propagation will be called seed, while pea or gram seeds used as food will not be referred to as seed. The seed used for raising a crop should be free from impurities, have high germination percentage and should be healthy; such seed would be considered as **good quality seed**. When such seed is of an improved variety, it is called **improved seed**. Improved seed is perhaps the most critical input in crop production since it would determine the ability of the crop to give the best performance by making the optimum use of other inputs like fertilizer, irrigation water, etc. The seed supplied to the farmers is certified seed, which is produced from other categories of seed through a step-wise multiplication scheme (Fig. 16.1). In this section we shall consider the various aspects of quality seed production in some detail.

16.3.1. Classes of Improved Seed

The scheme of seed production is designed to ensure a continuous production chain that is able to produce the required quantities of certified seed of a given variety every year. There are four classes of seed as follows: (i) nucleus seed or basic seed, (ii) breeder seed, (iii) foundation seed, and (iv) certified seed. Each of these seed classes is subdivided into stage I and stage II seed (Fig. 16.1). The stage II of a seed class is produced only when the stage I seed of that class is in short supply. Further, stage II certified seed can be produced only when the stage I certified seed was produced from stage I foundation seed. The quality of the foundation and certified seeds is certified by the Seed Certification Agency of the concerned state. The **nucleus seed** is the source of all other classes of seed. This seed is multiplied by selecting individual pods/spikes/plants and growing individual pod/spike/plant progenies, and selecting individual pods/spikes/plants from the true-to-the type progenies. This process is repeated every year. Nucleus seed stage I is produced by bulking the seeds from the true-to-the type progenies.

The **breeder seed** is progeny of nucleus seed and is produced under the supervision of the breeder who developed the variety or any other institution or qualified breeder recognized by the authorities. The quality of the breeder seed is indicated on a golden brown tag attached to the seed bag (Table 16.2). Multiplication of the breeder seed generates **foundation seed**. The quality of foundation seed has to meet the prescribed standards, and is certified by the Seed Certification Agency of the concerned state. The seed bags of foundation seed are labeled with white tags. The **certified seed** is the progeny of foundation seed, and is the seed distributed to the farmers for cultivation. The certified seed has to meet the prescribed standards of seed quality (Table 16.3). The quality of a seed lot of certified seed is certified by the Seed Certification Agency of the concerned state, and is listed on the azure blue tag of the seed bags.

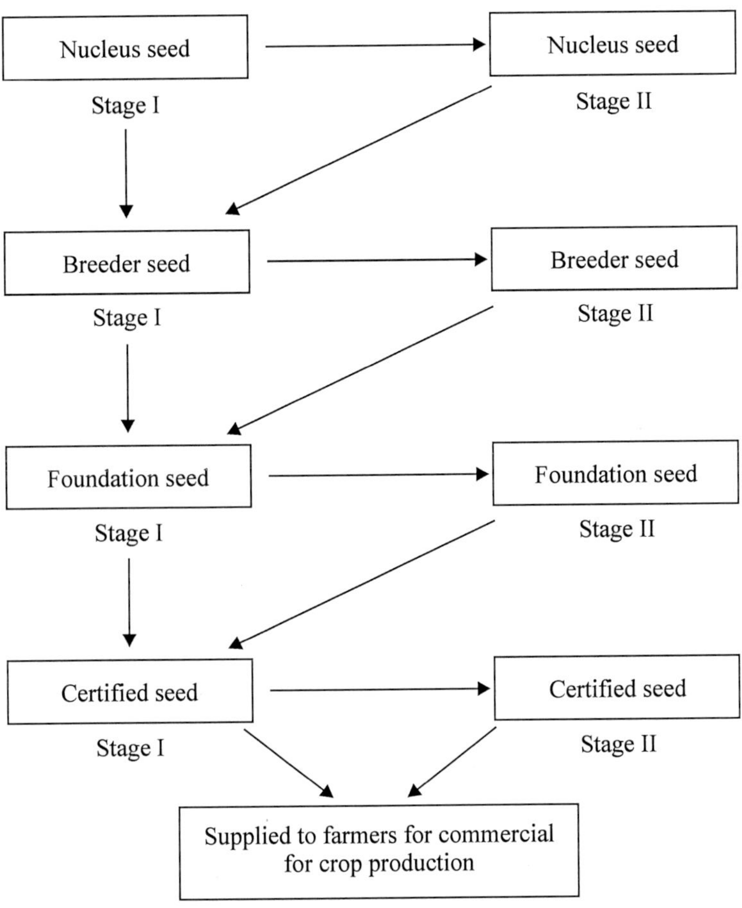

Fig. 16.1. A schematic representation of the four classes of improved seed and multiplication scheme used for their production. It should be noted that the stage II of any seed class is allowed to be produced only when that class of seed is in short supply.

Table 16.2: The tags used for the different classes of improved seed

Seed class	Tag color	Remarks
Nucleus seed	No tag used	Certificate issued by the concerned breeder
Breeder seed	Golden brown	Stage II seed can be produced when the supply of Stage I seed is not adequate*
Foundation seed	White	
Certified seed	Blue (shade ISI No. 104)	

* Stage II certified seed can be produced only when the Stage I certified seed was produced from Stage I foundation seed.

Table 16.3: The minimum quality standards required for the certified seeds of some selected crops

Crop	Pure seed (%)[1]	Other crop seed[2]	Other variety seed[2]	Total weed seed[2]	Total inert matter[3]	Moisture (%)[3]	Germination (%; minimum)
Maize	98	5	10	None	2	12	90**
Pearl millet	98*	10	-	10	2	12	75
Sorghum	98	5	10	5, 10**	2	12	75
Rice	98*	10	0.5%	10@	2	12	85
Wheat	98	10	-	10@	2	12	85
Cotton	98	5	-	5	-	10	65, 75**
Pigeon pea	98	5	10	10	2	9	75
Gram	98	None	5	None	2	9	80
Rapeseed and mustard	97	10	0.1%	5@	3	8	85
Groundnut	96	None	-	None	4	9	70
Peas	98	None	5	None	2	9	75
Soybean	98	None	5	5	3	12	70
Cauliflower	98	5	-	5	2	7	65
Tomato	98	5	-	None	2	8	70
Brinjal	98	None	-	None	2	8	70

[1] Minimum; by weight.

[2] Maximum number of seeds per kg.

[3] Maximum; by weight.

* Diseased seed (max.): pearl millet, ergot sclerotia 0.02%; rice, paddy bunt 0.1%; wheat, nematode galls of tundu/ear cockle or seeds affected by Karnal bunt 0.1% (all by number)

** For hybrid seed

@ objectionable weed seed (max.): rice, 2 seeds/kg; wheat, 2 seeds/kg; rapeseed and mustard, 5 seeds/kg.

Thus, the availability of quality improved seed to the farmers requires the following operations/functions: (i) development of new improved varieties, (ii) seed multiplication of the new variety in a step-wise manner, (iii) seed processing (drying, cleaning and grading of seed, followed by treatment with appropriate disinfectants and/or protectants, (iv) storage of the seeds in suitable containers like bags, (v) seed certification, (vi) publicity to generate awareness among the farmers, and (vii) marketing and distribution.

16.3.2. The Indian Seed Act (1966)

The Indian Seed Act (1966) was implemented in 1969, and amended in 1972, aims to regulate the quality of seed offered for sale. The Act requires the seed bags to be compulsorily labeled. The label attached to bags must contain the prescribed information on seed quality, and this information should accurately represent the quality of the seed in the container. In addition, the seed must conform to the prescribed norms of purity and germination. In addition, the seed offered for sale may be got certified by the concerned State Seed Certification Agency. Seed certification is not mandatory and left to the choice of the person/organization engaged in seed production.

16.3.3. Certified Seed Production

The production of certified seed of a crop includes cultivation of the seed crop, processing of the seed harvested from this crop and certification of the processed seed by the State Seed Certification Agency (SSCA).

Cultivation of the seed crop

The seed crop is grown in a field that is fertile, level, free from noxious weeds, and separated from other fields of the same crop at least by the minimum isolation distance. If the same crop was grown in this field in the previous year, the field should be irrigated at least three weeks before planting of the seed crop, and then ploughed. This is done to destroy any seeds of the previous crop that may have fallen in the field. The seed crop is generally grown from the foundation seed according to the recommended cultural practices to raise a good crop. The prescribed plant protection practices are followed to control insects in diseases. Effort is made to keep the crop free from weeds, particularly the objectionable weeds. Off-type plants or rogues are regularly removed; this is called **roguing**. A **rogue** is a plant of the same crop as the seed crop, but differs in morphological features from the plants of the seed crop. In case of hybrid seed production, special operations like detasselling of the pollen shedding plants of the female parent or emasculation of the flowers of the female parent, and harvesting of the male parent rows before the harvesting of the female parent need to be carried out. **Detasselling** is the removal of the tassels (the male inflorescence) from maize plants. **Emasculation** comprises removal of all the anthers from the flowers before they shed pollen; it is generally done one day prior to the expected date of anthesis of the concerned flower. **Anthesis** is the first opening of a flower. The seed crop is harvested at maturity and threshing is carried out on a clean floor so that the contamination of the seed by inert matter is minimized. In addition, care is to taken to avoid mechanical mixture with seeds of other crops, other varieties of the same crop, etc.

Seed processing

The seed obtained from the threshing floor is dried, cleaned, graded, treated, and bagged; these operations together constitute **seed processing.** The fresh seeds generally have much higher moisture content than prescribed for the seed. Therefore, they are **dried** either in the sun or in suitable seed drying facilities to reduce their moisture to the recommended level. In case of rabi crops, drying in the sun would be practicable, but the for kharif crops artificial drying may often be required. **Artificial drying** adds to the seed cost, and requires considerable care to avoid damage to the seeds by overheating and there is the risk of fire. The dried seeds are cleaned to remove the inert matter, and seeds of weeds and other crops (**cleaning**). The cleaned seeds are then subjected to **grading**, which comprises removal of the undersized and shriveled seeds. The cleaning and grading of seeds are generally done with the help of appropriate machines. For example, the air and screen machines carry out both cleaning and grading of seeds, and are extensively used in India.

The seeds obtained after grading are used for seed tests. The seeds are then treated with a suitable fungicide and an insecticide. For each crop, seed treatment with specific fungicides and insecticides are prescribed, and these need to be followed. These chemicals are either applied to the seeds as dry powder or as slurry. The seed treatments are used with the aim of (i) controlling seed-borne diseases, (ii) protecting the seed crop from soil-borne diseases, (iii) protecting the seeds from insects during storage, and (iv) protecting the seeds from soil-borne insects. The treated seeds are usually filled in 40 kg bags, and the label describing the kind and the quality of the seed is attached to the bags (**bagging and labeling**). The bags are finally stored in a clean dry storage space till the next crop season.

Seed certification

Seed certification is the issuance of a certificate by the concerned State Seed Certification Agency stating that the concerned seed lot meets the specified seed quality requirements. Seed certification is based on field inspection and seed tests.

Field inspection

Field inspection is the scrutiny of the seed crop field by a seed inspector of the concerned State Seed Certification Agency. The field inspection is done at the specific stages of crop growth, which differ to some extent for the different crops. For example, the following inspections are done in the case of hybrid maize seed crop: (i) before flowering, (ii) during flowering (two inspections), (iii) at maturity just before or during the harvest. The field inspections are done to monitor the isolation distance; the presence of rogues, noxious weeds, and plants of other crops; incidence of seed-borne and other diseases; ratio of male to female rows (for hybrid seed); etc.

Isolation distance is the minimum prescribed distance that should separate a seed crop from other fields of the same crop. The main aim of isolation distance is to prevent pollination of the seed crop by pollen from other plants of the crop growing in the neighboring fields. In case of self-pollinated crops, it also helps prevent mechanical mixtures, i.e., mixture with the seeds from the plants of the same crop species growing in other fields. The isolation distance varies with the crop species and depends mainly on the chances of cross-pollination in the concerned crop. For example, it is merely 3m in the case of crops like wheat, rice, barley, soybean, etc. through 200m in case of maize and pearl millet to 1000m in case of crops like cauliflower, radish, and turnip (Table 16.4).

Table 16.4: The minimum isolation distances for certified seed crops of some selected crop species

Crop	Isolation distance (m)
Maize hybrids and other varieties	200*
Hybrid sorghum:	
a) From sorghum fields	200
b) From Johnson grass fields	400
Hybrid pearl millet	200
Wheat, rice, barley, soybean	3
Hybrid rice and wheat	100
Pea, cowpea, rajma (common bean)	5
Rapeseed and mustard	25**
Tomato	25
Potato	5
Brinjal (eggplant)	150
Carrot	800
Cauliflower, radish, turnip	1,000

* Can be modified by growing the seed crop in a larger field and by planting 3-10 rows of the male parent of the hybrid around the field to ensure adequate pollination at the edges of the seed crop field.

** For self-compatible genotypes.

Seed tests

The determination of purity, germination, moisture content, etc. of a seed lot is called **seed test**. Seed tests are done in seed testing laboratories of the concerned state. Random samples, called **working samples**, of the seeds are carefully drawn for conducting the seed tests. The weight of a working sample depends mainly on the crop in question and the purpose of the test. For example, for physical purity test, the sample should be 15g in case of pearl millet, but it should be 900g for maize. But for genetic purity test, the working sample should be 150g in case of pearl millet and 1000g for maize.

Pure seed means the seed of the variety under certification; it is generally expressed as percent of the weight of the pure seed to the total seed weight. The seed lot may be contaminated by physical impurities as well as genetic impurities. The **impurities** in a seed lot are represented by seeds of other varieties of the same crop, other crops, and weeds; defective seeds (shrunken and broken seeds); and inert matter (sand, straw, stones, etc.). The amount of impurities in the sample is determined by a close examination of the seed sample. This constitutes the **physical purity test** since it is based on physical examination of the seed sample. Sometimes, the seeds of other varieties of the same crop may be identifiable on the basis of their morphological features like color, size, thickness, etc., but generally specific tests would be needed for their identification. The total weight of the different impurities in the seed sample is determined, and subtracted from the total weight of the seed sample to yield the weight of pure seed. Seed purity is then calculated by the following formula:

Seed purity (%) = (Weight of pure seed / Total weight of the seed sample) x 100

The tests designed for determination of the amount of seeds of other varieties of the same crop are called **genetic purity tests**. The genetic purity tests include examination of morphological features of the seeds, KOH bleach test (for sorghum), NaOH and phenol tests (both for wheat), peroxidase test (for soybean), electrophoretic patterns of seed proteins, including isozymes, and DNA markers (both applicable to all crops). The seeds may be germinated under controlled conditions and the seedlings may be evaluated for color and other morphological features. The specified amount of the test seeds may be planted in the field along with the authentic sample of the variety to be certified. The plants are evaluated for plant height, growth habit, flowering time, etc. to determine the percentage of plants belonging to the concerned variety; this is called **grow-out test**. Grow-out test is more suitable for self-pollinated crops than for cross-pollinated ones, and requires considerable field space and time.

Germination is a critical feature of the seed and is determined by germination test. The **germination test** determines the percentage of seeds of a test sample that would produce seedlings. Generally, seeds are germinated on wet filter papers kept in Petri dishes and incubated at 18-22°C either in dark or in diffuse light. The number of seeds that germinate to produce healthy seedlings is recorded and expressed as percent of the total number of seeds placed in the plate. From each seed lot at least four samples of 100 seeds are used for germination test. The germination test takes 7 days in most of the crops, but in some cases, it may take up to 25 days. In contrast **tetrazolium test** (TT) yields results in less than two days, and determines the percentage of viable seeds in the seed lot. These seeds may be expected to produce healthy seedlings on germination. For this test, the seeds are soaked overnight in tap water, the soaked seeds are split in two halves in such a way that each half has nearly half of the embryo. One half of each seed is treated with 1% solution of the reagent 2,3,5-triphenyl tetrazolium choloride (TTC) for 4 hr. The seeds are then washed in tap water, and the number of half seeds with red embryos is counted. The living cells reduce TTC to formazon, which gives the red color. The percentage of seeds with red colored embryos is the percentage of viable seeds. This method cannot be used for crops with small seeds that cannot be easily split into two parts.

The percentage amount (by weight) of water present in a seed sample is its **moisture content**. The moisture content of seeds determines its safe storage, and the damage caused by storage pests and molds. A simple way to estimate the moisture content of a seed sample is to determine the weight of the sample (W_1), and then dry it in an oven at 130°C for 90 min and determine its weight (W_2) again. The moisture content of the seed sample is then computed as follows:

$$\text{Moisture content (\%)} = [(W_1 - W_2)/ W_1] \times 100$$

Seed moisture content can also be determined by **moisture meters**, which measure the resistance of seeds to an electrical current. The resistance of seeds to the current decreases with an increase in seed moisture. The moisture meters have to be first calibrated for each crop. Their use requires some technical skill, but they are very efficient, extremely fast and are suited for handling of a large number of seed samples.

Based on the results from the field inspections and seed tests, the SSCA decides whether a given seed lot is suitable for certification as foundation or certified seed. If the seed is found suitable for certification, the agency issues the appropriate tags for the same. The SSCA carries out field inspections and seed tests only after a seed grower makes a written request to the agency for certification of the seed produced by him.

16.3.4. Maintenance of the Improved Varieties for Seed Production

The improved varieties used for certified seed production are maintained as nucleus or basic seed. The nucleus seeds of self-pollinated crops are maintained my mass selection. Every year, several hundred spikes/pods are selected and separate progeny rows are grown from them. Off-type progenies are eliminated, and the seeds from remaining progenies are harvested and bulked. In case, of the cross-pollinated crops, the hybrid and synthetic varieties are maintained in the form of their parental inbreds. The inbreds themselves are maintained by strict hand pollination and prompt and effective elimination of the off-type plants from the population. The varieties of asexually propagated plants are maintained as clones by asexual reproduction. Care is taken to avoid mechanical mixtures, and to promptly and effectively eliminate the off-type plants from the populations.

16.3.5. Agencies and Functionaries involved in Seed Production

In India, there are a number of bodies that perform specific functions in the production of quality seeds of crops. The Central Seed Committee advises the central and the state governments on all matters related to the Indian Seed Act. The Central Seed Laboratory carries out the functions assigned to it under the Indian Seed Act. The State Seed Laboratories conduct the seed tests for certification, and the tests are done by seed analysts using the samples collected by seed inspectors. The Central Seed Certification Board advises the central and the state governments on issues concerning seed certification.

The National Seeds Corporation (NSC) was established in 1961 to promote the development of seed industry in India, and to produce and supply the foundation seeds of the various crops. The phenomenal growth of the seed industry in India and the popularization of certified seed among the farmers are primarily due to the efforts of NSC. The NSC performs a variety of functions related to seed production, including maintenance of seed stocks of improved varieties, production and supply of foundation seed, interstate marketing of all classes of seed, import and export of seed, and production of certified seed where needed. In addition, it provides seed certification services in those states that do not have their own SSCA. The State Farms Corporation of India is involved in production and marketing of seeds; seed is marketed mainly through State Departments of Agriculture and State Seeds Corporations (SSCs). The SSCs themselves are concerned with the production and supply of certified seed within the concerned state. Finally, the SSCAs perform the various functions related to seed certification, including carrying out field inspections and seed tests, certification of the seed lots found to fulfill the requirements, guiding the seed producers in production, processing and distribution of seeds, and training the seed growers.

16.4. INTELLECTUAL PROPERTY RIGHTS

The word property means estate whether in lands, goods or money. This meaning defines **physical property**. **Intellectual property (IP)**, on the other hand, means an idea, a concept, a design, a process, etc. that can ultimately lead to generation of a useful product, which can be commercialized. Thus, intellectual properties originate from application of intellect and ingenuity, and can be converted into physical properties. The main difficulty with the intellectual properties is that they can be copied by others, and used for deriving economic benefits out of them. This will adversely impact the economic benefits to the authentic inventors of the concerned intellectual properties. This situation will act as disincentive to the inventors of the IPs, and would ultimately be detrimental to the economic progress of the society and the nation. Therefore, it would be highly desirable to develop and implement a legal system that will enable the inventors to derive exclusive benefits from their inventions/innovations. Such a system should also ensure that the society also benefits from the innovations in terms of improved products and lower costs.

The right of an individual to exclusively derive economic benefits from his IP(s) is known as **intellectual property right (IPR)**. An IPR held by a person prevents others from deriving economic benefits from the concerned IP unless authorized by the holder of the IPR. A legal system for the protection of IPRs began in Europe, and the first law on patents was passed in Venice in 1474. The protection of IPR can be achieved in one of the following ways: (i) trade secret, (ii) patent, (iii) plant breeder's rights, and (iv) copyright. The type of protection used for an IPR will mainly depend, among other things, on the nature of IP; this will become clear from the discussions in the following sections.

16.4.1. Trade Secret

A **trade secret** involves keeping the invention in as a closely guarded secret so that others are neither able to copy or improve upon it. This protection can be used for a formula, a microbial strain, a production process, a cell line, but it cannot be applied to a book, an equipment, etc. Coca Cola is the best guarded trade secret in the recorded history of human beings. A trade secret is for unlimited duration, does not involve legal formalities and procedures, and prevents others from improving upon the invention. But maintenance of trade secrets is costly, others may independently invent the same IP, and it hampers further improvements in the concerned fields.

16.4.2. Copyright

The copyright is the right granted to the creator of the copyrighted work to reproduce, publicize, adapt and translate the work and use it for commercial purposes. Copyright protection is granted to such intellectual properties as books, audio and video cassettes, computer software, cinematographic films, etc. The copyright protection prevents others from reproducing for commercial purposes either in whole or a part of the protected work. However, the information and/or the idea contained in the copyrighted work can be used by others in their works that would be commercialized. The holder of copyright can assign the same to some other person or organization for the purposes of commercial use. The Indian Copyright Act (1957) has been amended six times, the most recent amendment being the Copyright (Amendment) Act (2012). The duration of copyright protection for the literary, dramatic, musical and artistic works is the lifetime of the author plus sixty years beginning after the death of the author. For other types of works, the duration of protections is 60 years. India is the member of the important international conventions concerning copyright, including the Universal Copyright Convention of 1951, and the Agreement on Trade Related Aspects of Intellectual Property Rights (TRIPS).

16.4.3. Patent

A patent is the most stringent form of IPR protection. It is a certificate granted by a government to the inventor/innovator and gives him/her the exclusive right to manufacture and commercialize the patented item. The inventor/innovator has the right to exclude others from imitating manufacturing, using or selling the concerned invention for commercial purposes. However, other innovators can use the invention and improve upon it. A patent is valid for a specified period (around 20 years) in the country that has grated the patent. A patent is granted to an invention, an improvement in an invention, a production process and an idea or a concept. Further, a patent is awarded only when the following conditions are satisfied: (i) novelty, (ii) inventiveness, (iii) industrial application and usefulness, (iv) patentability, and (v) disclosure.

The invention should be new and it should not be already in the public knowledge (**novelty**). The invention should represent an innovation, and it should not be obvious to a person skilled in the concerned area (**inventiveness**). The invention should have either an immediate or a future industrial application, and this application should be useful to the society (**industrial application and usefulness**). Further, the invention should be patentable under the patent laws of the land in the context of their current interpretation (**patentability**). The issue of patentability is often complex and subject to variable interpretation. For example, microorganisms, plants, animals, and purified chemicals were granted patents in USA under the same patent laws that had earlier disallowed patents to these same subjects. Finally, the inventor has to disclose the invention to be patented in sufficient detail to allow a person with normal skill to reproduce it (**disclosure**). This is necessary to allow other inventors to improve upon the invention and thereby offer the society improved products. In the case of biological materials, already known organisms may be simply named. But in case of genetically modified organisms, the method of genetic modification has to be described in sufficient detail. Further, samples of microorganisms and cell lines are required to be deposited in the designated culture collections.

The procedure for obtaining a patent is, in very general terms, as follows. The inventor prepares the patent application on the prescribed proforma and submits it to the appropriate patent office. The patent examiners examine the application. If it is found to fulfill the requirements for the award of patent, it is published for public knowledge. If there is no objection to the application, the patent is awarded. But if there is an objection, the arguments in favor and against granting of the patent are heard by the competent authority of the patent office.

The authority would then decide either for awarding the patent or for rejecting the application. In case, the patent is not awarded, the invention stands disclosed without any protection whatsoever. It is highly desirable that the services of a competent patent attorney are used for filing and defending an application for a patent.

In India, the Act of protection of inventions was introduced in 1856 on the pattern of the British Patent Law of 1852. Subsequently, the Patents and Designs Act was passed in 1872. The Indian Patents Act was passed in 1970, and was later amended as Indian Patents (Amendment) Act (1999), which came in force from January 1, 1999. This Act was amended again in 2002 and 2006. The Indian Patents (Amendment) Act (1999) allows patenting of products, except for some specified medicines/drugs. The government of India may not disclose any information related to such patentable inventions that are relevant to the security of the country.

Efforts have been made to harmonize the patent laws of different countries, and to find acceptable means of extending the territorial validity of the patents. In this context, several treaties and conventions have been developed. The Paris Convention for the Protection of Industrial Property was the first concrete effort in this direction. This convention was signed in 1983, and India joined the convention in 1998. The European Patent Convention (EPC) began in 1978. The World Intellectual Property Organization (WIPO) administers the provisions of Paris Convention and other conventions on IPR. However, WIPO does not enforce these provisions. The TRIPs (trade related intellectual property rights) provisions of the GATT (General Agreement of Tariffs and Trade, 1994) is the most comprehensive multilateral agreement on IPR. It became effective in January, 1995, and is enforced by the World Trade Organization (WTO), Geneva. The member countries are obliged to implement all the articles of TRIPs. The TRIPs provisions cover a variety of IPs, including patents and plant variety protection. The member countries have the option to amend their own patent laws within the broad framework of the provisions of TRIPs.

16.4.4. Plant Breeder's Rights

There has been a prolonged debate on whether or not the IPR protection of plant varieties is justified. It has been argued that new plant varieties are produced essentially from the naturally existing lines of crops, and they are not equivalent to an industrial invention. The proponents of IPR protection of plant varieties point out that the development of a new plant variety requires considerable skill. In addition, it demands considerable intellectual input in planning and execution of the various operations for the creation of new combinations of the existing alleles of the genes involved in control of the relevant traits. Whatever we may feel about the validity of above arguments and counter arguments, nearly all the WTO member countries have agreed to provide IPR protection to plant varieties either by patent of an IPR regime comparable to patent. In USA, plant varieties can be protected by either patent or plant breeder's rights (PBR). In India, plant varieties are protected by a PBR system defined in the Protection of Plant Varieties and Farmer's Rights Act (2001; PPVFR, 2001). The UPOV (International Union for the Protection of New Varieties of Plants) convention (*http://www.upov.int/portal/index.html.en*), established in 1961, had 72 members in 2014. It encourages the member states to develop their PBR systems in conformity with the broad framework adopted by the convention. The UPOV Acts (1978) and (1991) have served as models for the PBR systems of many countries. The Indian PPVFR Act (2001) uses some provisions of the UPOV Act (1978) and some others of the UPOV Act (1991) (Table 16.5).

Table 16.5: Salient features of the UPOV Acts (1978) and (1991), the PVPFR Act (2001), and patents

Feature	UPOV Act (1978)	UPOV Act (1991)	PPVFR Act (2001)*	Patent
Protection coverage	Nationally defined species	All genera and species	Nationally specified species	Inventions
Requirements for protection	Distinctness Uniformity Stability	Novelty Distinctness Uniformity Stability	Novelty** Distinctness[@] Uniformity Stability	Novelty Inventiveness Non-obviousness Usefulness
Duration of protection	Min. 15 yr	Min. 20 yr	15 yr (18 yr for trees and vines)	17-20 yr (OECD[$])
Protection applies to	Propagating material	All material produced from the variety	All material produced from the variety	Commercial use of the protected subject
Breeder's exemption	Yes	Yes, except for the essentially derived varieties	Yes, except for the essentially derived varieties	No
Farmer's privilege	Yes	Left to the national laws	Yes	No

* Protection of Plant Varieties and Farmer's Rights Act (2001)

** A variety must not be in commercial use for one year in India or 4 years (6 years for trees and vines) outside India.

[$] OECD, Organization for Economic Cooperation and Development.

[@] In case of extant varieties, distinctness, uniformity and stability (DUS).

A PBR is granted by the government to the breeder, originator or owner of a variety. The PBR title gives its owner the right to prevent others from producing and commercializing the propagating material of the variety covered by the title. The PBR title-holder may authorize or license others to produce/sell the propagating material, i.e., the seed, of the variety. The terms for the licensing of the PBR title and/or the sale price of the seed should be reasonable. In case, the seed of the protected variety is not produced or is priced unreasonably high, the government may decide to grant license of the title in the public interest; this is called **compulsory licensing**. The PBR title offers protection to the variety concerned, but it does not protect the genetic components, the parental lines and the breeding methods used for the development of the variety. However, the parental inbreds of protected hybrids are also protected by the PBR title.

A variety has to satisfy the following four conditions for being registered for PBR protection: (i) novelty, (ii) distinctness, (iii) uniformity, and (iv) stability. A variety should not have been in commercial cultivation for more than one year for the grant of PBR (**novelty**). The variety must be distinguishable from other varieties of the crop for one or more readily identifiable characteristics (**distinctness**). In addition, the plants of the variety should be uniform in appearance under the conditions of its specified adaptation (**uniformity**). Finally, the variety must be stable in appearance and in the claimed features over years when grown under the environment of its adaptation (**stability**).

The PPVFR (2001) allows protection for farmer's varieties, extant (existing) varieties, and to new varieties. It recognizes a variety grown for less than one year in India, but four years (six years in case of vines) in foreign countries as **new varieties**. It does not permit the use of genetic use restriction technologies like the terminator technology. The duration of protection is for 15 years (18 years for trees and vines). It gives the farmer's the freedom to save, use, sow, resow, exchange, share or sell their farm produce, but it cannot be sold as **branded seed**. It also recognizes the farmer's rights; it provides for the establishment of a **gene fund** from the earnings of benefit sharing of the registered varieties, and for using this fund to reward the concerned farmers. A **Protection of Plant Varieties and Farmer's Rights Authority** has been established for implementation of the provisions of the PPVFR Act (2001). Plant variety registration began in 2007.

The PBR system allows breeders of the protected varieties to be used in breeding programmes for the development of new varieties; this is called **breeder's exemption**. However, under the UPOV Act (1991), essentially derived varieties are subject to the PBR protection granted to the initial variety, i.e., the variety from which it was derived. An **essentially derived variety** is defined as a variety predominantly derived from the initial variety, which retains the expression of the essential characteristics from the genotype or combination of genotypes of the initial variety. However, the decision whether a new variety is essentially derived or not is open to varied interpretations. The PBR systems generally allow the farmers to use their crop produce as seed for planting the crop in the next crop season; this is known as **farmer's privilege**. In addition, it is recognized that farmer families and farming communities have selected, developed and conserved the genetic resources used in the modern breeding programmes to develop the high yielding varieties. This contribution has been recognized by the FAO in form of farmer's rights, which seeks to return a part of the profit made from sales of the seeds of the improved varieties to the concerned farming families and communities that contributed the germplasm used for breeding the varieties. However, the farmer's right is yet to be implemented on this ground.

An effective system of PBR is needed to encourage the breeders for their achievements by rewarding them economically. It also encourages the private sectors to engage in plant breeding activities. A PBR regime in a country allows it the access to such varieties that are IPR protected. Finally, it increases competition among the plant breeding organizations, which ultimately benefits the farmers. However, PBR system tends to encourage monopolies, and discourage free exchange of germplasm. It may lead to dilution of the farmer's privilege, and increase seed costs, which will go against the poor farmers. The PBR title holder may produce less seed of the variety than the actual demand in an effort to maximize profits. Finally, the spread of new varieties may be slowed down in countries like India, where farmer to farmer seed exchange is a major vehicle for their wide adoption.

16.4.5. Protection of Biotechnological Inventions

The different biotechnological inventions can be broadly grouped into the following six categories: (i) methods or processes of generating specific products, (ii) the products themselves, (iii) the uses of various products and/or procedures, (iv) biological entities like strains of microorganisms, cell lines, and lines of plants and animals, (v) DNA molecules and proteins (protected as DNA and protein sequences), and (vi) specific methods like the techniques of genetic transformation, gene modification, etc. In countries like USA all the above are patentable, including products isolated from nature provided they differ in form or composition from that found in the nature. The products, production processes, techniques of genetic transformation are patentable in almost all the countries, including India. But genes and living forms are patentable in some countries other than USA, but they are not readily patentable in EPO and India. There has been considerable debate on whether life forms like microorganisms, plants and animals should be patented. The chief objections to the patenting of life forms have been moral, ethical and religious. But there has been an increasing inclination of the nations and societies to focus on the economic aspects of the debate. As a result, the patenting of life forms in gaining increasing acceptance among the nations.

A disturbing aspect concerning patents in biotechnology is the patenting of very broad subjects, e.g., all transgenic cotton (by W.R. Grace and Co.), all transgenic soybeans (by Calgene, Inc.), all transgenic plants with increased lysine content (by Dekalb Genetic Corporation), etc. Such broad coverage patents are inequitable because a single patent deprives others the opportunity for engaging in a broad spectrum of biotechnological activities. It is feared that such patents will encourage and facilitate monopolistic control of biotechnological modification of our major crops. Such a situation would be detrimental to food security and would even tend to limit the scientific and technological progress, especially in the developing countries. Many of the broad patents have been challenged in the courts of law and some of them have even been revoked.

Another negative aspect of IPR regime is the need for several different patented subjects in generating the desired biotechnology product like a transgenic plant with a specific trait. Often the access to few to several of the protected subject matters may not be readily available for various reasons. When this situation is encountered, either a new technology has to be created, or a merger/takeover has to be executed. Sometimes, the IPR situation of some of the techniques/materials may not be clear at the time the research for the development a product is initiated. It may be later discovered that one or more of technologies/materials used in the development of a useful product are not accessible. In such a case, an otherwise useful product has to be either shelved or it has to be developed all over again using an alternative strategy that is accessible to the researchers. The Endless Summer tomato, developed by the DNA Plant Technology, USA was ready for release for commercial cultivation in 1996, but it has not been released so far due to IPR problems.

16.4.6. Management of IPR

The benefit from an IPR would be realized only when it is used to generate the product on a commercial scale. Therefore, the holder of the IPR has to make every effort to commercialize the IPR. It may be added that generally additional research and development efforts are needed for commercial production using the IPRs. Therefore, the IPRs have to be properly and effectively managed, which would involve the following activities: (i) transfer of the IPR to a suitable organization/ company for commercialization on the most favorable terms and conditions, (ii) establishment of collaborations to facilitate the commercialization of IPRs, (iii) monitoring of IPR infringement and enforcing the IPR, (iv) periodical renewal of the IPRs, and (v) regular evaluation of the costs of and the benefits from the IPRs.

16.4.7. Benefits from IPR

1. IPRs safeguard the right of the creators of the IPs for commercial use of their intellectual and artistic creations.
2. IPRs encourage inventiveness and innovativeness by suitably rewarding the inventors and the innovators.
3. The disclosure requirement for patents promotes the spread of new technologies and allows improvements in them.

4. They encourage investment in research and development activities.
5. The consumers get a choice of a larger range of products.
6. IPRs promote all-round economic development by ensuring benefits to the inventors, investors, and the consumers.

16.4.8. Problems from IPR

1. IPR encourages mergers and takeovers, which ultimately promotes monopolies.
2. Often the IPR situations are quite complex. As a result, it may require an expert team to manage IPRs, which adds to the cost of production and takes up valuable resources.
3. Many perceive IPRs as a threat to food security.
4. IPRs might have an adverse effect on the biological diversity and ecological balance.
5. It may have adverse impact on the livelihood of the poor, particularly in the developing countries.

16.4.9. Geographical Indications

According to TRIPs of GATT (1994), **geographical indications** (GIs) are such indications which identify a good as originating in the territory of a member country, or a region or locality in that territory, where a given quality, reputation or characteristic of the good is essentially attributable to its geographical origin. Thus a GI recognizes that the given quality, reputation or characteristic of a specific good is primarily due to the specific region of its origin. Suitable laws have to be enacted by the concerned country to avail protection under the provisions of GI. Once a GI is awarded to a product as originating from a given region, the same product generated elsewhere cannot be called by the same name as that recognized in the GI. For example, basmati rice produced in a specific region of India has been granted GI. As a result of this GI, the produce from the same rice varieties grown in any other region cannot be called basmati. GIs are applicable to agricultural produce, natural products, manufactured goods (like Scotch Whiskey, Champagne, etc.), goods produced by handicraft, and even food products.

In India, the Geographical Indications of Goods Act (1999) came into force in September, 2003. The first GI was granted to Darjeeling tea. Since then, and up to September, 2010, over 130 different products/goods, including Bikaneri bhujia from Rajasthan, Banarasi brocade and sarees from Banaras (UP) had been tagged with GI.

16.4.10. International Convention on Biological Diversity (CBD)

The International Convention on Biological Diversity (*CBD*) recognizes the sovereign rights of the nations on their biological diversity and the genetic resources. The nations have the right to regulate the access of others to their genetic resources. They can link this access to sharing of the benefits from such access, and to transfer of relevant technologies. The CBD was signed on Dec. 29, 1983 by 173 countries, including India. The concept of CBD is counter to the International Undertaking on Plant Genetic Resources developed by FAO (Food and Agriculture Organization) of UNO, and signed in 1983 by India and 111 other nations. This undertaking recognizes the plant genetic resources to be 'common heritage of mankind'. The text of this undertaking is being modified to bring it in harmony with the CBD.

16.5. CONCLUSIONS

The new strains developed by plant breeders are systematically evaluated in replicated trials before they are released for cultivation. In India, these evaluations are done under the All India Coordinated Project for the concerned crop in a sequential manner: first, initial evaluation trials, then uniform regional trials, and minikit trials and adoptive research trials. The strains are also evaluated for disease and insect resistance, and the quality traits. Based on the results from the evaluation, superior strains are identified for release as new varieties by workshop of the All India Coordinated Project for the concerned crop. The seeds of the identified strains are multiplied through step-wise procedure beginning with nucleus seed, breeder seed, foundation seed and certified seed. The seed of any variety can be certified only when this variety is released by the competent committee and is duly notified, following its identification by the concerned All India Coordinated Project. The seed crop is cultivated following the recommended package of practices with proper isolation distance and other requirements for the seed crop. The seed crop is inspected by the State Seed Certification Agency to determine the suitability of seed crop for certification. The seed produced from a seed crop is dried, cleaned and graded, treated with disinfectants and protectants and then packaged in bags of suitable sizes. The seeds are subjected to tests for physical and genetic purity, germination, moisture content, etc. Based on the findings from field inspections and seed tests, the seed lots may be certified by the State Seed Certification Agency.

The biotechnological inventions and plant varieties are considered as intellectual properties. The intellectual properties are protected by intellectual property rights (IPRs). The IPR protection may be in the form of trade secret, patent, plant breeder's rights or copyright. The biotechnological inventions are protected by patent, while plant varieties are generally protected by plant breeder's rights (PBRs). The Indian Patents Act (1970) was amended as Indian

Patents (Amendment) Act (1999), which was amended twice, the latest amendment being effected in 2006. In India, plant varieties are protected by a PBR system under the Protection of Plant Varieties and Farmer's Rights Act (2001; PPVFR, 2001). The IPR system is believed to encourage innovation and economic benefits to inventors, the society and the nation.

PRACTICE QUESTIONS

1. Briefly describe the general procedure for the release of new crop varieties in India.
2. Define seed. Briefly describe the classes of seed produced in India and the general procedure for crop cultivation for certified seed production.
3. What is seed processing? Briefly describe the various steps in the processing of seed.
4. Explain the meaning of seed tests. Briefly describe the various seed tests carried out for seed certification.
5. Define intellectual property and intellectual property rights. Briefly describe the various forms of intellectual property rights, and discuss the significance of intellectual property rights.
6. What is a patent? Discuss the relevant features of patent protection and its usefulness in protection of biotechnological innovations.
7. What is plant breeder's rights protection of IPR? Discuss its features compared to patent protection. Briefly describe the salient features of the Indian plant breeder's rights law.
8. Write short notes on the following: (i) multi-location trials of new strains, (ii) variety identification and release, (iii) classes of improved seed, (iv) certified seed, (v) seed certification, (vi) seed tests, (vi) genetic purity of seed, (vii) field inspection of seed crop, (viii) isolation distance, (ix) intellectual property rights, (x) patent, (xi) plant breeder's rights, (xii) trade secret, (xiii) geographical indications, (xiv) International convention on biological diversity (CBD), (xvi) grow-out test for seed purity, (xvii) The Indian seed act (1966), (xviii) The plant variety protection and farmer's rights act (2001), (xix) seed processing.

FURTHER READINGS

Agrawal PK and Dadlani M 1987. Techniques in Seed Science and Technology. South Asian Publishers.

Gadwal VR 2003. The Indian seed industry: Its history, current status and future. *Curr. Sci.* 84: 399-406.

Ganguli P 1998. Gearing up for patents – The Indian Scenario. University Press, Hyderabad.

Ganguli P 1998. Patenting innovations: New demands in emerging contexts. *Curr. Sci.* 75: 433-439.

Indian Seed Sector (http://seednet.gov.in/material/IndianSeedSector.htm, also http://seednet.gov.in/Material/Prog-Schemes.htm; http://www.slideshare.net/sangacib/seed-produc; www.cicr.org.in/pdf/ legislation_seed_quality.pdf)

Joshi AK and Singh BD 2003. Seed Science and Technology. Kalyani Publishers, New Delhi.

Keayla BK 2000. Protection of Plant Varieties and Farmers Rights Bill, 1999: Some critical issues. *Jour. IPR* 5: 251-255.

Kochhar S 2011. Analysis of opportunities and challenges in IPR and agriculture in the Indian context. *Jour. IPR* 16: 69-73.

Kochupillai M 2011. The Indian PPV & FR Act 2001: Historical and implementation perspectives. *Jour. IPR* 16: 88-101

USDA 1961. Seed, Yearbook of Agriculture. Washington, DC, Government Printing Office.

Vijayalakshmi K 2013. A Manual on Seed Production and Certification. Centre for Indian Knowledge Systems, Chennai.

■■■■■

Subject Index

■■■■

Author Index